SEDIMENTARY PETROGRAPHY

PTR PRENTICE HALL
Sedimentary Geology Series
George deV. Klein, Series Editor
University of Illinois at Urbana–Champaign

CAROZZI, *Carbonate Rock Depositional Models:*
  A Microfacies Approach
CAROZZI, *Sedimentary Petrography*
FRASER, *Clastic Depositional Sequences*
WARREN, *Evaporite Sedimentology*

FUTURE TITLE

HARBAUGH, SLINGERLAND, AND FURLONG, *Simulating*
  Sedimentary Clastic Basins

SEDIMENTARY PETROGRAPHY

ALBERT V. CAROZZI

University of Illinois at Urbana–Champaign

PTR Prentice Hall, Englewood Cliffs, New Jersey 07632

Library of Congress Cataloging-in-Publication Data
CAROZZI, ALBERT V.
    Sedimentary petrography / Albert V. Carozzi.
        p.   cm.
    Includes bibliographical references and indexes.
    ISBN 0–13–799438–9
    1. Rocks, Sedimentary.   I. Title.
QE471.C362   1993   92–43987
552′.5—dc20   CIP

Editorial/production supervision: *Jane Bonnell*
Cover design: *Jerry Votta*
Buyer: *Mary Elizabeth McCartney*
Acquisitions editor: *Betty Sun*
Cover illustration: Coarsely crystalline mosaic of secondary anhydrite entirely replacing limestone under deep-burial diagenesis. Crossed nicols. Amapá Formation (Paleocene–Middle Miocene), Foz do Amazonas Basin, offshore northeast Brazil.

The publisher offers discounts on this book when ordered in bulk quantities. For more information, contact:

    Corporate Sales Department
    PTR Prentice Hall
    113 Sylvan Avenue
    Englewood Cliffs, NJ 07632
    Phone: 201-592-2863
    Fax: 201-592-2249

Printed in the United States of America

10 9 8 7 6 5 4 3 2 1

ISBN 0-13-799438-9

Prentice-Hall International (UK) Limited, *London*
Prentice-Hall of Australia Pty. Limited, *Sydney*
Prentice-Hall Canada Inc., *Toronto*
Prentice-Hall Hispanoamericana, S.A., *Mexico*
Prentice-Hall of India Private Limited, *New Delhi*
Prentice-Hall of Japan, Inc., *Tokyo*
Simon & Schuster Asia Pte. Ltd., *Singapore*
Editora Prentice-Hall do Brasil, Ltda., *Rio de Janeiro*

To Nadine for keeping up the challenge

CONTENTS

CHAPTER 2   RUDACEOUS ROCKS   43

CHAPTER 3   ARGILLACEOUS ROCKS   63

CHAPTER 4   VOLCANICLASTIC ROCKS   79

CHAPTER 8   PHOSPHORITES   173

CHAPTER 9   IRONSTONES   190

CHAPTER 10   EVAPORITES   209

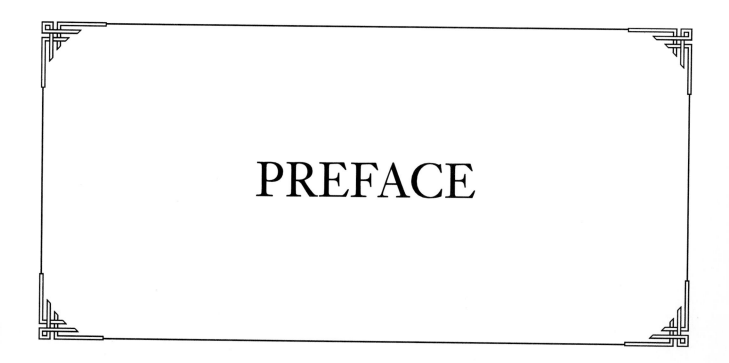

PREFACE

Today the emphasis of geological investigations is toward plate tectonics and geophysics in order to understand deep-seated processes, as well as toward the integration of stratigraphy into its natural dynamic framework of sea-level changes and plate deformation. At the same time, geochemistry attempts to understand better the depositional and diagenetic behavior of sedimentary rocks from their environments of deposition until deep burial conditions. In the near future, a return toward the petrographic study of sedimentary rocks seems inevitable, because it appears to have been by-passed without full investigation and integration into modern geological approaches.

This book is about sedimentary petrography rather than sedimentology or geochemistry. It focuses on fundamental observations under the petrographic microscope, their classification, and their theoretical and practical implications, rather than on speculations and far-reaching theories. I wish to stress the point, which is often underplayed, that the value of any theory or interpretation is a direct function of the quality of the observations on which it is based. Anything stated beyond this limit is pure unsupported speculation, which, although extremely appealing to many authors, does in reality hinder rather than promote any progress in geology.

ACKNOWLEDGMENTS

In the course of this work I have greatly benefited from my colleagues in the Department of Geology at the University of Illinois at Urbana–Champaign, who have generously given their time for critical reviews of the chapters pertaining to their respective expertise. My thanks go to Stephen P. Altaner (argillaceous rocks); David E. Anderson (iron-bearing rocks); Richard L. Hay (volcaniclastic rocks, siliceous rocks, and evaporites); George DeV. Klein (sandstones and conglomerates); and Philip A. Sandberg (limestones and dolostones).

The color plates of photomicrographs illustrating this book represent the best examples of textures of typical sedimentary rocks that I have studied together with some of my graduate students during more than 35 years. I am very grateful for their fundamental contribution to this critical aspect of the book. They are in alphabetical order: Sadeg H. Bakush, Renato T. Bertani, Nancy R. Black, William C. Dawson, Ibrahima Diaby, Frank U. H. Falkenhein, Sadat Feiznia, Milton R. Franke, Roger J. Kocken, Yaghoob Lasemi, Kathleen M. Marsaglia, Rasool Okhravi, Michael R. Owen, Jan L. Reichelderfer, J. G. William Soderman, Donald Von Bergen, and Breno Wolff.

Furthermore, I am very grateful to Louis F. Dellwig, Leo Ogniben, Richard C. Morris, B. Charlotte Schreiber, and Sharon A. Stonecipher, who provided me with original illustrations. Their contribution is acknowledged at appropriate places in the text. The advice of Gilles S. Odin on the use of his new proposed classification of glauconitic minerals is greatly appreciated.

The anonymous critical readers are thanked for their constructive criticisms and suggestions, which have improved the original version.

Sincere thanks are extended to Jessie Knox, who skillfully undertook the preparation of the color plates of photo-micrographs and the final redrafting of many figures.

I deeply appreciate the unwavering interest of executive editor Michael Hays, the devoted help of executive editor Betty Sun, and the efforts of production editor Jane Bonnell and others at PTR Prentice Hall who saw this book through press.

Finally, I would like to acknowledge again the debt of gratitude I owe to my wife Marguerite for her invaluable and constant encouragement and technical help during the intense but challenging times of preparing a synthesis of this kind.

Albert V. Carozzi
Urbana–Champaign and Geneva

SEDIMENTARY PETROGRAPHY

INTRODUCTION

This book represents an updated synthesis of a course on sedimentary petrography for advanced undergraduate and graduate students taught at the University of Illinois at Urbana–Champaign during more than thirty-five years. The same subject matter was also used, in a modified form, in short courses for training programs for major oil companies worldwide.

The order of topics in this book, which assumes a previous undergraduate knowledge of basic definitions, reflects the experience acquired during these many years of teaching. Answers to my repeated challenging questions to the students of "tell me," asked during lectures, and "show me," during laboratory sessions, are presented here in a highly systematic way. Each of the ten subject matters corresponding to individual chapters represents a self-sufficient didactical and practical package. Each package includes the following topics: introduction, classification, depositional environment, detailed petrography, diagenetic evolution, reservoir properties, color illustrations of typical examples, and pertinent updated references, which allow the student to pursue his or her interest in the appropriate specialized literature.

It seems logical to me to discuss first the association rudaceous rocks–sandstones–argillaceous rocks–volcaniclastic rocks because they are volumetrically the most important supergroup of rocks. The book begins therefore in Chapter 1 with the major types of sandstones presented in order of increasing mineralogical complexity and decreasing textural maturity. Why do I use this order from the simple to the complex, whereas the natural geological evolution of clastic constituents through time tends toward simplification of mineral composition and increase of textural maturity? This unorthodox approach originates from teaching experience, which has shown that features under the microscope, such as depositional fabrics, associations of minerals, and related diagenetic textures, can be better understood when proceeding from the simple to the complex, that is, from quartz arenites to glauconitic arenites, feldspathic arenites, lithic arenites, and finally wackes. Rudaceous rocks fall in place in Chapter 2 because most sandstone types build the interstitial matrix of many conglomerates. Argillaceous rocks, described in Chapter 3, represent the finest of the clastic rocks and are also, theoretically, recognizable fine-grained equivalents of the major sandstone families. Volcaniclastic rocks (Chapter 4) form a hybrid group closely related with clastic rocks and thus complete the review of the terrigenous rocks. Carbonate rocks are treated next, limestones in Chapter 5 and dolostones in Chapter 6. They are followed by siliceous rocks in Chapter 7, because many of their nodular chert types are intimately associated with carbonates. The succession of Chapter 8 on phosphorites and Chapter 9 on ironstones is a matter of personal preference; in fact, both display varieties interpreted as possible replacements of carbonates. Evaporites, in Chapter 10, are also closely associated with carbonates, and furthermore display puzzling similarities with the various types of laminations characteristic of Precambrian ironstones.

The reader will notice that the description of each

major rock type is given at first in generalized terms. This was done on purpose in an effort to present a "run-of-the-mill" petrographic picture derived from the examination of thousands of thin sections, a picture meant to be general or ideal and therefore devoid of references to literature. My teaching experience has shown that a student, or any researcher for that matter, needs to reach, in a first stage, a general mental picture of the rock being looked at under the microscope. Then, in a second stage, he or she can begin to "dissect" this initial picture with the use of the pertinent literature to determine to what extent a particular sample differs from the generalized description, and then to modify and gradually sharpen up the initial picture until a final qualitative and quantitative description is reached. The third stage consists of speculations about depositional and diagenetic processes. These speculations should furthermore include the fundamental contribution of geochemistry and of other appropriate forefront techniques of investigation.

Some readers might object to the repeated use of generalized statements, documented evidence, and speculations, which are played one against the other. My answer is that every scientist proceeds in this way every day—often without being aware of it—using a method of interplay between deductive and inductive reasoning. Speculations are the essence of scientific progress, even though at the time of their presentation they seem either supported or unsupported, acceptable or outrageous. Indeed, the history of science is filled with violently rejected speculations that were later vindicated. Eliminating speculations from this presentation would change this book into a soporific Cartesian or Linnaean catalog of rock types, which would be totally alien to the scientific philosophy of the author. As the famous Italian playwright Luigi Pirandello said, "Un fatto é come un sacco, che vuoto non si regge. Perché si regga, bisogna farci entrar dentro la ragione." ("A fact is like a sack that cannot stand straight when empty. For it to stand, reasoning should first be put into it," from his play *Six Characters in Search of an Author*.)

The uneven size of the ten chapters of this book expresses the various advances of knowledge during the past twenty years. This attempt at covering the entire spectrum of the petrography of sedimentary rocks at a level both relatively simple and practical, but at the same time based on the most recent developments, was no easy task. Simplification and scientific truth represent a delicate blend that can be easily tipped out of balance, and I take full responsibility for whatever shortcomings the reader may find in this synthesis.

To choose the most relevant and reliable papers to document the presentation and challenge the student toward further research can be compared metaphorically to a man swimming in the middle of a gigantic flood and grasping the most sturdy flotsam, which may bring him to the safety of the shore. This choice of papers is a highly critical and personal one. Still valuable older contributions, newer challenging works, even highly speculative ones must be selected. I am fully aware that some papers may have been overlooked

that the future will judge important, and apologies are extended to those who might have suffered from the inherent prejudice of a personal choice, albeit based on long experience.

During my career I have written three textbooks on sedimentary petrography, each entirely different in organization and content, showing how fast new advances and approaches occurred during the past forty years in the various disciplines of this important field of geological research. The first book was published in French in Geneva and Paris in 1953, written under the strong and lasting influence of Lucien Cayeux. The second, published in 1960 shortly after my arrival in this country, was reedited in 1972. It reflected the fundamental contributions of Francis J. Pettijohn, Paul D. Krynine, and Charles M. Gilbert. The third book is the present text, which attempts to show that a synthesis, regardless of its potential shortcomings, is an indispensable and opportune intellectual exercise, particularly at a time when extreme specialization threatens to lead toward the loss of the general overview indispensable to students and professionals alike. Students need a general picture for an adequate understanding of geology and as guide toward the choice of a future specialized field; for professionals, a general overview is equally critical for establishing a relationship between their specialized field and its broader geological context. Only the future will tell to what extent this attempt has fulfilled that purpose.

EXPLANATORY NOTES ON COLOR PLATES

The insert of 32 color plates in the middle of the book is fundamental because it illustrates the most important petrographic and textural aspects of the described sedimentary rocks.

All color photomicrographs are oriented for top and bottom and correspond to thin sections cut perpendicular to bedding. They were carefully chosen as the most spectacular examples among more than 15,000 thin sections in the author's collection preserved in the Department of Geology at the University of Illinois at Urbana–Champaign.

In the color plates, all photomicrographs are accompanied by a short caption and a linear scale. Full petrographic descriptions, stratigraphic positions, and geographic locations are given at the end of the appropriate sections in the text.

The symbols used in the color plates are:

  PL: plane-polarized light
  XN: crossed nicols
  CL: cathodoluminescence
  QP: quartz plate

Porosity is shown by a blue dye unless otherwise stated in the caption.

CHAPTER 1

SANDSTONES

INTRODUCTION AND CLASSIFICATION

Any attempt at establishing the composition and characteristics of the constituents of sandstones seems to be at first glance a pure descriptive process. However, it takes on immediately a genetic implication because the siliciclastic grains forming the framework are of detrital origin, whereas the origin of precipitated cement is chemical or diagenetic, and that of matrix is detrital or diagenetic. Inevitably, the integrated petrographic description provides data for interpretations on depositional environments, diagenetic sequences, and often reservoir properties (Wilson and Pittman, 1977). It also shows that sandstones, like all sediments, build mineralogically and texturally gradational natural sequences.

To understand these sequences, limits are drawn within them to subdivide the natural continuum into distinct units by using the most genetically significant and quantifiable parameters pertaining to grains, cement, and matrix. The distinction of these units, or types of sandstones in this particular case, is necessary to satisfy the need for a mentally logical but artificial order, for the purpose of communication and comparison, and for the final purpose of reconstructing the history of the rock from its origin to its present condition.

Whenever a genetic classification is reached, it should be considered a temporary human endeavor, continuously subject to modifications and improvements, because a final "correct" classification does not exist. The next problem is nomenclature, or terminology, because defined types should carry distinct designations, again for the purpose of communication, comparison, and genetic understanding.

The sedimentary petrographic literature shows that the selection of genetically significant parameters was a long and tedious process, dominated by the inevitable regionalism of geological studies, the evolution of sedimentological concepts, and human vanity. Further complication was introduced by the selection of rock terms to designate the genetically defined classes. Old terms were borrowed without scholarly research on their origin, and redefined many times to the extent of losing all real significance—such is the case of the terms arkose and graywacke—whereas many unnecessary terms were introduced for regional reasons or personal convenience.

The classification controversy has not been settled yet, but has been tempered by the time equalizing effect of daily usage by practicing sedimentary petrographers who have generally favored simple and practical classifications. Esoteric and sterile discussions on classification are increasingly being left aside in the present trend of research, which requires a clear petrographic background in order to pursue more rewarding diagenetic studies.

Sandstone classification relies essentially on two major criteria: (1) the mineralogical composition of framework grains, and (2) the texture. The first criterion is an expression of source-rock composition, tectonism, and weathering, and even of more complex conditions such as the final environment of deposition and subsequent diagenetic processes. Theoretically, all minerals forming the framework grains of sandstones, if not drastically affected by diagenesis, have the potential of providing data on the composition and tectonism of the source area and thus of becoming tools for the determination of provenance. In reality, for practical reasons and after many attempts, a choice was made to reach again a simple scheme. Three major types of grains were selected: quartz, feldspars, and sand-size rock fragments designated as lithics. They can conveniently form the apexes of a composition triangle QFL, a mode of representation borrowed from igneous petrography, on which are plotted the values of modal analyses required for a meaningful quantitative classification of sandstones. A controversy reappeared similar to that of the selection of the three major types of framework grains with respect to the criteria used to define fields within the triangular compositional diagram. Terminology excesses were again tempered by practical usage into the smallest possible number of fields for easy manipulation of data.

The second major criterion, to be combined with the mineralogical composition of framework grains, is the textural aspect of sandstones. As in the case of mineralogical composition, sandstones display a natural gradation of textures expressed by two depositional properties: (1) the sorting of framework grains, and, particularly, (2) the amount of interstitial argillaceous matrix (<30 μ fraction). Both properties were used by Folk (1954) to express "textural maturity," in fact an expression of mineralogical maturity, and to subdivide sandstones into immature, submature, mature, and supermature types. This proposed subdivision in which end terms are easily recognizable, but intermediate ones difficult to characterize, led eventually to the subdivision of all sandstones into two suites: *arenites* and *wackes* with the boundary at 10% matrix (Williams et al., 1982). Each suite was internally classified using the previously mentioned triangular QFL diagram.

The classification used in this book is a modified version of that proposed by Pettijohn et al. (1987), which in turn is a modification of the scheme presented by Dott (1964). Its primary criterion consists of the framework grains divided into quartz, feldspar, and lithics; the secondary criterion is the distinction of *arenites* with less than 15% matrix and *wackes* with more than 15% matrix, the upper limit of the matrix being set at 30 μ (**Fig. 1.1**) These figures make all arenites grain-supported rocks and provide a broad latitude for wackes to range from grain-supported to matrix-supported rocks. At this stage, it is understood that the term "*interstitial matrix*" and the related texture called *matrix-supported* are considered as purely descriptive because of their dual origin. Indeed, the origin of matrix can range from primary detrital deposition to complete early to late diagenetic reorganization in which unstable sand-size mineral and lithic constituents are altered into an interstitial material. Similarly, a matrix-supported texture can range from simultaneous deposition of the sand and clay fractions to a pure product of diagenetic processes. These distinctions are of critical importance in understanding the origin of wackes, discussed below.

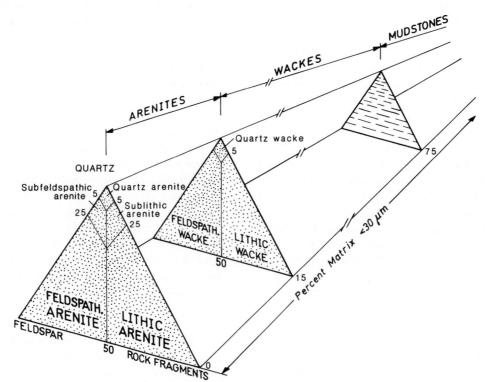

FIGURE 1.1 Classification of sandstones. Modified from Dott, Jr. (1964). Reproduced by permission of the Society of Economic Paleontologists and Mineralogists.

Among arenites, *quartz arenites* represent the end term, which consists predominantly of quartz and no more than 5% of either feldspar or lithic grains. Arenites that contain 25% or more feldspar grains exceeding lithic grains are designated as *feldspathic arenites;* those with 25% or more of lithic grains exceeding feldspars are called *lithic arenites.* Transitional classes are recognized as *subfeldspathic arenites,* which contain less feldspars than feldspathic arenites and few or no lithic grains, and, similarly, *sublithic arenites,* which show less lithic grains than lithic arenites and few or no feldspars.

Lithic arenites display a full spectrum of grain composition, including sedimentary, low- and high-rank metamorphics, plutonics, and volcanics combined, either in any proportions or predominating individually. This situation requires the use of appropriate adjectival modifiers, not only for the sake of accurate petrographic descriptions, but also for the potential importance of these rocks in provenance studies, such as siltstone–phyllite lithic arenite, or chert–dolomite lithic arenite. Because the classification adopted here takes into account only mineralogical and textural parameters of framework and matrix, it is critical that the mineralogy and texture of authigenic cements precipitated in the voids of the framework be clearly designated as, for instance, coarse sparite-cemented quartz arenite or hematite-cemented feldspathic arenite.

Among wackes, *quartz wackes* represent the end term, which consists predominantly of quartz and no more than 5% of either feldspar or lithic grains. Wackes with more than 5% feldspar grains exceeding lithic grains are designated as *feldspathic wackes,* and those with 5% or more of lithic grains exceeding feldspar grains are designated as *lithic wackes.* Contrary to arenites, transitional classes between the three major types of wackes do not seem to warrant particular terms at this time.

It would seem logical, at first glance, to begin the petrographic analysis of sandstones with immature types in terms of composition and texture, that is, the wackes that represent the least modified residues of parent rocks, and to proceed toward increasingly mature types, the arenites, among which quartz arenites are the nearest to the end product of the sedimentary evolution produced by transporting agents. However, wackes are characterized by a petrographic and textural complexity whose understanding appears greatly facilitated after the characteristics of arenites are described and interpreted. Consequently, description begins with quartz arenites and follows the path of increasing immaturity toward lithic wackes.

QUARTZ ARENITES

Framework

These arenites with 95% or more detrital quartz combine this predominant mineral with rare grains of feldspar, mainly microcline, orthoclase or albite, muscovite, and a few accessory minerals such as tourmaline and zircon. Occasionally, well-rounded grains of chert and altered silicic volcanic glass occur.

Monocrystalline quartz grains, with straight extinction, predominate over grains with undulatory or wavy extinction, and both are more abundant than polycrystalline grains. This situation results from the high degree of mineralogical maturity that these rocks reached through a history during which weaker varieties of quartz were gradually eliminated during transport.

Monocrystalline quartz is derived either from plutonic igneous rocks and high-rank metamorphic rocks, from reworked older sandstones, or from phenocrysts of silicic volcanic and pyroclastic rocks.

The distinction of monocrystalline straight and undulose quartz originating from plutonic or metamorphic rocks is very difficult because shape, extinction, nature of fluid-filled vacuoles, and mineral inclusions are properties difficult to assess because they overlap to the extent of becoming unreliable. High-temperature monocrystalline quartz of volcanic origin can be differentiated by its principal features: shape partially controlled by its original condition of bipyramidal phenocryst; limpid, strain-free aspect; and embayments and irregular inclusions of devitrified vitreous matrix due to partial resorption by the melt. Monocrystalline quartz grains originating from reworked older arenites show remnants of successive abraded quartz overgrowths. This feature seems to have been generally underestimated (Sanderson, 1984), because overgrowths occur only on certain parts of grain surfaces and in a random fashion; consequently the two-dimensional view in thin section is far from being representative of the real occurrence. Naturally, overgrowths tend to disappear with increasing maturity of the quartz arenite. Grains of quartz derived from phenocrysts in rhyolites may show overgrowths and intervening zones of inclusions generated during successive phases of accretion of the phenocrysts in the melt. In this case, superposed overgrowths consist of perfectly limpid high-temperature quartz identical to that of the original nucleus and no "dust line" separates nucleus from envelopes.

Polycrystalline quartz derived from plutonic igneous rocks tends to consist of randomly associated smaller and larger irregular grains, whereas associated sutured grains derived from high-rank metamorphic rocks display a more elongate shape and tend to show a subparallel crystallographic orientation. Another type of polycrystalline quartz of metamorphic origin consists of polygonized quartz mosaic in which component individuals form clear polygonal units with straight boundaries that tend to intersect at a 120° angle. This type of quartz is believed to have originated from annealing of highly strained quartz in the final stages of metamorphism. Polycrystalline quartz reworked from older quartzite grains can also consist of an irregular mosaic of individuals generated by completed overgrowth or pressure-solution processes. Rare grains of polycrystalline quartz originate from vein-filling quartz characterized by clear crystals arranged in a comblike structure of subparallel blades. Grains of chert displaying a fine-grained mosaic texture can be included in the group of polycrystalline quartz grains.

Regardless of their internal structure, many highly rounded quartz grains show a "frosted" external surface commonly attributed to the effect of saturation by micro-impacts during transport. However, this aspect is often of secondary origin and represents an irregular marginal replacement of quartz grains by a variety of diagenetic cements, such as carbonates, sulfates, and iron oxides.

Cements

The cementation of some quartz arenites, following the initial mechanical rearrangement between quartz grains, can begin by very early pressure solution (**Plate 1.A**), a process of reciprocal grain interpenetration recently modeled (Elias and Hajash, 1992; Stephenson et al., 1992), but generally occurs by quartz overgrowths generated by silica precipitated directly from circulating aqueous solution as well-ordered, low-temperature quartz. The overgrowth, or syntaxial rim, displays the same crystallographic and optical orientation as that of the underlying nucleating detrital grain. For monocrystalline grains, this similarity extends as far as reproducing the various types of undulatory extinction; for polycrystalline grains, the overgrowth changes orientation so as to match that of the individual crystals with which it is in direct contact, and this conformity extends even to microcrystalline chert grains.

SEM studies (Pittman, 1972) show that overgrowths begin either as growths with poorly defined crystal faces forming an interconnected anastomosing system at the surface of grains or as isolated growths with well-defined faces. Enlargement of these incipient overgrowths is followed by a partial envelopment of the nucleus and the formation of well-defined crystal faces when conditions of silica supply, time, and available space are favorable. This process continues by overlapping and merging of similarly oriented individuals or by envelopment of smaller individuals by one overgrowth that becomes predominant.

Overgrowths are generally in contact with their detrital substrate only at a few widely scattered points. Therefore, a large portion of the area between overgrowth and substrate is a capillary pore that forms the "dust line" visible under the microscope (**Plate 1.B**). This pore space is also partially infilled by minute mineral grains, coatings of clay minerals, iron oxides, organic matter, and even fluids, which contribute to outline the original shape of the detrital grain (Pittman, 1972). This shape may also be emphasized by other features, such as streaks of crystal inclusions or of fluid-filled vacuoles that occur only in the detrital grain and stop at its boundary with the overgrowth. In some instances, secondary quartz eventually fills the marginal capillary pore forming the "dust ring," and in the absence of mineral particles along it or of inclusions in the detrital core, it becomes impossible to separate grain from overgrowth under the petrographic microscope. However, recent techniques such as cathodoluminescence (Sippel, 1968) show the real boundaries between grains and quartz overgrowths. Backscatter electron imaging on the electron microprobe (Henry et al., 1986) reveals the complexity of overgrowths. They may differ from their detrital cores in trace element composition or in structural state, and they may display complex oscillatory zoning, disruption of zoning, and even zones of fluid inclusions. These features indicate that silica cementation developed in successive phases and that the chemical composition of circulating waters changed through time.

Initial overgrowth may begin selectively with a meniscus shape at the narrowest openings of capillary size between detrital grains. If intergranular space is available, overgrowths develop smooth, hexagonal and rhombohedral prismatic faces, but subsequent competition for space in pores leads to compromise boundaries ranging from irregular to straight or concavoconvex. Straight overgrowth faces commonly form 120° angles and develop spectacular triple junctions with adjacent grains. These variations of the shape of overgrowths result from their relative growth rate varying with the crystallographic orientation of the detrital grains, which is random with a given arenite, and which is also fastest along the *c* axis (Pittman, 1972).

The internal structure of detrital quartz grains affects the nature of the overgrowth and the volume of cement concentrated by individual grains. The preference is in decreasing order: straight quartz, quartz with undulatory extinction, polycrystalline quartz, and chert grains (James et al., 1986). Petrographic evidence seems to show that the development of overgrowth proceeds slower or starts later on polycrystalline grains and on chert grains than on monocrystalline grains, a situation also reproduced experimentally (Heald and Renton, 1966).

Whenever the overgrowth process is completed, the quartz arenite is changed into a tight mosaic of interlocking crystals. The texture is then called quartzitic, and the corresponding rock is a quartzite. This process is accompanied either by the almost complete disappearance of the dust rings, making the original shapes of the detrital quartz grains no longer visible (**Plate 1.C**), or by preservation of such rings when they are relatively thick (**Plate 1.D, E**). However, the quartzitic texture can develop also directly from a completed process of pressure solution (Skolnick, 1965) in which grains of quartz, originally with tangential point contacts, are differentially dissolved and display numerous straight, concavoconvex, and eventually sutured microstylolitic contacts. Silica liberated by this process is either reprecipitated in intervening spaces as small patches of syntaxial overgrowths or is altogether eliminated from the system.

Petrofabric studies on a possible crystallographic influence on the geometry of planar, concavoconvex, and sutured contacts due to pressure solution indicate that the geometry of all types of pressure-solution contacts are independent of the crystallographic orientation of opposing quartz grains (Hicks et al., 1986). With respect to the relative timing between quartz overgrowth and extensive pressure solution, observations (Houseknecht, 1988) indicate that most of the overgrowth cementation was completed before pressure solution began, although minor amounts of very early pressure solution between quartz grains, before any overgrowth, are observable in some instances after the initial mechanical rearrangement of the quartz grains. This relationship bears on the debated question of pressure solution providing silica for cementation. Indeed, in the general case of pressure solution occurring long after overgrowth, silica is released too late for cementation of that particular arenite. It is, however, available for cementing shallower arenites or for export out of the system.

Quartz overgrowth and pressure solution combine to such an extent in the generation of the quartzitic texture that, upon completion of the interlocking mosaic, it may be difficult to establish the respective parts ascribable to each process. Indeed, cathodoluminescence petrography (Sippel, 1968), by showing the outline of the original detrital grains and their surrounding overgrowths (**Plate 1.F**), reveals that cases of assumed pressure solution between grains are in reality compromise boundaries between overgrowths or result from pressure solution between overgrowths.

Flexible micaceous metaquartzites are called itacolumites, from the type locality in the Proterozoic of Itacolumi de Mariana in Minas Gerais, Brazil (Cayeux, 1929; Ginsburg and Lucas, 1949). These unusual rocks are characterized by mosaics of quartz grains ranging from extremely irregular to subrectangular; but in all instances individual grains are separated from each other by voids of uniform width, sometimes filled by short muscovite flakes (**Plate 2.A**). The flexibility of itacolumites results from a generalized but loose interlocking of the grains in which muscovite does not seem to play any important role. The generation of pervasive voids of uniform width between interlocking quartz grains, resulting in their free articulation, is attributed to a general and unusual contraction of quartz under conditions of retrometamorphism, since evidence of incomplete overgrowths or of irregular corrosion of the grains is wanting.

Whenever a quartzite is submitted to oriented tectonic stresses, as in overthrusts, the following sequence of textural changes of increasing intensity may be observed. Individual crystals of the mosaic display undulatory extinction, become elongate and reorient themselves along foliation planes, and develop eventually a mass extinction (**Plate 2.B**). Further deformation follows two distinct paths in relation to the ductility of the environment. The quartzite either recrystallizes

into highly irregular superindividuals of quartz with extreme undulatory extinction (**Plate 2.C**) or fractures into larger porphyroblasts surrounded by an irregularly granular matrix, generating a bimodality of pure structural origin (**Plate 2.D**). In both cases, the rock no longer displays any textural relationship to the original quartz arenite, and its complete reorganization makes it a real metaquartzite.

As stressed by Folk (1968), many quartz arenites have a bimodal texture, with each mode well sorted within itself. A diameter ratio of 6:1 or more may exist between the two grain sizes, with very few grains of intermediate size. Typically, the mixture consists of well-rounded and sorted coarse sand and fine to very fine sand from which the intermediate phase has been selectively removed (**Plate 2.E**). This is interpreted as the result of eolian desert processes as recognizable in Recent deserts all over the world. Most desert environments have developed by changes of climate upon older alluvial plains. The tendency of the wind is to remove the portion of the poorly sorted source sediment that is easy to set in saltation, that is, the fine sand, and to concentrate it into localized dunes of well-sorted unimodal sand (*erg*). Left behind on the desert floor are two fractions, the coarser grains that can only be rolled by the wind into ripples, irregular patches, or laminae, and fine sand or silt made of grains that have reciprocal cohesion on a flat surface and are too small to be easily moved by the wind. However, this small fraction can be occasionally removed by the wind and transported for long distances as dust storms. At any rate, the association of these two fractions builds the bimodal quartz arenites (*reg*).

Within the context of the problem of multicyclicity of quartz arenites, the association of widespread bimodal types and restricted unimodal ones indicates that such quartz arenites underwent an eolian deflation process whose imprint has survived within their subsequent fluvial or continental shelf environment of deposition.

Carbonates, mainly calcite, are common cements of quartz arenites. Calcite shows three major textures: (1) microcrystalline anhedral to subhedral aggregates of crystals (**Plate 2.F**); (2) one crystal twinned or not, filling a single pore; and (3) large poikilotopic patches that consist of interlocking single crystals enclosing many quartz grains (**Plate 3.A**). These various textures of calcite cement may be associated in a given rock or occur by themselves, but in some instances the size of the original pores appears to control the size of the calcite crystals. Commonly, boundaries of quartz grains and their overgrowth faces, both with high free-surface energy, preferentially display a variety of marginal replacement (commonly designated as *corrosion*) textures by the calcite cement. Dolomite and siderite cements seem to produce a less specific replacement. Marginal replacement features by calcite cement include small V-shaped notches and pits, regular to irregular embayments, and large depressions, all of which are clearly shown under the SEM (Burley and Kantorowicz, 1986). According to these authors, pits and notches are the smallest corrosion features and range in diameter between 1 and 2 μ; pits are very irregularly shaped, whereas notches have a V-shaped geometry and are more uniform in aspect. Overgrowths are still recognizable even when extensively pitted, and some larger rhomb-shaped notches may reach a size up to 10 μ.

Larger or coalesced pits are called *embayments*. They have a size commonly large enough to penetrate through overgrowths into detrital grains; they reach widths in excess of 20 μ.

The word *depression* describes considerably enlarged embayments that have removed such large amounts of overgrowths and of grains that the framework of the quartz arenite has been significantly reduced.

Marginal replacement features are particularly spectacular when the calcite cement displays a poikilotopic texture. The process may reach the extreme case of single quartz crystals becoming skeletal and consisting of irregularly shaped fragments "floating" in the calcite cement (**Plate 3.B**). These fragments have preserved the same optical orientation, which demonstrates that they originally belonged to a single grain. Even if one takes into account the misleading aspect of a randomly oriented thin section, which tends to give the impression that detrital grains of quartz do not form a grain-supported framework, quartz arenites with a poikilotopic calcite cement often show an "exploded" framework or displacive texture in which the cement has physically displaced or even broken the detrital grains of quartz. This situation appears to indicate a supersaturated environment, which also occurs in calcrete profiles (Buczynski and Chafetz, 1987; Saigal and Walton, 1988; Braithwaite, 1989).

A study of Recent calcite-cemented quartz arenites generated by the equatorial tree Iroko (*Chlorophora excelsa*) in the Ivory Coast (Carozzi, 1967) shows intense marginal replacement of quartz grains by the cryptocrystalline calcite cement reaching often the stage of skeletal grains. These conditions indicate that the replacement took place at the most over a period of 200 years and can therefore be considered as geologically instantaneous.

Cementation by patches of calcite and dolomite is frequent. The distinction of the two carbonates can only be reliably obtained by staining techniques. The supposed more idiomorphic rhombic habit of dolomite is a completely unreliable assumption, whereas that of siderite is a clearly established fact.

Other less common cements of quartz arenites include anhydrite (**Plate 3.C**), gypsum, barite, and celestite. All show the same three types of textures described above for calcite, as well as features of marginal replacement of detrital quartz grains. Opal occurs sometimes as cement of quartz arenites. It appears often in a concretionary form associated with small amounts of smectite (**Plate 3.D**). This amorphous form of silica seems to characterize Cenozoic quartz aren-

ites; with increasing geologic age it recrystallizes into fan-shaped microfibrous quartz. The presence of an opaline cement is often related to the diagenetic alteration of the products of silicic volcanism, particularly rhyolitic ashes into silica and smectite. Among clay mineral cements, fibrous rims of chlorite are frequent (**Plate 3.E, F**). Finally, cements of fan-shaped fibrous wavellite and of various types of zeolites although rare, are worth mentioning.

Depositional Environments and Provenance

Compared to all sandstones, quartz arenites display the highest concentration of well-sorted and well-rounded quartz grains combined with the most reduced suite of accessory minerals. The combination of these features presents a major genetic problem. Modern sands of this type were reported only in two equatorial fluvial environments (Franzinelli and Potter, 1983; Johnsson et al., 1988), but they occur in the geological record with an increasing frequency that appears to culminate in early Paleozoic and late Precambrian for reasons that remain unclear. In all instances, quartz arenites form widespread sheets, show large-scale cross-bedding, and occur interbedded with cratonic carbonates, but show little or no fossils. Their environment of deposition ranges from desert and fluvial to estuarine and continental shelf.

The major question raised by the origin of quartz arenites is their possible first-cycle derivation from weathering of granites and gneisses. According to studies by Suttner et al. (1981), such a direct derivation does not appear possible in areas of moderate to high relief located in temperate or drier climates, even if the final deposition environment consists of beaches or coastal dunes where high-energy conditions are favorable to the mechanical breakage of sand-sized grains by grain-to-grain collisions. The only realistic possibility of generating first-cycle quartz arenites is to consider the case of a long residence time of granitic and high-grade metamorphic material in a stable soil profile, implying low relief under a warm tropical climate, where it is known that intense chemical weathering can produce residual sands containing up to 95% quartz. This initial condition, which by itself would require the weathering of unusually large volumes of parent rock, should be followed by the effects of meandering alluvial systems with intermediate and extensive stages of storage in floodplains and coastal plains, together with further soil episodes, followed by a final setting of passive continental margins. The fluvial context conditions apparently occur today in the Orinoco river basin in Venezuela and Colombia (Johnsson et al., 1988) and in the Amazon basin in Brazil (Franzinelli and Potter, 1983). However, rates of sedimentation in passive margins, or even in cratonic basins, exceed by far the values required for keeping sand materials a sufficient time for effective shallow-water, grain-to-grain collision processes to further enhance by mechanical breakage the quartz content of the sediment. The combination of the above-mentioned conditions is highly constrained by tropical climate, low relief, and low sedimentation rate and does not seem able to account for the widespread sheets of quartz arenites; consequently, the unescapable conclusion is that such arenites are multicycle in origin.

Several modern techniques are being used to unravel the multiple origin of quartz grains by attempting to establish the provenance of these grains, a first step on the way to understanding multicyclicity. By using a combination of low and high degree of undulosity in monocrystalline quartz, and determining the amount of polycrystalline quartz and number of crystals per grain of polycrystalline quartz, Basu et al. (1975) and Basu (1985) discriminated between plutonic and low- and high-rank metamorphic source areas. However, these techniques are only applicable to first-cycle sands and sandstones, and parent rocks of sands derived from multiple sources cannot be distinguished. This approach is further complicated by diagenetic modifications and tectonic effects.

The shape of detrital quartz grains does not appear yet to be diagnostic of particular source rocks, although future progress may be expected from the application of new well-tested morphometric methods such as Fourier grain shape analysis (Hudson and Ehrlich, 1980). The determination of contents in trace and minor elements in individual quartz grains appears to be a promising tool to assign a particular grain to a specific rock type within a given basin when this sophisticated technique becomes widely applied. However, these approaches may well be restricted again to first-cycle sands and hindered by diagenetic effects.

The environmental interpretation of quartz grains has also been attempted by the study of the surface texture or roughness of the grains by using SEM microscopy in the secondary electron mode to image surface topography (Krinsley and Trusty, 1985). The technique is expensive and time consuming, and unfortunately some of the observed surface features that are of chemical or mechanical origin, or a combination of both, may represent several environments. At present, the following environmental combinations have been distinguished: weathering and youthful river; subaqueous littoral and shelf; turbidity currents and eolian processes; therefore only a combination of surface features can be very specific. Again, first-cycle sands can be partially understood in terms of environments of deposition, but recycling and diagenetic features make this technique inapplicable to helping solve the multicyclic character of quartz arenites.

The most promising technique for attempting to establish the provenance of monocrystalline quartz grains remains cathodoluminescence microscopy as shown by Owen and Carozzi (1986). Matter and Ramseyer (1985) introduced a subdivision of luminescing high-temperature quartz (originally mostly the β variety) into six types that were calibrated against specific types of igneous and metamorphic rocks, and hence have a reliable provenance significance. The types

are light blue to blue, bluish black, blue violet (mauve), violet, red, and brown.

Plutonic quartz and quartz phenocrysts of volcanic rocks display a wide variation of luminescence colors ranging from blue through mauve to violet. This similarity of colors results from the fact that phenocrysts of volcanic rocks crystallize early under plutonic and hypabyssal conditions. These phenocrysts are also identified by showing zoned or inhomogeneous distribution of luminescence colors. When quartz occurs as interstitial material in volcanic rocks, it luminesces red to red-brown because of its lower crystallization temperature combined with rapid rate of crystallization.

A bluish-black luminescence occurs in plutonic quartz crystals with strong undulatory extinction. The change from the original blue color reflects a reduced intensity of luminescence. Polygonization of quartz grains, which produces strain-free clear subgrains, generates a dull brownish-black color. Brown quartz is characteristic of regional metamorphic rocks, but if metamorphism was accompanied by high-temperature recrystallization, quartz grains revert to blue luminescence and can no longer be distinguished from plutonic quartz.

The same technique is applicable to polycrystalline quartz grains to distinguish their plutonic or metamorphic origin, as well as to fragments of fine-grained quartzites, metaquartzites, and silicic volcanic rocks.

In summary, the best approach to provenance problems raised by quartz grains is the combination of various techniques in their basinal context.

Diagenetic Evolution

The question pertains to the processes and the timing for an intracratonic well-sorted quartz sand of average porosity of 40% to become partially or entirely converted into a quartz arenite with a secondary quartz cement, mainly as overgrowth. A similar question for calcite cementation is discussed below.

The interbedding of quartz arenites with lithified carbonates and the intraformational reworking of cemented fragments of quartz arenites and of isolated grains with overgrowths show that lithification of quartz arenites by secondary quartz was an early diagenetic process that took place after a geologically short time on the order of a few tens of millions of years. Numerous cases exist as well in which quartz sands remained unlithified for tens of millions of years.

Blatt (1979) discussed at length the problem of cementation of quartz arenites and pointed out that the slow rate of movement of subsurface waters (on the order of 20 m per year) places critical restrictions on the time of cementation of a quartzose sand. To lithify such a sand, circulating waters must be supersaturated with respect to silica, which is to be precipitated into the pore space, and the number of pore volumes of water that must flow through the sand must be enormous. For a given well-sorted sand with a widespread extent

(65 km in length and width) and 20 m in thickness, Blatt's calculations indicate that its quartz cementation by horizontal flow of subsurface water is impossible within geologically reasonable periods of time. The only possible alternative is to assume that meteoric water with a silica concentration of 33 ppm SiO_2 first descended and then rose vertically during the cementation process immediately after deposition or when the depth of burial did not exceed a few hundred meters, and hence the flow distance was strongly reduced. Cementation could also have taken place at any subsequent time when tectonic conditions uplifted the buried quartz sand to near-surface conditions.

Petrographic evidence derived from the study of overgrowths shows that their precipitation occurred soon after deposition, at shallow depths, before any appreciable compaction of the quartz grains took place. The possible sources of silica remain widely debated. Solubility of quartz in water at 25°C is 6 ppm and rises slowly with increasing temperature and increasing pressure. Meteoric waters contain on the average 13 ppm dissolved silica; therefore, sufficient silica is present in them to cause cementation at a shallow depth of several hundred meters when solubility of quartz rises to 13 ppm. Although the means of chemically identifying the source of silica in overgrowths are lacking, it is reasonable to think that, under such conditions, silica in overgrowths originates from weathering processes, from dissolution of silicates in the shallow subsurface, and from dissolution of the quartz grains themselves, either in the uncemented portions of the same body of sand or in other contemporaneous or older sand bodies that remained unlithified, as pointed out above. However, increase in solubility of quartz with burial under increasing temperature and pressure leads to a content of dissolved silica in subsurface waters of several tens of parts per million, reaching at least 100 ppm in oil-field brines at depths of a few thousand meters. These conditions, which may occur in the deepest portions of intracratonic basins, obviously require additional sources of dissolved silica.

Two widespread processes can be considered. The first process releasing silica is the diagenetic conversion of smectite or mixed-layer illite/smectite to pure illite. Unquestionably, large amounts of silica are generated during this type of diagenetic reaction of clay minerals, but the problem lies in the timing between illitization and secondary quartz precipitation, because clay reactions should precede or be synchronous with quartz cementation. In the Gulf Coast, Boles and Franks (1979) showed that silica could be released at temperatures between 60° and 120°C to form quartz overgrowths in quartzose feldspathic lithic arenites. Modeling (Leder and Park, 1986) indicated that, under such conditions, a body of shale 1 km thick could release sufficient silica to cement completely by overgrowth an overlying 100-m-thick sandstone bed. Nevertheless, questions were asked by earlier authors (Yeh and Savin, 1977), as to whether the newly released silica had really left the shales.

The second process releasing silica is pressure solution. According to Houseknecht (1984), in the Pennsylvanian quartz arenites of the Arkoma Basin, pressure solution can be considered an important agent of mass transfer on a regional basis. Relatively fine grained quartz arenites tend to be silica exporters, that is, more quartz has been dissolved by pressure solution, whereas relatively coarse grained quartz arenites tend to be silica importers, that is, they show more quartz cement than pressure solution can account for. Temperature of burial expressed as thermal maturity also plays an important role in pressure solution if, at any locality, the amount of pressure solution increases with decreasing grain size. Regionally, the amount of pressure solution increases with relative thermal maturity, and vice versa. In this example, an internal mass balance on a regional scale can be implied. Arenites behaved as a more or less closed system in which silica dissolved by pressure solution in areas of high thermal maturity migrated down the temperature gradient within the arenites and precipitated as quartz cement in areas of low thermal maturity.

Other sources of silica, such as dissolution of feldspars and accessory minerals and kaolinization of feldspars, are of minor importance in the cementation of quartz arenites.

The abundance of calcite cement in quartz arenites also reveals texturally the effects of an early or shallow-depth meteoric cementation. Calcite shares the same timing with quartz since both cements may interfere, alternate, and precede or follow each other. Geochemical conditions of calcite cementation are easier to understand than those of silica because connate seawater in shallow marine sediments is saturated to slightly oversaturated with calcium carbonate, and meteoric waters contain abundant calcium carbonate. Furthermore, at depth, the effect of pressure solution is stronger on calcite than on quartz, and additional carbonate ions are released from adjacent arenites and carbonates.

In quartz arenites, cementation by authigenic clay minerals is very small and kaolinite or its polymorph, dickite, predominates. The necessary alumina is brought in by meteoric waters, and their extreme dilution favors precipitation of kaolinite rather than smectite or illite.

Cementation of quartz arenites by quartz and calcite, interpreted as an early meteoric water process ranging from the surface to shallow burial depth, appears further strengthened by the fact that these arenites are cyclically interbedded with cratonic carbonates for which freshwater phreatic cementation is a widely accepted fact. Therefore, it seems difficult to conceive that drastically different types of burial conditions and of circulating solutions should be required for cementing alternating clastic and carbonate units. As pointed out by McBride (1989), the idea of early cementation of quartz arenites by meteoric waters needs further geochemical testing using oxygen isotopes, but such techniques applied to the cementation of feldspathic arenites, as discussed below, favor the importance of meteoric burial diagenesis.

Reservoir Properties

Assuming that a quartz sand upon deposition possesses a primary porosity on the order of 40%, the following time succession of processes tends to change it into a tightly cemented quartz arenite: very early mechanical compaction expressed by tighter repacking of grains; rotation of grains and fracturing of grains; early, minor pressure solution between grains; and incomplete quartz overgrowth. These processes are responsible for a primary reduced porosity that is eventually entirely eliminated, either by complete overgrowth combined with pressure solution, or by incomplete overgrowth followed by precipitation of calcite and other carbonate cements, rarely by clay mineral cements. This was the assumed scenario for the disappearance of porosity with depth for quartz arenites, and other sandstones as well, until the studies of Schmidt and McDonald (1979a, b). These authors presented criteria for the recognition of secondary porosity, with emphasis on its widespread occurrence at depths, whereas earlier studies had concluded that porosity had been greatly reduced, if not completely obliterated, by compaction and various types of cementation.

Secondary porosity can be generated through dissolution of framework grains and/or authigenic cements, which can be either simple pore filling or in part replacive of framework grains (Shanmugam, 1985a,b). The process of dissolution develops under deep burial conditions by aggressive solutions rich in CO_2 originating from the decarboxylation of maturing organic matter in basinal shales that rise into adjacent sandstone bodies along basin margins. Besides ancillary types of secondary porosity, the most important in quartz arenites are hybrid pores (**Fig. 1.2**), which originate in two major ways. First, in the case of incomplete authigenic cementation by calcite, dolomite, or siderite followed by dissolution, hybrid pores are generated that combine features of residual primary porosity and secondary porosity. This process can theoretically reestablish final values of porosity identical to primary porosity. A second case involves complete authigenic cementation by calcite, dolomite, or siderite that has replaced the margins of framework grains. Upon dissolution of the cavity-filling and replacive authigenic cement, final values of porosity may be greater than primary porosity and then hybrid pores become oversized pores. These two possibilities emphasize the critical importance of burial secondary porosity, which can be combined with primary porosity in quartz arenites, particularly in view of the fact that secondary burial porosity can survive at depths of more than 6,000 m.

TYPICAL EXAMPLES

Plate 1.A. Porous quartz arenite weakly cemented by pressure solution shown by straight and concavoconvex contacts, with interstitial sericite films, between subrounded and well-rounded quartz grains. St. Peter Sandstone, Middle Ordovician, Klondike, Missouri, U.S.A.

Plate 1.B. Porous quartz arenite cemented by incipient quartz syntaxial overgrowths overlying "dust lines" and displaying subsequent partial dissolution of crystal faces. St. Peter Sandstone, Middle Ordovician, Ottawa, Illinois, U.S.A.

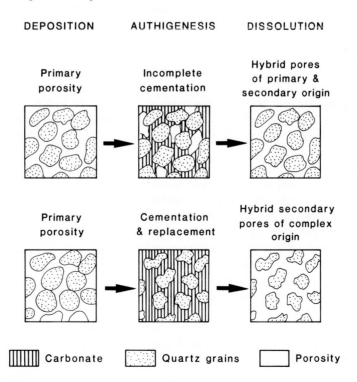

DEPOSITION   AUTHIGENESIS   DISSOLUTION

Primary porosity — Incomplete cementation — Hybrid pores of primary & secondary origin

Primary porosity — Cementation & replacement — Hybrid secondary pores of complex origin

▥ Carbonate   ⬚ Quartz grains   ☐ Porosity

FIGURE 1.2 Textural development of hybrid pores in sandstones. From Schmidt and McDonald (1979b). Reprinted by permission of the Society of Economic Paleontologists and Mineralogists.

Plate 1.C. Quartzite displaying a perfect mosaic of subhexagonal quartz grains interlocked by completed overgrowth with almost total destruction of original shapes of detrital grains. St. Peter Sandstone, Middle Ordovician, Ottawa, Illinois, U.S.A.

Plate 1.D. Quartzite displaying well-rounded grains of quartz with thin hematite coatings cemented by interfering syntaxial quartz overgrowths in optical continuity with their respective substrates. St. Peter Sandstone, Middle Ordovician, Oregon, Illinois, U.S.A.

Plate 1.E. Quartzite displaying well-rounded grains of quartz cemented by interfering syntaxial quartz overgrowths in optical continuity with their respective substrates and overlying "dust lines." St. Peter Sandstone, Middle Ordovician, Strawberry, Arkansas, U.S.A.

Plate 1.F. Same sample as Plate 1.E under cathodoluminescence, emphasizing the shapes of the detrital grains and their different provenance revealed by various shades of light blue, dark blue, and blue violet (mauve), whereas syntaxial overgrowths are light brown.

Plate 2.A. Itacolumite or flexible quartzite consisting of subrectangular grains of quartz elongated parallel to schistosity and articulated by means of open intergranular spaces of uniform width. Strong development of secondary muscovite flakes, parallel to the elongation of the quartz grains, plays no role in the flexibility, which is entirely due to the peculiar interlocking of the quartz grains. Proterozoic, Itacolumi de Mariana, Minas Gerais, Brazil.

Plate 2.B. Tectonized quartzite consisting of subrectangular to polygonal quartz grains. General tendency toward preferred orientation parallel to schistosity (upper-left corner to lower-right corner) is expressed by incipient mass extinction. An appreciable number of quartz grains show internal undulose extinction. Triassic, High Calcareous Alps, Arve Valley, France.

Plate 2.C. Highly tectonized quartzite consisting of completely regenerated and strongly interlocked quartz crystals with internal undulose extinction. General tendency toward preferred orientation parallel to schistosity (horizontal) is expressed by mass extinction. Triassic, High Calcareous Alps, Arve Valley, France.

Plate 2.D. Porphyroblastic quartzite consisting of irregularly shaped regenerated composite or single quartz crystals with internal undulose extinction set in a secondary matrix of microgranular undulose quartz, sericite, and chlorite. Triassic, High Calcareous Alps, Arve Valley, France.

Plate 2.E. Bimodal quartz arenite consisting of rounded and sorted, medium-sized quartz grains with straight extinction scattered in a groundmass of fine-sized quartz grains cemented by pressure solution and some interstitial sericite. Joachim Dolomite, Middle Ordovician, Montgomery City, Missouri, U.S.A.

Plate 2.F. Quartz arenite consisting of a framework of poorly sorted and irregularly shaped quartz grains cemented by pressure solution and interstitial cement of micro-

crystalline to incipiently poikilotopic calcite. Some quartz grains show irregular marginal replacement by calcite cement. Grand Tower Formation, Middle Devonian, Alto Pass, Illinois, U.S.A.

Plate 3.A. Quartz arenite consisting of a framework of coarse, subrounded to rounded and broken quartz grains with straight extinction. Well-developed interstitial cement of poikilotopic sparite with single crystals almost as large as the field of view. Moderate marginal replacement of quartz grains by the calcite cement in the shape of irregular embayments. Joachim Dolomite, Middle Ordovician, Brewer, Missouri, U.S.A.

Plate 3.B. Quartz arenite consisting of a framework of subangular quartz grains with predominant straight extinction and minor amount of pressure-solution contacts. Interstitial cement is fibrous calcite, with individual fibers perpendicular to quartz grain boundaries. Intense marginal replacement of quartz grains by calcite shown by original individual grains reduced to small irregular relicts with same optical orientation. Beachrock, Recent, Foz do Jacuipe, Salvador, Bahia, Brazil.

Plate 3.C. Quartz arenite consisting of a framework of subangular to subrounded quartz grains with straight or undulose extinction and appreciable amount of pressure-solution contacts. Interstitial cement is anhydrite (right half of picture) subsequently replaced by dolomite (left side of picture) consisting of a mosaic of fine subhedral to euhedral rhombs. Appreciable marginal replacement of quartz grains by anhydrite shown by fringes and embayments. Joachim Dolomite, Middle Ordovician, Isle de Bois Creek, Missouri, U.S.A.

Plate 3.D. Quartz arenite consisting of a framework of quartz and partially kaolinitized potassic feldspar grains with accessory lithoclasts of quartzites, ferruginous dolostone, and devitrified silicic volcanic glass. Interstitial cement is concretionary fibrous opal with argillaceous impurities often concentrated in the center of intergranular spaces. Ogallala Formation, Pliocene, Glade, Kansas, U.S.A.

Plate 3.E. Porous quartz arenite showing triangular intergranular space filled with a thin rim of green fibrous chlorite cement followed by a partially dissolved central single crystal of calcite cement. Basal Salina Formation, Lower Eocene, Talara, Peru.

Plate 3.F. Porous quartz arenite showing partially dissolved and collapsed thin rims of green fibrous chlorite cement. Basal Salina Formation, Lower Eocene, Talara, Peru.

GLAUCONITIC ARENITES

Framework

Peculiar arenites consisting of up to 90% of rounded green grains called glaucony are associated with quartz arenites and display all stages of transition to them. Odin and Matter (1981) stressed the fact that the term glauconite had been used to designate green grains as well as to name an assumed corresponding mineral species. In reality, the mineral species glauconite does not exist, and they suggested, to avoid further confusion, to discontinue the term glauconite and replace it by the term *glaucony* to designate the green grains. Indeed, glaucony grains are the reworked and transported end product of the submarine authigenic formation of glauconitic clay minerals, which form a sequence ranging from glauconitic smectite to glauconitic mica.

Numerous studies of Recent marine deposits (Odin, 1988) reported abundant glaucony on the sea floor and provided an understanding of the long debated conditions of its formation.

Glaucony presents a great morphological diversity. According to Odin (1988), four major factors control the habit of glaucony: (1) a substrate material that undergoes the authigenic process of verdissement or "greening"; (2) the porosity characteristics of the substrate because the authigenic phase begins to form in the confined microenvironment of pores; (3) the size of the substrate; and (4) the stage of evolution of the glauconitized material. The morphology of glaucony falls easily into two major categories: granular habit (**Plate 4.A**) and film habit (Odin and Matter, 1981). The granular habit is by far the most frequent and divides itself into five groups: mineral and lithic grains, carbonate and siliceous bioclasts, pelletoidal grains, completely evolved glaucony grains, and molds of calcareous and siliceous microfossils.

Individual minerals and lithic grains are an important group of granular substrates, which appear glauconitized regardless of their composition. The minerals are quartz (Hughes and Whitehead, 1987), feldspars, mostly K-feldspars (Dasgupta et al., 1990), muscovite, biotite, calcite, dolomite, and phosphates. Lithic fragments consist of all types of igneous, volcanic, and metamorphic rocks, as well as cherts (Hughes and Whitehead, 1987) and volcaniclastic rocks. Hyaloclastic grains of sideromelane glass from subaqueous basaltic eruptions may display ghosts of feldspar laths, flowage structure, and alteration textures preserved in subradiating and spherulitic aggregates of glaucony (Jeans et al., 1982). However, glauconitized minerals and lithic grains, contrary to pellets and bioclasts, rarely dominate by themselves a given deposit. The verdissement of mineral grains always begins along tiny fissures and particularly cleavage planes, which represent their major type of microporosity. In quartz grains, the process stops rather rapidly after filling the minute cracks, whereas in biotite and muscovite flakes the green authigenic clay grows between cleavage plates, splitting them apart and eventually replacing them. The final product appears greatly expanded compared to its original substrate and its accordionlike shape preserves residues of the former cleavages, a distinct high pleochroism, and unusual polarization colors under crossed nicols (**Plate 4.B**).

Pelletoidal grains are a frequent substrate for glauconitization. Most of them are ellipsoidal argillaceous (ka-

olinitic) or carbonate (micritic) fecal pellets of mud-eater or filter-feeding benthic organisms. The verdissement of these pellets begins with the filling of their microporosity and is completed by replacement of the substrate (**Plate 4.C**), which is often accompanied by the development of radial cracks forming the lobate glaucony grains (**Plate 4.D**). Carbonate and siliceous bioclasts are glauconitized in the same manner as internal molds, but at a much smaller scale. The green authigenic clay begins to fill the original test porosity or the secondary porosity due to biogenic perforations; partial to total replacement of the tests themselves follows, with frequent preservation of traces of their original microstructure.

Since the filling of intragranular porosity of all types of substrates is followed by the replacement of the substrates themselves, regardless of their mineralogical composition, the most important group of grains consists of individuals having reached their final stage of evolution with complete destruction of the original substrate. These grains predominate in glauconitic arenites because of their homogeneity and greatest resistance to abrasion during transport. Such grains are always slightly larger than associated detrital quartz grains because of the lower specific gravity of glaucony (**Plate 4.E**). In plane-polarized light, these glaucony grains display a wide range of shades of green, from pale green, often yellowish, to dark green, and pleochroism is weak. The range of colors expresses stages reached by the authigenic process when it was arrested by unfavorable environmental conditions; it has no relation to the age of formation of the mineral. Under crossed nicols, most fully evolved grains of glaucony appear cryptocrystalline to microcrystalline with an aggregate polarization. Whenever residual textures are inherited from the original substrate, whether bioclastic or micaceous, a preferred orientation of the crystallites is observable.

The chambers of calcareous and siliceous microfossils are either empty and filled by the green authigenic clay or the latter replaced an original calcareous or argillaceous filling (**Plate 4.F, Plate 5.A**). Both situations may be accompanied by partial to total replacement of the calcareous or siliceous walls of the chambers, generating enlarged molds with irregular shapes.

Two varieties of the film habit occur (Odin, 1988). The first develops at the surface of large carbonate bioclasts, lithic pebbles and boulders, and synsedimentary nodules of phosphates or chert. The green clay again forms in various types of marginal microporosity of these substrates, such as fissures and microborings resulting from submarine chemical, physical, or biogenic alteration. The thickness of the film depends on the intensity of the submarine alteration, ranging from a few millimeters to a few centimeters, with best development in carbonate substrates. The second variety of film habit covers extensive surfaces represented by hardgrounds on carbonate substrates and differs from the first only by its areal habit. Downward from these extensive films, glaucony may show a diffuse habit corresponding to the distribution of minute spots almost beyond petrographic resolution.

The various types of habit just described are different expressions of the same process acting on a porous substrate within which the authigenic mineral genesis occurs in a confined microenvironment. However, some relationship seems to exist between the size and nature of the substrate and observed stages of evolution of the green authigenic clay minerals (**Fig. 1.3**). For argillaceous fecal pellets, the evolution to pure dark-green clay is best reached at sizes between 200 and 500 μ; smaller and larger sizes are less evolved as if some optimal conditions of glauconitization existed at a specific distance from the outside of a glauconitized substrate (Odin, 1988). This situation probably expresses the best conditions for cation exchanges inside grains during this process. As mentioned above, not only size but also the mineralogical nature of the substrate plays a role, as for instance in quartz and micas.

The mineralogical variety of glaucony is shown by X-ray diffraction studies, which indicate that it consists of a family of glauconitic minerals that shows two end members: a potassium-poor glauconitic smectite and a potassium-rich glauconitic mica (Odin, 1988). The unique chemical and crystallographic features of this family are as follows: minor substitution of tetrahedral silicon, dominance of iron in the octahedral sites, and a high proportion of interlayer potassium.

Cements

Pure glauconitic arenites are cemented early by reciprocal deformation and interpenetration of glaucony grains involving compaction rather than pressure solution. Extreme cases

consist of patches of glaucony in which boundaries of individual grains have almost disappeared and now simulate concentration of interstitial matrix. Pure glauconitic arenites lack the required detrital quartz nuclei on which quartz cement could grow and thus escape cementation by quartz, although glaucony may be replaced by quartz (McBride, 1989). When glaucony grains are associated with overgrowths on quartz grains, they also appear strongly deformed against the quartz crystal faces. Otherwise, when calcite and dolomite cements are present, glauconitic arenites display the same textural properties as quartz arenites, including marginal replacement of glaucony grains by carbonates and sulfates, mainly anhydrite.

Depositional Environments and Provenance

Glaucony occurs in sedimentary rocks of all ages from the Precambrian to the Present. However, glauconitization reached peaks of development, basinwide and worldwide in scale, during certain time intervals such as Albian–Cenomanian, early Cenozoic, and Quaternary to Present.

Glaucony is restricted to the marine environment. Instances of occurrence in continental and restricted hypersaline environments consist of minerals that are not identical to glauconitic minerals as defined above, but are in fact ferric illites with a lower iron content (Kossovskaya and Drits, 1970).

SEM studies (Odin and Matter, 1981; Odin, 1988) show a definite textural evolution taking place during the genesis of glaucony or verdissement process close to the

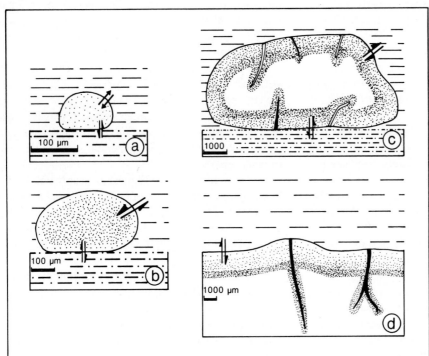

FIGURE 1.3 Glauconitization of substrates differing by size at the seawater–sediment interface: (a) small grains; (b) coarse grains; (c) lithoclasts; (d) hardground. Dots indicate the density of glauconitization and arrows represent cation exchanges. From Odin (1988). Reprinted by permission of the author and Elsevier Science Publishers.

water–sediment interface (**Fig. 1.4**). The first stage, or nascent stage, consists of the neoformed clay forming tiny, ill-defined globules, less than 0.5 μ in size, which eventually become attached to one another, forming "caterpillar" structures 2 to 3 μ in size, all of them attached to the pore walls, whereas small lepispheres, 3 to 4 μ in diameter, grow inside the pores. At this nascent stage, the green clay is iron rich and its K_2O content ranges from 2% to 4%. The destruction of the substrate generates new pore systems, which are filled by the second or slightly evolved stage. This stage consists of boxworks of small blades, 1 to 3 μ wide, that often relate to lamellae and cracks in micaceous substrates and to well-developed automorphic rosettes. The K_2O content is now 4% to 6%. Because the insides of the grains are more favorable for crystal growth than the surfaces, the larger and better organized crystallites develop in the center of grains, and radial cracks develop at the surface, giving the grains a lobate aspect. This widespread fissuration may break up individual grains into irregularly pie shaped segments that undergo their own subsequent evolution. Also at this stage, micaceous substrates expand into accordionlike grains. The K_2O content is now 6% to 8%, and the grains are considered evolved. Conditions can proceed further with K_2O content greater than 8%; the highly evolved grains have a smooth surface that has been restored by a new filling of the radial cracks. Under the SEM, the structure consists of well-developed lamellae, 5 to 10 μ long, slightly sinuous and showing a subparallel alignment.

At present, glauconitization appears to occur in all oceans; in terms of latitude, only the extreme coldest areas of the globe appear excluded; however, temperatures below 15°C appear most favorable, as well as a pH around 8. The Eh conditions are those of the oxidation–reduction boundary, which allow mobilization of the iron as Fe^{2+} from seawater and its stabilization inside the crystal structure as Fe^{3+}. In other words, the general environment of formation of glaucony is the open marine sea floor where various types of porous substrates are in direct contact with seawater. Glaucony forms between 60 and 1000 m depth of water, but mostly between 60 and 550 m. The environment should receive little or no detrital influx since the formation of glaucony occurs in contact with seawater and represents a hiatus in deposition. It is assumed (Odin, 1988) that a glaucony with a mean of 7% K_2O indicates a duration of submarine exposure of 10^4 to 10^5 years. Dark-green, highly evolved glaucony, possibly in part goethitized or phosphatized, may express 10^6 years of hiatus. The base of many transgressive series or eustatic rises of sea level, such as the worldwide Aptian–Cenomanian, is marked by glauconitic layers. Indeed, rises of sea level generate conditions favorable to glauconitization. First, they bring a variety of porous substrates (bioclasts, pellets, and minerals), originally produced in shallow waters unsuitable for glauconitization, into waters deeper than 60 m where the process can take place. Second, the clastic supply is greatly reduced by drowning of source areas, leaving the substrates in contact with seawater for long periods of time. However, a transgression or eustatic rise in sea level is not a prerequisite for glauconitization, be-

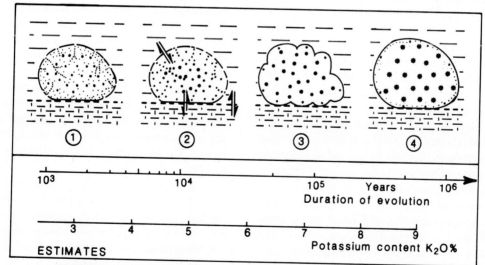

FIGURE 1.4 Glauconitization of a granular substrate: (1) nascent stage; (2) slightly evolved stage; (3) evolved stage; (4) highly evolved stage. Stars represent glauconitic minerals. The grain (about 5 mm in diameter) is shown at the boundary between seawater and sediment. After this evolution, the glauconitic minerals do not evolve anymore at the seawater–sediment interface and the green grain is relict. From Odin (1988). Reprinted by permission of the author and Elsevier Science Publishers.

cause a variety of other tectonic and sedimentological conditions can provide an adequate context for the process.

Diagenetic Evolution

Because glaucony is a diagenetic mineral, the evolution discussed here pertains only to changes taking place after its genesis that can be stopped at any stage when the environment is modified by two factors: sea level changes and shallow burial.

Shallowing brings glaucony grains into contact with an oxidizing environment or a renewed hiatus of deposition, in both cases the grains turn brown by development of goethite (**Plate 5.B, C**). Another form of alteration is the association glaucony–phosphates (Odin and Létolle, 1980), in which both minerals can replace each other. Although the two minerals form under different general environments and microenvironments, they share certain similarities, such as formation during a hiatus of sedimentation, at the sediment–water interface, and in interstitial waters. Still the problem of the association glaucony–phosphates remains unclear, and in the case of phosphatized glaucony, it could be related to deepening and upwelling conditions acting on previously formed glaucony.

Shallow burial under appreciable detrital influx stops glauconitization and is responsible for an early burial reorganization expressed by a crystallographic reordering of the previously formed glauconitic minerals. This observation (Odin, 1988) confirms the assumption that the final environment of equilibrium of these minerals could be interstitial water whose chemical composition is different from seawater.

Petrographic data on the effects of deep burial and tectonization on glaucony are lacking, but the geochemical aspects of these conditions were studied respectively by Odin (1982) and Conard et al. (1982). Weathering and reworking of glaucony were examined geochemically by Odin and Rex (1982). In ferricretes (hard iron crusts) of the Ivory Coast, West Africa, which can reach several meters thickness, Nahon et al. (1980) showed that their ferruginous ooids and pisoids in an iron-rich argillaceous matrix are of pedogenic origin. They result from the lateritic weathering of glaucony grains from Paleocene glauconitic argillaceous sandstones by successive centripetal concentrations and reorganizations of the iron oxides and hydroxides substituted with alumina (aluminous goethite). It is suggested that some types of older oolitic and pisolitic iron ores could represent paleoexposure crusts generated in tropical climates and derived from glauconitic facies.

Reservoir Properties

No fundamental differences exist between porosity development in glauconitic arenites and in quartz arenites. Primary porosity would occur regardless of the reciprocal proportion of glaucony and quartz grains. It would be reduced by a combination of quartz overgrowth, ductile deformation by compaction of glaucony grains, and finally by precipitation of carbonate and clay mineral cements.

Under burial conditions, secondary porosity can be generated by dissolution of framework grains and/or authigenic cements that were either simple pore filling or in part replacive of glaucony and quartz grains. This latter situation corresponds to the previously described two types of hybrid pores. However, in glauconitic arenites, an additional minor type of secondary porosity can develop through shrinkage of glaucony grains by recrystallization, dehydration, or alteration to goethite (Schmidt and McDonald, 1979a, b). The shrunken grains of glaucony appear almost free within the space they originally occupied after cementation. This type of porosity is by itself not accompanied by any increase of permeability. In some carbonates, this shrinkage porosity may be obliterated by precipitation of a rim of fibroradiated calcite, which, although it postdates the rigid interstitial calcite cement, has a texture indicating that it still represents a phase of submarine phreatic cementation, implying therefore also an early shrinkage of the glaucony grains (Lindström, 1980).

TYPICAL EXAMPLES

Plate 4.A. Glaucony quartz arenite consisting of a framework of highly pressure-welded glaucony grains separated by intergranular black pyrite films. Subangular to subrounded grains of quartz and a few incipiently altered potassic feldspars (extreme right of picture) are scattered between the glaucony grains and display the same intensity of pressure solution. Lion Mountain Sandstone, Cambrian, Kingsland, Texas, U.S.A.

Plate 4.B. Quartz glaucony arenite with microcrystalline calcite cement showing a large glauconitized muscovite flake. The highly expanded flake with an accordionlike shape displays residues of former cleavages and unusual polarization colors. Albian, High Calcareous Alps, Arve Valley, France.

Plate 4.C. Glaucony quartz arenite showing evolved glaucony pellets with partial pressure-solution contacts. Interstitial argillaceous calcite cement is stained by goethite and contains scattered subrounded grains of detrital quartz. Glaucony pellets are incipiently replaced along their margins by clear calcite rhombs. Hauterivian, High Calcareous Alps, Pilatus, Central Switzerland.

Plate 4.D. Glaucony quartz arenite showing highly evolved glaucony pellets with typical lobate shape due to development of radial cracks. Interstitial argillaceous calcite cement is stained by goethite and contains scattered subangular grains of detrital quartz. Albian, High Calcareous Alps, Arve Valley, France.

Plate 4.E. Glaucony quartz arenite consisting of a framework of subangular to subrounded grains of detrital glaucony associated with smaller angular grains of detrital quartz and brown phosphorite grains. Interstitial cement of coarse, slightly argillaceous sparite. Gallatin Formation, Cambrian, Sheridan, Wyoming, U.S.A.

Plate 4.F. Mold of textularid filled with a combination of glaucony and subsequent calcite replacing glaucony. Interstitial matrix is a fossiliferous argillaceous calcisiltite. Albo-Aptian, Clèdes, Aquitaine, France.

Plate 5.A. Crinoidal calcarenite showing glaucony filling microporosity of crinoid columnal and, in places, forming irregular patches by complete replacement of skeletal carbonate. Gallatin Formation, Cambrian, Sheridan, Wyoming, U.S.A.

Plate 5.B. Glaucony arenite consisting of small light-green pellets and large lobate pellets at all stages of alteration to opaque goethite. Interstitial cement is microcrystalline calcite. Scattered debris of *Discocyclina* and *Assilina* are filled or partially replaced by glaucony. Eocene, Fresco (Nagagrébo), Ivory Coast, Africa.

Plate 5.C. Altered glaucony arenite consisting of a framework of subrounded and accordionlike vermicular glaucony grains variably replaced by goethite. Interstitial material is an association of microgranular glaucony, clay minerals, and microcrystalline calcite stained by goethite. Cretaceous, Hazlet, New Jersey, U.S.A.

FELDSPATHIC ARENITES

Framework

As stated previously, feldspathic arenites contain 25% or more of K-feldspar grains exceeding lithic grains and less than 15% of interstitial detrital matrix. Subfeldspathic arenites contain less than 15% of K-feldspars and few or no lithic grains. Only feldspathic arenites are described here as characteristic representatives of the family.

Feldspathic arenites are by definition not only mineralogically less mature than quartz arenites, but also texturally immature as shown by a generally coarse grained texture, angular to poorly rounded grains, and the frequent presence of interstitial detrital matrix. Nevertheless, the predominant framework constituent is detrital quartz, although in some rare cases feldspars may exceed quartz. Quartz grains present all the petrographic types described in quartz arenites, but because of lower textural maturity, quartz grains with undulatory extinction and polycrystalline, which have survived mechanical destruction, occur in appreciable proportion compared to those with straight extinction.

When examining in feldspathic arenites other constituents besides quartz, it is necessary to deal, in a descriptive fashion, with the question of their pre- and postdepositional alteration, which is discussed again genetically when treating the section on the diagenetic evolution of these rocks. Feldspar grains with a few exceptions are mainly K-feldspars, generally microcline, orthoclase, and perthite, and sodic plagioclase (albite), which may be locally the predominant or even the only type of feldspar present in the "plagioclase feldspathic arenites." Calcic plagioclases are rare either because of original conditions or as a result of diagenetic albitization. K-feldspars can be perfectly fresh, typically clear, and colorless. They are distinguishable from the associated quartz by their cleavage, twinning, and refractive indices. Untwinned, clear potassic feldspar with diagenetic overgrowth may mimic quartz, and only staining techniques are reliable for their differentiation.

K-feldspars in feldspathic arenites display many types of diagenetic alteration. They range from incipient stages, with cloudy or turbid aspect, to advanced stages, appearing as subrectangular areas of aggregated secondary minerals that simulate either phyllitic lithic grains or patches of interstitial detrital matrix. The realization of the importance and variety of K-feldspar and sodic plagioclase alteration is critical for the determination of the real amount of feldspar present. Indeed, any underestimation has obvious consequences on classification, terminology, and interpretation of provenance, in particular for the evaluation of the tectonic setting of source areas (Helmold, 1985). Actually, processes of alteration that increase in intensity with time have three major aspects: (1) dissolution processes generate skeletal grains and final destruction, modifying the composition of the rock toward the pole of quartz arenites; (2) replacement processes by clay minerals generate concentrations mistaken for lithic grains, modifying the composition of the rock toward the pole of lithic arenites; and (3) replacement processes by other minerals such as quartzification and albitization, confusing the interpretation of provenance.

K-feldspars undergo several processes of alteration. Vacuolization is a microscopic preliminary stage of direct dissolution. It develops intraparticle microporosity, which gives to the grain a turbid aspect. This process appears to set up conditions favoring other types of alteration. Kaolinitization (**Plate 5.D**) is characterized by a deep brownish and cloudy aspect of the grains, accompanied by frequent preservation of relicts of original structures such as cleavage and twinning planes. Sericitization (**Plate 5.E**), chloritization, and muscovitization appear first as scattered flakes of these respective minerals along cleavage planes before involving the entire grain and changing it into a phyllitic-looking aggregate. Other processes are quartzification, albitization, calcitization (**Plates 5.F, 6.A**), anhydritization, zeolitization, and generation of authigenic K-feldspars (Milliken, 1989; Morad et al., 1989).

Sodic plagioclases undergo their own particular types of alteration following initial vacuolization: replacement by mixed-layered or illitic clays, sericitization, muscovitization, and zeolitization. Calcitization (**Plate 6.B**) and dolomitization are frequent processes that change the plagioclase grain into a fine mosaic of anhedral to subhedral crystals simulating marble or dolostone lithics. Replacement by aggregates of calcite and quartz is also common.

For both K-feldspars and sodic plagioclases, it is difficult to establish the precise timing of the various types of alteration, that is, to reconstruct diagenetic sequences for each basin, because they span the entire time interval between weathering in source rocks and additional processes during burial. The major obstacle is poor understanding of the capability of partially weathered feldspars to resist mechanical destruction during transportation. For the time being, the only available clue is simply that the alteration of a feldspar surrounded by a clear overgrowth took place in the predepositional environment, whereas an altered overgrowth is obviously postdepositional; all other instances are ambiguous.

Large flakes of detrital micas characterize numerous types of feldspathic arenites. They consist of relatively stable muscovite and unstable biotite, whose preservation ranges from fresh to various stages of alteration to chlorite and iron oxides and hydroxides. Mica flakes, because of their hydraulic behavior, are always much larger than associated grains of quartz and feldspars. They are generally aligned parallel to bedding, and sometimes imbricated. Compaction is responsible for distorted, crinkled, bent, and frayed flakes against more rigid adjacent quartz and feldspar grains.

Processes of alteration of muscovite are kaolinitization or direct dissolution. More spectacular is the partial to total dissolution of biotite flakes together with grains of hornblende, pyroxene, epidote, and chlorite into important concentrations of red iron oxide precursors, which later stabilize into hematite. In many feldspathic arenites, the "bleeding" from these altered grains and flakes imparts an intense reddish color to all framework constituents and interstitial matrix.

Minor accessory minerals in feldspathic arenites include apatite, sphene, zircon, tourmaline, rutile, epidote, garnet, magnetite, and ilmenite; but in many instances diagenetic dissolution has reduced the suite to the most stable types: zircon, tourmaline, and rutile.

Lithic grains are predominantly derived from igneous and high-rank metamorphic rocks such as granites, represented locally by intergrowths quartz–feldspars (myrmekites), pegmatites, gneisses, micaschists, and phyllites. They are associated with lithics of sedimentary rocks originating from intraformational reworking of the feldspathic suite itself, such as red shales, fined-grained feldspathic arenites, and occasionally older limestones and dolostones, and even chert.

Matrix and Cements

Cementation of many feldspathic arenites occurs by means of a variable combination of overgrowths on quartz and feldspar grains and pressure-solution processes (**Plate 6.C**).

An interstitial detrital matrix, whenever present in feldspathic arenites, consists of clay minerals that are almost beyond the resolution power of the petrographic microscope. Furthermore, their identification is complicated by frequent and abundant staining by authigenic iron oxides. However, under the SEM, the matrix constituents include kaolinite, illite, chlorite, sericite, and muscovite (**Plate 6.D**).

Ubiquitous films of iron oxides (**Plate 6.E**) are distributed throughout many types of feldspathic arenites with interstitial matrix or pressure solution when circulation of intrastratal fluids has occurred. These fluids also deposited within pore spaces authigenic clay mineral cements that, under the SEM, appear well crystallized and include vermicular kaolinite, dickite, illite, and chlorite.

Cementation of feldspathic arenites by calcite (**Plate 6.F**) and dolomite is more frequent than by anhydrite (**Plate 7.A**). Calcite cement has the same textural variations as in quartz arenites: microcrystalline aggregates, mosaic of anhedral to subhedral crystals, single crystals twinned or not, filling pore spaces, and large poikilotopic patches consisting of single crystals enclosing several quartz and feldspar grains. Marginal replacement of quartz and feldspar grains by the various types of carbonate cements (calcite, dolomite, and siderite) is widespread and again similar to that described in quartz arenites.

**Depositional Environments
and Provenance**

Feldspathic arenites, which occur throughout the entire geologic column, have an obvious provenance significance, a feldspar-rich source area characterized by predominant K-feldspar consisting of granites, gneisses, and other high-rank metamorphic rocks.

A first problem concerns the climatic and geomorphologic conditions under which K-feldspars can be released for transportation and final deposition without being entirely weathered into clay minerals in the source areas. Extremely arid or extremely cold climates that inhibit weathering processes were assumed at first, but feldspathic sands were also found in humid tropical and equatorial areas, indicating that climate is not the only critical factor. This was pointed out by Krynine (1935), who proposed that the generation of feldspathic sands required tectonically active high reliefs with related rapid erosion, rather than an arid climate. Indeed, under these conditions, a combination of fresh and altered feldspars is released, and if an equilibrium is reached between weathering processes and active erosion, feldspathic sands are generated within a wide range of climatic conditions.

A second problem pertains to the amount of transport that feldspar grains can withstand since they are more susceptible to mechanical destruction than quartz grains. At first, feldspathic sands were assumed to imply short transport, because feldspars indeed suffer destruction in short, high-gradient streams draining uplifted granitic source areas, but they do reach larger low-gradient streams in which they survive transport for thousands of miles (Russell, 1937). This means that feldspathic arenites span a wide variety of depositional environments.

The various environments of deposition of feldspathic arenites can be distinguished by a genetic interpretation of their textural properties. These environments range from in situ soil generation over granitic rocks to alluvial fans, fluviodeltaic conditions, and eventually marine turbidites.

Most feldspathic arenites are nothing more than the products of in situ weathering of granitic and gneissic rocks and are called *grus*. The distribution of quartz grains and feldspars, at various stages of vacuolization and kaolinitization, is so heterogeneous that patches with grain-supported and matrix-supported textures are randomly mixed. Local concentrations of framework grains with little interstitial argillaceous matrix permit cementation by a combination of quartz and feldspar overgrowths and pressure solution, forming the *regenerated granites*. These particular types of residual deposits, which can be incipiently stratified and contain pebbles of both quartz and the underlying granite, still display a textural heterogeneity that makes them strikingly different from a typical granitic texture.

Residual feldspathic arenites may undergo limited transport downslope by mass movements and form alluvial fans along the margins of tectonically active fault-bounded grabens, often referred to as *granite wash*. Under these conditions, incipient sorting is developed, and a combination of interstitial detrital clay matrix and clay mineral cements deposited by circulating waters is common. These fluids, if alkaline and oxidizing, initiate also the diagenetic alteration of biotite flakes and hornblende crystals into reddish iron oxides, which show a mottled or uniform appearance. These iron oxides, which can be considered as an early diagenetic cement, display the temporal sequence limonite–goethite–hematite. They were observed to form today in the subsurface of hot, arid to semiarid climates in first-cycle feldspathic arenites (Walker, 1967; Walker et al., 1978; Hubert and Reed, 1978).

In braided streams and lower fluvial environments, which emerge from tectonic grabens, feldspathic materials undergo further sorting and rounding of the grains, accompanied by stratification and cross-bedding of the deposits. Equivalent feldspathic arenites display a decreasing amount of matrix and a related increase of porosity and permeability, favoring the precipitation of a variety of well-crystallized clay mineral cements together with calcite and iron oxides produced by the alteration of biotite and hornblende. The proportion of quartz increases as a consequence of the elimination of altered feldspars combined with their partial or complete replacement by the calcite cement.

The above-mentioned association of cements occurs also in deltaic environments, and marine platforms of continental trailing edges and in related turbidites. Calcite cement also characterizes freshwater lacustrine feldspathic arenites, being associated with dolomite, anhydrite, and gypsum in saline lakes and playas. Zeolite cements (natrolite, analcite, heulandite, stilbite, and clinoptylolite) occur mainly in feldspathic arenites deposited in saline lakes submitted to the influence of subaqueous volcanism in the context of early phases of the development of rift valleys.

Feldspathic arenites are a favorable context for analyzing the role of feldspars as provenance indicators. Since some of the properties of feldspars, such as chemical composition, zoning, twinning, and structural state, have genetic implications, they provide clues on the provenance of individual grains (Helmold, 1985).

The determination of the chemical composition of feldspars using the electron microprobe is now a rapid and precise procedure (Trevena and Nash, 1981). These analyses indicate that potassium content is a useful parameter for establishing the origin of both alkali feldspars and plagioclases. Detrital grains of alkali feldspars more sodic than Ab 50 are derived almost exclusively from volcanic parent rock. Alkali feldspars more potassic than Or 88 are derived from plutonic and metamorphic rocks. The potassium content of volcanic plagioclases increases markedly as sodium content increases. This situation permits discrimination between a volcanic and a metamorphic or plutonic origin for the more sodic varieties of detrital plagioclases.

Chemical zoning in alkali feldspars and plagioclases is well known and has been used for plagioclases as a possible provenance indicator (Pittman, 1963, 1970), an application further refined by means of the electron probe (Trevena and Nash, 1981). For practical purposes, it is convenient to consider three situations: progressive zoning, oscillatory zoning, and absence of zoning. Progressive zoning is generally revealed, under crossed nicols, as a broad wave of extinction with or without sharp boundaries. Oscillatory zoning appears, under crossed nicols, as a fine structure of successive thin bands of alternating extinction. Most moderately to strongly zoned plagioclases belong to the high-potassium volcanic–plutonic domain, while weakly zoned to unzoned types belong to the low-potassium metamorphic–plutonic domain. All the plagioclases with oscillatory zoning concentrate in the high-potassium volcanic–plutonic trend, with most of them in the volcanic field. Plagioclases with progressive zoning or unzoned fall in the low-potassium metamorphic–plutonic field, with most of the unzoned plagioclases in the metamorphic field. A higher degree of resolution of provenance on the basis of chemical composition and zoning of feldspars has not yet been reached.

Feldspar twinning, besides being a common property for identification, has also been used as a provenance indicator. For plagioclases, Gorai (1951) divided the twins into two groups: A-twins, which include the albite, pericline, and acline laws, and C-twins, which include all the other laws. He attempted to establish the distribution of the two groups among plutonic, volcanic, and metamorphic rocks (gneisses, schists, and hornfels). The results indicate that A-twins occur in all three types of rocks, but are usually less abundant in hornfels. C-twins are restricted to volcanic and plutonic rocks, except a few present in hornfels. The relative frequency of twinned and untwinned plagioclases in plutonic and volcanic rocks varies with composition. C-twins are more abundant in calcium-rich (> An 50) plagioclases, whereas both A-twins and untwinned crystals are more abundant in sodic plagioclases. No relationship seems to exist between the frequency of twinned plagioclases and the composition of plagioclases in gneisses and schists.

The above-described provenance considerations based on zoning and twinning of feldspars have to be carefully evaluated as a function of mechanical abrasion during transport (Helmold, 1985). Indeed, abrasion reduces grain size and modifies the frequency of feldspar grains presenting certain types of zoning or twinning; thus it becomes responsible for erroneous interpretations of the nature of source rocks. A typical example is the destruction of plagioclase by stream transport (Pittman, 1969). Increasing distance of transport corresponds to an increase in both the A-twin to C-twin ratio and in the untwinned to twinned ratio for all size fractions. The downstream increase of the A-twin to C-twin ratio was interpreted by Pittman (1969) as the result of the selective destruction of C-twins. He proposed that plagioclase grains twinned according to the Albite–Carlsbad A law would break along the 010 cleavage, which is parallel to the composition plane of the Carlsbad twin, thus producing two daughter A-twins out of one parent C-twin. The same author interpreted the downstream increase of the untwinned to twinned ratio as resulting from breakage of A- and C-twins along their composition planes, hence releasing untwinned grains.

Assuming the above-mentioned distribution of A- and C-twins of plagioclases in source rocks as established by Gorai (1951), and a process of transportation destroying selectively most or all C-twins, the source area would be erroneously interpreted as metamorphic instead of igneous (volcanic and/or plutonic).

The ratio of unzoned to zoned plagioclases may also increase with the distance of transport (Helmold, 1985) if, for instance, mechanical abrasion of coarsely zoned plagioclases concentrates unzoned grains in the finer fractions. The resulting interpretation of provenance would be the same as that due to the destruction of plagioclase twins. Indeed, since plagioclases in volcanic and plutonic rocks are characterized by frequent zoning, their absence would falsely suggest a metamorphic provenance.

Twinning of potassic feldspars has frequently been used in provenance studies to separate rocks containing microcline with its typical cross-hatch twins from others rich in orthoclase or sanidine (Plymate and Suttner, 1983), with the reservation that microcline may be untwinned and that orthoclase and sanidine may show their particular twins (Carlsbad twins).

The structural state of feldspars is also a potential indicator of provenance. Studies concentrated on potassic feldspars (Suttner and Basu, 1977; Plymate and Suttner, 1983) mostly because such data are easily obtained by using standard X-ray diffraction techniques, which by themselves are not sufficient to unravel the structural state of plagioclases. In potassic feldspars, the structural state is controlled by the degree of Al–Si ordering in the tetrahedral sites, which is a complex function of the equilibrium temperature of the atomic structure, the cooling rate, the total chemistry of the natural environment of formation, tectonic stresses, and other factors. All these factors contribute to define what is called the *genetic regime* (Suttner and Basu, 1977), which is the assemblage of all interacting physical and chemical variables responsible for a particular feldspar structural state. In essence, high equilibration temperatures result in increased Al–Si disorder and the formation of potassic feldspars with enhanced disordered structure, whereas low equilibration temperatures generate more ordered structures. Slow rates of cooling provide the required time for Al–Si ordering and for the generation of well-ordered potassic feldspars. Testing of the use of structural data of detrital K-feldspars as provenance indicators in Holocene fluvial sands derived from known plutonic, volcanic, and metamorphic source rocks (Suttner and Basu, 1977) yielded positive results that are applicable to older feldspathic arenites. For instance, K-feldspars derived from volcanic sources have disordered structures similar to high sanidine, expressing high equilibration temperatures and rapid cooling of the source rocks. Detrital K-feldspars derived from plutonic rocks range from moderately ordered (orthoclase) to well-ordered structure (microcline). Feldspars from younger plutons are more disordered, whereas those of older plutons are well ordered. This situation indicates that older and deeply eroded plutons release well-ordered K-feldspars from their slowly cooled interiors, whereas younger and less dissected plutons release more disordered K-feldspars from their rapidly chilled margins and upper parts. All detrital K-feldspars of high- and low-rank metamorphic sources are well-ordered maximum microcline, indicating that their final reequilibration temperature falls within the field of the deeper zones of granitic plutons.

Therefore, it appears possible to derive from the structural state of detrital K-feldspars valuable data on the source rocks that provided detrital particles to a given basin. This approach would also permit distinguishing mineralogical provinces within such a basin.

Detrital mica in feldspathic arenites may afford data on its actual environment of final deposition rather than on provenance or processes of transport because of its relative softness compared to associated quartz and feldspar grains. Preliminary SEM studies (Park and Pilkey, 1981) were completed on mica flakes from a variety of sedimentary environments, ranging from source area soil profiles, glacial till, and fluvial deposits, to marine continental margins. The flakes display distinct combinations of surface textures (mainly mechanical scratches), of types and extent of biological degradation, of grain roundness, and of "book" thickness. For instance, glacial grains are scratched, soil profile micas are very angular, fluvial grains are rarely scratched but display distinct degrees of biological degradation, and finally flakes from marine continental margins are rounded. Further quantification of the properties of mica flakes may provide data useful for paleoenvironmental determinations.

Feldspathic arenites may offer a rather rich suite of heavy minerals that, when complete, consists of more than 30 species, many of which have characteristic parageneses. Heavy minerals played for many years a key role in provenance studies, but without a sufficiently critical look at the limitations of the technique. These restrictions result from several potential factors: effects of weathering in the source areas, abrasion during transport, hydraulic controls at the time of deposition, and diagenetic intrastratal dissolution (Morton, 1985). Conclusions concerning provenance become highly unreliable without an adequate understanding of these factors and of the methods used to reduce or minimize their effects.

The initial control on heavy mineral assemblages in a sandstone lies naturally in the composition of the source rock. The purpose of provenance studies is precisely to attempt to isolate this initial control from subsequent modifications that took place during transportation, sedimentation, and diagenesis. A typical example of the problems encountered is the question of multicycle sands, which, during their evolution toward mineralogical and textural maturity, tend to concentrate the heavy minerals that have the highest chemical and mechanical stability. A measure of maturity can be expressed by the zircon–tourmaline–rutile (ZTR) index (Hubert, 1962). However, intense source area weathering and appreciable intrastratal dissolution also increase the ZTR index, and their effects have to be thoroughly evaluated before drawing provenance conclusions. Furthermore, the use of rutile in the ZTR is limited (Force, 1980), because most rutile is derived from high-rank metamorphics and is rare in most igneous rocks, except in alkalic ones such as alkalic anorthosites and kimberlites.

A frequently neglected aspect of provenance pertains to heavy minerals from direct pyroclastic origin. This neglect is part of the general underestimation of the contribution of volcanic processes to most sediments. Heavy minerals of pyroclastic origin are generally euhedral and build distinct mineral suites including zircon, apatite, sphene, titanite, and biotite, which characterize many bentonites, with sometimes amphiboles and pyroxenes (Weaver, 1963).

Unquestionably, weathering and soil-forming processes can modify the mineralogy of source rocks; but even in the areas of most intense weathering, high relief, tectonic uplifting, and related rapid erosion are sufficient to release fairly representative suites of heavy minerals, which are eventually deposited in feldspathic arenites. Therefore, as a whole, the damaging effect of weathering on heavy minerals in source rocks can be considered as relatively minor (Morton, 1985). Hence it is possible to compare heavy mineral suites of different age, but derived from the same source area, and obtain data on variations through time of the intensity of the weathering processes.

Abrasion effects on heavy minerals during fluvial transportation and even in beach environments can likewise be considered as minor. Such effects become appreciable only if sands are submitted to high-energy conditions for a long period of time and, as previously discussed for quartz arenites, rates of subsidence usually prevent such exposure.

Hydraulic effects during deposition are produced by variations in grain density, grain shape, and entrainment conditions, which are responsible for fluctuations in the proportion of heavy minerals between samples that affect provenance interpretations. Several techniques are available to counteract the effects of hydraulic differentiation, such as the use of hydraulic ratios, or only of minerals that display variations independent of grain size as shown by scatter plots of mean grain size versus mineral proportions, or the study of variations of one or several minerals, thus eliminating density effects.

Intrastratal dissolution strongly affects the distribution of heavy minerals in arenites and can easily reduce a rich suite into the stable zircon–tourmaline–rutile assemblage. This dissolution, recognizable by etched grain surfaces and skeletal grains, begins in the domain of meteoric waters and continues under moderate burial conditions unless early cementation by calcite or quartz protected heavy minerals from its action. Orders of general stability of heavy minerals have been established for various conditions of groundwater circulation (Morton, 1985). Nevertheless, destruction of mineral species by intrastratal dissolution may practically eliminate the possibility of establishing the original nature of an assemblage and heavily jeopardize provenance interpretations. Under such conditions, Morton (1985) suggested undertaking "varietal study," which deals only with the varieties of one particular mineral or mineral group. This approach greatly reduces the effects of intrastratal dissolution because it concentrates only on grains of a given density and stability. Techniques for varietal studies include standard optical methods pertaining to color, form, habit, or inclusions; electron microprobe for establishing minute variation of composition; cathodoluminescence for revealing properties related to certain types of parageneses; and, finally, radiometric dating methods. Although many of these techniques are still in the initial stage of development, they promise a reliable use of heavy minerals as provenance indicators.

Diagenetic Evolution

Among the various products of diagenesis of feldspathic arenites mentioned above during the discussion of framework, matrix, and cements, three important ones deserve further analysis: albitization of K-feldspars and calcic plagioclases, quartzification of K-feldspars and plagioclases, and generation of hematite, which directly relates to the origin of red beds.

Albitization of both K-feldspars and calcic plagioclases in feldspathic arenites has been recently recognized as a widespread process (Morad, 1986; Saigal et al., 1988; Aagaard et al., 1990; Morad et al., 1990; Ramseyer et al., 1992). Failure to recognize it is responsible for erroneous interpretation of the nature of the source area as consisting of albite-rich igneous or metamorphic rocks (Helmold, 1985).

Diagenetic albitization of K-feldspars during burial can proceed either by direct replacement or through one or several prior replacements by other minerals, such as anhydrite, calcite, and dolomite, which are in turn replaced by albite (Walker, 1984). Direct replacement may occur in a diffuse manner or be controlled by cleavage planes, with the final product being grains of pure or almost pure albite that are pseudomorphs of the parent grains of K-feldspars. Typically, these pseudomorphs are clear or clouded with submicroscopic fluid inclusions and/or hematite. Although many pseudomorphs are untwinned, they commonly display "chessboard" twinning. The development of twinning seems to depend on the manner in which albite nucleates. If it nucleates on optically continuous relicts of K-feldspar, the final product is untwinned and optically continuous pseudomorphs. However, commonly albite displays chessboard twinning characterized by a blocky to lath-shaped pattern produced by short, discontinuous twin lamellae. This type of twinning is interpreted (Walker, 1984) as the product of nucleation of albite on submicroscopic relicts of unreplaced twinned microcline. It is very similar to that observed in chessboard albite formed by metasomatic replacement of K-feldspar in igneous and metamorphic rocks and should not be confused with it.

In many cases, K-feldspar is not replaced directly by albite, but first by anhydrite (**Fig. 1.5**), which in turn is selectively replaced by albite, which presumably nucleates on relicts of K-feldspar remaining in the anhydrite. This situation

DIAGENETIC ALBITIZATION OF POTASSIUM FELDSPAR

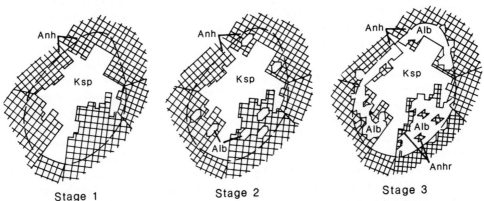

Stage 1   Stage 2   Stage 3

FIGURE 1.5 Replacement of potassium feldspar by anhydrite, followed by replacement of the anhydrite by albite. Diagrams show successive stages of the replacement sequence. In stage 1, a detrital grain of potassium feldspar (Ksp) has been deeply replaced by anhydrite (Anh), which also forms the interstitial cement. In stage 2, tiny crystals of albite (Alb) have replaced the anhydrite within the area formerly occupied by potassium feldspar. The albite crystals, although separated, are aligned, and the alignment provides evidence that the crystals have nucleated on relicts of the original potassium feldspar that remained in the anhydrite in stage 1. In stage 3, the albite crystals have grown in size and have become merged together, forming optically continuous albite that has replaced nearly all the anhydrite that replaced the feldspar in stage 1. Many relicts of optically continuous anhydrite (Anhr) still remain in the albite. Growth of the albite generally has terminated at the boundary of the original feldspar grain; however, a few crystals of albite have grown beyond that border by replacing the interstitial anhydrite cement. From Walker (1984). Reprinted by permission of the Society of Economic Paleontologists and Mineralogists.

is shown by initial isolated crystals of albite being aligned, although no relicts of the parent K-feldspar are visible; the alignment is interpreted as reflecting the orientation of the parent mineral. In more advanced cases, all or almost all the intermediate-phase anhydrite is replaced by albite, which may extend beyond the boundaries of the original grain as overgrowths. At this stage, all areas of albite are optically continuous or may display chessboard twinning, and the relicts of anhydrite within the albite show optical continuity with the parent anhydrite. The final product of this type of albitization is a complete albite pseudomorph of the original K-feldspar grain, which cannot be distinguished from that resulting from direct replacement.

Another intermediate replacement process involves partial or complete replacement of K-feldspar by calcite or dolomite, both minerals tending to be automorphic against the feldspar and in optical continuity with the intergranular carbonate cement whenever present. Textural features are the same as those described in the replacement involving anhydrite, and the final product is again an almost complete or complete pseudomorph of the original K-feldspar undistinguishable from the product of direct replacement.

As mentioned earlier, albitization of K-feldspar, if not recognized, can be responsible for erroneous interpretations of provenance. Indeed, partial replacement generates perthitic intergrowths of albite and K-feldspar that could be misinterpreted as detrital grains of perthite and antiperthite of igneous or metamorphic origin. However, authigenic albitization is recognizable by the following features: presence of relicts, high purity of albite, common occurrence of chessboard twinning, unusual abundance of albite grains relative to associated grains of K-feldspar, and even unusual total elimination of K-feldspar grains.

Sodium required for albitization is generally of intrastratal origin, either from the diagenetic alteration of detrital grains of sodium-rich plagioclases by dissolution or replacement by clay minerals or from circulating connate-marine pore waters originating from compaction of shales associated or adjacent to bodies of feldspathic arenites, or under certain circumstances from adjacent evaporites.

The diagenetic replacement of detrital feldspars (mostly microcline and albite) by quartz occurs in several ways and spans a variety of shallow to deep burial conditions (Morad and AlDahan, 1987). In a first type, quartz penetrates feldspar grains along fractures, cleavages, and twinning planes, in all instances forming veinlets with sharp contacts and pinching toward the center of grains. This type of quartz is in optical continuity with precursors such as interstitial quartz cement, quartz overgrowths, or quartz grains adjacent to the feldspar. SEM examination shows that quartz precipitated along fracture planes of feldspars consists of fine crystals ($<10\,\mu$ in size). These crystals are an association of individuals showing a combination of trigonal pyramids and trapezohedrons and others that are double-terminated

hexagonal pyramids. Quartz crystals developed along traces of cleavages and twinning planes are always double-terminated hexagonal pyramids with their c axis parallel to these planes.

A second type is an association of illite and quartz scattered in patches throughout the feldspar grain, which may be locally bounded by cleavage and twinning planes. In this case, SEM examination shows that the quartz crystals are aggregates of numerous crystals or a few discrete individuals with preferred orientation along their c axis. The latter crystals are usually aligned parallel to the cleavage planes of the replaced feldspar.

A third type involves fine crystalline quartz cement replacing the margins of detrital feldspar or a few crystals of quartz of variable size filling dissolution voids within feldspar grains. These quartz crystals are oriented at random to each other and to cleavage and twinning planes of the replaced feldspar.

A fourth and last type consists of pervasive replacement of feldspar by quartz. The feldspar forms etched relicts among randomly oriented quartz crystals ranging in size from 10 to $30\,\mu$.

In summary, quartz replaces feldspar in two ways (Morad and AlDahan, 1987): (1) by precipitation of silica along fractures, cleavage planes, and twinning planes of feldspars, which act as voids for precipitation, eliminating this type of intracrystalline porosity. Sources of silica are probably intrastratal from pressure solution of immediately adjacent quartz grains. (2) By precipitation of silica almost simultaneously with dissolution of feldspars, or derived from their replacement by authigenic clay minerals. In both instances, the patchy to pervasive texture generates intracrystalline porosity within feldspar grains.

Generation of diagenetic hematite and its precursor limonitic oxide is characteristic of many feldspathic arenites that belong to the red bed facies (Turner, 1980; Hubert and Reed, 1978). Important amounts of iron are released by pervasive intrastratal dissolution of Fe-silicates, such as pyroxene, amphibole, epidote, chlorite, and biotite, as well as ilmenite and magnetite, all of which display various stages of etching until final destruction. Augite characteristically develops "cockscomb" terminations that indicate preferential dissolution along lattice planes transverse to the c axis (Walker et al., 1978). Intrastratal dissolution of iron-bearing silicate grains also generates clay minerals and calcite in addition to iron oxides, all of which contribute to the diagenetic reorganization of the original sediment. In numerous instances, etched iron-bearing silicate grains appear under the microscope as "bleeding out" a red stain in the shape of irregular halos of iron oxide, which appears amorphous or consists of crystals too fine to be resolved even under the SEM. These pervasive halos stain indiscriminately all framework grains, matrix, and cement, contributing effectively to the red bed aspect (Walker, 1967). In other instances, hematite appears as tabular concentrations between cleavage laminae of biotite flakes, and when the parent mineral is completely destroyed, hematite forms a drusy lining of voids intact or collapsed by compaction.

Extensive generation of red staining in feldspathic arenites requires the original presence of Fe-silicate grains and postdepositional conditions favoring intrastratal dissolution, which may reach total destruction, and Eh pH values favoring the formation of ferric oxide, initially as limonite. Further requirements are an absence of subsequent reduction of the ferric oxide and sufficient time for limonite to convert into hematite. The wide distribution of red beds indicates that such requirements of meteoric to burial water circulation were frequently met during geologic time and that periods on the order of 10^6 years are sufficient to form most of the red color in given instances (Blatt, 1979).

Feldspathic arenites, because of their tectonic context of deposition in relatively restricted basins surrounded by uplifted recharge areas, can be excellent examples of meteoric burial diagenesis. Furthermore, their mineralogical composition offers enough possibilities of chemical reactions to generate diagenetic sequences of cementing minerals during the various stages of burial. A typical example is represented by fan-delta feldspathic arenites of the Pennsylvanian of the southern Anadarko Basin (Dutton and Land, 1985). In the distal portion of the fan delta where feldspathic arenites were reworked in marine conditions, the earliest diagenetic events were the formation of chlorite ooids and precipitation of fibrous submarine calcite cement. They were followed during early burial, in a meteoric fluid environment, by precipitation of iron-poor calcite sparite. The earliest cement in the nonreworked arenites that did not have an internal source of calcium carbonate was pore-lining authigenic chlorite. With increasing burial, precipitation of authigenic quartz, feldspar, and kaolinite cements occurred. Apparently, the dissolution of detrital feldspars would be sufficient to account for these cements. Upon further burial and temperatures on the order of 65°C, organic matter reactions liberated CO_2 and ferrous iron. Dissolution of the iron-poor calcite was followed by precipitation of iron-rich calcite, followed by ankerite. Because of the occurrence of overlying Permian evaporites, the last diagenetic events were partial albitization of detrital plagioclases by sodium-rich fluids and precipitation of anhydrite and celestite from strontium and sulfate solutions. Oxygen isotope data indicate that the entire sequence of diagenetic cements precipitated from meteoric fluids that became increasingly hot with burial, but remained relatively constant in isotopic composition.

Reservoir Properties

Primary porosity in feldspathic arenites is gradually reduced and even obliterated by processes similar to those described for quartz arenites, that is, mechanical compaction, pressure solution, and overgrowths on quartz and feldspar grains, fol-lowed by precipitation of interstitial and partially replacive carbonate, sulfate, and clay mineral cements, whose importance in porosity control of feldspathic and lithic arenites has been recently widely documented and emphasized (Houseknecht and Pittman, 1992). However, the abundance of feldspars and the presence of carbonate and sulfate cements are conducive to the generation of secondary burial porosity when feldspathic arenites are buried in basins along passive plate margins and submitted to compaction-driven fluids.

Feldspathic arenites display many of the most characteristic features of secondary porosity (**Fig. 1.6**) reported by Schmidt and McDonald (1979a, b) and Shanmugam (1985a, b), which are briefly described below.

Partial dissolution of framework grains and cement constituents (**Plate 7.B**) generates irregularly shaped pores that often show patches of remnant materials adjacent to them that have corroded margins. Such secondary pores may occur entirely within cements and could be confused with incomplete cementation of primary porosity. However, their secondary nature is obvious when they show cement remnants with the same optical orientation or if they cut across growth zonations of the cement.

Moldic pores derived from completely dissolved grains or sometimes bioclasts characteristically display original shapes of the precursors, such as a subrectangular aspect for feldspar grains, unless such cavities were deformed or collapsed under the action of subsequent compaction.

Inhomogeneity of packing is a very common feature of feldspathic arenites. This texture consists of areas of loosely packed grains with high intergranular porosity adjacent to areas of tightly packed grains with little or no interstitial porosity. This texture reflects an original inhomogeneous distribution of soluble cement. The tight packing originates from mechanical compaction and pressure solution in areas devoid of cement. Elsewhere, early cement was present and prevented compaction and grain interpenetration. Subsequent dissolution of the cement, which is cavity filling or partially replacive, generated secondary porosity between the loosely packed grains.

Oversized pores are among the most spectacular aspects of secondary porosity. They have a diameter that is several times larger than that of adjacent grains. Most oversized pores result from the dissolution of several adjacent grains, together with their overgrowths whenever present, and the surrounding cavity-filling or partially replacive cement. In the frequent case of feldspar grains surrounded by very delicate rims of clay minerals, these rims may subsist outlining the shapes of the original grains or may be at various stages of collapse at the bottom of the pore.

Elongate pores result from the burial dissolution of relatively uniform zones along the margins of tightly packed grains that were replaced by authigenic cements. They should not be confused with primary pores, which may occur in feldspathic arenites particularly rich in micaceous flakes.

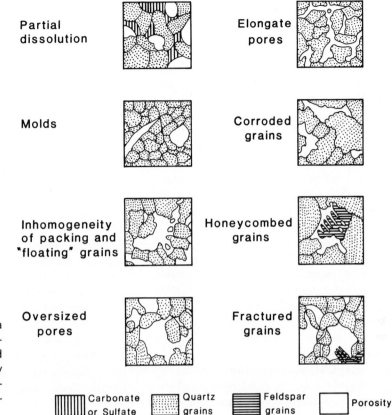

FIGURE 1.6 Petrographic criteria for recognition of secondary sandstone porosity. From Schmidt and McDonald (1979b). Reprinted by permission of the Society of Economic Paleontologists and Mineralogists.

Corroded grains, often associated with oversized pores, result mainly from dissolution of authigenic cements that unevenly replaced the margins of detrital grains, including their overgrowths whenever present. Incipiently corroded grains should not be confused with similar aspects resulting from depositional processes. Obviously, corroded overgrowths are the best criteria of secondary origin, as well as highly corroded grains that could not have survived transportation. The same reasoning applies to intragranular pores, which range from nonfabric selective to cases highly controlled by cleavage and twinning planes. In the latter case, the term *honeycombed grains* is used. For feldspar (**Plate 7.C**), honeycombed texture is generated either by direct differential dissolution or by removal of areas previously replaced by authigenic soluble minerals such as carbonates and sulfates.

Finally, open grain fractures often occur in pillars of grains that apparently support the roofs of adjacent oversized pores or of areas of inhomogeneity of packing. These fractures express failure of the framework after formation of such large pores. The secondary origin of open grain fractures is particularly obvious when fractures cut across grains that were previously reciprocally welded by pressure solution or display syntaxial overgrowths.

By the combination of the various types of secondary porosity just described, feldspathic arenites may reach a sec-

ondary porosity of more than 40%, which is far in excess of the primary one, making them extremely important reservoirs (Schmidt and McDonald, 1979a, b; Shanmugam, 1985a, b; Couto Anjos and Carozzi, 1988). Surdam and Crossey (1987) provided a process-oriented integrated diagenetic model that combined the generation of secondary porosity in sandstones, shale diagenesis, and the effects of maturing organic matter, some chemical aspects of which were recently reevaluated (Stoessell and Pittman, 1990).

TYPICAL EXAMPLES

Plate 5.D. Feldspathic arenite showing brownish cloudy grain of kaolinitized potassic feldspar with relict structures. Fountain Formation, Pennsylvanian, Lyons, Colorado, U.S.A.

Plate 5.E. Feldspathic arenite showing grain of twinned potassic feldspar partially and randomly sericitized with preservation of cleavage traces. Fountain Formation, Pennsylvanian, Lyons, Colorado, U.S.A.

Plate 5.F. Feldspathic arenite showing grain of potassic feldspar altered to calcite in bands controlled by cleavages separating zones altered to microvermicular kaolinite. Basal Salina Formation, Lower Eocene, Talara, Peru.

Plate 6.A. Feldspathic arenite showing grain of untwinned potassic feldspar altered marginally by calcite and internally by microvermicular kaolinite. Basal Salina Formation, Lower Eocene, Talara, Peru.

Plate 6.B. Feldspathic arenite showing partially calcitized plagioclase grain with relicts controlled by orientation of cleavages. Basal Salina Formation, Lower Eocene, Talara, Peru.

Plate 6.C. Feldspathic arenite consisting of subrectangular grains of quartz, partially altered potassic feldspars and plagioclases, chloritized biotite, and lithoclasts of quartzites (extreme right). Intense pressure solution of all constituents associated with patches of iron oxides and chlorite matrix give rise to the "regenerated granite" texture. Fountain Formation, Pennsylvanian, Lyons, Colorado, U.S.A.

Plate 6.D. Feldspathic arenite consisting of poorly sorted subangular grains of quartz, potassic feldspars deeply altered to a mosaic of quartz–sericite–chlorite, abundant muscovite flakes, rare calcitized plagioclases, and lithoclasts of quartzites, micaschists, and gneisses. Interstitial matrix of sericite–chlorite–quartz is combined with moderate pressure solution between all constituents. Carboniferous, Reyran Basin, Estérel, France.

Plate 6.E. Feldspathic arenite consisting of poorly sorted subangular grains of quartz, fresh and kaolinitized potassic feldspars, chloritized biotite flakes, and lithoclasts of cherts, quartzites, and micaschists. Moderate pressure solution between all constituents is combined with a highly pervasive opaque hematite cement and minor amounts of sericite and chlorite matrix. "Brownstone," Triassic, Portland, Connecticut, U.S.A.

Plate 6.F. Feldspathic arenite consisting of poorly sorted subangular grains of quartz, potassic feldspars at various stages of alteration to chlorite–sericite–calcite aggregates, chloritized biotite, rare glaucony, and lithoclasts of chertified silicic volcanic glass, micaschists, and ferruginous dolostone. Interstitial cement of sparite filling pores and replacing margins of quartz and feldspar grains. Flysch, Oligocene, High Calcareous Alps, Arve Valley, France.

Plate 7.A. Feldspathic arenite consisting of subrounded grains of quartz and potassic feldspars deeply altered to aggregates of calcite–kaolinite and opaque iron oxides. Interstitial cement is a combination of early syntaxial overgrowths on quartz grains and subsequent subrectangular laths of anhydrite. Mulichinco Formation, Lower Cretaceous, Mendoza Basin, Argentina.

Plate 7.B. Porous feldspathic arenite showing subangular grains of quartz and subrounded grains of kaolinitized potassic feldspars. Incomplete interstitial cementation is by rims of fibrous chlorite on quartz grains and fine microporous vermicular kaolinite in central portions of pores. Basal Salina Formation, Lower Eocene, Talara, Peru.

Plate 7.C. Porous feldspathic arenite showing subangular quartz grains and partially dissolved (honeycombed) feldspar grain with argillaceous residues along relict cleavage traces. Basal Salina Formation, Lower Eocene, Talara, Peru.

LITHIC ARENITES

Framework

As defined above, lithic arenites contain 25% or more lithic grains exceeding feldspar grains. Rock particles display an extraordinary variety, which includes all types of resistant sedimentary, metamorphic, igneous, volcanic, and volcaniclastic rocks, but average lithic arenites are mixtures of low-rank metamorphic rocks and sedimentary rocks. Only a few examples are illustrated here: graphitic schist (**Plate 7.D**), micaschists (**Plate 7.E**), marble (**Plate 7.F**), chert (**Plate 8.A**), radiolarite (**Plate 8.B**), fracture filling (**Plate 8.C**), volcanic (**Plate 8.D**), volcaniclastic (**Plate 8.E**), basaltic glass (**Plate 8.F**), serpentine (**Plate 9.A**), porphyritic andesite (**Plate 9.B**), and myrmekite (**Plate 9.C**).

The general character of lithic grains is to display a fine-grained texture at the sand-size level because coarser-textured rocks would break up into individual mineral grains. Hence, the clear dominance of debris derived from shales, siltstones, phyllites, slates, and micaschists, which are ubiquitous rocks capable of generating, by reworking, the entire range of fine to coarse sand particles. The remarkable abundance of shale fragments is caused by their common calcareous composition and hence better resistance to abrasion. The fine-grained character applies also to volcanic glasses, but lithic arenites derived from volcanic rocks possess unique properties described below.

In most lithic arenites, quartz grains are well represented, and because they are reworked from older rocks, they display inherited features such as straight extinction and good rounding when derived from quartz arenites, or angularity, wavy extinction, and polycrystallinity when derived from low-rank metamorphic rocks. A derivation from rhyolites is shown by broken and embayed phenocrysts of strain-free high-temperature quartz, with enclosed or attached blebs of aphanitic glassy matrix (bubble-wall texture).

The predominant derivation of lithic arenites from older sedimentary rocks and low-rank metamorphic rocks is responsible for the relatively small amount of feldspar they contain. Plagioclases, with more abundant sodic than calcic types, largely predominate over K-feldspars, a separation further enhanced by burial albitization of calcic plagioclases and K-feldspars (Land and Milliken, 1981). In lithic arenites derived from rhyolitic volcanics, sanidine is common, whereas young lithic arenites derived from andesitic and basaltic rocks display zoned calcic plagioclase. With increasing age, however, sodic plagioclase becomes the rule again as the result of calcic varieties being destroyed by diagenesis or albitized.

Detrital micas, with muscovite more common than biotite, are frequent constituents of lithic arenites, particularly those derived from metamorphic sources. Mica flakes are frequently deposited parallel to bedding or concentrated along certain bedding planes and are particularly susceptible to deformation by compaction between more resistant constituents. An extremely varied suite of heavy minerals occurs in lithic arenites, often partially destroyed by diagenesis as in feldspathic arenites, but nevertheless expressing provenance in combination with the lithic grains. Plant fragments are also often reported.

Certain types of lithic grains deserve additional comments. Chert grains can be predominant in certain lithic arenites (**Plate 8.A**). They represent redeposited residual products of extensive chemical weathering of carbonate rocks containing bedded or nodular cherts and indicate a relatively local provenance from low-relief areas. The possible confusion between chert grains and clasts of devitrified rhyolitic glass can be avoided by the following observations: chert grains display a great variety of microcrystalline to microfibrous textures, with abundant impurities representing the various types of replaced carbonate, minute inclusions of calcite and dolomite, small quartz grains, and "ghosts" of silicified bioclasts. Clasts of devitrified volcanic glass contain small microlites of feldspars, collapsed vesicles, and "ghosts" of flow structures (**Plate 8.E, F**).

Grains of carbonate rocks in lithic arenites express rapid erosion and high reliefs because, under normal conditions, carbonates are removed by dissolution processes. Carbonate grains are mainly various types of calcilutites and calcisiltites together with dolostones, marbles, and dolomitic marbles (**Plate 7.F**) consisting of fine polycrystalline aggregates. Interesting and unusual lithic grains are carbonate or siliceous vein fillings with characteristic concretionary structures or comblike crystals that grew perpendicularly to the original walls of the fracture in the country rock (**Plate 8.C**). These lithics indicate that erosion reached terranes highly fractured by orogenic events and subsequently rehealed before being dismantled by erosion.

Lithic grains derived from volcanic rocks show very characteristic aspects. Those coming from silicic magmas are clasts of uncollapsed or collapsed vesicular pumice (**Plate 10.C**), rhyolitic glass with flow structure, and massive vitreous groundmass with euhedral microlites of sanidine. Volcanic glass, which is optically amorphous, varies in aspect from colorless to yellow, red, brown, and black, depending on the degree of oxidation of the iron it contains and the presence of impurities. In general, volcanic glass is assumed to grade from colorless to dark brown or black with increasing mafic nature of the magma, but this relationship suffers numerous exceptions, shown in particular by light-tan basaltic glass. Lithic grains of chert in volcanic lithic arenites are derived from radiolarian cherts (radiolarites) and show typical remains of Radiolarians (**Plates 8.B, 9.E**). They are generally red, but colorless, green, brown, and black varieties occur also (Shanmugam and Higgins, 1988). Lithic grains of andesites and basalts show a highly developed microlitic texture and abundant phenocrysts of zoned plagioclase (**Plate 9.B**), augite, hypersthene, hornblende, and oxyhornblende.

Identification of volcanic lithic clasts is complicated by their chemical instability and reactivity, which makes them very susceptible to diagenetic alteration. For instance, rhyolitic glass devitrifies into microcrystalline aggregates of quartz–cristobalite–tridymite resembling chert, but still often displaying scattered microlites of sanidine. Other types of alteration of intermediate and basaltic glass to smectites and zeolites are texturally even more destructive.

Matrix and Cements

Because of their moderate to excellent sorting, lithic arenites are essentially devoid of interstitial detrital matrix. Certain types remain weakly consolidated and highly porous or display only weak pressure solution. Others are tightly cemented by extensive pressure solution shown by microstylolitic contacts and strong ductile deformation of the softer grains of shales and low-rank metamorphics between the other more resistant constituents. In these situations, care should be taken not to confuse the various types of deformed soft grains that give the illusion of filling some interstitial pores with a possible real interstitial matrix of clay minerals. The latter would be more uniformly distributed and of similar composition throughout a particular rock.

In lithic arenites derived from sedimentary and low-rank metamorphic rocks, the most frequent cements are calcite and quartz, by themselves or associated, occasionally anhydrite and barite, and a subordinate amount of authigenic clay minerals. Calcite cement displays the following wide spectrum of textures: microcrystalline mosaic or randomly associated anhedral to subhedral crystals (**Plate 9.D**); cavity-filling texture in which blades of calcite grown perpendicularly to the pore walls merge toward the center into a medium- to coarse-grained equant mosaic; coarse mosaic of randomly associated anhedral crystals (**Plate 9.E**); and, finally, poikilotopic texture in which large crystals of calcite interlock irregularly, each enclosing many framework grains (**Plate 9.F**). The texture of calcite cement seems to reflect rather the effects of initial porosity and the conditions of circulation of solutions than the effects of the composition of lithic grains, except in the rare cases of abundant clasts of marbles, which may nucleate in part the cement crystals.

Quartz cementation also displays little relationship to the composition of lithic clasts because it occurs predominantly as widespread overgrowths around quartz grains. Only in the presence of chert grains is a microfibrous quartz cement developed.

In lithic arenites derived from andesitic and basaltic rocks, cements are more varied, as would be expected. For instance, chlorite is common, showing several generations of concentric rings around grains and clasts or forming a bright green, microfibrous rim grading into a lighter-colored, microgranular central area. Other instances are fibrous rims of chlorite with central portions of pores filled with microvermicular kaolinite (**Plate 10.A**) or nontronite (**Plate 10.B**). In still other cases, concentric rings of various clay minerals, such as smectites, may be ubiquitous in filling most of the interstitial porosity with well-developed crystals or leaving the central parts of the pores open.

Depositional Environments and Provenance

Lithic arenites are among the most common immature sandstones. They occur throughout the geological column and characterize several important ancient environments. Indeed, they occur as alluvial deposits that filled postorogenic frontal basins, generally called molassic basins; as fluvial deposits over large cratonic masses; in major deltas and related beach and shallow-water shelf deposits along passive continental margins; and, finally, as certain types of turbidites that belong to the flysch facies in continental collision orogenic belts.

The present-day importance of sands equivalent to lithic arenites was demonstrated by Potter (1978). He reported that the sands of 36 rivers that represent 45% of continental surfaces draining into the various oceans have the composition of predominantly moderately to well-sorted, angular to subangular lithic arenites, poor in clay and silt. These sands have the following average grain composition: quartz, 60%; feldspars, 11%; and rock fragments, 19%.

Lithic arenites reflect provenance in the most effective manner because rock particles carry by themselves the obvious proofs of their origin more than any kind of monomineralic grains does. The combination of both types of indicators affords a high degree of precision in recognizing classes or provenance as a function of plate tectonics (Dickinson and Suczek, 1979; Dickinson et al., 1983). The most significant framework compositional variation of lithic arenites can be displayed as ternary plots on triangular diagrams. The three poles represent recalculated proportions of key categories of grain type determined by modal point counts. Two types of diagrams are used: (1) QFL diagrams in which the poles are Q, total quartzose grains including polycrystalline lithic grains of chert and quartzite; F, monocrystalline feldspar grains (plagioclase + K-feldspar); and L, igneous, metamorphic, and sedimentary unstable polycrystalline lithic grains, and (2) QmFLt diagrams in which the poles are Qm, monocrystalline quartz grains; F, monocrystalline feldspar grains (plagioclase + K-feldspar), and Lt, total polycrystalline lithic grains including quartzose varieties.

According to Dickinson and Suczek (1979), mean compositions of sandstone suites derived from various source areas controlled by plate tectonics tend to cluster in distinct and separate fields on QFL and QmFLt triangular diagrams. Three main types of provenance were therefore distinguished: continental blocks, magmatic arcs, and recycled orogens. The positions of the boundary lines between subdivisions on both types of triangular diagrams are empirical and susceptible to adjustments as more data become available to refine this working tool. At present, Dickinson et al. (1983) have distinguished a number of subfields among the three main types of source areas as follows (**Fig. 1.7**). Within continental blocks, source areas correspond to stable shields, platforms, or uplifts, marking plate boundaries or zones of intraplate deformation. Three types of quartz–feldspathic sands, poor in lithics, are distinguished within continental blocks: the most quartz rich are derived from stable craton interiors of low relief, more feldspathic sands form an intermediate group, and feldspathic sands originate from deep erosional action on basement uplifts.

Within active magmatic arcs, source areas consist of volcanics overlying igneous and metamorphic belts and granitic plutons in the roots of arcs. The released sands, which are lithic–feldspathic and feldspathic–lithic, plot across the central and lower parts of both types of triangular diagrams. Three types are distinguished: the most lithic are volcanic sands derived from undissected arcs, less lithic types form an intermediate group called transitional arc, and the most quartz–feldspathic are volcanic–plutonic sands originating from dissected arcs where erosion has penetrated batholiths beneath volcanic covers. Possible feldspathic sands generated under these conditions would be impossible to recognize by petrographic studies alone from those derived from cratonic uplifts.

In recycled orogens, source areas consist of sedimentary and metamorphic rocks, with a minor amount of volcanic rocks undergoing erosion as the result of orogenic uplift of folded and thrusted belts under various tectonic settings. The released sands are quartzose–lithic to lithic–quartzose, but low in feldspars because igneous rocks are not major contributors, and can be divided on the QmFLt plot into three types: quartzose recycled, transitional recycled and lithic recycled. These types cannot be distinguished on the QFL plot because many lithics are chert grains that plot together with quartz at the same pole. The quartz-rich varieties may have been recycled several times after their original cratonic source, whereas the lithic-rich types have abundant chert grains coming either from the reworking of radiolarian cherts of subduction zones or from cherty carbonates of the shelf sequences.

On a more regional scale, such as within individual basins, provenance modeling using mineral and lithic grains is a very effective approach for unraveling provenance and its evolution in time and space in feldspathic and lithic arenites.

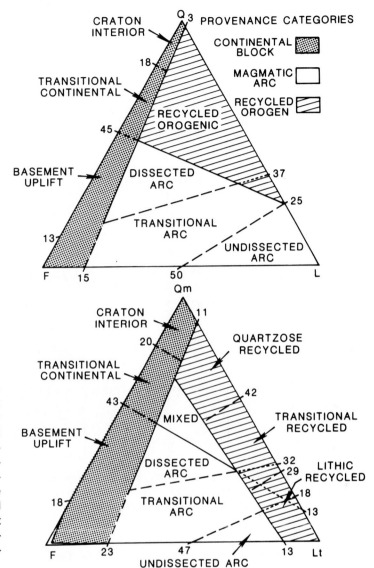

FIGURE 1.7 QFL (upper) and QmFLt (lower) plots for framework modes of sandstones showing provisional subdivisions according to inferred provenance type. Numbered ticks on legs of triangles denote positions of empirical provenance division lines in percentage units measured from nearest apical pole. Modified from Dickinson et al. (1983). Reproduced by permission of the authors and the Geological Society of America.

It is a deductive technique for identifying best statistical matches between compositions of sandstones and modeled sandstone compositions from hypothetical source sections (see, for instance, Ingersoll et al., 1987).

Another aspect of provenance studies in lithic arenites is the use of the occurrence of specific types of lithic or monomineralic grains to subdivide thick and apparently uniform flysch (submarine fan–turbidites) sequences into petrographic (lithostratigraphic) units for the purpose of making practical subdivisions into units usable for correlation in the absence of index fossils (**Fig. 1.8**). Such studies disclose also disconformities and hiatuses not otherwise recognizable, episodes of volcanic activity (rhyolite grains), and traces of orogenic events (grains of vein and fracture filling). They also demonstrate, in the absence of reliable structural criteria, if such sequences are right-side up or overturned, or top and bottom of beds when they are in a vertical position.

The same approach, called *inverted stratigraphy,* permits the reconstruction of the geological composition of source areas and following the progress of erosion destroying them. Other economically interesting aspects are determining relationships between paleoplacers (gold, diamond) and their sources within the terranes undergoing erosion. The principle of inverted stratigraphy can easily be applied to lithic arenites involved in low-grade metamorphism.

Diagenetic Evolution

The diagenetic evolution of lithic arenites, whether deposited in cratonic areas or submitted to shallow burial conditions where meteoric waters are the predominant circulating fluids shows many similarities to that of feldspathic arenites. These comparisons pertain to the various types of feldspar alteration, carbonate and quartz cementation, and authigenic

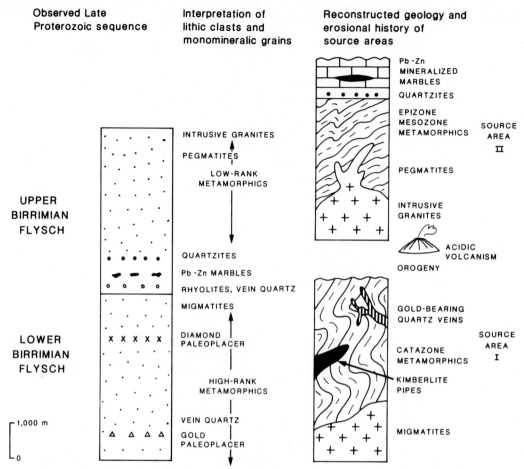

FIGURE 1.8   Examples of interpretation of lithic constituents in sandstones and their use in reconstructing geological composition and erosional history of source areas. Late Proterozoic, Ivory Coast, Africa.

clay cements. Whenever thick lithic arenites are involved in actively subsiding areas, their diagenetic evolution becomes more complex and many types of diagenetic sequences of cementation develop as a function of numerous factors that change from basin to basin (Leder and Park, 1986; McBride, 1989). At present, it is difficult to assess the relative importance of these factors, which are initial porosity and permeability, compaction, burial rate, age, basin size and geometry, composition of fluids, effects of organic matters, fluid drive mechanisms, and geothermal gradient. More case histories are needed to reach reasonable generalizations and predictions.

A fundamental problem is the understanding of hydrologic regimes operating in the subsurface because diagenetic reactions require a certain amount of mass transfer as well as an assortment of chemical reactions (Galloway, 1984; Surdam and Crossey, 1987). If, in some instances, diffusion of components appears sufficient to explain diagenetic changes, particularly during early stages, advection of components into or out of a given system appears required by observations. Several types of advectional flow are to be considered:

meteoric water flow at relatively shallow depth, as discussed previously for quartz and feldspathic arenites and some types of lithic arenites, and compaction-driven flow and convectional flow for deeply buried systems of lithic arenites. Convectional flow or recirculation of basinal fluids appears as a major transport mechanism in thick basinal sequences of lithic arenites because it explains the extremely high amounts of water and their very important upward migration required to introduce diagenetic cements, particularly quartz, and the acids to generate secondary porosity (Wood and Hewett, 1984; McBride, 1989). Of great importance is the determination of the zones of potential mixing of the various types of meteoric, compactional, and convectional flow, where the numerous reactions of diagenetic cementation are more likely to occur and which have shifting positions in the subsurface during basinal history.

A generalized diagenetic sequence was proposed for volcanic lithic arenites deposited in basins along active continental margins such as the arc-trench system of western North America during its Mesozoic–Cenozoic evolution

(Galloway, 1979, 1984; Burns and Ethridge, 1979). This diagenetic sequence can be divided into four major stages. In the first stage, calcite pore-filling cement forms locally as sparite, and its irregular distribution appears related to depositional conditions. The second stage develops with increasing depth of burial on the order of 300 to 1,200 m. Chemical alteration of unstable grains and aphanitic volcanic fragments mobilizes silica and aluminum, which are combined as clay rims around detrital grains. The most common clay minerals are chlorite and smectite, locally illite and chlorite. Clay rims forming near grain-to-grain contacts are deformed and extruded, with increasing burial causing mechanical compaction of the lithic arenite. Stage 3 occurs at burial depths between 1,000 and 3,000 m and corresponds to a second episode of cementation. The remaining open-pore spaces are infilled either by authigenic zeolite (laumontite) or by a second layer of well-crystallized phyllosilicate, usually chlorite or smectite, in places kaolinite. The clay rims developed during stage 2 are preserved as halos around detrital grains and separate them from the pore-filling cements of stage 3. The relative time of formation of phyllosilicates and laumontite varies because they are rarely found together; it probably depends upon the local variations of pore fluid chemistry. Stage 4, which characterizes still greater depths of burial, shows the formation of a variety of replacement minerals and minor cements, which increases the complexity of the fabric and mineralogy of lithic arenites. Calcite replacement of feldspar grains, volcanic lithics, and matrix occurs in an appreciable but spotted manner, as does replacement by albite, chlorite, and epidote. Development of quartz and feldspar overgrowths generates a welded texture with many sutured and convexoconcave intergranular contacts.

The diagenetic sequence just described should be visualized as resulting from a combination of thermal factors related to burial depth with variations in fluid composition (Surdam and Boles, 1979). Indeed, early diagenetic conditions are characterized by hydration and carbonatization reactions, such as changes of volcanic glass and plagioclase to zeolites, whereas later diagenetic conditions consist mainly of dehydration reactions related to fracturing, such as heulandite changing to laumontite, analcite, albite, or prehnite, or laumontite changing to pumpellyite and prehnite.

Another implication of the generalized diagenetic sequence of volcanic lithic arenites has direct implications for the origin of the diagenetic matrix of wackes (Galloway, 1974). If deeper burial conditions are considered together with greater compressional stresses and changes of fluid composition, complex reactions between the phyllosilicates formed in stages 2 and 3, the replacement products of stage 4, and further alteration of essentially all types of lithic grains, pyroxenes, amphiboles, and heavy minerals would produce an intergranular diagenetic matrix consisting mainly of chlorite, sericite, epidote, and zeolites. This matrix is extremely similar to that of wackes, and its proposed origin is discussed in more detail below.

Reservoir Properties

Lithic arenites, deposited in cratonic areas or submitted to shallow burial conditions where meteoric waters are the predominant circulating fluid possess a primary intergranular porosity ranging from 30% to 40%. They display the same diagenetic features as feldspathic arenites, which involve here alteration of lithic grains, including chert (Shanmugam and Higgins, 1988) and feldspars, and precipitation of carbonate, quartz, and clay mineral cements. Buried in basins along passive plate margins and in the presence of compactional and convectional fluid flow, lithic arenites develop all the characteristics of secondary porosity of feldspathic arenites (**Plate 10.C, D, E, F**) under similar conditions, making them excellent potential reservoirs (Kaiser, 1984; Loucks et al., 1984; Shanmugam, 1985a, b).

Arc-derived volcanic lithic arenites, subjected to the generalized diagenetic sequence described above, do not achieve very remarkable reservoir properties, and their changes of average porosity are a direct function of the diagenetic evolution they underwent (Galloway, 1979). The initial lithic sands have a primary intergranular porosity on the order of 30% to 40% (**Fig. 1.9**). Precipitation of chalky calcite pore-filling cement of diagenetic stage 1 reduces porosity to about 10%, but only in very localized places, so most of the sands remain unaffected by this process. Lithic arenites showing the clay rims of diagenetic stage 2 retain relatively high porosities from 15% to 30% depending on the degree of mechanical compaction. These authigenic clays fill only about 2% of intergranular pore space, but the actual porosity reduction is even smaller because these rims are microporous. The formation of laumontite or phyllosilicate pore-filling cements of diagenetic stage 3 reduces porosity to an average of 10%. At the beginning of diagenetic stage 4, porosity is further reduced and may be entirely eliminated. If the loss of porosity appears to be a somewhat continuous process directly related to the succession of diagenetic stages, the loss of permeability, on the contrary, occurs mainly during the formation of clay rims of diagenetic stage 2, mostly as the result of the platy clay minerals acting as baffles extending into pore throats and having a very efficient retarding effect on fluid flow. At this stage, permeability is reduced to a few tens of millidarcies, and upon reaching diagenetic stage 3, permeability is reduced to only a few millidarcies.

In essence, the diagenetic sequence defines a porosity economic basement at the beginning of diagenetic stage 3 of pore-filling cementation and an even shallower permeability economic basement within diagenetic stage 2 of clay rim formation. Under these conditions, arc-derived volcanic lithic arenites are poor prospects for commercial production of hydrocarbons.

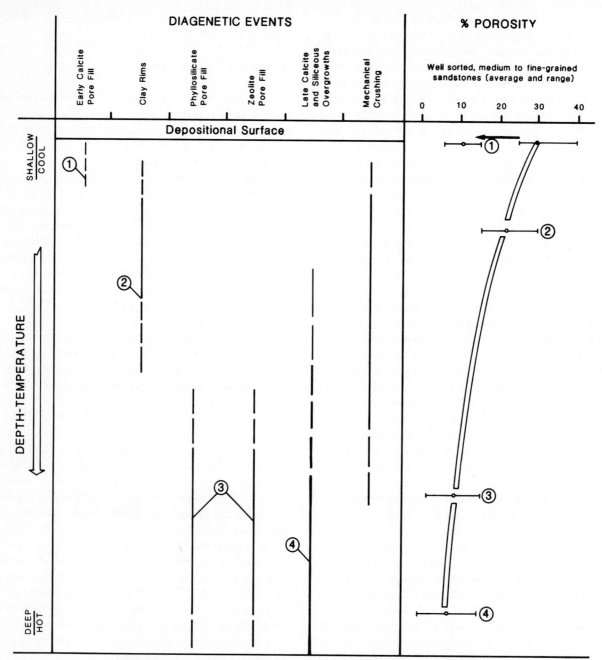

FIGURE 1.9 Sequential development of diagenetic features in arc-derived sandstones based on paragenetic sequence and/or first occurrence in wells. The vertical bars denote interpreted relative depth range of major diagenetic events (specific events defining stages are numbered). Right-hand curve shows successive porosity decrease for sandstones in successive diagenetic stages. Circles indicate average porosity for all measured samples within each stage; bars show the range of measurements. Porosity curve is calibrated for fine- to medium-grained sandstones. From Galloway (1979). Reprinted by permission of the Society of Economic Paleontologists and Mineralogists.

TYPICAL EXAMPLES

Plate 7.D. Porous lithic arenite showing quartz grains and pressure-welded lithoclast of dark brown well-laminated graphitic schist. Honeycombed feldspar grain (lower-left corner) shows angular dissolution pattern. Basal Salina Formation, Lower Eocene, Talara, Peru.

Plate 7.E. Lithic arenite showing pressure solution between grains of quartz, lithoclasts of quartzites, ferruginous dolostones, and, in the center, elongated muscovite–quartz schist. Basal Salina Formation, Lower Eocene, Talara, Peru.

Plate 7.F. Lithic arenite consisting of a poorly sorted assemblage of subrounded to well-rounded lithoclasts of various types of light-colored, coarsely crystalline dolomitic marbles, dark brown bituminous shales, fine-grained metaquartzites and cherts, and quartz grains with straight or undulose extinction. Intense pressure solution with iron oxide films and no visible cement, although some calcite may be present. Flysch, Oligocene, High Calcareous Alps, Central Switzerland.

Plate 8.A. Lithic arenite showing an assemblage of subrounded lithoclasts of light-colored cherts of variable crystallinity, dark brown cherty siltstones, and grains of quartz and fibrous chalcedony. Intense pressure solution between all constituents with interstitial concentrations of opaque iron oxides. Kootenai Formation, Lower Cretaceous, Alberta, Canada.

Plate 8.B. Porous lithic arenite showing a lithoclast of argillaceous radiolarite with intense pressure-solution contacts against adjacent quartz grains. Basal Salina Formation, Lower Eocene, Talara, Peru.

Plate 8.C. Porous lithic arenite showing a lithoclast of symmetrically banded fibrous quartz, probably a fracture filling, which has been preferentially dissolved along some zones. Moderate pressure solution against adjacent quartz grains; one of them (upper-right corner) shows euhedral syntaxial overgrowth. Basal Salina Formation, Lower Eocene, Talara, Peru.

Plate 8.D. Lithic arenite with interstitial calcite cement showing a subrounded and broken lithoclast of unaltered hypohyaline volcanic rock with pale green chlorite or zeolite-filled amygdules. Basal Salina Formation, Lower Eocene, Talara, Peru.

Plate 8.E. Porous lithic arenite showing a lithoclast of volcaniclastic rock containing large embayed quartz grains partially set in a dark brown, highly weathered, microlitic silicic glass matrix. Moderate pressure solution between all constituents accompanied by fine fracture porosity. Basal Salina Formation, Lower Eocene, Talara, Peru.

Plate 8.F. Lithic arenite consisting of predominant lithoclasts of various types of tan to brown basaltic glass with massive or well-developed flow structure and less abundant lithoclasts of silicic to intermediate glass altered to

pale green chlorite and yellow zeolites. Intense pressure solution between all volcanic constituents and scattered grains or phenocrysts of quartz and patches of opaque iron oxides. Cretaceous, Two Medicine, Montana, U.S.A.

Plate 9.A. Lithic arenite consisting predominantly of angular to subrounded lithoclasts of various types of fresh to deeply altered serpentine, quartzites, cherts, and calcisiltites, associated with grains of quartz, muscovite flakes, and concentrations of opaque iron oxides. Moderate pressure solution between all constituents and irregularly distributed patches of sparite cement. "Macigno," Paleogene, Narni, Tuscany, Italy.

Plate 9.B. Lithic arenite consisting predominantly of subangular to subrounded lithoclasts of various types of fresh to deeply altered porphyritic andesites, massive andesitic tuffs, basalts, and less abundant lithoclasts of fine metaquartzites and micaschists associated with quartz phenocrysts. Moderate pressure solution between all constituents and widespread sparite cement. Middle Miocene, Argao Canyon, Cebu, Philippines.

Plate 9.C. Metalithic arenite showing a large lithoclast consisting of a myrmekitic intergrowth of quartz and microcline set in a highly metamorphic quartz–feldspar–sericite groundmass. Birrimian, Bouaké, Ivory Coast, Africa.

Plate 9.D. Lithic arenite consisting of an assemblage of coarse-grained and subangular lithoclasts of cherts, andesites, schists, quartzites, and porphyritic volcanics associated with angular grains of quartz, fresh feldspars, chloritized biotite, and muscovite. Moderate pressure solution between all constituents and widespread interstitial, finely crystalline, argillaceous calcite mixed with some opaque iron oxides, chlorite, and sericite. Upper Cretaceous, Cache Creek, California, U.S.A.

Plate 9.E. Lithic arenite consisting predominantly of subangular to subrounded lithoclasts of light-colored radiolarian cherts of variable texture, dark bituminous shales, laminated argillaceous calcisiltites, and quartzites, associated with quartz grains. Appreciable pressure solution between all constituents and widespread interstitial coarse calcite cement. Flysch, Oligocene, High Calcareous Alps, Central Switzerland.

Plate 9.F. Lithic arenite consisting of an assemblage of large lithoclasts of feldspathic quartz arenite (upper-left and upper-right corners of picture), silicic intrusive (lower-left corner), associated with smaller grains of metaquartzites and undulose quartz. Minor pressure solution between all constituents, which are set in a highly developed poikilotopic calcite cement. Basal Salina Formation, Lower Eocene, Talara, Peru.

Plate 10.A. Porous lithic arenite showing partially dissolved felspar grain (lower-left corner of picture) with a thin fibrous rim of chlorite cement and microporous vermicular kaolinite cement (white) filling the surrounding pore space. Basal Salina Formation, Lower Eocene, Talara, Peru.

Plate 10.B. Lithic arenite consisting of lithoclasts of various types of porphyritic andesites with large fibrous rims of green chlorite cement and clear nontronite filling central portions of pore spaces. Cretaceous, Two Medicine, Montana, U.S.A.

Plate 10.C. Porous lithic arenite showing large lithoclast of altered amygdaloidal pumice. Intense pressure-solution contacts with adjacent quartz grains. All amygdules, except one calcite filled, have been preferentially dissolved, generating intragranular porosity. Basal Salina Formation, Lower Eocene, Talara, Peru.

Plate 10.D. Porous lithic arenite showing large lithoclast of partially dissolved volcanic rock with intragranular porosity. Its groundmass shows brown and green hydrothermal alteration and contains resorbed plagioclase phenocrysts. Basal Salina Formation, Lower Eocene, Talara, Peru.

Plate 10.E. Porous lithic arenite showing almost complete dissolution of feldspar grain into a fine honeycombed texture with argillaceous residues emphasizing relict cleavage traces. Basal Salina Formation, Lower Eocene, Talara, Peru.

Plate 10.F. Porous lithic arenite showing oversized pore resulting from dissolution of two adjacent feldspar grains, with barely visible outlines, and surrounding cement. Basal Salina Formation, Lower Eocene, Talara, Peru.

WACKES

Framework

The sand-size fraction of wackes is made up essentially of quartz, feldspar, and lithic grains. As defined above, varieties consisting predominantly of quartz with no more than 5% feldspar or lithic grains are designated as quartz wackes, those with more than 5% feldspar exceeding lithic grains are called feldspathic wackes, and those with more than 5% lithic grains exceeding feldspar are called lithic wackes. The characteristic of wackes is the presence of an interstitial matrix of less than 30 μ grain size, and ranging from a minimum of 15% to as high as 60%. This variation implies that the texture of wackes ranges from grain supported to matrix supported.

Quartz grains are variable in shape, but predominantly angular. Most of the grains display undulatory extinction and intense marginal replacement by the matrix, giving them ragged to hazy boundaries. The same marginal replacement characterizes feldspar grains, which are commonly phenocrysts, broken or entire, of sodic plagioclase and often zoned (albite, oligoclase, andesine). Albite predominates, generally clear and fresh with its typical multiple twinning; in some instances it is altered to sericite, chlorite, or epidote. Calcic plagioclases have been albitized or replaced by laumontite.

K-feldspars are rare to absent and may also have been albitized. Detrital micas are occasionally present as muscovite, biotite, and chloritized biotite. Heavy minerals are mostly zircon and tourmaline.

Recognizable lithic grains belong to the major suites of sedimentary to low-rank metamorphic rocks, such as mudstones, slates, siltstones, phyllites, micaschists, polycrystalline quartz, and quartzites; silicic volcanic rocks, such as fine-grained rhyolitic glass and rhyolites with flow structure and microlites of feldspar; and fine-grained intermediate to basaltic volcanic rocks, predominantly andesites. The presence of chert and quartz pseudomorphs after shardlike fragments, together with well-formed phenocrysts of sanidine and plagioclases, could indicate products of contemporaneous volcanism (Condie et al., 1970). Occasional grains of serpentine, as well as grains of sedimentary chert and calcilutites, may be encountered.

Matrix and Cements

The matrix is a microcrystalline to fine-grained intergrowth of chlorite, sericite, quartz, feldspars, and micas whose relatively dark color is further enhanced by scattered pigments of carbonaceous or graphitic material and minute pyrite crystals. The matrix is normally heterogeneous and displays relicts of almost completely replaced unstable volcanic lithic and monomineralic grains. Other patches of matrix have a distinct composition, indicating that they are lithic grains of metamorphic and unstable volcanic rocks deformed by compaction and tectonic stresses, squeezed between more resistant constituents. Clearly, the designation of matrix covers a complex origin, which prompted Dickinson (1970) to introduce an appropriate terminology. His attempt does not solve the petrographic complexity resulting from mineralogical convergence, but clarifies the problem in recognizing how many kinds of origin are involved under the general term of matrix. The term *protomatrix* is used for a matrix of depositional origin consisting mainly of smectite, illite, and possible pyroclastic contribution of glass and zeolites. The term *orthomatrix* is applied to a recrystallized protomatrix with no visible textural reorganization. It consists of sericite, chlorite, epidote, and biotite. Wackes with protomatrix and orthomatrix are relatively rare. The examples from the Miocene–Pliocene of Timor (Audley-Charles, 1967) show a protomatrix ranging from 5% to 30% and consisting entirely of detrital silt- to clay-size argillaceous materials. This matrix shows very slight marginal replacement of quartz and feldspar grains and no relicts of partially or entirely altered lithic grains. The relative rarity of wackes with protomatrix or orthomatrix results from two factors. First, the simultaneous deposition of a sand and clay fraction, even by high-density turbidity currents, remains controversial and, according to Kuenen (1966), theoretical and experimental data indicate that the coarsest wackes could have at the best 10%

protomatrix. Second, such a protomatrix would be recognizable only in young, weakly buried wackes, because burial would change protomatrix into orthomatrix, which is extremely difficult to distinguish from a matrix of authigenic origin.

The matrix of recognizable authigenic origin is called *epimatrix*. It is a typical heterogeneous interstitial material that, upon investigation by techniques such as X-ray diffraction, SEM, and electron microprobe (Morad, 1984), indicates a development in place from the chemical diagenetic alteration of a great variety of constituents. These include quartz, feldspars, and biotite, lithic grains of shales, siltstones, phyllites, micaschists, and silicic intermediate, and basaltic volcanics, with a particular contribution from andesites. The resulting matrix of illite, mixed-layer illite/smectite, sericite, chlorite, muscovite, epidote, and zeolites contains skeletal grains of quartz, biotite, and feldspars, displaying the effects of intense marginal replacement. Often the epimatrix forms fibrous coatings of irregular thickness with fibers oriented perpendicularly to the margins of the corroded grains. Whenever the epimatrix originates from the alteration of unstable volcanic grains, it varies enough in mineralogical composition to still reveal the relict outline of the grains. However, instances are known in which the parent grain has completely disappeared, contributing only to the heterogeneous texture of the epimatrix. The epimatrix can become more texturally complex by subsequent zeolitization and recrystallization under tectonic stresses.

Shannon (1978) studied lower Paleozoic quartz–lithic–feldspathic, grain-supported wackes from Ireland, in which the lithic grains are mudstones, shales, siltstones, low-rank metamorphics, and some igneous rocks, but no volcanics. He showed the important contribution that the alteration of quartz grains can make to the generation of the epimatrix. Detrital quartz grains have very irregular boundaries resulting from their marginal replacement by the diagenetic chlorite matrix. Chlorite flakes frequently form a zone of marginal infiltration, 0.03 to 0.06 mm wide, surrounding the quartz grain. The inner part of this zone consists of small chlorite flakes; the outer part tends to be more quartz rich and is formed by rounded to elongated matrix-size recrystallized quartz grains mixed with a smaller amount of interstitial chlorite. The latter eventually grades into the interstitial chlorite. This unusual marginal structure of quartz grains is interpreted as follows: the inner part is a corrosion front, and the outer part, which is quartz rich, represents the recrystallized original margin of the quartz grain. In the same suite of wackes studied by Shannon (1978), detrital andesine is moderately to extensively sericitized with hazy boundaries. When totally replaced, it becomes barely recognizable because of its subrectangular shape and a faintly banded organization of the sericite, which is a relict of multiple twinning. Rock fragments are equally sericitized to a variable extent and highly deformed by compaction between more resistant constituents.

The epimatrix may also contain authigenic silicate minerals precipitated from solutions in its microporosity (Morad, 1984). These minute pore fillings consist of mixed illite–smectite and chlorite without the preferred orientation of the previously described coatings around the skeletal residues of various minerals. They occur as concentrations of pseudohexagonal flakes and can at the most penetrate into detrital grains along microfractures. The cations obviously came from adjacent altering minerals. In some instances, finely crystalline authigenic quartz occurs between illite flakes, as well as minute pseudorhombic crystals of authigenic potassic feldspars.

Dickinson (1970) also introduced the term *pseudomatrix* to characterize discontinuous patches of interstitial materials that result from the compactional deformation of still-recognizable weak grains. These grains consist of shales, phyllites, and glassy materials with microlites and were changed to aggregates of sericite, chlorite, epidote, and zeolites. Thus they have a composition similar to epimatrix, but are recognizable as former framework constituents. Their identification contributes to a better understanding of the spectrum of original lithic grains. Practically, two main criteria can be used to recognize pseudomatrix: (1) the pseudofluidal internal texture of the lithic fragments, which makes them conform to the shape of adjacent more rigid grains and penetrate in all adjacent narrow pore throats; and (2) the fact that the composition of these crushed grains is homogeneous inside each one and different from one another.

Morad (1984) described further stages of alteration of pseudomatrix. For instance, coarsely crystalline chlorite was found to occur as patches within finely crystalline chloritic pseudomatrix, with contacts ranging from sharp to transitional. It appears as if the coarse chlorite developed through local fixation of iron and magnesium derived from biotite constituents. In other cases, mainly in coarse-grained wackes, well-crystallized kaolinite extensively replaced extremely fine and poorly crystallized illitic–chloritic pseudomatrix. It is suggested that kaolinite resulted from the leaching of iron, magnesium, and potassium from the pseudomatrix. The remaining silica and alumina after the dissolution combined with a required additional amount of alumina, derived from the dissolution of detrital feldspars, crystallized into kaolinite.

The diagenetic origin of the interstitial matrix of wackes is supported by the results of experiments of hydrothermal nature simulating deep burial conditions (Whetten and Hawkins, 1970). These authors used moderately to poorly sorted sands of the Columbia River, which are chemically similar to wackes, also, in their high Na to K ratio. They consist of a small amount of detrital quartz and plagioclase, rare K-feldspar, and a predominant quantity of unstable grains of altered volcanic glass and porphyritic volcanics, and some mafic minerals, all of which are mainly derived from andesitic volcanics, with a small contribution from plutonic and metamorphic rocks.

Depending on the experiment, Whetten and Hawkins (1970) placed samples in distilled water or in artificial seawater in silver tubes and heated them externally by resistance furnaces. Tests were run for 14 to 60 days at an average temperature of 250°C and a water pressure of 1 kbar. New minerals, not present in the original sands, were synthesized from the hydration and alteration of partially to completely devitrified volcanic lithic grains. These complex reactions etched the grain surfaces and generated a meshlike to fibrous coating of authigenic minerals such as smectite, chlorite, clinoptilolite, heulandite, and riebeckite. For technical reasons, it was not possible to ascertain if quartz, feldspars, and mafic minerals, such as hornblende and pyroxene, originally present in the materials submitted to experimentation, participated or not in the reactions. However, many wackes show that these minerals can be replaced by phyllosilicates and zeolites, and there are no reasons why they would not behave in the same way as lithic volcanics during the experimentation. At any rate, these experiments by Whetten and Hawkins (1970) demonstrate the initial stages of the formation of an epimatrix at the expense of framework grains. The physical conditions of these experiments were approximately equivalent to a burial on the order of 3 to 4 km with a geothermal gradient of 60° to 80°C/km. The duration of the experimental runs was too short to obtain equilibrium at the pressure and temperature used, and equilibrium conditions had to be considerably overstepped to produce the mineral phase formed; therefore, the geothermal gradients may be in excess by as much as a factor of 2 compared to those of natural environments.

In summary, most wackes have derived their authigenic matrix from the alteration of unstable metamorphic and volcanic rock fragments, with a contribution from grains of quartz, feldspars, pyroxenes, and amphiboles. This authigenic alteration increases with depth of burial and tectonic stresses. It may eventually reach a near complete replacement of detrital grains by a chlorite–sericite epimatrix interspersed with microcrystalline quartz, feldspar, calcite, and zeolite (Lovell, 1972). In a given context, the ratio of still recognizable rock fragments to matrix could be used as an index of relative depth of burial (Lovell, 1972).

The presence of abundant calcite cement in some wackes is of critical interest for itself and its implication on the generation of epimatrix. The question does not pertain to small patches of calcite scattered in epimatrix, which may represent recrystallized, small carbonate lithic grains or patches of calcite precipitated by pore fluids during burial to be discussed later, but to early cavity-filling calcite cement, occurring not only in lithic arenites interbedded with wackes of many turbidite sequences, but also in certain parts of the wackes themselves (Brenchley, 1969). He described lithic volcanic wackes from the Ordovician of North Wales that

consist of angular grains of quartz, feldspar, and aphanitic volcanics with microlites of feldspar scattered in an epimatrix of needlelike, pale green chlorite ranging from 40% to 60%. This matrix shows relics of almost entirely chloritized sand grains of presumably unstable volcanics. Adjacent parts of the same wacke bed, which contain shell debris, are well-sorted quartz–feldspar–lithic arenites with an early poikilotopic calcite cement filling an interstitial porosity on the order of 30% and accompanied by practically no matrix. The calcite cement has replaced some of the grains, in particular aphanitic volcanics whose feldspar microlites remain scattered within the poikilotopic calcite.

Because both wackes with epimatrix and calcite-cemented lithic volcanic arenites occur in adjacent parts of the same bed, the conclusion is that both were deposited as well-sorted lithic sands and that early calcite pore-filling cementation protected unstable lithic grains from diagenetic alteration or growth of authigenic phyllosilicates, by reducing porosity and permeability, which elsewhere enhanced diagenetic alteration. The latter, in the absence of early calcite, proceeded toward an advanced stage of destruction of lithic grains into a chloritic material. This epimatrix is responsible for changing an original grain-supported texture into a diagenetic matrix-supported texture (Galloway, 1974).

In the description of the framework of wackes, the abundance of albite was mentioned as the major reason for the high Na to K ratio of all wackes. The question can be raised if this feldspar was originally detrital albite or resulted from the complete diagenetic albitization of calcic plagioclases or in some cases of perthitic K-feldspars (Ogunyomi et al., 1981). There are many reasons to believe that albitization of plagioclases and the generation of a chloritic epimatrix are the most important diagenetic processes to which wackes owe their typical texture and composition.

Albitization of plagioclases is a diagenetic burial process that does not necessarily require abnormal salinities or extremely high sodium concentrations of the circulating brines, which are derived from diagenetically modified seawater (Boles, 1982). In the Cenozoic of the Gulf Coast, the process occurs in feldspathic–lithic arenites at temperatures ranging from 110° to 120°C and at depths between 2,500 and 2,870 m. The process of albitization of plagioclases is the conversion of the anorthite component of calcic plagioclase through an equal-volume replacement reaction during which Na^+ ions are incorporated from solutions and Ca^{2+} ions released into solution until final completion. However, in many instances, mineral by-products are formed such as laumontite, calcite, and kaolinite (dickite). Upon formation of laumontite, calcium-bearing plagioclases can be partially albitized and partially replaced by laumontite. Authigenic pure calcite (Fe–Mg free) develops as inclusions within albitized grains. It originates from the Ca^{2+} released during albitization combined with carbonate ions from pore fluids. Authigenic kaolinite (dickite) acts as a sink for alumina released during albitization (Boles, 1982). Authigenic albite is very pure, has a low albite structure, is calcium free, and rarely contains residual potassium from the original replaced grains.

Authigenic albite appears in three major ways (Helmold and Van de Kamp, 1984): (1) clear filling and replacement along fractures and cleavage planes of detrital plagioclase grains, suggesting that such zones of weakness favor fluid penetration; (2) clear albite overgrowths with euhedral faces that are optically continuous with albitized plagioclase grains, but are optically discontinuous with calcium-bearing plagioclases as a result of the difference in chemical composition; and (3) completely albitized plagioclase grains characterized by a murky brown aspect resulting from numerous "dusty" inclusions often associated with calcite inclusions and intragranular porosity.

On the basis of the above criteria, one can recognize that albite in many wackes is, to a large but variable extent, diagenetic in origin. In an example from the Great Valley Sequence of California, Dickinson et al. (1969) described albitization in buried feldspathic–lithic arenites and wackes, starting at a depth of 2,250 m and reaching completion at 6,000 m. It is associated with chloritization of biotite and, as in the case of the generation of a chloritic matrix from unstable lithics, it is very reduced or absent when feldspathic–lithic arenites have an early calcite cement.

Dolomite is another type of cement reported in significant amount (average 6.7%, maximum 14%) within the epimatrix of some late Precambrian wackes of South Africa (Condie et al., 1970; Reimer, 1972). In these wackes, dolomite is irregularly distributed in patches and veinlets and often in well-formed rhombs. The fact that dolomite often penetrates and embays detrital grains demonstrates its authigenic origin, which is also shown by the absence of any grains of detrital carbonate that might have recrystallized during diagenesis. The most probable explanation for the generation of dolomite is its introduction and precipitation from calcium–magnesium bicarbonate-rich pore fluids during an early stage of burial when some original porosity was still present, which consequently was reduced or eliminated. Calcium was apparently derived from the alteration of calcic plagioclases and partial albitization of plagioclases in the wackes and intercalated shales. Indeed, the amount of dolomite in a given sample is negatively correlated with the average An content of its plagioclase population. Also, the amount of dolomite varies inversely with the relative amount of plagioclase in the size fraction larger than 0.03 mm (Reimer, 1972). Magnesium was mainly derived from the original mafic minerals and rock fragments. These diagenetic changes reached their maximum intensity under an overburden of 500 to 1,000 m and were virtually completed at about 3,000 m. Because the overburden of the investigated wackes was about 5,500 m, the temperature was about 165°C. The simultaneous occurrence of siderite and pyrite suggests that these diagenetic processes took place at a pH of about 6.5 to 9.5 and an Eh of −0.2, to −0.4. Reimer (1972) suggested that, while calcite is the usual cement found in wackes, dolomite is preferentially formed in those containing a high amount of magnesium, expressing abundant mafic detrital minerals.

Depositional Environments and Provenance

A valid subdivision of wackes as a function of their tectonic setting is possible on the basis of framework quartz content and chemical composition (Crook, 1974). He recognized the following general correlation:

> Quartz-poor wackes (volcanic lithic and volcanic-lithic wackes): Q < 15%; average percent SiO_2, 58; average K_2O to Na_2O << 1.0; deposited in island-arc systems (**Plate 11.A, B, C, D**)
>
> Quartz-intermediate wackes (feldspathic wackes): Q, 15% to 65%; average percent SiO_2, 68 to 74; average K_2O to Na_2O < 1.0; deposited on leading edges with subduction (**Plate 11.E, F**)
>
> Quartz-rich wackes (quartz wackes): Q > 65%; average percent SiO_2, 89; average K_2O to Na_2O > 1.0; deposited along trailing edges (**Plate 12.A, B**)

Contrary to a long-held belief, the mineralogical composition of wackes varied through geological time. For instance, McLennan (1984) studied the characteristics of Archean wackes and compared them with those of the Phanerozoic. The comparison shows a drastic change in upper-crustal composition during late Archean toward a more felsic character related to a period of growth and differentiation of the continental crust. In terms of QFR and of alkali elements (K, Na) relationships, most Archean wackes are similar to Phanerozoic quartz-intermediate wackes. But several important differences prevent direct comparisons. First, felsic volcanic rock fragments are much more abundant than andesitic rock fragments in Archean wackes, whereas the reverse is frequent in Phanerozoic cases. Second, the predominance of plagioclase over K-feldspar in Archean wackes cannot be taken as evidence of a volcanic source because early Archean granitic rocks are typically sodium rich (tonalites, trondhjemites). The abundance of quartz in most Archean wackes (Q typically > 40) indicates a major granitic component. In summary, many Archean wackes seem to have been derived from the well-documented igneous bimodal suite of many Archean terranes, that is, mafic volcanics, Na-granites, and felsic granites. Geochemical and petrographic differences between Archean wackes and younger equivalents do not allow a direct comparison of tectonic settings, but indicate only that many Archean wackes were deposited along evolving continental margins.

The most characteristic Phanerozoic wackes originated mainly from the destruction of volcaniclastic and volcanic rocks, with an average composition of andesites, redeposited by turbidity currents, in rapidly subsiding zones of active magmatic arcs. In these zones, characterized by elevated geothermal gradient and strong tectonic stresses, the unstable lithic and mineral constituents underwent intense diagenetic transformations, reaching weak metamorphism, generating the variable amount of epimatrix that is the typical trademark of all wackes.

Provenance of wackes is obvious from the examination of the lithic and mineral constituents still recognizable in spite of the intense diagenesis, with the reservation that only a minimum number of the original constituents might have been preserved. To a certain extent, the provenance problem of wackes is simplified by the fact that they are restricted to orogenic belts at the exclusion of any other structural domain. In the triangular diagrams (**Fig. 1.7**) proposed by Dickinson et al. (1983), wackes belong to the three types occurring within active magmatic arcs.

Studies of provenance of wackes forming thick and apparently monotonous flysch (submarine fan–turbidites) sequences permit, as in the previously discussed case of lithic arenites, subdividing them into petrographic (lithostratigraphic) units, as long as they belong to relatively undisturbed structural units and to the same sedimentary petrographic province. Korsch (1978) studied such a sequence of late Paleozoic wackes of Eastern Australia. Dacitic volcanism provided most of the lithic constituents, with a small contribution from nonvolcanic sources. He divided the investigated sequence into four petrographic units on the basis of the presence or absence of detrital hornblende and the relative ratio of volcanic rock fragments to feldspar. Vertical petrographic variations within the entire sequence indicate that silicic volcanic sources were predominant throughout the entire time of deposition, but that the contribution of intermediate volcanics, silicic plutonics, low-rank metamorphics, and sedimentary sources increased toward the top. The presence or absence of minor constituents such as zircon, epidote, chlorite, tourmaline, and various textures in volcanic rock fragments delineated lateral variations within one subdivision.

In another example of uninterrupted sequence of early Precambrian wackes of the Sheba Formation (Fig Tree Group) of South Africa, Reimer (1971) used volcanic quartz as a provenance indicator. These wackes, among the oldest known, have the following average composition in percent: quartz, 26; chert fragments, 14; feldspar, 13; matrix (< 0.03 mm), 34; other rock fragments, mica, and opaques, 5; dolomite, 8. In this study, volcanic quartz is characterized by these features: almost complete absence of monomineralic inclusions, almost exclusive straight extinction, idiomorphic shape or outline controlled by crystal faces, and irregular resorption structures.

The stratigraphic variation of the components of these wackes was expressed by two indices: a ratio of nonundulatory to undulatory quartz and a ratio of porphyry and tuff lithics to lava fragments. Both ratios show a similar cyclic pattern and a general decrease upward, indicating a change in composition of the source areas, at least with respect to silicic volcanic rocks that were part of the underlying volcanic Onverwacht Group.

Provenance studies on wackes in flysch series involved in complex structures resulting from multiple thrusting, as in the Alpine orogenic belt, are much more difficult. However, if in some cases provenance remains obscure, it is possible to use the petrographic characterization of units as a tool to unravel structures.

Diagenetic Evolution

The diagenetic evolution of wackes, originally porous sands, consists essentially of the above-described alteration of unstable mineral grains and rock fragments, which produced their typical epimatrix. This alteration does not require any further elaboration here. Small amounts of calcite and dolomite cements were also deposited by circulating pore fluids. Whenever large-scale early calcite cementation took place, the original lithic sands remained volcanic lithic arenites.

Reservoir Properties

Primary porosity of the original lithic sands has been practically destroyed by the very process of generation of the epimatrix of wackes. A small amount of secondary porosity could result, in coarse-grained wackes, from burial dissolution of altered unstable grains of feldspars, mafics, and volcanic rocks, from dissolution of clay minerals formed by alteration of pseudomatrix, and from dissolution of patches of calcite and dolomite cements formed during burial as byproducts of epimatrix generation. At any rate, permeability would be negligible and porosity so low as to make most wackes unfavorable producers of hydrocarbons, which, at best, would require stimulation.

TYPICAL EXAMPLES

Plate 11.A. Lithic wacke consisting predominantly of lithoclasts of weathered micaschists, phyllites, and metaquartzites associated with grains of quartz with straight or undulose extinction, altered plagioclases, rare potassic feldspars. Well-developed epimatrix of chlorite, sericite, epidote, graphite, and pyrite. Rounded quartz grain (lower right of picture) displays strong marginal replacement by epimatrix material. Proterozoic, Brasiliano cycle, Tucano Basin, Brazil.

Plate 11.B. Lithic wacke consisting predominantly

of lithoclasts of metaquartzites, gneisses, micaschists, and phyllites associated with grains of quartz with straight or undulose extinction and altered plagioclases. Small development of epimatrix of chlorite, epidote, sericite, graphite, and pyrite combined with intense pressure solution between grain constituents and epimatrix. Proterozoic, Brasiliano cycle, Tucano Basin, Brazil.

Plate 11.C. Volcanic–lithic wacke consisting of a coarse-grained and poorly sorted assemblage of lithoclasts of various types of porphyritic andesites, andesitic tuffs, quartzites, schists, and gneisses associated with abundant isolated and broken phenocrysts of variably altered plagioclases, also of andesitic origin, grains of quartz, augite, hornblende, chloritized biotite, and muscovite flakes. Almost opaque interstitial orthomatrix consists of chlorite, epidote, and unidentifiable association of clay minerals and pyrite. Taveyannaz Sandstone, Eocene, High Calcareous Alps, Arve Valley, France.

Plate 11.D. Volcanic–lithic wacke consisting of a coarse-grained and poorly sorted assemblage of lithoclasts of various types of porphyritic andesites, andesitic tuffs, and quartzites associated with large and broken phenocrysts of zoned plagioclases, also of andesitic origin, grains of quartz with undulose extinction, augite, hornblende, chloritized biotite, and muscovite flakes. Moderately developed interstitial orthomatrix consists of chlorite, epidote, and pyrite. Taveyannaz Sandstone, Eocene, High Calcareous Alps, Arve Valley, France.

Plate 11.E. Feldspathic wacke consisting of a poorly sorted assemblage of large broken phenocrysts of fresh and variably altered twinned plagioclases, rare potassic feldspars, quartz grains with straight or undulose extinction, deeply chloritized biotite, and muscovite flakes. Subordinate lithoclasts of micaschists, cherts, and porphyritic volcanic rocks. Well-developed interstitial orthomatrix consists of

chlorite, epidote, and pyrite, which has appreciably replaced the margins of all grain constituents. Upper Cretaceous, Cache Creek, California, U.S.A.

Plate 11.F. Feldspathic–quartz wacke consisting of a poorly sorted assemblage of predominant fresh to variably altered plagioclases, quartz grains with straight or undulose extinction, deeply chloritized biotite, and muscovite flakes. Subordinate and small lithoclasts of metaquartzites, weathered micaschists, and phyllites. Well-developed interstitial epimatrix consists of sericite, chlorite, epidote, graphite, and pyrite. Quartz grain at center of picture shows deep embayment, expressing the generalized marginal replacement of grain constituents by epimatrix material. Proterozoic, Brasiliano cycle, Tucano Basin, Brazil.

Plate 12.A. Quartz wacke consisting predominantly of poorly sorted, subangular to subrounded, and broken grains of quartz with straight or undulose extinction, rare weathered plagioclase grains, chloritized biotite flakes, and muscovite flakes. Scattered small lithoclasts of opaque iron ore. Well-developed interstitial orthomatrix of quartz, illite, chlorite, sericite, pyrite, and bituminous material, which has appreciably replaced the margins of quartz grains. Cabeças Formation, Upper Devonian, Maranhão Basin, Brazil.

Plate 12.B. Metawacke consisting of entirely reorganized, large, polycrystalline quartz pseudoclasts with undulose extinction, rare grains of plagioclases, and subordinate lithoclasts of metaquartzites and dark schists. All grain constituents display extremely irregular margins due to intense pressure solution involving the interstitial reorganized matrix of quartz with abundant secondary muscovite and sericite. Original bimodality of this metamorphic quartz wacke is barely recognizable. Ocoe Series, Proterozoic, Gatlinburg, Tennessee, U.S.A.

REFERENCES

AAGAARD, P., EGEBERG, P. K., SAIGAL, G. C., MORAD, S., and BJØRLYKKE, K., 1990. Diagenetic albitization of detrital K-feldspars in Jurassic, Lower Cretaceous, and Tertiary reservoir rocks from offshore Norway. II. Formation water chemistry and kinetic considerations. *J. Sed. Petrology*, 60, 575–581.

AUDLEY-CHARLES, M. G., 1967. Greywackes with a primary matrix from the Viqueque Formation (Upper Miocene–Pliocene), Timor. *J. Sed. Petrology*, 37, 1–11.

BASU A., 1985. Reading provenance from detrital quartz. In G. G. Zuffa (ed.), *Provenance in Arenites*. NATO ASI Series, Series C, vol. 148. D. Reidel Publishing Co., Dordrecht, pp. 231–247.

———, YOUNG, S. W., SUTTNER, L. J., JAMES, W. C., and MACK, G. H., 1975. Reevaluation of the use of undulatory extinction and polycrystallinity in detrital quartz for provenance interpretation. *J. Sed. Petrology*, 45, 873–882.

BLATT, H., 1979. Diagenetic processes in sandstones. In P. A. Scholle and P. R. Schluger (eds.), *Aspects of Diagenesis*. Soc. Econ. Paleontologists and Mineralogists, Special Publ. 26, pp. 141–157.

BOLES, J. R., 1982. Active albitization of plagioclase, Gulf Coast Tertiary. *Amer. J. Sci.*, 282, 165–180.

———, and FRANKS, S. G., 1979. Clay diagenesis in Wilcox sandstones of southwest Texas: implications of smectite diagenesis on sandstone cementation. *J. Sed. Petrology*, 49, 55–70.

BRAITHWAITE, C. J. R., 1989. Displacive calcite and grain breakage in sandstones. *J. Sed. Petrology*, 59, 258–266.

BRENCHLEY, J. P., 1969. Origin of matrix in Ordovician greywackes, Berwyn Hills, North Wales. *J. Sed. Petrology*, 39, 1297–1301.

BUCZYNSKI, C., and CHAFETZ, H. S., 1987. Siliciclastic grain breakage and displacement due to carbonate crystal growth: an example from the Lueders Formation (Permian) of north-central Texas, U.S.A. *Sedimentology,* 34, 837–843.

BURLEY, S. D., and KANTOROWICZ, J. D., 1986. Thin section and S.E.M. textural criteria for the recognition of cement-dissolution porosity in sandstones. *Sedimentology,* 33, 587–604.

BURNS, L. K., and ETHRIDGE, F. G., 1979. Petrology and diagenetic effects of lithic sandstones: Paleocene and Eocene Umpqua Formation, southwest Oregon. In P. A. Scholle and P. R. Schlunger (eds.), *Aspects of Diagenesis.* Soc. Econ. Paleontologists and Mineralogists, Special Publ. 26, pp. 307–317.

CAROZZI, A. V., 1967. Recent calcite-cemented sandstone generated by the equatorial tree Iroko (*Chlorophora excelsa*), Daloa, Ivory Coast. *J. Sed. Petrology,* 37, 597–600.

CAYEUX, L., 1929. *Les roches sédimentaires de France—Roches siliceuses.* Mém. Carte géol. dét. France. Imprimerie Nationale, Paris, 774 pp.

CONARD, M., KREUZER, H., and ODIN, G. S., 1982. Potassium-argon dating of tectonized glauconies. In G. S. Odin (ed.), *Numerical Dating in Stratigraphy,* Part One. John Wiley & Sons, New York, pp. 321–332.

CONDIE, K. C, MACKE, J.E., and REIMER, T. O., 1970. Petrology and geochemistry of early Precambrian graywackes from the Fig Tree Group, South Africa. *Geol. Soc. Amer. Bull.,* 81, 2759–2776.

COUTO ANJOS, S. M., and CAROZZI, A. V., 1988. Depositional and diagenetic factors in the generation of Santiago arenite reservoirs (Lower Cretaceous), Araças oil field, Recôncavo Basin, Brazil. *J. South Amer. Earth Sci.,* 1, 3–19.

CROOK, K. A. W., 1974. Lithogenesis and geotectonics: the significance of compositional variation in flysch arenites (graywackes). In R. H. Dott, Jr., and R. H. Shaver (eds.), *Modern and Ancient Geosynclinal Sedimentation.* Soc. Econ. Paleontologists and Mineralogists, Special Publ. 19, pp. 304–310.

DASGUPTA, S., CHAUDHURI, A. K., and FUKUOKA, M., 1990. Compositional characteristics of glauconitic alteration of K-feldspars from India, and their implications. *J. Sed. Petrology,* 60, 277–281.

DICKINSON, W. R., 1970. Interpreting detrital modes of graywacke and arkose. *J. Sed. Petrology,* 40, 695–707.

———, and SUCZEK, C. A., 1979. Plate tectonics and sandstone compositions. *Amer. Assoc. Petroleum Geologists Bull.,* 63, 2164–2182.

———, OJAKANGAS, R. W., and STEWART, R. J., 1969. Burial metamorphism of the late Mesozoic Great Valley Sequence, Cache Creek, California. *Geol. Soc. Amer. Bull.,* 80, 519–526.

———, BEARD, L. S., BRAKENRIDGE, G. R., ERJAVEC, J. L., FERGUSON, R. C., INMAN, K. F., KNEPP, R. A., LINDBERG, F. A., and RYBERG, P. T., 1983. Provenance of North American Phanerozoic sandstones in relation to tectonic setting. *Geol. Soc. Amer. Bull.,* 94, 222–235.

DOTT, R. H., Jr., 1964. Wacke, graywacke and matrix—What approach to immature sandstone classification? *J. Sed. Petrology,* 34, 625–632.

DUTTON, S. P., and LAND, L. S., 1985. Meteoric burial diagenesis of Pennsylvanian arkosic sandstones, southwestern Anadarko Basin, Texas, *Amer. Assoc. Petroleum Geologists Bull.,* 69, 22–38.

ELIAS, B. P., and HAJASH, A., Jr., 1992. Changes in quartz solubility and porosity due to effective stress: an experimental investigation of pressure solution. *Geology,* 20, 451–454.

FOLK, R. L., 1954. The distinction between grain size and mineral composition in sedimentary-rock nomenclature. *J. Geol.,* 62, 344–359.

———. 1968. Bimodal supermature sandstones: product of the desert floor. *Proc. XXIII Internat. Geol. Congress,* Prague, 8, 9–32.

FORCE, E. R., 1980. The provenance of rutile. *J. Sed. Petrology,* 50, 485–488.

FRANZINELLI, E., and POTTER, P. E., 1983. Petrology, chemistry, and texture of modern river sands, Amazon river system. *J. Geol.,* 91, 23–29.

GALLOWAY, W. E., 1974. Deposition and diagenetic alteration of sandstone in Northeast Pacific arc-related basins: implication of graywacke diagenesis. *Geol. Soc. Amer. Bull.,* 85, 379–390.

———, 1979. Diagenetic control of reservoir quality in arc-derived sandstones: implications for petroleum exploration. In P. A. Scholle and P. R. Schluger (eds.), *Aspects of Diagenesis.* Soc. Econ. Paleontologists and Mineralogists, Special Publ. 26, pp. 251–262.

———, 1984. Hydrogeologic regimes of sandstone diagenesis. In D. A. McDonald and R. C. Surdam (eds.), *Clastic Diagenesis.* Amer Assoc. Petroleum Geologists, Memoir 37, pp. 3–13.

GINSBURG, L., and LUCAS, L., 1949. Présence de quartzites élastiques dans les grès armoricains de Berrien (Finistère). *C. R. Acad. Sciences, Paris,* 228, 1657–1658.

GORAI, M., 1951. Petrological studies on plagioclase twins. *Amer. Mineralogist,* 36, 884–901.

HEALD, M. T., and RENTON, J. J., 1966. Experimental studies on sandstone cementation. *J. Sed. Petrology,* 36, 977–991.

HELMOLD, K. P., 1985. Provenance of feldspathic sandstones—The effect of diagenesis on provenance interpretations: a review. In G. G. Zuffa (ed.), *Provenance of Arenites.* NATO ASI Series, Series C, vol. 148. D. Reidel Publishing Co., Dordrecht, pp. 139–163.

———, and VAN DE KAMP, P. C., 1984. Diagenetic mineralogy and controls on albitization and laumontite formation in Paleogene arkoses, Santa Ynez Mountains, California. In D. A. McDonald and R. C. Surdam (eds.), *Clastic Diagenesis.* Amer. Assoc. Petroleum Geologists, Memoir 37, pp. 239–276.

HENRY, D. J., TONEY, J. B., SUCHECKI, R. K., and BLOCH, S., 1986. Development of quartz overgrowths and pressure solution in quartz sandstones: evidence from cathodoluminescence, backscattered electron imaging, and trace element analysis on the electron microprobe (abstract). *Geol. Soc. America Abstr. Program,* No. 18, p. 635.

HICKS, B. D., APPLIN, K. R., and HOUSEKNECHT, D. W., 1986. Crystallographic influences on intergranular pressure solution in a quartzose sandstone. *J. Sed. Petrology,* 56, 784–787.

HOUSEKNECHT, D. W., 1984. Influence of grain size and temperature on intergranular pressure solution, quartz cementation, and porosity in a quartzose sandstone. *J. Sed. Petrology,* 54, 348–361.

———, 1988. Intergranular pressure solution in four quartzose sandstones. *J. Sed. Petrology,* 58, 228–246.

———, and PITTMAN, E. D. (eds.), 1992. *Origin, Diagenesis, and Petrophysics of Clay Minerals in Sandstones.* Soc. Econ. Paleontologists and Mineralogists, Special Publ. 47, 288 pp.

HUBERT, J. F., 1962. A zircon–tourmaline–rutile maturity index and the interdependence of the composition of heavy mineral assemblages with the gross composition and texture of sandstones. *J. Sed. Petrology,* 32, 440–450.

———, and REED, A. A., 1978. Red-bed diagenesis in the East Berlin Formation, Newark Group, Connecticut Valley. *J. Sed. Petrology,* 48, 175–184.

HUDSON, C. B., and EHRLICH, R., 1980. Determination of relative provenance contributions in samples of quartz sand using Q-mode factor analysis of Fourier grain shape data. *J. Sed. Petrology,* 50, 1101–1110.

HUGHES, A. D., and WHITEHEAD, D., 1987. Glauconitization of detrital silica substrates in the Barton Formation (upper Eocene) of the Hampshire Basin, southern England. *Sedimentology,* 34, 825–835.

INGERSOLL, R. V., CAVAZZA, W., and GRAHAM, S. A., 1987. Provenance of impure calclithites in the Laramide foreland of southwestern Montana. *J. Sed. Petrology,* 57, 995–1003.

JAMES, W. C., WILMAR, G. C., and DAVIDSON, B. G., 1986. Role of quartz type and grain size in silica diagenesis, Nugget Sandstone, south-central Wyoming. *J. Sed. Petrology,* 56, 657–662.

JEANS, C. V., MERRIMAN, R. J., MITCHELL, J. G., and BLAND, D. J., 1982. Volcanic clays in the Cretaceous of southern England and northern Ireland. *Clay Minerals,* 17, 105–156.

JOHNSSON, M. J., STALLARD, R. F., and MEADE, R. H., 1988. First-cycle quartz arenites in the Orinoco river basin, Venezuela and Colombia. *J. Geol.,* 96, 263–277.

KAISER, W. R., 1984. Predicting reservoir quality and diagenetic history in the Frio Formation (Oligocene) of Texas. In D. A. McDonald and R. C. Surdam (eds.), *Clastic Diagenesis.* Amer. Assoc. Petroleum Geologists, Memoir 37, pp. 195–215.

KORSCH, R. J., 1978. Petrographic variations within thick turbidite sequences: an example from the late Paleozoic of Eastern Australia. *Sedimentology,* 25, 247–265.

KOSSOVSKAYA, A. G., and DRITS, V. R., 1970. Micaceous minerals in sedimentary rocks. *Sedimentology,* 15, 83–101.

KRINSLEY, D., and TRUSTY, P., 1985. Environmental interpretation of quartz grain surface textures. In G. G. Zuffa (ed.), *Provenance of Arenites.* NATO ASI Series, Series C, vol. 148. D. Reidel Publishing Co., Dordrecht, pp. 213–229.

KRYNINE, P. D., 1935. Arkose deposits in the humid tropics, a study of sedimentation in southern Mexico. *Amer. J. Sci.,* 29, 353–363.

KUENEN, Ph. H., 1966. Matrix of turbidites, experimental approach *Sedimentology,* 7, 267–297.

LAND, L. S., and MILLIKEN, K. L., 1981. Feldspar diagenesis in the Frio Formation (Oligocene), Texas Gulf Coast. *Geology,* 9, 314–318.

LEDER F., and PARK, W. C., 1986. Porosity reduction in sandstone by quartz overgrowth. *Amer. Assoc. Petroleum Geologists Bull.,* 70, 1713–1728.

LINDSTRÖM, M., 1980. Glauconite shrinkage and limestone cementation. *J. Sed. Petrology,* 50, 133–138.

LOUCKS, R. G., DODGE, M. M., and GALLOWAY, W. E., 1984. Regional controls on diagenesis and reservoir quality in Lower Tertiary sandstones along the Texas Gulf Coast. In D. A. McDonald and R. C. Surdam (eds.), *Clastic Diagenesis.* Amer. Assoc. Petroleum Geologists, Memoir 37, pp. 15–45.

LOVELL, J. P. B., 1972. Diagenetic origin of graywacke matrix minerals: a discussion. *Sedimentology,* 19, 141–143.

MATTER, A., and RAMSEYER, K., 1985. Cathodoluminescence microscopy as a tool for provenance studies in sandstones. In G. G. Zuffa (ed.), *Provenance of Arenites.* NATO ASI Series, Series C, vol. 148. D. Reidel Publishing Co., Dordrecht, pp. 191–211.

MCBRIDE, E. F., 1989. Quartz cement in sandstones: a review. *Earth-Science Reviews,* 26, 69–112.

MCLENNAN, S. M., 1984. Petrological characteristics of Archean graywackes. *J. Sed. Petrology,* 54, 889–898.

MILLIKEN, K. L., 1989. Petrography and composition of authigenic feldspars, Oligocene Frio Formation, South Texas. *J. Sed. Petrology,* 59, 361–374.

MORAD, S., 1984. Diagenetic matrix in Proterozoic graywackes from Sweden. *J. Sed. Petrology,* 54, 1157–1168.

———, 1986. Albitization of K-feldspar grains in Proterozoic arkoses and graywackes from southern Sweden. *N. Jb. Mineralogie,* Monatsheft 4, 145–156.

———, and ALDAHAN, A. A., 1987. Diagenetic replacement of feldspars by quartz in sandstones. *J. Sed. Petrology,* 57, 488–493.

———, MARFIL, R., AND DE LA PEÑA, J. A., 1989. Diagenetic K-feldspar pseudomorphs in the Triassic Buntsandstein sandstones of the Iberian Range, Spain. *Sedimentology,* 36, 635–650.

———, BERGAN, M., KNARUD, R., and NYSTUEN, J. P., 1990. Albitization of detrital plagioclase in Triassic reservoir sandstones from the Snorre field, Norwegian North Sea. *J. Sed. Petrology,* 60, 411–425.

MORTON, A. C., 1985. Heavy minerals in provenance studies. In G. G. Zuffa (ed.), *Provenance of Arenites.* NATO ASI Series, Series C, vol. 148. D. Reidel Publishing Co., Dordrecht, pp. 249–277.

NAHON, D., CAROZZI, A. V., and PARRON, C., 1980. Lateritic weathering as a mechanism for the generation of ferruginous ooids. *J. Sed. Petrology,* 50, 1287–1298.

ODIN, G. S., 1982. Effect of pressure and temperature on clay mineral potassium–argon ages. In G. S. Odin (ed.), *Numerical Dating in Stratigraphy,* Part One. John Wiley & Sons, New York, pp. 307–319.

——— (ed.), 1988. Green marine clays. Oolitic ironstone facies, verdine facies, glaucony facies and celadonite-bearing facies—A comparative study. *Develop. Sedimentology* 45, Elsevier Publishing Co., Amsterdam, 445 pp.

———, and LÉTOLLE, R., 1980. Glauconitization and phosphatization environments: a tentative comparison. In Y. K. Bentor (ed.), *Marine Phosphorites.* Soc. Econ. Paleontologists and Mineralogists, Special Publ. 29, pp. 227–237.

———, and MATTER, A., 1981. De glauconiarum origine. *Sedimentology,* 28, 611–641.

———, and REX, D. D., 1982. Potassium–argon dating of washed, leached, weathered, and reworked glauconies. In G. S. Odin (ed.), *Numerical Dating in Stratigraphy,* Part One. John Wiley & Sons, New York, pp. 363–385.

OGUNYOMI, O., MARTIN, R. F., and HESSE, R., 1981. Albite of secondary origin in Charny sandstones, Québec: a re-evaluation. *J. Sed. Petrology,* 51, 597–606.

OWEN, M. R., and CAROZZI, A. V., 1986. Southern provenance of upper Jackfork sandstones, southern Ouachita Mountains: catho-doluminescence petrology. *Geol. Soc. Amer. Bull.,* 97, 110–115.

PARK, Y. A., and PILKEY, O. H., 1981. Detrital mica: environmental significance of roundness and grain surface textures. *J. Sed. Petrology,* 51, 113–120.

PETTIJOHN, F. J., POTTER, P. E., and SIEVER, R., 1987. *Sand and Sandstone,* 2nd ed. Springer-Verlag, New York, 571 pp.

PITTMAN, E. D., 1963. Use of zoned plagioclase as an indicator of provenance. *J. Sed. Petrology,* 33, 380–386.

———, 1969. Destruction of plagioclase twins by stream transport. *J. Sed. Petrology,* 39, 1432–1437.

———, 1970. Plagioclase feldspar as an indicator of provenance in sedimentary rocks. *J. Sed. Petrology,* 40, 591–598.

———, 1972. Diagenesis of quartz in sandstones as revealed by scanning electron microscopy. *J. Sed. Petrology,* 42, 507–519.

PLYMATE, T. G., and SUTTNER, L. J., 1983. Evaluation of optical and X-ray techniques for detecting source-rock-controlled variation in detrital potassium feldspar. *J. Sed. Petrology,* 53, 509–519.

POTTER, P. E., 1978. Petrology and chemistry of modern big river sands. *J. Geol.,* 86, 423–449.

RAMSEYER, K., BOLES, J. R., and LICHTNER, P. C., 1992. Mechanism of plagioclase albitization. *J. Sed. Petrology,* 62, 349–356.

REIMER, T. O., 1971. Volcanic quartz as indicator mineral in gray-wackes. *Sedimentology,* 17, 125–128.

———, 1972. Diagenetic reactions in early Precambrian gray-wackes of the Barberton Mountains Land (South Africa). *Sed. Geol.,* 7, 263–282.

RUSSELL, R. D., 1937. Mineral composition of Mississippi River sands. *Geol. Soc. Amer. Bull.,* 48, 1307–1348.

SAIGAL, G. C., and WALTON, E. K., 1988. On the occurrence of displacive calcite in Lower Old Red Sandstone of Carnoustie, Eastern Scotland. *J. Sed. Petrology,* 58, 131–135.

———, MORAD, S., BJØRLYKKE, K., EGEBERG, P. K., and AAGAARD, P., 1988. Diagenetic albitization of detrital K-feldspars in Jurassic, Lower Cretaceous, and Tertiary clastic reservoir rocks from offshore Norway. I. Textures and origin. *J. Sed. Petrology,* 58, 1003–1013.

SANDERSON, I. D., 1984. Recognition and significance of inherited quartz overgrowths in quartz arenites. *J. Sed. Petrology,* 54, 473–486.

SCHMIDT, V., and MCDONALD, D. A., 1979a. The role of secondary porosity in the course of sandstone diagenesis. In P. A. Scholle and P. R. Schluger (eds.), *Aspects of Diagenesis.* Soc. Econ. Paleontologists and Mineralogists, Special Publ. 26, pp. 175–207.

———, and ———, 1979b. Texture and recognition of secondary porosity in sandstones. In P. A. Scholle and P. R. Schluger (eds.), *Aspects of Diagenesis.* Soc. Econ. Paleontologists and Mineralogists, Special Publ. 26, pp. 209–225.

SHANMUGAM, G., 1985a. Significance of secondary porosity in interpreting sandstone composition. *Amer. Assoc. Petroleum Geologists Bull.,* 69, 378–384.

———, 1985b. Types of porosity in sandstones and their significance in interpreting provenance. In G. G. Zuffa (ed.), *Provenance of Arenites.* NATO ASI Series, Series C, vol. 148. D. Reidel Publishing Co., Dordrecht, pp. 115–137.

———, and HIGGINS, J. B., 1988. Porosity enhancement from chert dissolution beneath Neocomian unconformity: Ivishak Formation, North Slope, Alaska. *Amer. Assoc. Petroleum Geologists Bull.,* 72, 523–535.

SHANNON, P. M., 1978. The petrology of some lower Paleozoic greywackes from south-east Ireland: a clue to the origin of the matrix. *J. Sed. Petrology,* 48, 1185–1192.

SIPPEL, R. F., 1968. Sandstone petrology: evidence from luminescence petrography. *J. Sed. Petrology,* 38, 530–554.

SKOLNICK, H., 1965. The quartzite problem. *J. Sed. Petrology,* 35, 12–21.

STEPHENSON, L. P., PLUMLEY, W. J., and PALCIAUSKAS, V. V., 1992. A model for sandstone compaction by grain interpenetration. *J. Sed. Petrology,* 62, 11–22.

STOESSELL, R. K., and PITTMAN, E. D., 1990. Secondary porosity revisited: the chemistry of feldspar dissolution by carboxylic acids and anions. *Amer. Assoc. Petroleum Geologists Bull.,* 74, 1795–1805.

SURDAM, R. C., and BOLES, J. R., 1979. Diagenesis of volcanic sandstones. In P. A. Scholle and P. R. Schluger (eds.), *Aspects of Diagenesis.* Soc. Econ. Paleontologists and Mineralogists, Special Publ. 26, pp. 227–242.

———, and CROSSEY, L. J., 1987. Integrated diagenetic modeling: a process-oriented approach for clastic systems. *Ann. Rev. Earth Planet. Sci.,* 15, 141–170.

SUTTNER, L. J., and BASU, A., 1977. Structural state of detrital alkali feldspars. *Sedimentology,* 24, 63–74.

———, ———, and MACK, G. H., 1981. Climate and origin of quartz arenites. *J. Sed. Petrology,* 51, 1235–1246.

TREVENA, A. S., and NASH, W. P., 1981. An electron microprobe study of detrital feldspar. *J. Sed. Petrology,* 51, 137–150.

TURNER, P. (ed.), 1980. Continental red beds. *Developments in Sedimentology,* No. 29. Elsevier Scientific Pub. Co., Amsterdam, 562 pp.

WALKER, T. R., 1967. Formation of red beds in modern and ancient deserts. *Geol. Soc. Amer. Bull.,* 78, 353–368.

———, 1984. Diagenetic albitization of potassium feldspars in arkosic sandstones. *J. Sed. Petrology,* 54, 3–16.

———, WAUGH, B., and GRONE, A. J., 1978. Diagenesis in first-cycle desert alluvium of Cenozoic age, southwestern United States and northwestern Mexico. *Geol. Soc. Amer. Bull.,* 89, 19–32.

WEAVER, C. E., 1963. Interpretative value of heavy minerals from bentonites. *J. Sed. Petrology,* 33, 343–349.

WHETTEN, J. T., and HAWKINS, J. W., 1970. Diagenetic origin of graywacke matrix materials. *Sedimentology,* 15, 347–361.

WILLIAMS, H., TURNER, F. J., and GILBERT, C. M., 1982. *Petrography. An Introduction to the Study of Rocks in Thin Sections.* 2nd ed. W. H. Freeman and Co., San Francisco, 626 pp.

WILSON, M. D., and PITTMAN, E. D., 1977. Authigenic clays in sandstones: recognition and influence on reservoir properties and paleoenvironmental analysis. *J. Sed. Petrology,* 47, 3–31.

WOOD, J. R., and HEWETT, T. A., 1984. Reservoir diagenesis and convective fluid flow. In D. A. McDonald and R. C. Surdam (eds.), *Clastic Diagenesis.* Amer. Assoc. Petroleum Geologists, Memoir 37, pp. 99–110.

YEH, H. W., and SAVIN, S. M., 1977. Mechanism of burial metamorphism of argillaceous sediments. 3. O-isotope evidence. *Geol. Soc. Amer. Bull.,* 88, 1321–1330.

CHAPTER 2

RUDACEOUS ROCKS

INTRODUCTION AND CLASSIFICATION

Rudaceous rocks are coarse clastic rocks with an average grain size ranging from 2.0 to 2.5 mm to 10 m, the upper size limit for boulders. They consist of more than 25% rock fragments or pebbles and an interstitial material ranging from sandstones and calcarenites to siltstones–claystones and calcilutites, and thus are megascopically bimodal. When these rocks have a predominance of rounded pebbles, they are called *conglomerates,* and when the clasts are predominantly angular, the term of *breccia* is used. However, a complete gradation exists between the two end terms, and the designation of breccioconglomerate is used for such intermediate types. Conglomerates are by far more frequent than breccias and represent the consolidated equivalent of numerous types of gravels.

The coexistence of pebbles and an interstitial material requires that any petrographic description be undertaken at several levels in decreasing order of grain size. For instance, in the case of the interstitial material consisting of a sandstone, the description should deal with the pebbles, the monomineralic or lithic sand-size grains, and the matrix and/or cement. This procedure avoids the lack of precision introduced when the term matrix is applied indiscriminately to whatever interstitial sediment is smaller than the pebbles. The deposition of pebbles and interstitial material is generally synchronous and attributable to the same sedimentological process. However, cases occur of subsequent infiltration

of very fine detrital interstitial material into the pores of a completely or partially open pebble framework, as well as precipitation of diagenetic cements.

Sundell and Fisher (1985) attracted attention to the lack of a systematic descriptive terminology for very coarse rudaceous rocks consisting of elements larger than boulders, and thus seriously complicating their understanding. Because extremely large rock fragments may reach a maximum length of more than 10 km, such as tectonic plates, these authors introduced a dual classification that includes a nongenetic term combined with a genetic modifier in an attempt to bridge a possible continuum from sedimentary rocks to tectonically derived rocks. The following terms are proposed: *megaboulder* (10 to 100 m of maximum length diameter) with megamictite as the equivalent rock name; *block* (100 to 1000 m), equivalent rock name teramictite; *megablock* (1 to 10 km), equivalent rock name oromictite; and *tectonic plate* (> 10 km). The genetic modifiers pertain either to the process of mixing, such as sedimentary, tectonic, magmatic, volcanic, meteor impact, glacial, and biologic processes, or to the environment of mixing, such as continental, subaerial, reef, karst, talus, and so on. The proposed rock names require that at least two large clasts of the designated size limit be present among any proportion of smaller particles (considered as matrix) within a continuous rock body. Preferably, the two large clasts should occur within an area having sides no longer than 10 times the maximum length of the considered size limit.

The composition of conglomerates can be estimated by pebble count and represented by grouping rock types into four major classes: extrusive (E), plutonic (P), sedimentary (S), and metamorphic (M). Generally, these classes are plotted on triangular diagrams in which sedimentary and metamorphic rocks are associated in the same pole (Pettijohn, 1975), because the distinction of supracrustal and plutonic provenances affords clues on uplift and depth of erosion, both of which are functions of tectonism. The composition of the interstitial material expresses the various types of sandstones, since conglomerates are the rudaceous equivalents of quartz arenites, feldspathic arenites, lithic arenites, and of the three types of wackes.

Rudaceous rocks are distinctly less abundant and more restricted than associated sandstones. The effect of the degree of maturity of sandstones on related rudaceous rocks is that the latter increase in frequency with the immaturity of interstitial and associated sandstones. Quartz arenites have about 1% conglomerates, feldspathic arenites, 5%, and lithic arenites and wackes, 10% to 20%. Furthermore, as a result of the rapid destruction of certain types of less resistant pebbles during transportation (to be discussed below), mature conglomerates tend to be associated with submature and even immature sandstones, in which case the best interpretation of provenance has to be derived from the study of the interstitial materials.

The contribution of petrography to the study of rudaceous rocks is somewhat restricted, because of the size of the pebbles that are usually identified megascopically, and hence reduced to the study of the mineralogy and texture of the interstitial materials. However, finer-grained types called microconglomerates whose pebble size ranges up to a few centimeters allow enough constituents over the surface of a thin section to be valid representatives of the coarser ones.

Conglomerate and breccias of sedimentary origin can be classified into categories that take into account three major features: origin of the pebbles, presence or absence of a pebble framework, and lithological composition of the pebbles. This approach is mostly descriptive, but also genetic in a very broad sense. Indeed, attributing a given conglomerate to a particular agent of deposition or a specific environment of deposition requires extensive textural studies of the conglomerate in its basinal and tectonic context, which belong to sedimentology and are not discussed here (see Koster and Steel, 1984; Walker, 1984).

Pebbles have two possible origins. They result either from the mechanical destruction of rocks from outside the basin of final deposition, and the conglomerates they form are designated as *extraformational*, a feature they share with sandstones, or the pebbles are derived from reworking processes inside the basin of deposition and the corresponding conglomerates (or often breccias) are called *intraformational*. In the latter case, fragmentation processes such as syneresis, desiccation, submarine reworking, and storms are essentially contemporaneous with deposition.

Presence or absence of a pebble framework has important genetic significance. *Orthoconglomerates* have an intact framework of pebbles bound together by less than 15% of interstitial materials as defined above, that is, any type of arenite or calcarenite with its matrix and/or cement. They were deposited by traction processes in highly turbulent waters of fluvial and marine environments or resedimented by high-energy turbidity currents (Walker, 1984). *Paraconglomerates,* also called diamictites or pebbly mudstones, are devoid of a pebble framework. Pebbles are scattered within more than 15% (up to 70%) interstitial materials, which represent wackes with an argillaceous proto- to orthomatrix. Paraconglomerates were deposited by mass transportation processes such as glacial action, mudflows, and debris flows in freshwater and marine environments.

The lithological composition of pebbles is expressed by the terms of *monomictic* and *oligomictic* when conglomerates and breccias consist of pebbles of one or a few rock types. These rudaceous rocks are either intraformational or very mature deposits such as chert conglomerates. *Polymictic* conglomerates have pebbles representing many rock types among which one or several may predominate. They are generally extraformational in origin. The terminology pertaining to the lithological composition of pebbles is very relative and depends on the technique used. For instance, a calcareous intraformational conglomerate or calcirudite appears megascopically homogeneous, whereas, at the microscopic level, it consists of an association of five or six distinct microfacies, indicating that an entire carbonate platform has contributed to its formation, for instance by storm action.

All the above-described types of conglomerates and breccias result from several mechanical processes that transported and deposited clasts of extraformational or intraformational origin. The generation of other rudaceous rocks implies little transport and they owe their origin to diagenesis. They are designated as *pseudobreccias* and *pseudoconglomerates* to emphasize their particular mode of formation. They involve a small amount of vertical displacement by gravity and are produced by early diagenetic processes in carbonate rocks associated with evaporites. These processes are dissolution of soluble constituents and subsequent collapse of unsupported portions, thus forming a breccia with sometimes monomictic clasts fitting together in a puzzlelike fashion, with interstitial filling of the dissolved portions by a variety of precipitated cements.

Other breccias, which also occur in carbonates, consist of monomictic pseudoclasts in an apparent puzzlelike texture that hides in reality a complicated three-dimensional shape, which implies no movement at all, except a fissuration of the rock. They are produced by incomplete late diagenetic processes such as neomorphism, replacements, and mineralizations controlled by fracture patterns. For instance, in the case of incomplete dolomitization, the pseudoclasts represent the nondolomitized original limestone and the cement consists of the replacive dolomite. However, in rare cases of marginal reaction between pseudoclasts and diagenetic cements, irregularly rounded pseudoclasts are generated and the rock is designated as a pseudoconglomerate.

In summary, the following classification is used:

Rudaceous rocks of sedimentary origin
  Extraformational
  Orthoconglomerates (mono-, oligo- to polymictic)
  Paraconglomerates or diamictites (polymictic)
  Intraformational
  Conglomerates and breccias (mono- to oligomictic)
Rudaceous rocks of diagenetic origin
  Pseudobreccias and pseudoconglomerates (monomictic)

Rudaceous rocks produced by volcanic processes are described under the heading of pyroclastic rocks; those of tectonic or cataclastic origin, such as fault breccias, thrust breccias, and meteorite impact breccias are not discussed here.

ORTHOCONGLOMERATES

Petrography and Depositional Environments

Monomictic to oligomictic types are mature concentrations of generally well rounded pebbles of the materials most resistant to abrasion and decomposition, such as metaquartzites (**Plate 12.C**), quartzites, vein quartz, chert, arenaceous siltstones (**Plate 12.D**), and occasionally even limestones (**Plate 12.E**). They are residual deposits from the destruction of large volumes of rocks and generally build rather small deposits intercalated as lenticular bodies in equally mature quartz arenites and in less mature feldspathic arenites. In the case of open framework, quartz is a frequent interstitial cement, with cavity-filling textures consisting of prismatic to bladed crystals growing from pebble margins and grading toward the center of the pores into an irregular mosaic of anhedral crystals. These very mature conglomerates occur throughout the geological column in high-energy fluvial and coastal marine environments, as well as in resedimented channel deposits of some turbidite sequences (Seiders, 1983).

On the other hand, polymictic types build much thicker accumulations either close to their source areas, as alluvial fans and braided stream deposits, or farther away, as turbidite resedimented conglomerates. They are the coarse equivalents of many feldspathic and lithic arenites. Their composition is highly variable and includes virtually all types of metastable igneous, metamorphic, and sedimentary rocks, although usually three or four lithologies tend to predominate because of the rapid destruction of certain types of less resistant pebbles during transport (Lindholm et al., 1979).

Provenance

Since pebbles of orthoconglomerates are readily identifiable clasts of bedrock, they are generally favored in provenance studies over monomineralic and lithic grains of the interstitial sandstones. However, the latter are always less mature and, in spite of the numerous unanswered questions raised by the determination of their provenance discussed earlier, they should be investigated with priority, or at least in combination with the pebbles. A typical example of the use of sand-size lithic grains of the matrix of an orthoconglomerate is provided by a study of the freshwater upper fan conglomerates deposited during the Early Cretaceous along the major active fault that formed the eastern border of the Recôncavo Basin in Brazil (Carozzi et al., 1976). On the basis of the vertical distribution in cores of interstitial lithic grains of sandstone, limestone, chert, metaquartzite, and gneiss belonging to the cover and the substratum of the eastern basement margin, it was possible to divide the conglomerates into three well-characterized compositional units (**Fig. 2.1**). By means of these petrographic subdivisions, correlation between wells was possible, as well as the reconstruction of the progress of erosion along the eastern marginal source areas. Nevertheless, the question remains as to what extent do pebble populations in conglomerates represent the relative amount of the different types of bedrock exposed in the source area? Clearly, the percentage of pebble types in a conglomerate is not an exact representation because the following factors play a role: inherent capacities of different rocks to release a certain number of clasts and of a given original size; durability of the various types of clasts to abrasion; distance of transport; competence and energy of the transporting agent; and size and proportion of associated sediments (Pettijohn, 1975).

The capacity of certain rocks to release clasts is a function of their selective destruction by erosion, which expresses topography, climate, and structural conditions. For instance, beds of jointed quartzites yield angular blocks of a size controlled by the joint system, granite disintegrates into residual feldspathic sands with a few pebbles, and schists split up into irregular platy chips. The resistance to abrasion and the effect of the distance of transport are shown by schist chips that are so weak as to survive only in immature clast populations transported a short distance. Whenever clasts of schists occur in a given orthoconglomerate, their volumetric significance in the source area is certainly underestimated. Conversely, pebbles of quartzite, chert, or rhyolite are so resistant to abrasion that after becoming somewhat rounded

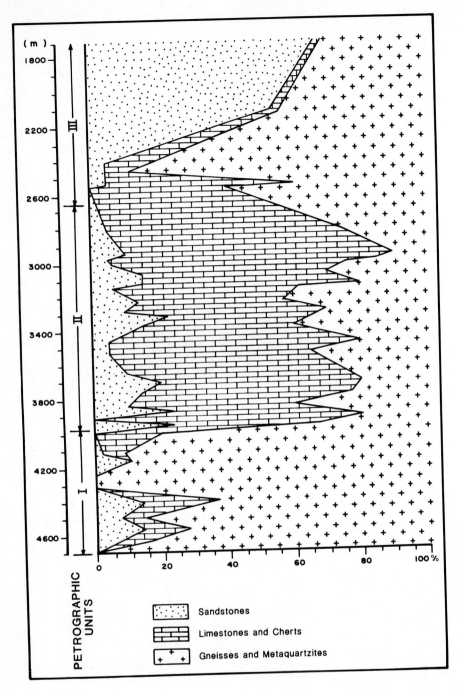

FIGURE 2.1 Vertical variation of lithic constituents in the arenaceous matrix of conglomerates and their use in reconstructing the geological composition and erosional history of source areas. Salvador Formation, Lower Cretaceous, Recôncavo Basin, Brazil. From Carozzi et al. (1976).

their size no longer decreases. Thus their assumed importance in source areas is overestimated. Their extreme abrasion durability permits them to be recycled from older conglomerates, and the overrepresentation is continued and even increased through successive sedimentary cycles. In summary, it is generally agreed that the composition of fluvial conglomerates is modified rather rapidly with increasing distance of transport downstream from an original broad spectrum of pebble lithologies to a ubiquitous restricted suite of the most resistant constituents: vein quartz, quartzites, and chert.

In reality, the problem of the interpretation of conglomerate composition is more complex than can be expressed by the effects of original release, abrasion durability, and distance of transport. Additional controlling factors include processes that occur during transportation. Among them, two appear particularly important: (1) differential weathering of various types of pebbles during temporary alluvial storage, and (2) resistance to abrasion of individual types of pebbles when transported in polylithologic assemblages. The question of differential weathering during temporary storage was discussed by Bradley (1970) in his study on the abrasion of granitic gravel of the lower Colorado River in Texas. For a distance of travel of 160 miles, the pebbles decreased in size by about 50%, and the gravel changed in composition from a dominantly granite–gneiss–aplite assemblage to a dominantly pegmatite–graphic granite assemblage. Bradley studied abrasion and its relationship to weathering by running experiments on fresh and weathered samples of the lower Colorado River granitic gravel. The degree of weathering varied from slight to moderate and involved to the highest degree biotite followed by feldspar. Abrasion of fresh granitic gravel during a simulated travel of 160 miles produced only a 10% reduction of size. Abrasion of weathered gravel produced a reduction of more than 50% and was clearly related to lithology. The biotite-bearing rocks (granite, gneiss, and some aplite) were the least durable; the biotite-free rocks (pegmatite and graphic granite) were the most durable. The decrease in size and the change of lithology of the lower Colorado River gravels in Texas are best explained by abrasion of clasts that have slightly weathered during periods of temporary alluvial storage.

The resistance to abrasion of various types of pebbles when transported in polylithologic assemblages was studied by Abbott and Peterson (1978). They found experimentally that in a first phase of transport the abrasion rate of a particular pebble type is slowed when it grinds against less durable clasts. In a second phase of transport, after elimination of the less durable clasts, the abrasion rate increases when it grinds against more durable clasts. In summary, the presence of relatively low durability clasts increases the distance of transport necessary to achieve a given attritional loss for high durability clasts.

In spite of the numerous incertitudes just mentioned, lithologic provenance modeling is a very effective approach for establishing the provenance, and its evolution in time and space, of orthoconglomerates. As previously stated for feldspathic and lithic arenites, it is a deductive technique for identifying best statistical matches between compositions of conglomerates and modeled conglomerate compositions from hypothetical source sections (Graham et al., 1986; DeCelles, 1988).

Diagenetic Evolution

Since the composition of the interstitial materials of orthoconglomerates represents the various types of quartz arenites, feldspathic arenites, and lithic arenites, the major diagenetic evolution of conglomerates resides in these materials and is identical to that previously described when discussing arenites. Diagenetic changes involving the framework of pebbles are rather limited to intrastratal dissolution and pressure solution at the contact points between pebbles. The features generated involve not only carbonates, but also the most resistant lithologies, such as vein quartz, quartzites, and cherts. They are either stylolitic contacts or distinct pits

and concave circular depressions in which less soluble pebbles have penetrated more soluble ones. Sometimes, pebbles appear deformed by sets of microfaults subsequently healed by calcite or quartz. All these features, which imply minute mechanical readjustments of the framework, require apparently only a slight overburden, since they occur in completely undeformed conglomerates and even in Pleistocene interglacial and postglacial outwash gravels.

Reservoir Properties

Orthoconglomerates have unusual reservoir properties that result from their depositional bimodality, expressed by a rigid framework of pebbles of zero porosity and an interstitial material considered here as a sand. The sand content is less than or equal to the porosity the conglomerate would have had if the sand content were zero, that is, 25% to 35%. If the sand does not fill originally the entire megapore space system between pebbles, a partial open framework texture exists with geopetal cavities in the upper part of the voids. Furthermore, if the sand fills completely or not the megapore system between pebbles, it escapes repacking into a tighter fabric by mechanical or chemical compaction. These original high-porosity and high-permeability conditions are of short duration because, upon burial, circulating solutions preferentially flow through these types of deposits and initiate the diagenetic cementation of the uncompacted sand into an arenite, as well as the filling of any open space by carbonate or silica cements.

Assuming that the depositional porosity of 25% to 35% corresponding to the interstitial sand fraction (which in fact corresponds only to 12% for the whole rock) is obliterated by diagenetic cementation, any potential reservoir can only be generated by burial secondary porosity, as previously discussed for quartz, feldspathic, and lithic arenites. However, the presence of the rigid pebble framework is an obstacle that increases the tortuosity of flow pathways. Consequently, the efficiency of secondary porosity generation is decreased and values rarely reach primary porosity or become even higher as in many quartz, feldspathic, and lithic arenites. In essence, potential reservoirs for hydrocarbons in orthoconglomerates result from a complex combination of the effects of bimodality and diagenesis, which have to be carefully evaluated in each particular case (Clarke, 1979).

TYPICAL EXAMPLES

Plate 12.C. Quartzite orthoconglomerate consisting of a framework of subrounded lithoclasts of metaquartzites, tectonized quartzites, and quartzites with an interstitial sand-size matrix of grains of quartz, altered potassic feldspars, abundant sericite, chlorite, and iron oxides. Pottsville Formation, Pennsylvanian, St. Clair, Pennsylvania, U.S.A.

Plate 12.D. Siltstone orthoconglomerate consisting of a framework of subangular lithoclasts of dark brown banded and finely arenaceous siltstone with an interstitial sand-size matrix of angular grains of quartz, potassic feldspars, biotite, muscovite and sericite flakes, and iron oxides. Appreciable generalized pressure solution has deformed the siltstone lithoclasts. Franciscan Group, Upper Jurassic, Bartlett Springs, California, U.S.A.

Plate 12.E. Limestone orthoconglomerate consisting of a framework of subangular lithoclasts of several types of limestones: pelletoidal calcarenites, argillaceous calcisiltites, pelletoidal calcisiltites, and massive or banded calcilutites. Interstitial cement is impure microgranular calcite and fibrous chalcedony; the latter also occurs as microconcretions inside some lithoclasts. Flysch, Oligocene, High Calcareous Alps, Arve Valley, France.

PARACONGLOMERATES

Petrography and Depositional Environments

Paraconglomerates are polymictic conglomerates consisting of pebbles scattered within 15% to 70% interstitial materials. These materials are the best developed examples of the three major types of wackes. However, in comparison with wackes, only protomatrix and orthomatrix have been described so far in paraconglomerates. Epimatrix seems to be very rare, because these rocks appear to be rather impervious to circulating solutions capable of appreciable diagenetic textural reorganizations, but pseudomatrix can occur when compaction processes are strongly developed.

According to the sedimentological context, paraconglomerates represent either tilloids deposited from mass transportation by debris flows and mudflows not involving glacial ice or tillites that are demonstrated glacial deposits. Tilloids are relatively minor deposits in the geological column and little petrography has been done on them; tillites are by far better known in that respect (**Plates 12.F, 13.A**). It is clear that the glacial or non-glacial origin of many paraconglomerates remains controversial and cannot be solved by petrographic studies because of converging textures.

The tillites of the Precambrian Gowganda Formation of Canada (Lindsey, 1969) and of the late Paleozoic of the Lesser Himalaya (Jain, 1981) are so characteristic as to provide data applicable to the entire spectrum of these rocks. A typical nondeformed and unmetamorphosed tillite is a nongraded, nonbedded massive rock. The scattered pebbles can belong to the entire spectrum of igneous, metamorphic, and sedimentary rocks; however, the assemblage, as in all conglomerates, is not so rich as that displayed by the sand-size lithic grains of the interstitial materials. Continental ice sheets are powerful erosional agents that pluck clasts from

their substratum at any place along their path; hence long-distance transportation is not necessarily implied. The power of glacial erosion is shown by the fact that clasts are fresh to slightly weathered and have their original shape controlled by factors inherent to the rock themselves, such as joint systems and bedding. Complex attrition processes take place beneath ice sheets, under conditions comparable to pressurized debris flows and mudflows, between water-saturated silty materials and clasts that abrade them into pebbles of characteristic shapes. These shapes include the following varieties: faceted and striated, tubular, wedge shaped, parallel sided, pentagonal, flat-iron shaped, and even disc shaped. Unquestionably, the subrounded to rounded shape of some pebbles may be inherited when they are derived from recycled interglacial outwash gravels or reworked from older tills. The extent of these recycling processes remains debatable. Surface features of pebbles, besides striations, chatter marks and impact marks, are not characteristic or lacking.

The orientation of the more elongate pebbles within the interstitial materials is generally parallel to the direction of ice flow and consistent with observed striations, grooves, chatter marks, and other directional features of underlying tillite pavements (Lindsey, 1969; Pettijohn, 1975; Potter and Pettijohn, 1977). Care should be exercised not to confuse such depositional orientation with subsequent tectonic reorientation of the elongate pebbles.

Petrographic study of the interstitial materials shows a sand-size fraction of quartz, feldspar, and lithic grains in the proportions characteristic of the three types of wackes. Quartz grains are angular to subrounded, rarely rounded to well rounded. Recycling is shown by abraded overgrowths. All types of quartz occur, emphasizing the immature character of the deposits, such as quartz with straight extinction, undulatory extinction, polygonized, polycrystalline, and euhedral volcanic. Feldspar grains are mainly plagioclases, orthoclase, microcline, perthite, and myrmekite. They are angular to subrounded, fresh or with the various types of alteration described in feldspathic arenites. Lithic fragments are a highly polymictic association of igneous, metamorphic, and sedimentary rocks. The heavy mineral suites are also very rich. The protomatrix ($<30 \mu$) may form up to 70% of the total constituents. It consists of variable proportions of smectite and illite, with frequent additions of minute granules of calcite and dolomite when the ice moved over carbonate terranes.

At the scale of the interstitial materials of tillites, an orientation of the elongate sand grains parallel to the ice movement has also been reported from oriented samples of the Gowganda tillite (Dreimanis, 1959). In Wisconsinan tills from Ohio and Pennsylvania, Sitler and Chapman (1955) reported a microfoliation roughly horizontal in till outcrops, which they attributed to rotation and packing due to intergranular movement of the till, probably facilitated by the presence of interstitial water. Microfoliation is parallel to the orientation of the elongate sand grains. In tills composed largely of a silty matrix, flakes of micaceous clay minerals and chlorite can be seen weaving between slightly coarser silt-size granules of quartz, feldspars, and heavy minerals in a meshlike pattern, forming a general felted texture. Both flakes and elongate granules are parallel to each other, giving under crossed nicols a characteristic mass extinction.

Microfoliation decreases in perfection with increasing admixture of fine sand constituents. In moderately sandy tills, microfoliation is still clearly defined, but highly irregular in distribution and trend. It is well developed in portions relatively free of grains; elsewhere, foliated streaks bifurcate as they approach sand grains. They form around them a thin shell of concentric foliation and may even build a type of augen structure in which the opposed triangular areas at both ends of the structure show a random felty texture before merging into the general foliation. In very sandy tills, the orientation of the silty flakes is strongly influenced by the shape of the sand grains. Each grain appears also encased by a thin shell of concentric foliation, very pronounced at the contact with the grain and grading outward into the general trend. But, if sand grains are very numerous, the flakes are used almost entirely in making up the concentric foliation shells, and in the intergranular areas most of the preferred general microfoliation is lost.

Sitler and Chapman (1955) also described a veining structure in which small and thin veins or bands, rich in silty flakes, intersect the till. Silt flakes are oriented parallel to the vein walls, and veins are either parallel to microfoliation or cut it at a large angle. Veins clearly postdate microfoliation and are attributed to shear processes within the till. The above-described microfabrics, particularly microfoliation (not to be confused with a possible subsequent tectonic foliation), remain to be studied in detail in tillites since they represent additional diagnostic features of glacial origin.

Some of the best arguments in favor of the glacial origin of tillites lie in the existence of an underlying striated pavement and in associated varvites (**Plate 14.F**). These are finely laminated shales or siltstones that periodically contain layers with scattered, sand-size, angular grains of quartz, feldspar, or lithic clasts. Laminae are distorted next to these sands grains, bent down beneath them, and arched over them in a very characteristic fashion. This texture is produced by the dropping of grains from rafting ice into the unconsolidated laminated muds during periods of spring and summer thawing. This process also involves pebbles and boulders called *dropstones*, which form in such a case laminated pebbly mudstones considered, in the field, as the best evidence of glaciolacustrine or glaciomarine sedimentation.

Provenance

Pebbles in tillites form a polymictic association by far richer than that of any orthoconglomerates because continental ice moves over a great variety of terranes. As mentioned above, glacial mass transportation has abrasion effects on the various lithologies, which are limited to a characteristic shaping of the clasts, but with virtually no destruction of weaker lithologies. In the examples studied by Jain (1981), each tillite appears characterized by a specific suite of pebble lithologies, which therefore becomes a useful tool for correlation and provenance studies. Some variations were observed in the proportion of the various lithologies, but no clear pattern of variation has been reported either laterally or in the direction of ice flow. At the sand-size level of the interstitial materials, lithic grains display richer suites than pebbles, which makes them even more useful. Jain (1981) quoted tillites containing up to 18 different types of lithic grains. As for pebbles, each tillite appears characterized by a given association of quartz grain types and of lithic grains. For both constituents, internal variations in their relative proportions were observed, but no clear trend emerged either laterally or in the direction of ice flow. Similar observations were made on the rich suites of heavy minerals. Characteristic assemblages of given tillites were based on varieties of garnet, zircon, tourmaline, rutile, and staurolite. Recycling is shown by abraded overgrowths on tourmaline grains.

However, use of heavy minerals as provenance indicators in tillites is also subject to some reservations introduced mainly by the effects of reworking during interglacial episodes and of intrastratal dissolution over long periods of burial (Gravenor and Gostin, 1979). These authors studied pebbles in Pleistocene tills of North America and Europe, late Paleozoic tillites of South Africa and Australia, and late Precambrian tillites of Australia, and they concluded that in all three cases ice sheets eroded fresh igneous, metamorphic, and sedimentary rocks from the shields and terranes over which they moved, after having rapidly removed unconsolidated mantles of weathered materials. Therefore, it is acceptable to assume that in all three cases the original heavy mineral suites were similar and dominated by the most common heavy minerals of igneous and metamorphic rocks: amphiboles, with lesser amounts of garnet, pyroxenes, epidote, tourmaline, and zircon.

In reality, however, and although tillites are at present relatively impervious compared to sediments above and below, heavy mineral suites indicate obvious changes in mineralogy through time. Gravenor and Gostin (1979) found that the heavy minerals of Pleistocene tills are dominated by amphiboles, with lesser amounts of garnet, pyroxenes, epidote, tourmaline, and zircon. The late Paleozoic tillites are dominated by garnet (80% to 99%) and late Precambrian tillites by zircon (up to 70%) and tourmaline, with traces of garnet.

At first glance, the most obvious explanation for the changes in mineralogy is the removal of the less chemically stable minerals from dissolution by intrastratal solutions through geological time, particularly in the light of the fact that little removal of material by such a process has taken place in the oldest Pleistocene tills over a period of 700,000 years or more. However, the answer does not appear that simple.

In late Paleozoic tillites, half of the garnets are well rounded and show delicate surface chattermark trails typical of glacial action. They obviously were not submitted to intrastratal dissolution since deposition. Their high degree of rounding is of interest since garnet is very resistant to abrasion. The combination of rounding and chattermarks indicates that these grains were not plucked out directly by the ice from shield rocks of Gondwanaland, but were first eroded from these rocks, sorted and abraded in beach or fluvial environments during interglacial episodes, and subsequently picked up by readvances of the ice, which developed the chattermark trails on grain surfaces. This sequence of events could have been repeated several times during a glaciation, which is estimated to have lasted as long as 75 my. Under these conditions, other less resistant heavy minerals, such as hornblende, pyroxene, and epidote, must have been lost primarily by the same processes of sorting and mechanical abrasion, which left the surviving grains with greater surface, making them more susceptible to subsequent intrastratal solution. The other half of the garnets shows clear evidence of extensive in situ dissolution by their "rectangular" etching pattern. If such an action continued, the final result would be a suite of heavy minerals dominated by tourmaline and zircon.

The most obvious change of mineralogy between late Paleozoic and late Precambrian tillites is the almost complete loss of garnet, and sometimes of tourmaline, leaving up to 80% zircon. Tourmaline and zircon are furthermore highly rounded. If the same line of reasoning used to explain the late Paleozoic suite is applied to the late Precambrian situation, one must assume that the major loss of amphibole, pyroxene, and epidote, as well as the rounding of tourmaline and zircon, occurred through extensive reworking of the sediments during interglacial episodes. The subsequent loss of garnet and the etching of tourmaline and zircon were caused by intrastratal dissolution.

Gravenor and Gostin (1979) speculated on the nature of intrastratal solutions. In some late Paleozoic tillites, carbonate clasts and even carbonate matrix occur next to etched garnets; late Precambrian tillites also contain carbonate pebbles and boulders with relatively fresh surfaces. Well-rounded apatite grains occur in tillites of both ages. As apatite is quite unstable under acid conditions, its presence and that of carbonate clasts indicate that the solutions that circulated for a long period of time in the tillites, dissolving garnet and tourmaline, were neutral or alkaline. Although it is known that garnet dissolves under acid conditions, it can apparently also dissolve by extended contact with alkaline solutions. Some late Precambrian tillites, in spite of their extensive loss of stable minerals by intrastratal solution, contain fresh microcline and a small amount of fresh to altered plagioclase; possibly the circulating solutions

were rich in calcium and sodium and decreased the solubility of feldspars.

Even when taking into account the above-mentioned restrictions, the correlation and provenance of tillites can be established by means of a great variety of constituents, ranging in size from pebbles to heavy minerals, because mass transportation by ice preserves to an unusually high degree the composition of the original materials.

Diagenetic Evolution

The diagenetic evolution of tillites involves only the change of the protomatrix of smectites and illite to an orthomatrix of sericite, chlorite, epidote, and zeolites, with almost no visible textural reorganization. Most of the tillites described by Jain (1981) have an orthomatrix consisting of poorly birefringent, randomly oriented, and incipiently recrystallized sericite associated with patches of calcite, dolomite, siderite, and quartz, derived from the authigenic neomorphism of minute detrital granules of the protomatrix. The only visible textural reorganization is a marginal rim of fibroradiated chlorite, indicating marginal replacement of quartz and feldspar grains. Similarly, the margins of unstable lithics, such as phyllites, become completely gradational with the sericite matrix. Under tectonic stresses, the sericite orthomatrix becomes an aggregate of coarser flakes by neomorphism and displays a preferred orientation along foliation planes that wrap around grains. Sericite becomes even coarser under greenschist facies metamorphism.

Reservoir Properties

Primary porosity does not exist in tillites since their interstitial materials are wackes with abundant proto- to orthomatrix. A remote possibility of secondary porosity can occur through burial dissolution of altered unstable grains of feldspars, mafics, and volcanic rocks and from dissolution of patches of authigenic calcite, dolomite, and siderite, but practically no permeability would exist after such possible dissolutions.

TYPICAL EXAMPLES

Plate 12.F. Paraconglomerate or glacial diamictite consisting of a well-developed pyritic and bituminous groundmass of illitic material in which are scattered randomly oriented lithoclasts of dark bituminous shales and metaquartzites, subangular to rounded sand-size grains of quartz, fresh potassic feldspars, plagioclases, and opaque grains of iron ore. Stratification is poorly developed and deformed by ice action. Cabeças Formation, Upper Devonian, Maranhão Basin, Brazil.

Plate 13.A Paraconglomerate or glacial diamictite

consisting of a well-developed argillaceous groundmass of illitic–smectitic material in which are scattered randomly oriented and subrounded lithoclasts of arenaceous limestone, dark brown shale, quartzite, dolostone, opaque iron ore, associated with subangular sand-size grains of quartz, fresh and weathered potassic feldspars, and plagioclases. Kansan till, Pleistocene, Banner, Illinois, U.S.A.

INTRAFORMATIONAL CONGLOMERATES AND BRECCIAS

Petrography and Depositional Environments

Intraformational conglomerates and breccias result from penecontemporaneous fragmentation, related or not to subaerial exposure and desiccation, followed by an almost immediate redeposition of the clasts below normal wave base in marine shelf environments (**Plate 13.B, C**). These rocks represent a single or complex, relatively short, sedimentologic event producing beds that vary in thickness from a few centimeters to about one meter. They may be very widespread and repeated in a cyclic pattern within a given sedimentary sequence.

Two major types exist, the first among carbonate rocks, the second among argillaceous rocks. Modern studies (Einsele and Seilacher, 1982) showed that most intraformational conglomerates and breccias belong to the category of *tempestites*, or storm deposits, which are the effects of high-energy storm waves reworking and redepositing weakly consolidated carbonate and argillaceous sediments of shallow marine platforms. Thin and relatively fine grained beds represent one-event situations, whereas thick and coarser ones result from the superposition of several erosive and depositional events.

Among carbonate rocks, the most common tempestites are grain-supported flat pebble conglomerates, as in the Ordovician Pogonip Group of Nevada, in which a crudely graded bedded framework of flat intraclasts is associated with an interstitial material consisting of smaller intraclasts, lithic pellets derived from the abrasion of larger intraclasts, and all types of entire tests or bioclasts at various stages of disarticulation. All these constituents are set in a cavity-filling submarine to phreatic sparite cement, which grades upward into a calcisiltite to calcilutite matrix as another expression of the general graded bedding texture.

Carbonate intraclasts reach a size up to 10 cm or more. They often display a wide spectrum of shapes, ranging from extremely irregular, indicating a subaqueous semiconsolidated condition at the time of reworking, to subrectangular flat and abraded, representing original desiccation chips formed during episodes of subaerial exposure. Occasional oxidation aureoles and perforations confirm the latter condi-

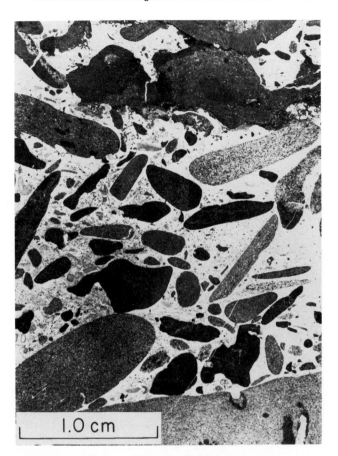

FIGURE 2.2 Carbonate lithoclastic tempestite from the Pogonip Group (Lower Ordovician) of the Arrow Canyon Range, Clark County, Nevada. Larger lithoclasts show a highly imbricated texture with frequent development of umbrella effects. Lithoclasts range in shape from extremely irregular to abraded flat desiccation chips displaying oxidation aureoles and perforations from subaerial exposure on tidal flats. Interstitial matrix consists of smaller lithoclasts, lithic pellets derived from the abrasion of larger clasts, bioclasts of crinoids, gastropods, pelecypods, *Nuia*, and trilobites set in a cavity-filling sparite cement. The predominant lithologies of the lithoclasts indicate derivation from the reworking of adjacent microfacies by high-energy storm events.

tions (**Fig. 2.2**). With respect to bioclasts, some tempestites consist entirely of stromatolite flat to concavoconvex chips, while others are rich in disarticulated valves of large brachiopods or pelecypods as in the Ordovician Galena Group of Iowa (Bakush and Carozzi, 1986). These components, as an effect of the crude graded bedding, tend to concentrate in the lower half of the tempestite, usually with their curvature convex upward in an equilibrium position and with an "umbrella effect" beneath them consisting of a geopetal association of internal sediment overlain by cavity-filling submarine to phreatic sparite cement. In other instances of more tumultuous storm deposition, graded bedding is absent or incipient, intraclasts and bioclasts are randomly mixed, and disarticulated large shells may be imbricated (edgewise structure) or even concentrated at any place near the top of the tempestites (**Figs. 2.3, 2.4**). Intense bioturbation, which

may in many cases follow storm deposition and which consists mostly of escape burrows, contributes considerably to the disorganization of original depositional textures.

In most carbonate tempestites, the base is sharp and erosional, sole marks include large scours, gutter casts, and tool marks; the top is sharp, but straight or gradational into the overlying fair-weather subtidal carbonates (Carozzi,

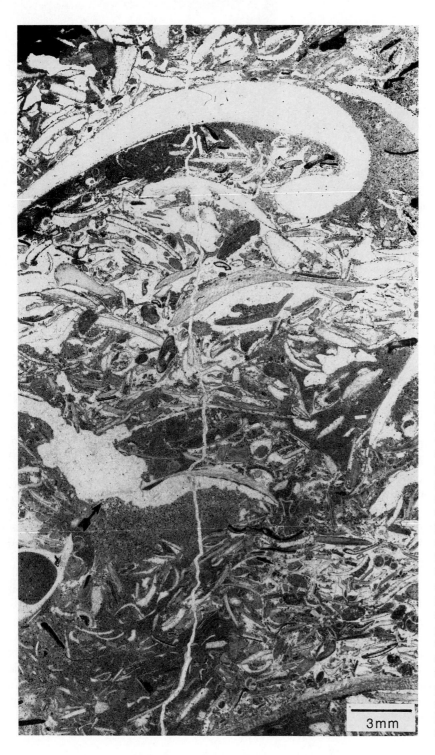

FIGURE 2.3 Carbonate bioclastic tempestite of the Galena Group (Middle Ordovician), Guttenberg, Iowa. Moderately bioturbated and disorganized grain-supported biocalcirudite with sparite cement and calcisiltite matrix. Bioclasts of brachiopods and pelecypods predominate over those of gastropods, bryozoans, crinoids, and trilobites. The bioclasts display an irregular texture where the fine to medium components are concentrated in the lower half of the photomicrograph with calcite cement (at lower-right corner) and calcisiltite matrix (at lower-left corner). Larger shells of brachiopods and pelecypods are in equilibrium position at top of photomicrograph with calcite cement and rare calcisiltite matrix. Note dissolution feature in large cavity with typical geopetal internal sediment overlain by sparite cement (arrow). Plane-polarized light. From Bakush and Carozzi (1986).

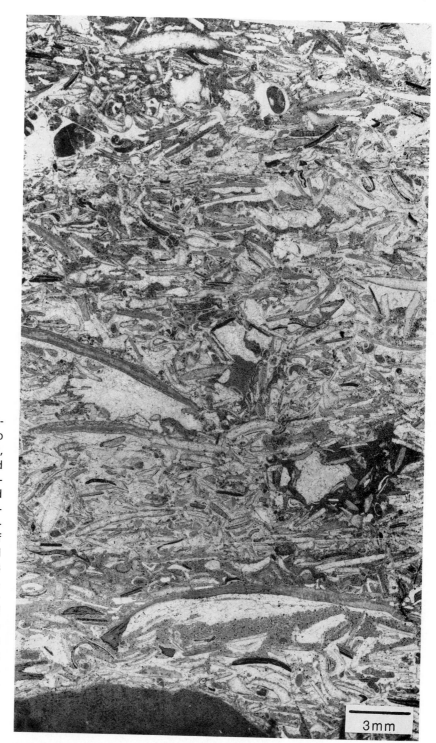

FIGURE 2.4 Carbonate bioclastic tempestite of the Galena Group (Middle Ordovician), Guttenberg, Iowa. Moderately bioturbated and disorganized grain-supported biocalcirudite with sparite cement and pelletoidal calcisiltite matrix. Bioclasts of brachiopods and pelecypods predominate over those of gastropods, crinoids, trilobites, and bryozoans. Bioclasts show various orientations either perpendicular, inclined at various angles, or parallel to bedding. Note the surface of mud at bottom and geopetal internal sediment overlain by sparite cement under brachiopod shells (middle left and bottom). Burrows are partially filled with sparite cement and calcisiltite matrix. Plane-polarized light. From Bakush and Carozzi (1986).

3mm

1989). Reconstruction of storm paths can be reached by two major approaches. In the first, which applies to the case of a well-documented carbonate platform depositional model, one notices that tempestite intraclasts consist of a certain number of microfacies whose relative position is well identified in the model. By comparing the distribution of such intraclasts at different sampling locations of the tempestite bed, it is possible to reconstruct the storm path across the carbonate platform (Carozzi, 1989). The second approach, discussed in detail by Aigner (1982), is based on the fact that storm effects, such as storm waves and storm-induced currents, decrease in intensity toward deeper offshore areas. Therefore, as in turbidites, a lateral succession of types of deposits is observable between proximal and distal areas.

Proximal tempestites are relatively thick bedded and coarse grained calcirudites with sparite cement, with intraclasts and bioclasts mixed in variable proportion; they commonly form composite and amalgamated beds. Distal tempestites are fine-grained litho- and biocalcarenites, with a matrix of calcilutites and calcisiltites predominating over sparite cement.

In terms of basin analysis, since storms are instantaneous and isochronic events, tempestites and their paleoecological impact represent useful tools for high-resolution regional stratigraphy.

Among siliciclastic rocks, intraformational shale clast conglomerates and breccias are widespread in formations consisting of alternating sandstones and shales of fluvial and deltaic origin. Petrographically, they consist of a framework of subrectangular to weakly abraded tabular chips or flakes of shale, several centimeters in size, embedded in a matrix of quartz, feldspathic, or lithic arenites. These shale clast conglomerates and breccias typically occur at the base of sandstone units, and the origin of the clasts is local. Desiccation processes during temporary episodes of subaerial exposure, rather than subaqueous syneresis, appear responsible for the fragmentation, while episodes of flood are the major reworking and redeposition agents.

Some particular types of shale clast conglomerates and breccias occur as discontinuous lenses and streaks at the base of turbiditic wackes in flysch sequences and represent rip-up clasts produced by the erosive action of turbidity currents.

In most intraformational conglomerates and breccias,

the shape of the intraclasts suggests a semiconsolidated to consolidated state of the reworked and redeposited materials. However, situations occur where two types of alternating sediments, such as fine pelagic carbonate muds and glauconitic quartzose sands, may be constantly reworked during a certain span of time by waves and bottom currents. This process generates a complex mixture of soft lumps of pelagic mud and of glauconitic sand of extremely irregular shapes and extended apophyses with equally irregular reciprocal inclusions. An example of such an intraformational conglomerate by mixed sedimentation was described by Carozzi (1956) at the base of the Upper Cretaceous transgression in the Calcareous Alps of High Savoy, France (**Fig. 2.5**). The deepening- and thickening-upward trend shows deposition at the base of the transgressive sequence of glauconitic sands in a highly agitated environment alternating with low-energy pelagic calcilutite muds, which increase upward in thickness until becoming predominant. However, scattered quartz and glaucony grains within the pelagic calcilutite muds indicate that the clastic supply never stopped entirely. If intraformational reworking had not been continuously active, the deposited sequence would consist of alternating beds of calcilutite and glauconitic quartz arenites, with the former gradually replacing the latter (**Fig. 2.5G**). Instead, the reworking, decreasing in intensity upward, of soft carbonate mud and loose glauconitic sand by oscillatory movements of the waves and later by submarine currents generated the curiously shaped pebbles. It is obvious that during each episode

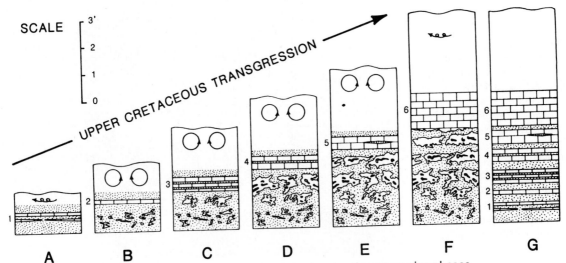

FIGURE 2.5   Diagrammatic sketches showing the successive phases of development of an intraformational conglomerate by mixed sedimentation. Upper Cretaceous, Roc-de-Chère, autochthonous, High-Savoy, France. Sketches A to F correspond to the deposition of the main sandy beds just before reworking, with accompanying deformation of their limestone mud intercalations. Circles with arrows indicate wave action and curled arrows current action. Sketch G shows the inferred succession of sandy and limestone beds if no reworking had occurred. From Carozzi (1956). Reprinted by permission of the Society of Economic Paleontologists and Mineralogists.

FIGURE 2.6 Megascopic and microscopic illustrations of tempestite chert breccia in Burlington Limestone (Middle Mississippian), western margin of Illinois Basin. **A.** General view of chert breccia. Lighter-colored chert breccia with a matrix of crinoidal calcarenite is interbedded between darker crinoidal calcisiltites. Top is even, whereas base is irregularly wavy. Concentration of angular chert clasts in lower half of bed gives rise to a crude graded bedding. Scale in decimeters. **B.** Typical chert nodule with whitish patina (arrow) and not shattered as observed at both ends of chert breccia bed. Scale in inches. **C.** Examples of variation in size and shape of chert clasts that have been extracted from enclosing crinoidal calcarenite. Scale in inches. **D.** Typical angular clast of chert partially embedded in crinoidal calcarenite, with conchoidal character of fracture surfaces. Scale as in C. **E.** Texture of groundmass of chert breccia. Coarse-grained, poorly sorted crinoidal calcarenite consisting of 70% crinoidal fragments in an interstitial matrix of silt-size debris of crinoids and bryozoans. Scale at lower right. Plane-polarized light. **F.** Texture of chert nodules. Silicified crinoidal calcisiltite with streaks of sand-size bioclasts of crinoids, bryozoans, and brachiopods. Observe cryptocrystalline nature of quartz in groundmass. Plane-polarized light. From Carozzi and Gerber (1978). Reprinted by permission of the Society of Economic Paleontologists and Mineralogists.

of reworking the underlying semiconsolidated clasts might have been distorted again. The upward decrease in intensity of the reworking, because of general deepening of the environment, is shown by the fact that clasts tend to display less complicated shapes grading into larger subrectangular masses of relatively undisturbed calcilutite and finally into continuous beds of calcilutite upon cessation of the intraformational reworking process.

Storms and other high-energy instantaneous events can generate various types of lag deposits, which result from the reworking and rapid settling as a basal concentrate of large biogenic constituents such as whole shells or lithic clasts in sandstone and calcarenite beds. The latter display a coarse texture and a crude graded bedding of the clasts, an erosive basal contact of variable relief, or simply a sharp contact. These lag deposits tend to be of local extent and not usable for basinal correlations. A typical example corresponds to the storm-generated coquinoid sandstones from the Upper Jurassic of Wyoming and Montana (Brenner and Davies, 1973).

An unusual chert intraformational conglomerate occurs in a section of Burlington Limestone (Middle Mississippian), near Hannibal, Missouri (Carozzi and Gerber, 1978; Carozzi, 1989). It is of local occurrence and could indicate the touchdown of a tornadolike system that behaved like a funnel, rather than the passage of a storm front sweeping a shallow carbonate platform. In addition, this chert intraformational conglomerate demonstrates a geologically instantaneous submarine formation of completely indurated chert nodules, including their external whitish crust or patina, within unconsolidated crinoidal calcarenitic sands.

The conglomeratic bed of crinoidal calcarenite extends over a distance of 30 m and displays ellipsoidal chert nodules with their whitish external crust or patina, which, instead of being regularly lined up parallel to bedding as in overlying and underlying beds, appear extensively disturbed and brecciated (**Fig. 2.6A to D**). This bed shows neither evidence of interruption of sedimentation nor of subaerial exposure. The only visible sedimentary structure is a general crude graded bedding of the chert clasts throughout the thickness of the bed and a concentration of most clasts in its lower half.

Some chert nodules were shattered in situ, with opening of systems of fractures into which the enclosing crinoidal calcarenitic sands were injected (**Fig. 2.7A**). Other nodules were broken into two parts by a major transverse fracture, with subsequent rotation of the individual segments (**Fig. 2.7B**). Still other nodules display either truncated edges (**Fig. 2.7C**) or shattering so that the nodules were reduced to concentrations of numerous polygonal fragments and slivers (**Fig. 2.7A, D**).

In thin section, the groundmass of the chert conglomerate consists of a coarse-grained, poorly sorted, crinoidal calcarenite, 70% of which is crinoidal fragments in an inter-

stitial matrix of silt-size bioclasts of crinoids and bryozoans. This texture indicates a high-energy environment of deposition (**Fig. 2.6E**). The chert fragments display the texture of a silicified calcisiltite with streaks of sand-size bioclasts of crinoids, bryozoans, and brachiopods deposited under lesser energy conditions similar to those of the beds overlying and underlying the chert conglomerate (**Fig. 2.6F**).

The inferred succession of events that generated this intraformational chert conglomerate is as follows: (1) synsedimentary formation of completely indurated chert nodules, including their peripheral patina, by submarine geologically instantaneous replacement of the enclosing unconsolidated crinoidal calcisiltite (**Fig. 2.8, stage 1**); (2) transit of a high-energy event that broke the nodules by reciprocal impacts and dispersed the debris over a short distance (**Fig. 2.8, stage 2**); (3) final deposition of the clasts with crude graded bedding within a poorly sorted, coarse-grained crinoidal calcarenite (**Fig. 2.8, stage 3**); and (4) renewed deposition of the undisturbed overlying crinoidal calcisiltite.

Provenance and Diagenetic Evolution

Intraformational conglomerates and breccias are megascopically monomictic and rarely oligomictic, consisting for instance of clasts of limestone, shale, or chert, and are intrabasinal in origin. Whenever limestone clasts, for instance, are studied in detail petrographically, they may represent several microfacies with well-established locations within a platform carbonate model; hence at such a scale of observation one could talk about the rock being polymictic. It is really a matter of semantics. As mentioned above, the distribution of these microfacies within a given body of intraformational conglomerate or breccia can afford data on its provenance and, in the case of tempestites, provide information on the path of the storm, as well as defining proximal and distal facies of a storm depositional system. It is not uncommon to find in intraformational conglomerates and breccias of storm origin extrabasinal constituents such as grains of detrital quartz and of other minerals. The presence of these minerals indicates that local reworking and redeposition events have either interfered with local extrabasinal supplies or carried along with them materials originating from extrabasinal areas where storm activities began.

The diagenetic evolution of intraformational conglomerates and breccias is identical to that of the enclosing lithologies, whether siliciclastic, argillaceous, or carbonate, and reference should be made to the appropriate sections of this book.

Reservoir Properties

Intraformational conglomerates and breccias have the same reservoir properties as those of the enclosing coarse-grained sandstones or carbonates and again reference should be

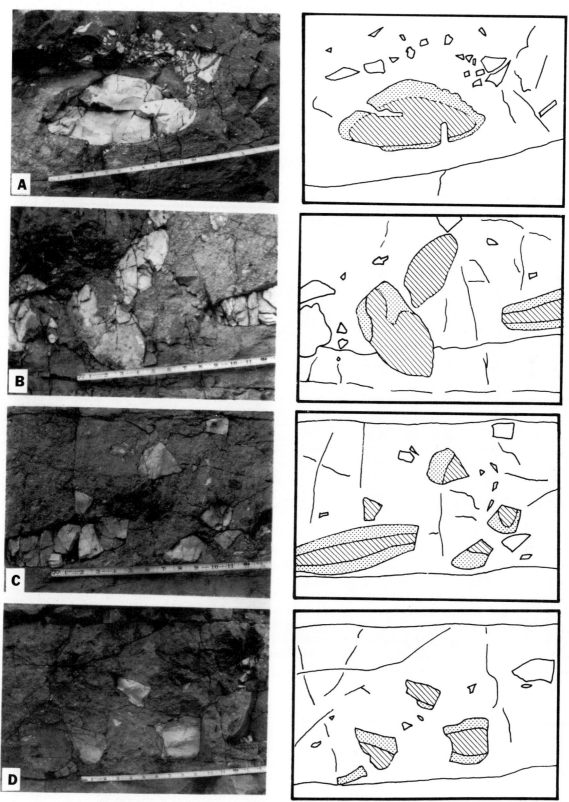

FIGURE 2.7 Examples of shattering of chert nodules in tempestite chert breccia, Burlington Limestone (Middle Mississippian), western margin of Illinois Basin. Central part of nodules represented by hachured pattern, marginal patina by dotted pattern. Scale in inches. **A.** Nodule shattered in place with double penetration of crinoidal calcarenite in open cracks. **B.** Rotated half nodules. **C.** Angular clasts and nodule with truncated edges. **D.** Angular clasts showing truncation of marginal patina. From Carozzi and Gerber (1978). Reprinted by permission of the Society of Economic Paleontologists and Mineralogists.

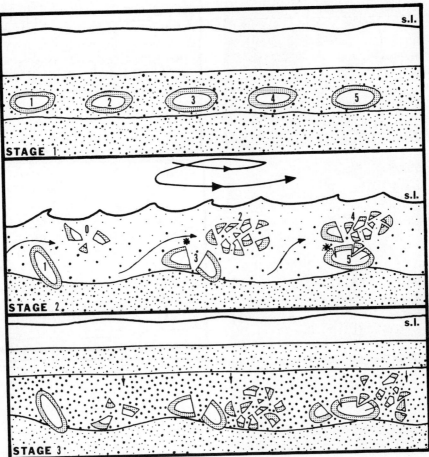

FIGURE 2.8 Schematic diagram (not to scale) illustrating the assumed process of formation of tempestite chert breccia, Burlington Limestone (Middle Mississippian), western margin of Illinois Basin. **Stage 1.** Chert nodules with marginal patina forming within unconsolidated crinoidal calcisiltite. **Stage 2.** Shattering of chert nodules by moderate horizontal displacement and reciprocal impact within the slurrylike crinoidal calcarenite churned by high-energy tornadolike event. **Stage 3.** Settling of chert clasts with crude graded bedding within crinoidal calcarenite. Renewed deposition of undisturbed overlying crinoidal calcisiltite. From Carozzi and Gerber (1978). Reprinted by permission of the Society of Economic Paleontologists and Mineralogists.

made to the appropriate sections of this book. Intraformational conglomerates and breccias consisting of shaly clasts generally display a cavity-filling calcite cement that upon secondary burial dissolution can present the same potential reservoir capacities as its equivalent in sandstones and carbonate rocks. In terms of horizontal extent of potential reservoirs in intraformational conglomerates and breccias, those displaying the best-developed characteristics of a tempestite system can be extensive with the most favorable conditions in their composite and amalgamated proximal areas. Other types of tempestites with a thin sheet geometry and a channel lag to shoestring shape tend to be of local extent. In all instances, the cyclic repetition of storm-controlled in-

traformational conglomerates and breccias is of added interest in their exploration as potential coarse-grained reservoirs.

TYPICAL EXAMPLES

Plate 13.B. Intraformational limestone conglomerate consisting of a framework of flat pebbles of slightly arenaceous and pelletoidal calcisiltite, derived from desiccation chips, some of which have a superficial oolitic coating, and well-developed normal ooids with interfering concentric and fibroradiated internal structure. Moderate pressure solution between all constituents and well-developed, cavity-filling sparite mosaic cement. Hamburg Oolite, Lower Mississippian, Hamburg, Illinois, U.S.A.

Plate 13.C. Intraformational limestone conglomerate consisting of a framework of superposed flat pebbles of dolomitized calcilutite derived from desiccation chips. Interstitial material shows smaller, sand-size to silt-size lithic grains of similar calcilutite and cavity-filling sparite cement. Joachim Dolomite, Middle Ordovician, Isle de Bois, Missouri, U.S.A.

PSEUDOBRECCIAS AND PSEUDOCONGLOMERATES

Petrography and Depositional Environments

In this category of rudaceous rocks, the composition of the clasts is exclusively monomictic at all scales of observation. The generally angular shape of the clasts implies little transport under the action of gravity. The most representative types are carbonate collapse breccias (**Plate 13.D**) developed in evaporitic calcilutites, calcisiltites, dololutites, and dolosiltites interbedded with beds of halite. The latter was subsequently dissolved after complete induration of the associated carbonates, which often show calcite, dolomite, and silica pseudomorphs after gypsum and anhydrite, and collapsed by lack of physical support in the spaces previously occupied by halite (Swennen et al., 1990). This process accounts for the completely random assemblage of angular clasts of variable size, ranging from slender slivers to rectangular and irregular chunks. Locally, a jigsaw texture may be observed expressing the intersections of the networks of barely opened fissures, which released the clasts that themselves have undergone essentially no displacement.

Collapse breccias fill cavities of irregular shape among otherwise undisturbed similar fine-grained carbonates, but frequently the shape of the cavities tends to be conical, with the apex downward terminating into a single open fracture, like a miniature sinkhole (**Fig. 2.9**) as in the Devonian Wapsipinicon Formation of Iowa (Kocken and Carozzi, 1991).

The late diagenetic origin of carbonate collapse breccias is demonstrated by their interstitial cement, which consists of freshwater phreatic sparite overlain by several generations of geopetal concentration of vadose silt. In many instances, gypsum, anhydrite, barite, or fluorite take the place of calcite.

Pseudoconglomerates were described in hemipelagic chert units of the Archean Onverwacht Group of the southern Barberton greenstone belt of South Africa and interpreted as products of submarine exhalative activity (Stanistreet and Hughes, 1984; Paris et al., 1985). The cavities containing these pseudoconglomerates vary considerably in shape and size; some are concordant with bedding of the enclosing cherts, whereas others are discordant. They

range in size from thin fractures to elongate cavities greater than 50 cm in diameter. The pseudoconglomerates grade in places marginally into angular breccias with frequent edgewise and imbricate structures. These brecciated zones are in places many meters wide and crosscut bedding for tens of meters. The clasts of the pseudoconglomerates are identical to those of the host rock, and their shape depends on whether the chert was massive or had primary laminations along which fracturation could occur. The interstitial material between clasts is microquartz, which in places shows residual textures of an earlier concretionary silica cement. The size of the clasts ranges from less than 2 mm to more than 20 cm. Clasts within smaller cavities generally show only minor displacement and rotation; they are commonly elongate and angular to subrounded. Larger cavities contain disorganized clasts, many of which are subspherical and well rounded.

Obviously, pseudoconglomerates derived from monomictic breccias. The clasts of the breccias are interpreted as formed by a process of roof and wall collapse during hydraulic disaggregation of the cherts within exhalative systems. Clasts in larger cavities were moved together during continued fluid flow and became rounded by reciprocal attrition, forming pseudoconglomerates. The cavities filled with the latter connect spatially with vein and breccia networks, and all gradations are visible between veins, stockwork veins, breccia-filled cavities with diffuse margins, and pseudoconglomerate-filled cavities. This succession is interpreted by Stanistreet and Hughes (1984) and Paris et al. (1985) as expressing the progressive stages of development of pseudoconglomerates.

The enclosing cherts represent deep-water turbidites and hemipelagic muds that were totally replaced by chert, with preservation of many typical depositional structures. These rocks overlie silicified and carbonated mafic volcanics, including pillow lavas, some of which were also completely silicified. In such a context, the process of chertification, the veins, the pseudoconglomeratic and pseudobrecciated structures, and the cavity-filling silica cement are interpreted as products of hydraulic fracturing and fluid flow immediately below the vent areas of submarine exhalative systems, and therefore they are almost penecontemporaneous with sedimentation.

The last category of pseudobreccias displays monomictic pseudoclasts that typically show a jigsaw texture implying no movement at all of all the components. These have in fact strange and very complicated shapes that interconnect in three dimensions. These pseudobreccias, which characterize carbonate rocks (**Plate 13.E**), result from an initial combination of fracture surfaces and bedding planes along which late diagenetic processes, such as neomorphism, replacement (often dolomitization), and all types of mineralization, have remained incompleted and were controlled by the complex network of fractures and bedding. Consequently, the portions of the original rocks are pseudoclasts, simulating angular clasts,

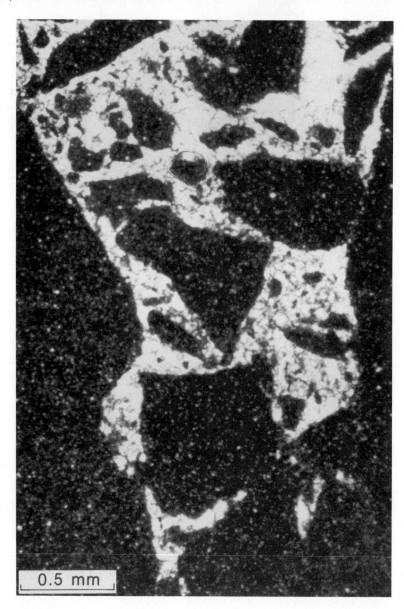

0.5 mm

FIGURE 2.9 Collapse breccia due to dissolution of halite and consisting of partially dolomitized calcisiltite clasts with interstitial sparite cement. Wapsipinicon Formation (Middle Devonian), Fayette, Iowa.

and the interstitial cement results from neomorphism, replacement, or mineralization. If any displacement at all has occurred, it was limited to the initial opening and perhaps widening of fractures and bedding planes. In certain instances, a marginal reaction took place between pseudoclasts and diagenetic cements, which has somewhat rounded the edges of the pseudoclasts, making them often hazy, and thus generating a pseudoconglomeratic texture.

Provenance and Diagenetic Evolution

The complete in situ generation of pseudobreccias and pseudoconglomerates eliminates any provenance problem. Their diagenetic evolution is the same as that of the enclosing carbonates and cherts. In those associated with carbon-

ates, late diagenetic processes range from freshwater vadose and phreatic to burial conditions. For the above-described case of those associated with hemipelagic cherts, replacement and submarine brecciation appear penecontemporaneous.

Reservoir Properties

Pseudobreccias of collapse nature in carbonate rocks fill scattered and irregularly shaped cavities within fine-grained carbonates. Their various types of interstitial cements could undergo burial dissolution, developing irregularly scattered and unconnected potential reservoirs. The diagenetic evolution of pseudobreccias and pseudoconglomerates produced by neomorphism, replacement, or mineralization is a direct function of the late diagenetic history of the host rock.

TYPICAL EXAMPLES

Plate 13.D. Limestone collapse breccia with typical puzzlelike texture of the angular and irregularly shaped clasts of dolomitized calcilutite. Interstitial cracks are filled by vadose silt, geopetal in places, and poikilotopic sparite cement. Joachim Dolomite, Middle Ordovician, Matson, Missouri, U.S.A.

Plate 13.E. Pseudobreccia by dolomitization. Diagenetic selective replacement isolated irregular, but often subrectangular, pseudoclasts of dark homogeneous calcilutite. Slender apophyses join the pseudoclasts, excluding any mechanical action. Replacive coarse and irregularly crystalline dolomite mosaic often shows fringes of subhedral rhombic crystals perpendicular to margins of pseudoclasts. Some perfect dolomite rhombs are scattered inside the pseudoclasts. Mifflin Formation, Middle Ordovician, Kentland, Indiana, U.S.A.

REFERENCES

ABOTT, P. L., and PETERSON, G. L., 1978. Effects of abrasion durability on conglomerate clast populations: examples from Cretaceous and Eocene conglomerates of the San Diego area. *J. Sed. Petrology,* 48, 31–42.

AIGNER, T. 1982. Calcareous tempestites: storm-dominated stratification in Upper Muschelkalk limestones (Middle Trias, SW. Germany). In G. Einsele and A. Seilacher (eds.), *Cyclic and Event Stratification.* Springer-Verlag, New York, pp. 180–198.

BAKUSH, S., and CAROZZI, A. V., 1986. Subtidal storm-influenced carbonate ramp model: Galena Group (Middle Ordovician) along Mississippi River (Iowa, Wisconsin, Illinois, and Missouri), U.S.A. *Archives Sciences, Genève,* 39, 141–183.

BRADLEY, W. C., 1970. Effect of weathering on abrasion of granitic gravel, Colorado River (Texas). *Geol. Soc. Amer. Bull.,* 81, 61–80.

BRENNER, R. L., and DAVIES, D. K., 1973. Storm-generated coquinoid sandstone: genesis of high-energy marine sediments from the Upper Jurassic of Wyoming and Montana. *Geol. Soc. Amer. Bull.,* 84, 1685–1698.

CAROZZI, A. V., 1956. An intraformational conglomerate by mixed sedimentation in the Upper Cretaceous of the Roc-de-Chère, autochthonous chains of High Savoy, France. *J. Sed. Petrology,* 26, 253–257.

———, 1989. *Carbonate Rock Depositional Models: A Microfacies Approach.* Prentice Hall, Englewood Cliffs, N.J., Advanced Reference Series, 604 pp.

———, and GERBER, M. S., 1978. Synsedimentary chert breccia: a Mississippian tempestite. *J. Sed. Petrology,* 48, 705–708.

———, DE ARAUJO, M. B., DE CESERO, P., DOS REIS FONSECA, J., and DA SILVA, V. J. L., 1976. Formação Salvador: um modelo de deposição gravitacional subaquosa. *Boletim Técnico da PETROBRAS,* 19, 47–79.

CLARKE, R. H., 1979. Reservoir properties of conglomerates and conglomeratic sandstones. *Amer. Assoc. Petroleum Geologists Bull.,* 63, 799–809.

DECELLES, P. G., 1988. Lithologic provenance modeling applied to the Late Cretaceous synorogenic Echo Canyon Conglomerate, Utah: a case of multiple source areas. *Geology,* 16, 1039–1043.

DREIMANIS, A., 1959. Rapid macroscopic fabric in drill-cores and hand specimens of till and tillite. *J. Sed. Petrology,* 29, 459–463.

EINSELE, G., and SEILACHER, A. (eds.), 1982. *Cyclic and Event Stratification.* Springer-Verlag, New York, 536 pp.

GRAHAM, S. A., TOLSON, R. B., DECELLES, P. G., INGERSOLL, R. V., BARGAR, E., CALDWELL, M., CAVAZZA, W., EDWARDS, D. P., FOLLO, M. F., HANDSCHY, J. W., LEMKE, L., MOXON, I., RICE, R., SMITH, G. A., and WHITE, J., 1986. Provenance modelling as a technique for analysing source terrane evolution and controls on foreland sedimentation. In P. A. Allen and P. Homewood (eds.), *Foreland Basins.* Internat. Assoc. Sedimentologists, Special Publ. 8. Blackwell Scientific Publications, Boston, 425–436.

GRAVENOR, C. P., and GOSTIN, V. A., 1979. Mechanisms to explain the loss of heavy minerals from the Upper Palaeozoic of South Africa and Australia and the late Precambrian of Australia. *Sedimentology,* 26, 707–717.

JAIN, A. K., 1981. Stratigraphy, petrography, and paleogeography of the late Paleozoic diamictites of the Lower Himalaya. *Sed. Geol.,* 30, 43–78.

KOCKEN, R. J., and CAROZZI, A. V., 1991. Microfacies of Middle Devonian transgressive carbonates of central Midcontinent, U.S.A.: shallowing-upward sequences in a deepening-upward eustatic trend. *Archives Sciences, Genève,* 44, 371–415.

KOSTER, E. H., and STEEL, R. J. (eds.), 1984. *Sedimentology of Gravels and Conglomerates.* Canadian Soc. Petroleum Geologists, Memoir 10, 441 pp.

LINDHOLM, R. C., HAZLETT, J. M., and FAGIN, S. W., 1979. Petrology of Triassic–Jurassic conglomerates in the Culpeper Basin, Virginia. *J. Sed. Petrology,* 49, 1245–1262.

LINDSEY, D. A., 1969. Glacial sedimentology of the Precambrian Gowganda Formation, Ontario, Canada. *Geol. Soc. Amer. Bull.,* 80, 1685–1702.

PARIS, I., STANISTREET, I. G., and HUGHES, M. J., 1985. Cherts of the Barberton greenstone belt interpreted as products of submarine exhalative activity. *J. Geol.,* 93, 111–129.

PETTIJOHN, F. J., 1975. *Sedimentary Rocks,* 3rd ed. Harper & Row, New York, 628 pp.

POTTER, P. E., and PETTIJOHN, F. J., 1977. *Paleocurrents and Basin Analysis,* 2nd ed. Springer-Verlag, New York, 425 pp.

SEIDERS, V. M., 1983. Correlation and provenance of Upper

Mesozoic chert-rich conglomerates of California. *Geol. Soc. Amer. Bull.,* 94, 875–888.

SITLER, R. F., and CHAPMAN, C. A., 1955. Microfabrics of till from Ohio and Pennsylvania. *J. Sed. Petrology,* 25, 262–269.

STANISTREET, I. G., and HUGHES, M. J., 1984. Pseudoconglomerates and a reexamination of some paleoenvironmental controversies. *Geology,* 12, 717–719.

SUNDELL, K. A., and FISHER, R. V., 1985. Very coarse grained fragmental rocks: a proposed size classification. *Geology,* 13, 692–695.

SWENNEN, R., VIAENE, W., and CORNELISSEN, C., 1990. Petrography and geochemistry of the Belle Roche breccia (lower Visean, Belgium): evidence for brecciation by evaporite dissolution. *Sedimentology,* 37, 859–878.

WALKER, R. G. (ed.), 1984. *Facies Models,* 2nd ed., Geological Association of Canada, Geoscience Canada, Reprint Series 1, 317 pp.

CHAPTER 3

ARGILLACEOUS ROCKS

INTRODUCTION AND CLASSIFICATION

The fine-grained clastic rocks, extremely abundant in the geological column in all environments of deposition and commonly called argillaceous rocks, consist predominantly of "clay minerals," which are hydrous layer silicates representing a large proportion of the phyllosilicates.

The average grain size of these rocks is more in the silt range (4 to 62 μ) than in the clay range (< 4 μ), and the phyllosilicates are associated with minor amounts of detrital granules of quartz, feldspars, carbonate minerals, pyrite, and, in particular instances, carbonaceous, bituminous, and siliceous matter. Picard (1971) attempted a classification of fine-grained clastic rocks and sediments using thin-section petrography. It emphasized texture and composition of the silt-size components combined with the composition of clay minerals. This attempt, which is essentially an extension of the conventional sandstone terminology to fine-grained rocks, also used a triangular representation with poles represented by sand, silt, and clay (**Fig. 3.1**). Lewan (1978) defined fine-grained clastic rocks as containing more than 45% by volume of microscopic material < 5 μ in size. This value was chosen on the basis that most silt- or sand-size grains will be matrix supported when microscopic material exceeds 40% to 50% of the rock, depending on sorting, angularity, shape, and packing of its grains. Thus the intermediate 45% value implies that the very fine grained clastic sedimentary rocks are usually matrix supported. Lewan distinguished two categories: shales with more than 65% by volume microscopic material (fissility is de-emphasized) and mudstones with 45% to 65% by volume microscopic material. Determination of the percentage of microscopic material was done in thin section by point counting or visual estimates and the 5-μ value was chosen for practical reasons because below that grain size results are unreliable. Semiquantitative X-ray diffraction was used to determine mineralogy and to complete the proposed terminology. Spears (1980) advocated a classification of fine-grained clastic rocks based on quartz content, still retaining the importance of fissility and the terms mudstone to massive siltstone when fissility is absent, and shale to flaggy siltstone when fissility is present. The boundary mudstone–shale versus siltstone was set at 40%, a value apparently of widespread application.

However, these two classifications do not escape the drawbacks pointed out by Weaver (1980, 1989), that the widely used terms of claystone, mudstone, siltstone, and shale have been continuously redefined to the extent of being ambiguous. Furthermore, confusion arises from the assumption of a fixed relationship between grain size and mineralogy and from the use of structural terms such as fissility for shales as basic parameters. Weaver pointed out that the term "clay minerals" and to a certain extent the word "clay" have been used as synonyms for the term "phyllosilicates," but with the connotation of fine grain size (finer than 1/256 mm or 3.9 μ). This means that a mineral group is defined on the basis of both mineralogy and size, which is an anomaly.

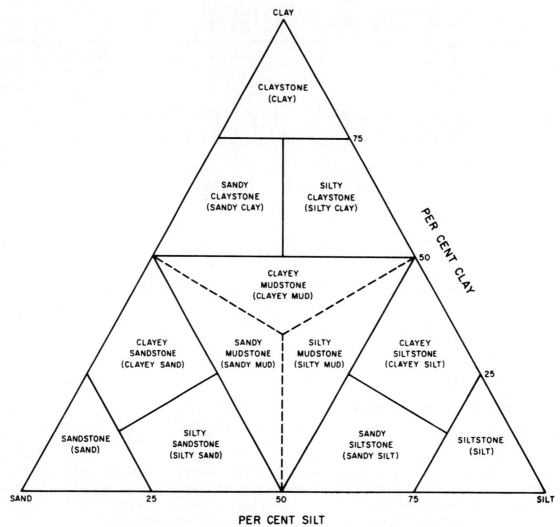

FIGURE 3.1 Textural classification of fine-grained clastic rocks and sediments. From Picard (1971). Reprinted by permission of the Society of Economic Paleontologists and Mineralogists.

There is indeed no common term available to describe a fine-grained clastic rock consisting predominantly of phyllosilicates without any implication of a particular grain size. For instance, the term claystone is defined on the basis of grain size rather than mineralogy, but at the same time it is considered to have a high phyllosilicate content, which in turn is assumed to be clay sized. In actual practice, the identification of claystone is usually based on the predominance of phyllosilicates rather than on size, but closer examination of the phyllosilicates may show that they are in reality silt sized; hence many so-called claystones are siltstones.

Weaver (1980, 1989) also stressed the fact that sedimentary petrologists have tended to use textural data from unconsolidated sediments to classify phyllosilicate-rich rocks, assuming therefore that the grain size of unconsolidated materials is the same as that of rocks. This attitude ignores one of the most important properties of phyllosilicate-rich rocks: their diagenetic evolution upon burial, which

drastically changes their mineralogical and textural properties. An aspect of this question needs to be addressed here. In unconsolidated clays, Recent marine muds, and weakly buried phyllosilicate-rich rocks (with 60% or more phyllosilicates), these minerals occur as dispersed particles finer than 4 μ (often 2 μ) and thus form a clay fraction ranging from 50% to 80% of the deposits. Under the effect of increased temperature and pressure related to burial, the average grain size of all phyllosilicates increases to silt sized, reducing the clay fraction to 10% to 40%. Claystones are thus converted into siltstones, while major mineralogical changes take place, such as the conversion of smectite into mixed-layer illite/smectite and eventually into complete illite, while chlorite may also increase in amount and size, becoming even coarser than illite.

A related problem is presented by fissility, which is included in the definition of shale and expresses the property of the rock to split along relatively smooth surfaces parallel to bedding. Fissility is produced by depositional laminae, concentration of organic matter in distinct streaks, and mainly by the parallel orientation of the sheet phyllosilicates, which itself can be of depositional, diagenetic, or metamorphic origin. Ingram (1953) stated that fissility, in addition to being a function of particle orientation and bedding, is also influenced by weathering, temperature, and water content. All fine-grained clastic rocks with oriented phyllosilicates are massive under overburden pressure. Fissility develops upon removal of overburden, exposure to weathering fluids, and, for those consisting of swelling phyllosilicates, upon appreciable loss of water by dehydration. Weaver (1989) made the interesting conclusive statement on this subject by stressing that shales do not exist in the subsurface, only potential shales occur.

This brief review of the nomenclature problems of fine-grained argillaceous rocks indicates the need for a new term including the entire family of phyllosilicates, with no connotation of grain size. Weaver (1980, 1989) proposed the general term of *physil* as an abbreviated form of the term phyllosilicate. This designation has no size implication, it is similar to the terms quartz or feldspar, and includes both the "clay minerals" and the coarser micas and chlorites, since in fact many physils in argillaceous rocks are silt sized or coarser. The terms clay (claystone) and silt (siltstone) would be used like sandstone as simple grain-size designations, and the term shale would be restricted to field usage for fissile siltstones and claystones.

According to Weaver (1980, 1989), rocks containing more than 50% physils should be called *physilites,* a term that would include all rocks commonly designated as claystones, siltstones, and shales. The corresponding suggested terms are physil claystones and physil siltstones, modified by a prefix such as massive, laminated, or fissile in the case of shales. If the content is less than 50% physils, the adjective physilitic would be applied, as in physilitic claystone, physilitic siltstone, or even physilitic sandstone. For slates and schists, the designation of metaphysilites could be used.

Unquestionably, two- and three-word descriptions are not as popular as traditional single words, as ambiguous as they might be, but they would reduce considerably the widespread confusion and lack of precision that plague at present the terminology of fine-grained argillaceous rocks. A similar problem occurs in the classification of limestones to be discussed below (Chapter 5), and only future studies can spell out the success or demise of new terms.

This book is not the most suitable place to try out the new terminology of physils; the task really belongs to physil mineralogists themselves, particularly since this chapter deals only with the petrography of shales (bentonites and tonsteins are treated with volcaniclastic rocks), in which investigations under the petrographic microscope can still provide some important descriptive and genetic data. It is clear to everyone that the study of the nature and composition of clay minerals is done at present by means of a vast array of modern techniques, among which the most important are X-ray diffraction, differential thermal and gravimetric analysis, SEM and TEM observation,

FIGURE 3.2 Kaolinite under SEM. Mesaverde Formation, Sweetwater County, Wyoming. Well Latham 1-14, depth 10,254.6 ft. Courtesy of Sharon A. Stonecipher, Marathon Oil Company, Exploration and Production Technology, Littleton, Colorado.

infrared spectroscopy, microprobe analysis, and microgeochemistry (Velde, 1985; Moore and Reynolds, 1989; O'Brien and Slatt, 1990; Bennett et al., 1991).

For the particular purpose of this chapter, it is sufficient to consider a few of the phyllosilicates viewed under SEM with their characteristic authigenic shapes particularly well developed when deposited as pore lining and pore filling in sandstones by migrating fluids during burial (Wilson and Pittman, 1977; Weaver, 1989). Kaolinite (**Fig. 3.2**) most commonly occurs as pore-filling packets or small stacks of pseudohexagonal plates randomly arranged with respect to each other. Individual plates range in size from 3 to 20 μ. Occasionally, large vermicular stacks or books can develop. Chlorite (**Fig. 3.3**) occurs in a variety of forms in which indi-

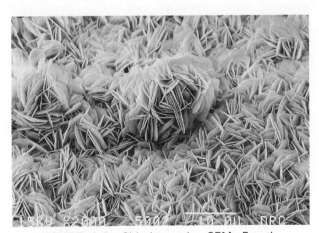

FIGURE 3.3 Chlorite under SEM. Frontier Formation, Spearhead Ranch Field, Wyoming. Well Spearhead Ranch No. 4, depth 12,534.3 ft. Courtesy of Sharon A. Stonecipher, Marathon Oil Company, Exploration and Production Technology, Littleton, Colorado.

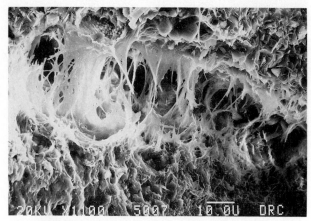

FIGURE 3.4 Illite under SEM. Frontier Formation, Spearhead Ranch Field, Wyoming. Well Spearhead Ranch No. 3, depth 12,786.6 ft. Courtesy of Sharon A. Stonecipher, Marathon Oil Company, Exploration and Production Technology, Littleton, Colorado.

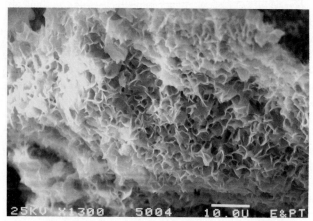

FIGURE 3.6 Montmorillonite under SEM. Grayling Sand, Trading Bay Field, Alaska. Well TBU M-1, depth 7,038.95 ft. Courtesy of Sharon A. Stonecipher, Marathon Oil Company, Exploration and Production Technology, Littleton, Colorado.

vidual flakes, ranging in size from 2 to 10 μ, grow perpendicular to the surface of sandstone grains in a face-to-edge orientation, forming a honeycomb structure that can grade laterally into fan-shaped clusters and rosettes, 5 to 10 μ in size, where face-to-face orientation of the flakes is locally more predominant. Illite (**Figs. 3.4, 3.5**) appears as freestanding to aggregated tape-shaped fibers of variable length (2 to 70 μ), width (1 to 2 μ), and thickness (0.01 to 0.02 μ), which either line pore walls or completely bridge pores. There hairy illites have a profound influence on the reduction of permeability in sandstone reservoirs. Montmorillonite (**Fig. 3.6**) and other smectites form highly wrinkled or honeycomblike pore linings in which individual flakes are

not resolvable. Mixed-layer illite/smectite have morphologies intermediate between the two end-members.

The application of the above-mentioned modern laboratory techniques throughout the geological column opened the important field of clay sedimentology, which traces the history of phyllosilicate-rich sediments, including weathering processes (Nahon, 1991), transportation from land to sea, and early and late diagenesis, which furthermore has fundamental implications, as mentioned above, on the petrophysical properties of the various types of arenites (Houseknecht and Pittman, 1992). The final synthesis is clay stratigraphy and basinal interpretation. Potter et al. (1980), Chamley (1989), and Weaver (1989) have presented exhaustive treatments of these new approaches.

SHALES

Microstructures

In thin sections cut perpendicular to bedding, shales show a broad spectrum of sedimentary structures (Potter et al., 1980). These structures have to be examined at all times with the understanding that their shapes have been strongly influenced by postdepositional compaction and other diagenetic effects.

Primary Sedimentary Structures. These structures formed by hydraulic processes are prominent in shales. They express microscopically processes of flocculation, pelletization, suspension and settling, traction transport, and gravity flow. Consequently, primary structures represent an important source of data for the environmental interpretation of shales.

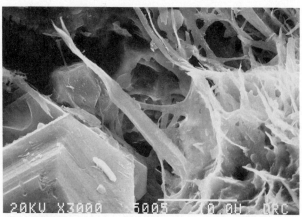

FIGURE 3.5 Illite under SEM. Frontier Formation, Spearhead Ranch Field, Wyoming. Well Spearhead Ranch No. 3, depth 12,786.8 ft. Courtesy of Sharon A. Stonecipher, Marathon Oil Company, Exploration and Production Technology, Littleton, Colorado.

The various types of horizontal stratification are clearly displayed both at the scale of beds (greater than 5 mm in thickness) and at the scale of laminae (smaller than 5 mm in thickness). Horizontal stratification and its varieties (**Plate 13.F**) are produced by a combination of factors, among which are differences in grain size, composition, and fabric, which is mostly the arrangement and perfection of the orientation of platy phyllosilicate particles. Beds and laminae in shales are depositional units with sharp or gradational boundaries representing episodic events in the amount and nature of the clastic supply. These events are in turn controlled by seasonal variations, reworking and redeposition processes related to storm effects, changes of water chemistry that lead to the precipitation of minor amounts of certain minerals such as calcite or gypsum, and variations in organic productivity of planktonic organisms providing calcite, silica, phosphate, and organic matter. *Parallel horizontal stratification* is characterized by distinct and continuous laminae, parallel to each other, with sharp or gradational boundaries. This type of stratification is the most typical structure of shales; it indicates episodic deposition from suspension in quiet water. *Parallel discontinuous stratification,* in which laminae terminate abruptly or taper off, results from the combination of episodic deposition from suspension and some bottom current action. *Wavy or lenticular stratification,* which can be parallel, nonparallel, or discontinuous, represents episodic traction transport forming minute ripples of silt and sand interbedded with fine mud deposited from suspension. This type of stratification grades into microscopic ripple marks and flaser bedding with internal cross laminae in silt- and sand-size materials, which are mostly phyllosilicate floccules and fecal pellets. *Varved stratification* is a typical expression of seasonally controlled sedimentation, particularly when associated with tillites. The process in this case is deposition by suspension of a light-colored silt rapidly deposited during the summer and grading into a darker, organic-rich finer layer slowly settled during the winter. *Graded bedding* is the expression of the distal deposition of turbidity currents (**Plate 14.A**) and is often accompanied by a variety of sole marks along the bottom surface of the graded beds, even at the microscopic scale. Finally, laminae and beds of shales can be internally massive or structureless, a texture indicating either very uniform and continuous deposition, rapid or slow, from suspension or the final product of complete homogenization by bioturbation of any other primary structure.

Burrowed, Bioturbated, and Mottled Structures.

These structures are extremely variable in shales, where they are produced by infaunal activity. Commonly, they appear as tunnels and shafts of complex shapes and orientation, spiral concentrations of fecal pellets and of unassimilated bioclasts (**Plate 14.B**). Biogenic structures are particularly susceptible to extensive deformation and destruction by compaction. For instance, a tunnel with original circular section is often reduced to a flat ellipsoidal shape or even to a simple horizontal streak; individual fecal pellets are merged into massive mud; tunnels and shafts can eventually produce complex mottled textures of darker and lighter areas.

Intraclastic Structures. These structures consist of angular to rounded, discoidal to spherical, or very irregularly shaped clasts derived either from the finer or the coarser laminae of shales among which they are dispersed or form distinct layers. The clasts are randomly oriented or imbricated with oriented long axes unless compaction has modified their original aspect. Intraclasts in shales are locally derived from syneresis (**Plate 14.C**) or desiccation fracturation of semiconsolidated laminae or by their hydraulic erosion.

Early Compactional and Deformational Structures. These particular structures develop during or immediately after deposition and indicate environmental changes between depositional events triggered by gravity, downslope or vertical compression, differences in density of the various constituents, internal fluid movements, and desiccation processes upon temporary subaerial exposure. *Mud cracks* are produced by shrinkage, either from desiccation due to subaerial exposure or from subaquatic contraction or syneresis, which is a continuation of the flocculation process by means of which phyllosilicate particles are attracted to each other and force interstitial water out. No reliable criteria are available for distinguishing the two major processes of formation of mud cracks, except a critical examination of the environmental evolution of the shale sequence (Plummer and Gostin, 1981). *Load casts* result from soft sediment gravity displacements of laminae of sand and silt into underlying finer-grained muds. Load casts are frequently recognizable under the microscope in distal turbidites, where they are associated with *convolute lamination* and *synsedimentary microfolds,* often recumbent and sometimes folded. Whenever convolutions and microfolds are asymmetrical, they can represent subaquatic failures and could be used for the determination of paleoslope directions. Finally, fluidal texture may develop in situations of appreciable compaction when finer-grained layers are symmetrically molded around coarser particles, either mineral grains or bioclasts.

Diagenetic Structures. These structures develop during early to late diagenesis, and in the first case they can be simultaneous with compaction. The majority of diagenetic structures appear to have formed very early, most probably because fine-grained argillaceous sediments are still characterized by appreciable permeability, high content of slowly migrating fluids, reactive phyllosilicates, and low oxygen levels. The most common diagenetic structures in shales are *concretions*, *nodules*, and *spherulites*, which range in size from a few microns to several meters. They are formed by a wide spectrum of minerals: calcite, aragonite, dolomite, gypsum, anhydrite, barite, hematite, limonite, pyrite, marcasite, and often phosphates, all these minerals being usually associated with organic matter. In essence, diagenetic structures are not only an expression of the geochemical environment of the shales, but their very early formation can provide data for estimating the generally very high degree of compaction of the enclosing shales.

Precipitation of the various diagenetic minerals can occur in different ways, such as syngenetically at the water–sediment interface, by displacement of the host sediment, by early filling of the pore space of the surrounding shale, by late displacement of the host rock, and finally by very late void and fracture filling. The character of the bedding and lamination of the surrounding shale is often the best clue to establish the mode of precipitation (Raiswell, 1971). Concretions and nodules may occur in well-defined, widespread, and thin stratigraphic units and could assume the significance of marker beds. Minor, but often spectacular as seen under the microscope, are early diagenetic individual crystal growths. These are crystals of gypsum, anhydrite, halite, and zeolites or their casts and late diagenetic growths of certain phyllosilicates such as vermicular books of kaolinite and dickite, which displace hosting shales. Finally, color banding, often related to the diffusion of iron oxides, represents very late diagenesis under weathering conditions.

Compaction

Compaction in shales is of great importance in many theoretical and practical aspects (Heling, 1970; Rieke and Chilingarian, 1974). Petrographically, its effects are critical in the evolution of fabric and porosity from the initially deposited phyllosilicate flakes to the final shale condition.

Phyllosilicates settling as individual flakes, floccules, and aggregates of various types, charged with various coatings of absorbed organic matter, oxides, and hydroxides, coalesce into a loose network that has a porosity between 70% and 90% and is the site of extensive water and ion exchange. With further accumulation, the overburden pressure causes compaction by rearrangement and deformation of the particles. The decrease of porosity is rather rapid for the first 500 m and then becomes slower. The rate of porosity decrease depends on numerous factors, mainly type and size of phyllosilicate flakes, absorbed cations, composition and pH of interstitial circulating fluids, temperature of burial, and time. The combination of time and temperature tends to increase particle size and to decrease surface area (Weaver, 1989).

One of the main textural effects of compaction is fissility of shales, which is generally considered proportional to the degree of preferred orientation of phyllosilicate flakes. O'Brien (1970) conducted SEM and TEM investigations of shale surface morphology and produced actual detailed pictures of a variety of shales. He confirmed the close correlation between phyllosilicate flake orientation and fissility. Shales with the best fissility have the highest degree of preferred orientation. Randomness of flakes prevails in nonfissile claystones and siltstones. Depositional conditions in seawater and freshwater, which are supposed to control the possibility of whether phyllosilicates are deposited as individual flakes in a parallel manner or as floccules consisting of face-to-face bonded flakes organized in an edge-to-face arrangement, remain unclear, probably because of the complex influence of organic matter, pelletizing infauna, and related bioturbation (Weaver, 1989). Experimental data confirm the fact that overburden produces an increase in parallel particle orientation (Bowles et al., 1969), but again many buried argillaceous rocks do not display parallel orientation of phyllosilicate flakes. This situation can be attributed to several factors: early cementation by carbonate or silica, bioturbation, and occurrence of fine sand-size quartz and feldspar grains inhibiting or limiting the amount of orientation. On the other hand, increasing organic matter content acting as a dispersive agent favors parallel orientation (Weaver, 1989).

Color

Shales occur in a great variety of colors. The significance of shale color, which is an obvious feature of these rocks, has not been investigated in as much detail as that of sandstones. Indeed, color could be used in principle for correlation purposes, for environmental interpretations, and for detecting variations in the supply of phyllosilicate types, but conclusions are often ambiguous (Potter et al., 1980; Weaver, 1989).

The color of shales can be detrital in origin, representing the original color of the various types of phyllosilicates or the pigments of hematite brought in from red soils. It can be depositional and reflect the physical–chemical conditions of the environment, such as the state of the organic matter and of the iron. Color can be also early to late diagenetic when circulating burial fluids change the mineralogical assemblage, and even very late diagenetic upon weathering.

In essence, the color of shales is largely controlled by iron and organic matter. The intrinsic color of the various phyllosilicates, whitish when kaolinitic or greenish when illitic or chloritic, is only a factor when organic matter and iron oxides are present in small amounts.

The color of shales is independent of the total amount of iron present, but highly controlled by the Fe^{3+} to Fe^{2+} ratio. High ratios generate reddish colors; low ratios generate greenish colors. McBride (1974), in a study of continental shales, found a succession of red, yellow, green, and gray corresponding to a decreasing Fe^{3+} to Fe^{2+} ratio, but with a constant total iron. When this ratio is high in an oxidizing environment, whether depositional, diagenetic, or related to weathering, the intense red color is due to hematite pigments forming a film on phyllosilicates or concentrations in interstitial spaces. In all instances, iron is originally leached from various iron-bearing silicates and ilmenite. However, the degree of oxidation of the iron can be so easily modified that the red color expresses essentially the most recent environment in which oxygenated fluids were present in the shales; hence the ambiguity in the environmental interpretation of red beds.

Potter et al. (1980) pointed out that the oxidation state that controls the behavior of iron in shales is in turn controlled by the amount of organic matter, which is probably the most critical factor. Organic matter is measured as amount of organic carbon in the various shales. The preservation of appreciable organic matter (3% to 10%) in anoxic environments, which are also characterized by a low Fe^{3+} to Fe^{2+} ratio, shifts the color of shales to darker greenish and eventually to black upon precipitation of metastable Fe monosulfides changed to pyrite during early diagenesis. This change essentially eliminates iron as a color factor, because the hues of black are related to the amount and size of organic matter.

Petrography of Common Shales

Common shales with variable degrees of fissility are also texturally siltstones. A typical example of their mean grain-size distribution is 3% sand, 59% silt, and 38% clay, yet they may contain 60% phyllosilicates, mostly in the silt-size fraction, together with 31% quartz and 4% feldspar granules of silt to sand size. Iron oxides represent 0.5%, carbonate minerals 4%, other minerals (phosphates, sulfates, zeolites) 2%, and organic matter 1% (Shaw and Weaver, 1965). Unquestionably, phyllosilicates control the shale properties of density, plasticity, fissility, compactibility, and swelling, as well as the reaction to weathering.

In common laminated fissile shales, the groundmass of constituent phyllosilicates, such as illite, kaolinite, smectite, and chlorite, cannot be resolved under the petrographic microscope. However, larger flakes of chlorite, muscovite, and biotite, which display a strong tendency toward uniform orientation parallel to the laminae, contribute predominantly to the general mass extinction characteristic of shales. The boundaries between the larger micaceous flakes and the other finer-grained phyllosilicates forming the groundmass are often quite distinct, and the micaceous books show split and frayed ends. The general tendency toward parallel orientation of the phyllosilicates is often hindered by the appreciable amount of bending of the flakes around dispersed detrital granules of quartz and feldspars. These can also occur in distinct laminae in which they predominate over phyllosilicates. Quartz is in angular to subrounded grains of silt to fine sand size; even at this microscopic scale, secondary overgrowths are frequently displayed around isolated grains, whereas the same process inside quartz-rich lenses or laminae develops an interlocking microquartzitic texture. Quartz grains transported in suspension or by traction processes in distal turbidites have various origins; whereas a portion is certainly of direct eolian origin, the predominant fraction originates from weathering processes, particularly those operating in tropical zones. This origin appears more important than glacial grinding in producing silt-size quartz (Nahon and Trompette, 1982). Other varieties of silica occur in small amount in common shales, such as chalcedony, opal-CT, and amorphous silica, all of which are of biogenic origin.

Besides quartz, feldspar grains can be observed, although always less abundant than quartz. Both plagioclase and orthoclase, fresh or little altered, may be found, and plagioclase is said to be more frequent than potassic feldspar, although there should be shale equivalents to all the previously described sandstone types. Part of the feldspars may be of diagenetic origin, but its distinction from detrital grains is problematic because of the minute size of the individuals. In deeply buried shales, during the conversion of mixed-layer illite/smectite to illite, a considerable amount of potassic feldspars may be lost, thus affecting, to a yet unknown amount, the feldspar abundance in shales (Boles and Franks, 1979) and accounting perhaps for the observed predominance of plagioclase over potassic feldspars.

Calcite is a rare to frequent constituent of common shales. It occurs as irregular grains and aggregates, which may represent a biochemical precipitate cementing agent or minute bioclasts. At any rate, no carbonate mineral should exist in marine shales below the calcite compensation depth unless brought in by turbidity currents or if it is a diagenetic precipitate. The same restriction applies to the presence of dolomite in shales, which is even less common than calcite, although it may act as a cementing agent. Siderite and ankerite are mostly concentrated in diagenetic concretions and, where related to iron sulfides, indicate strongly reducing conditions at the water–sediment interface or in interstitial waters. Pyrite and marcasite cannot be distinguished in thin section from each other within shales. Pyrite appears as minute grains or larger lenticular to framboidal aggregates. In many instances, perfect cubical forms are displayed as well as a variable alteration to iron oxides and hydroxides. The latter form by themselves coatings around phyllosilicate flakes. A varied suite of accessory heavy minerals occurs in common shales, potentially best preserved inside nodules and concretions; these minerals are mostly angular and display little or no alteration.

A great number of common shales contain small amounts of discrete and structured organic particles, which are mostly palynomorphs or small coaly fragments (vitrinite). These constituents, often concentrated in streaks parallel to the laminae, are used in correlation and in establishing the proximity of shorelines. Their reflectance and color are tools for defining the thermal history of shaly basins. Amorphous algal organic material called kerogen is also scattered among shale laminae. It systematically changes color with increasing burial temperature until its final conversion to graphite. Kerogen is an indicator of the gas and oil potential of a basin and of its thermal history.

Petrography of Calcareous Shales

In this type of very common shale, phyllosilicates consist of illite, smectite, kaolinite, and chlorite, with well-developed platy structure. Accessory components are angular to subangular grains of detrital quartz, feldspars, muscovite, and biotite.

Calcite typical of these shales displays several forms: minute bioclasts sometimes with still recognizable internal structures; isolated grains that appear detrital; and spherulites and minute lenses of microcrystalline to cryptocrystalline calcite, parallel to the laminae, which together with well-developed rhombs, are certainly of authigenic origin. It is often difficult to establish the real origin of the carbonate particles, some of which may represent the products of biochemical precipitation in marine or freshwater environments. Spherulites of siderite and micronodules of finely crystalline pyrite occasionally occur in calcareous shales.

The increasing abundance of carbonate particles tends to decrease the fissility of shales, providing a transition toward massive argillaceous calcisiltites and calcilutites. The organic matter that stains many thin sections of calcareous shales consists of unfigured colloidal material that cannot be resolved under the petrographic microscope.

Petrography of Siliceous Shales

Siliceous shales have an extremely high content of silica, which may reach 85%. Besides the phyllosilicate constituents, which are smectite, illite, kaolinite, and chlorite, the coarser angular granules consist of quartz, orthoclase, and sanidine.

Silica appears as an amorphous to cryptocrystalline material scattered at random among phyllosilicates or forming distinct laminae. It is possible to identify all transitions between amorphous opal-CT and cryptocrystalline quartz. A large amount of dark unfigured organic matter characterizes many types of siliceous shales; it occurs both in phyllosilicate laminae and in siliceous ones.

The occurrence of euhedral sanidine, the angularity and freshness of detrital quartz, and the occasional presence of a few cuspate shards of faintly brownish silicic volcanic glass, totally replaced by smectites, indicates that most of the silica is derived from the diagenetic alteration of silicic volcanic glass (Goldstein and Hendricks, 1953). Occasional Radiolaria, diatoms, and sponge spicules do not seem to have played an important genetic role in most siliceous shales. However, particular siliceous shales, also called porcelaneous shales or porcelanites, are the indurated equivalents of diatomites and spiculites, as in the case of the Monterey Formation (Miocene) of California. For these rocks, it is assumed (Bramlette, 1946) that biochemical silica was precipitated when abundant microorganisms used silica originally of pyroclastic origin.

Petrography of Bituminous Shales

High amounts of kerogen or sapropelic algal organic matter characterize bituminous shales or oil shales, which may be of marine or freshwater origin (Bradley, 1931; Spears and Amin, 1981). Their pronounced laminated texture is due to laminae consisting mainly of phyllosilicates and carbonates alternating with other layers in which the bituminous organic matter is concentrated (**Plate 14.D**). The phyllosilicate-rich laminae consist of smectite, illite, and kaolinite associated with a variable amount of angular to subangular detrital granules of quartz, orthoclase and plagioclase, and flakes of muscovite. In certain bituminous shales, the detrital granules are regularly distributed in the phyllosilicate-rich laminae, indicating deposition by distal basinal currents. In other instances, these same granules lack sorting and are scattered at random among the phyllosilicates. Such an occurrence, combined with the predominance of euhedral sanidine over orthoclase and the presence of shards of silicic volcanic glass, implies that a considerable proportion of the clastic granules are of pyroclastic origin and have undergone eolian transport.

Authigenic components are characteristically abundant and varied in bituminous shales. Among them, calcite, dolomite, analcite, apophyllite, quartz, opal, and feldspars predominate. Calcite and dolomite are the predominant authigenic minerals both in rich and lean bituminous shales. Both carbonates occur in shapes ranging from equidimensional grains to perfectly developed rhombs. Analcite and apophyllite display typical crystallographic shapes, but may show distortions due to compactional effects; they are both the expression of the influence of volcanic ash (Bradley, 1931). Silica is present as minute bipyramidal crystals and as patches of impure opal-CT. The layers of bituminous shales with the highest organic matter content seem to be the most favorable for the development of authigenic feldspars, in particular albite (Moore, 1950). Albite appears as well-developed crystals, sometimes with a darker central core surrounded by one or several marginal zones of different extinction, and hence of slightly different composition.

Pyrite occurs in various amounts in bituminous shales, in subspherical, framboidal, or irregular grains, as well as in radial aggregates of elongate rodlets, but cubes and other crystal forms are less common.

Two kinds of organic matter can be distinguished in thin sections of bituminous shales. One is entirely structureless or unfigured, translucent, and of lemon yellow to reddish brown color depending on its degree of thermal maturation. It constitutes the groundmass of the bituminous shales and contains variable amounts of finely divided mineral matter, often segregated into more or less well-defined sublaminae and streaks (**Plate 14.D**). The other kind of organic matter consists of complete or fragmentary figured material of marine or freshwater algal origin, among which the best known are the spherical unicellular varieties of algae called *Tasmanites* (**Plate 14.E**). These spherical bodies are often the centers for early pyritization that filled completely their interior, preserving their original shape from subsequent compaction. The juxtaposition of spherical individuals with completely flattened nonmineralized ones gives a measure of the high degree of compaction these particular bituminous shales have undergone. Besides compaction, smectite to illite conversion, and thermal maturation, bituminous shales do not seem to have undergone other mineralogical changes during diagenesis except for the diagenetic growth of calcitic and siliceous concretions frequently coated with pyrite (Curtis, 1980). As mentioned above, whenever bituminous shales have a typical varved texture combined with "dropgrains," they represent the best argument in favor of the glacial origin of associated tillites (**Plate 14.F**).

Petrography of Carbonaceous Shales

Phanerozoic carbonaceous shales represent the association of phyllosilicates and humic organic matter in the state of gas-yielding carbonaceous matter, rather than as sapropelic algal matter susceptible to yielding liquid hydrocarbons.

The phyllosilicate-rich laminae consist of illite, smectite, kaolinite, and chlorite in variable proportions. Detrital granules consist of quartz, orthoclase and plagioclase feldspars, and muscovite flakes. A very small amount of carbonate is usually present as idiomorphic rhombs of calcite, either scattered throughout the matrix or concentrated in laminae, in thin lenses or aggregates. Authigenic pyrite is an abundant constituent of carbonaceous shales and occurs in irregular masses and as single crystals or aggregates of individuals: many of the smallest crystals are commonly surrounded by amorphous carbonaceous material. Pyrite aggregates show shapes and relict structures indicating replacement of minute plant debris. The carbonaceous and graphitic material that characterizes these shales is either concentrated in distinct laminae or scattered among phyllosilicates as detrital, irregular thin shreds and flakes, often surrounded by thin coatings of pyrite.

In the Precambrian, before the advent of continental plants, the formation of carbonaceous shales was apparently entirely different than during Phanerozoic times. Schieber (1986) studied the carbonaceous silty shales of the Newland Formation, Mid-Proterozoic of Montana, and interpreted them as remnants of benthic microbial mats formed in shallow subtidal conditions basinward of a carbonate platform with algal laminites and flat-pebble conglomerates. These shales consist of irregularly shaped, wavy to crinkly carbonaceous and pyritic silty laminae with minute lenses of silt-size granules alternating with laminae of dolomitic argillaceous shale draping over the carbonaceous ones.

The carbonaceous silty laminae consist of superposed crinkly to wavy carbonaceous mats that show patterns of particle trapping, false cross-bedding, filamentous structures, and a mechanical strength during episodes of intraformational reworking and deformation. These features indicate a tough, leathery membrane, rather than an accumulation of a mixture of mud and detrital carbonaceous matter; hence these laminae are interpreted as the remnants of benthic microbial mats or stromatolites, rather than the products of settling during seasonal blooming of planktonic microbial matter. During the Mid-Proterozoic, cyanobacteria (blue-green algae) were probably the sole contributors of organic matter. Although microbial mats occur today in intertidal settings, in the Proterozoic they also occupied subtidal environments. In fact, because of the absence of grazing animals, they may have occupied all the available niches with sufficient light and a low rate of sedimentation. Schieber (1986) suggested that microbial mats may have acted as membranes that separated reducing underlying sediments from overlying oxygenated waters, thus allowing the accumulation of highly carbonaceous shales in shallow subtidal conditions.

DIAGENESIS OF SHALES

It appears appropriate to discuss briefly the diagenesis of shales before considering the question of their depositional environments and provenance because, during burial diagenesis, phyllosilicate types undergo appreciable changes. These changes occur mainly as an effect of temperature and, to some degree, the chemical composition of the system, rather than pressure. Therefore, the implications of such changes are fundamental in reconstructions of environments of deposition and provenance of shales.

The main modification is the alteration of smectite to illite and chlorite by means of an intermediate stage of mixed-layer illite/smectite (Hower et al., 1976). This diagenetic process consists of the incorporation of K^+ ions into the smectite structure, accompanied by a loss of interlayer water. In areas of average geothermal gradient, the destruction of smectite begins at temperatures of 70° to 90°C, corresponding to depths of 2 to 3 km. At slightly higher temperatures and greater depths, kaolinite is replaced by illite and chlorite, and if the original sediment is illitic, the degree of illite crystallinity increases (Kübler, 1968). Still deeper, under conditions of incipient metamorphism, illite is replaced by sericite, which coexists with chlorite.

Boles and Franks (1979) stressed the importance of the by-products of the illitization process in the cementation of sandstones interbedded or adjacent to shales undergoing this type of diagenesis (**Fig. 3.7**). Silica and calcium released at

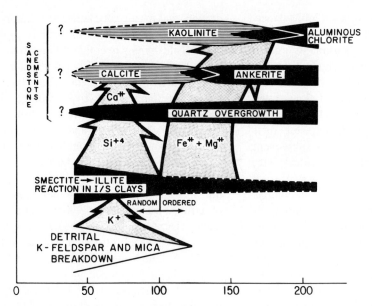

FIGURE 3.7 Schematic diagram of the influence of illite/smectite clay reactions on Wilcox sandstone cements. Vertical arrows depict ion transfer between illite/smectite clay reactions and phases in sandstones. No sample control at temperatures less than 60°C. From Boles and Franks (1979). Reprinted by permission of the Society of Economic Paleontologists and Mineralogists.

temperatures as low as 60°C form quartz overgrowths and early calcite cements. Iron and magnesium released at temperatures of about 100°C react with kaolinite to produce high-aluminum chlorite. They also react with calcite to generate ankerite.

Mineralogical changes in shales also take place through geologic time (Dunoyer de Segonzac, 1970). The percentage of smectite dramatically decreases in time, while that of illite increases; similarly, kaolinite shows a smaller decrease, whereas chlorite increases. Consequently, lower Paleozoic and Precambrian shales tend to consist predominantly of illite and chlorite. This evolution toward a more stable composition of shales with time is another expression of the burial diagenetic alteration discussed above. Another possible factor is the advent of continental plants during late Devonian and Carboniferous times, which influenced weathering processes and related soil formation and from that time on contributed to the generation of a greater variety of phyllosilicates than before.

DEPOSITIONAL ENVIRONMENTS AND PROVENANCE OF SHALES

The depositional environments and provenance of shales can be approached in three major ways: (1) study of the basinward distribution of the various types of phyllosilicates; (2) study of the basinward distribution of detrital quartz, feldspars, and accessory minerals; and (3) study of the directional properties of shales to establish paleocurrent patterns.

Definite patterns of basinward distribution of the various types of phyllosilicates were observed in shallow intracratonic marine basins, located in areas of relatively low geothermal gradient, where diagenetic processes have not affected the nature of major phyllosilicates. These patterns are particularly clear in deltaic environments. However, their origin remains disputed as to whether it results from flocculation of the phyllosilicates upon reaching seawater or from an original size distribution of these platy minerals. The basinward trend can be a change, as in Chesterian shales and sandstones of the Illinois Basin (Smoot, 1960), as follows: high-kaolinite sandstone facies of lower deltaic distributaries, low-kaolinite sandstone facies of delta front, mixed-layer shale facies of prodelta, and illite shale facies of basinal environment. In a generalized fashion (Parham, 1966), the basinward trend can be a change from predominant kaolinite to illite and chlorite, with smectite present over the entire range.

In still other instances, such as the bituminous shales of two members of the Curuá Formation (Famennian–Frasnian) of the Middle Amazon Basin (Carozzi, 1979), the basinward trend is a change from kaolinite to a mixed kaolinite–illite assemblage and finally to illite (**Figs. 3.8, 3.9**). In both cases, kaolinite lobes, corresponding to the major axes of influx during shale deposition, extend to various distances toward the depocenter of the basin and coincide with those of the underlying and overlying sandstone formations, representing well-defined delta systems. Therefore, it is possible to establish the genetic relationship between shales and associated sandstones and in particular to determine if a given shale represents the end term of the underlying deltaic sandstone system or the beginning stage of the overlying one.

In Desmoinesian cyclothems of Iowa and Missouri, the major environmental variation within a deltaic system is the change from normal marine salinities in deltaic marine environments to brackish and freshwater conditions in marshy delta plains and upper interdistributary bays and within adjacent interdeltaic environments. Brown et al. (1977) showed that lateral changes in phyllosilicate types from marine to nonmarine shales, as proved by limestone

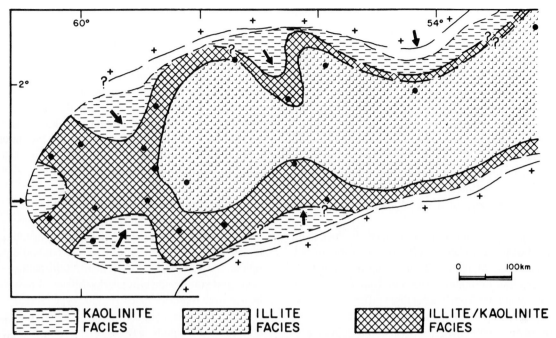

KAOLINITE FACIES   **ILLITE FACIES**   **ILLITE/KAOLINITE FACIES**

FIGURE 3.8 Distribution of clay mineral facies in black shales, Barreirinha Member, Curuá Formation, Upper Devonian (Frasnian) of the Middle Amazon Basin, Brazil. From Carozzi (1979). Reprinted by permission of the *Journal of Petroleum Geology* and Scientific Press Ltd.

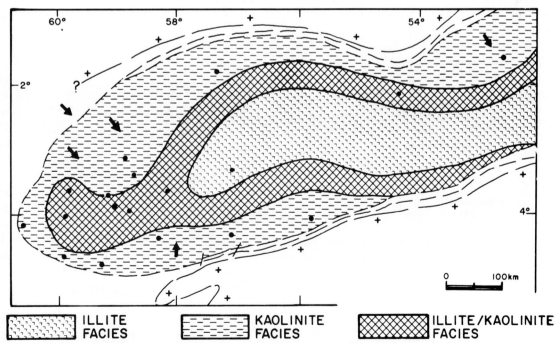

ILLITE FACIES   **KAOLINITE FACIES**   **ILLITE/KAOLINITE FACIES**

FIGURE 3.9 Distribution of clay mineral facies in black shales, Curiri Member, Curuá Formation, Upper Devonian (Famennian) of the Middle Amazon Basin, Brazil. From Carozzi (1979). Reprinted by permission of the *Journal of Petroleum Geology* and Scientific Press Ltd.

pinchouts, coincide with a predictable decrease of illite and an increase of kaolinite, mixed-layer illite/smectite, and percentage of expansible layers within the mixed-layer phyllosilicate (**Fig. 3.10**). It appears that progradation of deltas supplied a fresh influx of predominant illitic clays from distant sources. Clays were deposited mainly in prodelta environments, but some settled in adjacent embayments. During repeated exposure on intertidal flats or as a result of the effect of acidic marshes, illitic clays degraded rapidly to mixed-layer illite/smectite with maximum expansible layers. This variation of phyllosilicates is an appropriate tool to establish, together with other sedimentological and paleoecological data, the cyclic variations of delta building and abandonment.

With respect to establishing provenance by means of suites of phyllosilicates, Carozzi (1980) described, in the shallow intracratonic Maranhão Basin of northeastern Brazil, the Pimenteiras Formation (Eifelian–Givetian), which received clastic supplies from at least three different sources. The latter could be identified on the basis of the relative proportions of various combinations of kaolinite, illite, and chlorite (**Fig. 3.11**).

Shales of flysch facies deposited as distal turbidites and hemipelagic muds in relatively deep water can also be investigated in terms of provenance and tectonic setting. Bhatia (1985) undertook such a study on Paleozoic flysch shales of eastern Australia. He established a maturity index based on the ratio of total phyllosilicates to phyllosilicates + feldspar + quartz, and divided the shales into tectic, phyllic–tectic, and phyllic on the basis of increasing maturity. The

latter also corresponds in the element geochemistry of these rocks to an increase of the K_2O to Na_2O ratio. This is attributed to the increase in K with the enrichment of phyllosilicates and the depletion of Na due to the decrease in abundance of feldspar in source areas and loss during transport.

Bhatia was able to distinguish four major tectonic settings for the investigated flysch shales. Shales of oceanic island-arc type have a low maturity index (tectic type) and are characterized by a high feldspar and low phyllosilicate content and a low K_2O to Na_2O ratio. These shales are derived predominantly from calc-alkaline andesitic rocks of oceanic island arcs.

Shales of continental island arcs and active continental margins (Andean type and strike-slip basins) have an intermediate maturity index (phyllic–tectic type) and are characterized by higher phyllosilicate to feldspar and K_2O to Na_2O ratios. They are derived from quartz-feldspathic (felsic volcanic and granite–gneissic) rocks. Shales of passive margins have a high maturity index (phyllic type) and are characterized by a high phyllosilicate content, a low feldspar content, and extremely high K_2O to Na_2O ratios, suggesting recycling from older metamorphic and sedimentary terranes.

Immobile trace elements (such as La, Th, Zr, Nb, Y, and Sc) appear very useful in discriminating tectonic settings. An increase occurs in La, Th, Nb and in the Nb to Y, Rb to Sr, and Ba to Sr ratios, and a decrease takes place in the Zr to Th and Zr to Nb ratios from oceanic island arc to continental island arc and to active continental–passive margin flysch shales.

The second approach to establishing depositional envi-

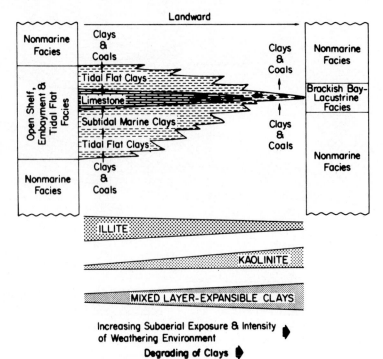

FIGURE 3.10 Lateral relationships of clay minerals within a marine-to-nonmarine pinchout, Desmoinesian Series, northern Missouri and southern Iowa. From Brown et al. (1977). Reprinted by permission of the Clay Mineral Society.

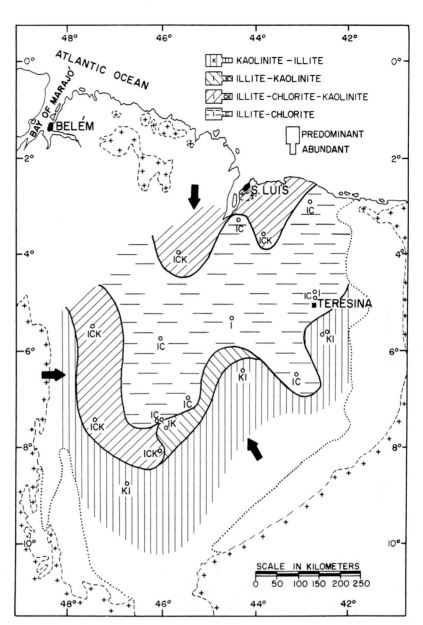

FIGURE 3.11 Distribution and provenance of clay mineral assemblages in the shales of the Pimenteiras Formation, Eifelian–Givetian, Maranhão Basin, Brazil. From Carozzi (1980). Reprinted by permission of the *Journal of Petroleum Geology* and Scientific Press Ltd.

ronments and provenance of shales is based on the study of the basinward distribution of detrital quartz, feldspars, and accessory minerals. The gradient of decrease of the percentage of mean grain size of detrital quartz in epicontinental shales can be used to determine increasing distance from shorelines (Blatt and Totten, 1981). Quartz was released from shales by the sodium bisulfate fusion technique, which becomes, however, inaccurate in proportion to the percentage of very calcic plagioclase in the sample. Fortunately, this problem is only encountered in shaly basins along convergent plate boundaries. Therefore, the only significant problem would be encountered in deeply buried shales when the conversion of mixed-layer illite/smectite to illite produces quartz as a by-product (Boles and Franks, 1979). This quartz

is appreciable in amount and of small grain size, but quantitative data are lacking.

The mechanism that produced the direct relationship between grain size and percentage of quartz and distance to shorelines consists of dilute, fine-grained turbidity currents initiated by subaqueous debris flows. This situation means that the rate of decrease of size and percentage should be related to the slope of the shaly depositional basin so that future refinements of the technique could be very useful in basinal slope estimates, with possible reservations related to the disruptive action of storms and hurricanes.

An earlier discussion (Chapter 1) dealt with the difficulties of provenance studies based on the mineralogy of sandstones, particularly with respect to feldspars, which are commonly dissolved, altered to clay minerals or albitized, and to heavy minerals often reduced to a ZTR assemblage. Blatt (1985) pointed out that shales have the potential of greater preservation of feldspars and accessory minerals because of their permeabilities, which are typically orders of magnitude lower than those of sandstones. However, caution is in order because intrastratal solution also occurs in shales, and if the basinward distribution of orthoclase is usable for establishing distance to a known granitic source area in fluvial shales of the Vanoss Formation (Pennsylvanian) of Oklahoma (Blatt and Caprara, 1985), selective dissolution of plagioclase in these shales prevents the use of this mineral for provenance studies. Apparently, solution phenomena are more common in shales than generally believed; very slow but large-scale movement of solutions, pressure solution, and diffusion of various ions along fissility planes are to be considered (Blatt, 1985). Furthermore, shales are not immune to diagenetic losses of nonclay minerals with burial. There is a general agreement that the abundance of potassium feldspar decreases with increasing depth in shales of different ages and various tectonic, environmental, and sedimentological settings, but the explanation of this decrease remains disputed. It may not be always the effect of illitization during which sodium is released by smectites, causing albitization of orthoclase, because calcite replacement of orthoclase produces the same results.

The behavior of accessory minerals in shales indicates that they are certainly better preserved than in associated sandstones, but again the effects of intrastratal dissolution and phyllosilicate diagenesis remain unclear (Blatt, 1985). At any rate, shales deserve more attention in provenance studies than they have received.

Directional properties of shales usable for establishing patterns of paleocurrents are best displayed in those deposited as distal turbidites. These properties are as follows: orientation of cross laminae in microscopic ripple marks and flaser bedding; of flat intraclasts; of silt-size elongate quartz grains in thin sections cut parallel to bedding of silt laminae (Piper, 1972); and of delicate and elongate rhabdosomes of graptolites in the pelitic parts of graded beds (zone *e* of Bouma), which have an orientation in agreement with the underlying turbidites as determined from sole markings and internal structures (Moors, 1959).

However, most shales were deposited by very weak currents unrelated to turbidites and even more sensitive indicators are required. A variety of them are still available, such as the orientation of entire fossils or of their bioclasts, a question dealt with in great detail by Seilacher (1973), of fecal pellets (Hakes, 1976), and fine charcoal debris and woody fragments (Zangerl and Richardson, 1963). All these depositional indicators have proved reliable as long as subsequent bioturbation or slumping processes have not appreciably modified or destroyed original conditions (Potter et al., 1980). Finally, Colton (1967) and Jacob (1973) reported orientation of early diagenetic concretions parallel to paleocurrent directions.

TYPICAL EXAMPLES

Plate 13.F. Arenaceous shale consisting of a weakly laminated to irregularly textured groundmass of clay minerals in which are scattered minute angular grains of detrital quartz, pyrite flecks, and rare rodlets of carbonaceous material. Green River Formation, Eocene, Bridger Basin, Wyoming, U.S.A.

Plate 14.A. Shale with interbedded microturbidites consisting of silt-size angular grains of quartz, potassic feldspars, muscovite flakes, and intraformational shale pellets, set in an interstitial cement of argillaceous microcrystalline calcite and pyrite aggregates. Microturbidites show relatively sharp basal contacts and incipient graded bedding. Green River Formation, Eocene, Bridger Basin, Wyoming, U.S.A.

Plate 14.B. Shale with irregularly textured groundmass in which silt-size angular grains of detrital quartz and potassic feldspars, muscovite flakes, and fragments of smooth ostracod shells are concentrated in a variety of incomplete spiral structures due to intense infauna bioturbation. McKenzie Formation, Middle Silurian, Mifflintown, Pennsylvania, U.S.A.

Plate 14.C. Shale with laminated texture emphasized by streaks of minute angular grains of quartz and pyrite flecks. Left portion of bed is ruptured and uplifted by effect of synersis crack system. Overlying massive siltstone, with abundant angular quartz grains, intraformational shale pellets, and pyrite concentrations, has filled in the intervening space produced by the dislocation. Wills Creek Formation, Upper Silurian, Huntingdon, Pennsylvania, U.S.A.

Plate 14.D. Bituminous shale displaying an irregularly laminated structure consisting of thin, tan to yellow bands of almost pure bituminous material alternating with thicker, dark-colored bands consisting of an intimate mixture of clay minerals, bituminous material, and cryptocrystalline calcite in which are scattered angular, silt-size grains of quartz and potassic feldspars, muscovite flakes, and minute fragments of thin ostracod shells. Green River Formation, Eocene, Bridger Basin, Wyoming, U.S.A.

Plate 14.E. Tasmanite consisting almost entirely of compaction-flattened spherical algae preserved as almost pure bituminous matter, with internal dark streak representing original central cavity. Interstitial matrix is an association of minute quartz grains, sericite and muscovite flakes, dark clay minerals, and pyrite flecks. Minute white calcite concretions are scattered throughout the matrix. High amplitude of compaction is shown by a noncompacted and early pyrite-infilled single algal sphere in right portion of figure. Tasmanite bed, Permian, Latrobe, Tasmania.

Plate 14.F. Glacial varvite consisting of a layer of dark bituminous shale overlain and underlain by coarse bituminous siltstones consisting of abundant angular grains of

detrital quartz and potassic feldspars, flakes of chloritized biotite and muscovite, and intraformational bituminous shale pellets set in a matrix of bituminous clay minerals. The three larger and angular quartz grains at top of lower siltstone are interpreted as "dropgrains" from melting, floating ice. Cabeças Formation, Upper Devonian, Maranhão Basin, Brazil.

REFERENCES

BENNETT, R. H., BRYANT, W. R., and HULBERT, M. H. (eds.), 1991. Microstructure of fine-grained sediments. From mud to shale. *Frontiers in Sedimentary Geology,* Springer-Verlag, New York, 582 pp.

BHATIA, M. R., 1985. Composition and classification of Paleozoic flysch mudrocks of eastern Australia: implications in provenance and tectonic setting interpretation. In R. Hesse (ed.), Sedimentology of siltstone and mudstone. *Sed. Geol.,* Special Issue, 41, 249–268.

BLATT, H., 1985. Provenance studies and mudrocks. *J. Sed. Petrology,* 55, 69–75.

———, and CAPRARA, J. R., 1985. Feldspar dispersal patterns in shales of the Vanoss Formation (Pennsylvanian), south-central Oklahoma. *J. Sed. Petrology,* 55, 548–552.

———, and TOTTEN, M. W., 1981. Detrital quartz as an indicator of distance from shore in marine mudrocks. *J. Sed. Petrology,* 51, 1259–1266.

BOLES, J. R., and FRANKS, S. G., 1979. Clay diagenesis in Wilcox sandstones of southwest Texas: implications of smectite diagenesis on sandstone cementation. *J. Sed. Petrology,* 49, 55–70.

BOWLES, F. A., BRYANT, W. R., and WALLIN, C., 1969. Microstructure of unconsolidated and consolidated marine sediments. *J. Sed. Petrology,* 39, 1546–1551.

BRADLEY, W. H., 1931. Origin and microfossils of the oil shale of the Green River Formation of Colorado and Utah. *U.S. Geol. Survey Prof. Paper 168,* 58 pp.

BRAMLETTE, M. N., 1946. The Monterey Formation of California and the origin of its siliceous rocks. *U.S. Geol. Survey Prof. Paper 212,* 57 pp.

BROWN, L. F., JR., BAILEY, S. W., CLINE, L. M., and LISTER, J. S., 1977. Clay mineralogy in relation to deltaic sedimentation patterns of Desmoinesian cyclothems in Iowa–Missouri. *Clays and Clay Minerals,* 25, 171–186.

CAROZZI, A. V., 1979. Petroleum geology in the Paleozoic clastics of the Middle Amazon Basin. *J. Petroleum Geol.,* 2, 55–74.

———, 1980. Tectonic control and petroleum geology of the Paleozoic clastics of the Maranhão Basin, Brazil. *J. Petroleum Geol.,* 2, 389–410.

CHAMLEY, H., 1989. *Clay Sedimentology.* Springer-Verlag, New York, 623 pp.

COLTON, G. W., 1967. Orientation of carbonate concretions in the Upper Devonian of New York. *U.S. Geol. Survey Prof. Paper 575B,* 57–59.

CURTIS, C. D., 1980. Diagenetic alteration in black shales. *J. Geol. Soc. London.* 137, 189–194.

DUNOYER DE SEGONZAC, G., 1970. The transformation of clay minerals during diagenesis and low-grade metamorphism: a review. *Sedimentology,* 15, 281–346.

GOLDSTEIN, A., JR., and HENDRICKS, T. A., 1953. Siliceous sediments of Ouachita facies in Oklahoma. *Geol. Soc. Amer. Bull.,* 64, 421–442.

HAKES, W. G., 1976. Trace fossils and depositional environment of four clastic units, Upper Pennsylvanian megacyclothems, northeast Kansas. *Univ. Kansas Paleontol. Contrib.,* Art. 63, 46 pp.

HELING, D., 1970. Micro-fabrics of shales and their arrangement by compaction. *Sedimentology,* 15, 247–260.

HOUSEKNECHT, D. W., and PITTMAN, E. D. (eds.), 1992. *Origin, Diagenesis, and Petrophysics of Clay Minerals in Sandstones.* Soc. Econ. Paleontologists and Mineralogists, Special Publ. 47, 288 pp.

HOWER, J., ESLINGER, E. V., HOWER, M. E., and PERRY, E. A., 1976. Mechanism of burial metamorphism of argillaceous sediment: 1. Mineralogical and chemical evidence. *Geol. Soc. Amer. Bull.,* 87, 725–737.

INGRAM, R. L., 1953. Fissility of mudstones. *Geol. Soc. Amer. Bull.,* 64, 869–878.

JACOB, A. J., 1973. Elongate concretions as paleochannel indicators, Tongue River Formation (Paleocene), North Dakota. *Geol. Soc. Amer. Bull.,* 84, 2127–2132.

KÜBLER, B., 1968. Évaluation quantitative du métamorphisme par la cristallinité de l'illite. *Bull. Centre Recherches S.N.E.A.P., Pau,* 1, 259–278.

LEWAN, M. D., 1978. Laboratory classification of very fine-grained sedimentary rocks. *Geology,* 6, 745–748.

MCBRIDE, E. F., 1974. Significance of color in red, green, purple, olive, brown, and gray beds of DiFunta Group, northeastern Mexico. *J. Sed. Petrology,* 44, 760–773.

MOORE, D. M., and REYNOLDS, R. C., 1989. *X-ray Diffraction and the Identification and Analysis of Clay Minerals.* Oxford University Press, New York, 352 pp.

MOORE, F. E., 1950. Authigenic albite in the Green River oil shales. *J. Sed. Petrology,* 20, 227–230.

MOORS, H. T., 1959. The position of graptolites in turbidites. *Sed. Geol.,* 3, 241–261.

NAHON, D., 1991. *Introduction to the Petrology of Soils and Chemical Weathering.* John Wiley & Sons, Inc., New York, 313 pp.

———, and TROMPETTE, R., 1982. Origin of siltstones: glacial grinding versus weathering. *Sedimentology,* 29, 25–35.

O'BRIEN, N. R., 1970. The fabric of shale—an electron microscope study. *Sedimentology,* 15, 229–246.

———, and SLATT, R. M., 1990. *Argillaceous Rock Atlas.* Springer-Verlag, New York, 141 pp.

PARHAM, W. E., 1966. Lateral variations of clay mineral assemblages in modern and ancient sediments. *Proc. Intern. Clay Conf.,* Jerusalem, Israel, 1, 135–145.

PICARD, M. D., 1971. Classification of fine-grained sedimentary rocks. *J. Sed. Petrology,* 41, 179–195.

PIPER, D. J. W., 1972. Turbidite origin of some laminated mudstones. *Geol. Magazine,* 109, 115–126.

PLUMMER, P. S., and GOSTIN, V. A., 1981. Shrinkage cracks: desiccation or synaeresis? *J. Sed. Petrology,* 51, 1147–1156.

POTTER, P. E., MAYNARD, J. B., and PRYOR, W. A., 1980. *Sedimentology of Shale, Study Guide and Reference Source.* Springer-Verlag, New York, 306 pp.

RAISWELL, R., 1971. The growth of Cambrian and Liassic concretions in the Upper Lias of N.E. England. *Chem. Geol.,* 18, 227–244.

RIEKE, H. H., and CHILINGARIAN, G. V., 1974. *Compaction of Argillaceous Sediments.* Developments in Sedimentology 16, Elsevier Publishing Co., New York, 424 pp.

SCHIEBER, J., 1986. The possible role of benthic microbial mats during the formation of carbonaceous shales in shallow Mid-Proterozoic basins. *Sedimentology,* 33, 521–536.

SEILACHER, A., 1973. Biostratonomy: the sedimentology of biologically standardized particles. In R. N. Ginsburg (ed.), *Evolving Concepts in Sedimentology,* Johns Hopkins University Studies in Geology No. 21, Johns Hopkins University Press, Baltimore, 159–177.

SHAW, D. B., and WEAVER, C. E., 1965. The mineralogical composition of shales. *J. Sed. Petrology,* 35, 213–222.

SMOOT, T. W., 1960. Clay mineralogy of Pre-Pennsylvanian sandstones and shales of the Illinois Basin. Part III. Clay minerals of various facies of some Chester formations. *Illinois State Geol. Survey Circular No. 293,* 19 pp.

SPEARS, D. A., 1980. Towards a classification of shales. *J. Geol. Soc. London,* 137, 125–129.

———, and AMIN, M. A., 1981. Geochemistry and mineralogy of marine and non-marine Namurian black shales from the Tansley borehole, Derbyshire. *Sedimentology,* 28, 407–417.

VELDE, B., 1985. *Clay Minerals. A Physico-chemical Explanation of Their Occurrence.* Developments in Sedimentology 40, Elsevier Publishing Co., New York, 427 pp.

WEAVER, C. E., 1980. Fine-grained rocks: shales or physilites. *Sed. Geol.,* 27, 301–313.

———, 1989. *Clays, Muds, and Shales.* Developments in Sedimentology 44, Elsevier Publishing Co., New York, 819 pp.

WILSON, M. D., and PITTMAN, E. D., 1977. Authigenic clays in sandstones: recognition and influence on reservoir properties and paleoenvironmental analysis. *J. Sed. Petrology,* 47, 3–31.

ZANGERL, R., and RICHARDSON, E. S., JR., 1963. The paleoecologic history of two Pennsylvanian black shales. *Fieldiana Geol. Memoirs,* 4, 352 pp.

CHAPTER 4

VOLCANICLASTIC ROCKS

INTRODUCTION AND CLASSIFICATION

According to the definition of Fisher (1961, 1966), volcaniclastic rocks include all clastic volcanic materials formed by any process of fragmentation, dispersed by any kind of transport agent, deposited in any environment, or mixed in any significant proportion with nonvolcanic sediments.

The awareness of the direct or indirect volcanic contribution to the geological column has been growing in recent years and was recently reviewed exhaustively (Fisher and Schminke, 1984; Cas and Wright, 1987; Fisher and Smith, 1991). This contribution is extremely varied in aspect, ubiquitous in all depositional environments, and volumetrically enormous. Nevertheless, it seems to have often remained undetected for three major reasons: (1) the study of volcanic material seems peripheral to both igneous and sedimentary petrographers and hence remains neglected; (2) volcanic material of relatively small size is dispersed among other constituents of sedimentary rocks and only detailed petrographic work is able to disclose its presence; and (3) the various types of relatively rapid diagenetic alteration of volcanic material, such as zeolitization, and replacement by clay minerals tend also, if not recognized, to obscure this kind of contribution, whose understanding is critical for a complete reconstruction of sedimentary environments.

The classification of volcaniclastic rocks proposed by Fisher (1961, 1966) has reached general agreement (Schmid, 1981; Fisher and Schminke, 1984). The following five major subdivisions are recognized. *Autoclastic* fragments are produced by fragmentation in volcanic vents from mechanical friction within flowing lava, by internal gaseous explosions disrupting partially indurated portions of lava flows, and finally by gravity crumbling of lava spines and domes. These various processes produce autoclastic materials ranging from breccia (> than 64-mm fragments) to dust size (< than $\frac{1}{16}$ mm). Under these conditions, autoclastic products can contain nonvolcanic rocks derived either from the walls of the volcanic vent or from the substratum over which lava flows.

Alloclastic fragments result from the various types of fragmentation of preexisting volcanic rocks by igneous processes beneath the surface of the earth, with or without intrusion of fresh magmatic material, but involving the country rock. Autoclastic and alloclastic fragments are not of direct concern to sedimentary petrographers.

Pyroclastic fragments, also designated as pyroclasts or tephra (Schmid, 1981), are of great importance. They are particles, such as individual crystals, crystal fragments, glass fragments, and rock fragments, expelled from volcanic vents without reference to the type of eruption or the origin of the particles. The shapes of pyroclastic fragments are generated during the disruption process or transport to the primary place of deposition and should not have been changed by subsequent redeposition processes, which would make them reworked pyroclasts.

Hydroclastic fragments represent a particular variety of pyroclasts generated by steam explosions at the interface between erupting magma and seawater or freshwater and also by rapid quenching and mechanical granulation of the lava when it comes in contact with water or water-saturated sediments.

Finally, *epiclastic* fragments are particles produced by weathering processes and mechanical erosion of indurated older volcanic rocks, and not by explosions. These kinds of lithoclasts are particularly abundant in volcanic lithic arenites and volcanic lithic wackes, but may also occur scattered in carbonate rocks or in any other kind of sedimentary rocks.

The interpretation of the origin of volcaniclastic rocks presents some critical problems resulting mainly from reworked pyroclastic fragments becoming mixed with other types of clasts, in particular epiclastic fragments. Fisher and Schminke (1984) stressed the importance of studying carefully the physical aspect, composition, and percentages of clasts for successfully approaching this problem. Indeed, pyroclastic fragments being produced so to speak instantaneously by volcanic processes are, when correctly identified, indicators of penecontemporaneous volcanism. However, they are often submitted to reworking episodes by rain, running water, or wind action, which can occur as early as a few hours after initial fallout on the slopes of volcanoes. If pyroclastic fragments consist of glass shards and pumice that have undergone little alteration, their redeposition in streams, lakes, and marine environments, adjacent to volcanoes or at a certain distance from them, strongly impairs or even destroys their power of resolution as indicators of penecontemporaneous volcanism. This situation is particularly critical in older deposits. As mentioned above, epiclastic fragments produced by weathering and erosion of volcanic rocks are not markers of penecontemporaneous volcanism although, if large enough, they may contain inside themselves smaller pyroclastic debris. A critical feature of epiclastic deposits is their small content or even lack of individual shards and pumice fragments, because they were altered either by weathering to clay minerals and zeolites or destroyed by mechanical abrasion. However, this feature is by no means entirely reliable because certain pyroclastic deposits may consist entirely of lithic fragments (Fisher and Schminke, 1984), and if reworked, slightly rounded, and redeposited in various environments, they mimic epiclastic debris. All these problems can be in part resolved for older deposits by setting these pyroclastic rocks in their depositional, stratigraphic, and basinal context.

PYROCLASTIC ROCKS

Magmatic eruptions are caused by the rapid release of gases or frothing of the magma when it reaches the ground surface. They range from the extrusion of high-viscosity lavas ac-

companied by large-scale explosions producing ashes, to the fountaining of low-viscosity lavas with small-scale release of ashes (Heiken, 1972). Regardless of the composition of the magma, pyroclastic rocks fall into three major categories: *pyroclastic fall deposits, pyroclastic flow deposits,* and *pyroclastic surge deposits* (Wright et al., 1980). Petrographic emphasis in this chapter is on the first two categories.

Pyroclastic fall deposits are produced when material is explosively ejected from the volcanic vent into the atmosphere, producing an eruption column in the form of a convective plume. The expansion of the plume releases the pyroclasts, which fall back under the action of gravity and wind, in a downwind direction and at a variable distance from the vent. Fall deposits are generally well sorted, sometimes show internal stratification due to pulsations of eruptions, and display mantle bedding of rather uniform thickness over restricted areas while draping over most of the topography (Wright et al., 1980).

Pyroclastic flow deposits result from the lateral movement of pyroclasts as a gravity-controlled, hot, and high-concentration gas–solid dispersion, in some cases partly fluidized. Deposits are topographically controlled, filling valleys and depressions. They are massive and devoid of internal stratification, and their poor sorting is attributed to high particle concentration rather than to internal turbulence. The dominant flow mechanism is probably laminar (Wright et al., 1980).

Pyroclastic surge deposits involve the lateral movement of pyroclasts as expanded, turbulent, low-concentration gas–solid dispersion. Their deposits mantle topography, but tend to accumulate thickest in depressions. They typically show unidirectional sedimentary bedforms (cross-stratification, dunes, planar lamination, antidunes, pinch and swell structures, chute and pool structures). They are better sorted than pyroclastic flow deposits. A complete gradation certainly occurs between pyroclastic flows of high concentration and high density and pyroclastic surges of low concentration and low density. Nevertheless, the two types should be kept separate (Wright et al., 1980).

Major Constituents

Many features described in pyroclastic rocks also occur in hydroclastic ones, which are described in a subsequent section. Three main varieties of pyroclastic ejecta are distinguished on the basis of origin: (1) *juvenile particles,* also called essential particles, are derived directly from the erupting magma itself; they consist of dense or variably vesicular particles of chilled lava or of phenocrystals that grew in the magma before the eruption, and which are also called pyrogenic crystals; (2) *cognate particles* or accessory particles, which represent fragments from comagmatic volcanic rocks erupted previously by the same volcano; and (3) *accidental particles* derived from the volcano substratum, and which therefore can be of any composition, either volcanic, igneous, metamorphic, or sedimentary. Pyroclasts and unimodal well-sorted pyroclastic deposits are designated by numerous genetic and descriptive criteria not yet unified (Wright et al., 1980), but the fundamental parameter is average grain size (Schmid, 1981).

Bombs.   Bombs are grossly rounded pyroclasts with an average diameter exceeding 64 mm (fine bombs are defined as having a size between 64 and 256 mm; coarse bombs have a size > 256 mm). They consist almost exclusively of juvenile material; their shape (ellipsoidal, discoidal, or irregular) indicates that they were ejected in a partly molten state, shaped during flight, and further modified upon impact, if still plastic at that time. Bombs are mostly produced by explosions of magmas of silicic and intermediate composition; many of them have chilled external crusts with fracture patterns due to continued expansion of gases inside the still plastic cores. Basaltic bombs usually display little cracking, although their thin and iridescent glassy surfaces may show patterns of fine hairlike cracks due to stretching of the external surface over an internal plastic core upon landing.

Consolidated aggregates consisting of predominantly (more than 75% by volume) reciprocally welded and deformed bombs, with or without any interstitial matrix of lapilli or ashes, are designated as *agglomerates* (Schmid, 1981).

Blocks.   Blocks are angular to subangular pyroclastic fragments of cognate and accidental origin derived from the volcanic apparatus or its basement. Their average diameter exceeds 64 mm (fine blocks are defined as having a size between 64 and 256 mm; coarse blocks have a size > 256 mm). Blocks of juvenile origin can also develop from the dislocation of penecontemporaneous lava domes or from the rupture of indurated bombs upon impact.

Consolidated aggregates consisting predominantly of blocks (more than 75% by volume), with or without an interstitial matrix of lapilli or ashes, are designated as *pyroclastic breccias* (Schmid, 1981).

Lapilli.   Lapilli are angular to subrounded pyroclasts with an average diameter ranging between 2 and 64 mm. Subrounded lapilli are generally of juvenile origin, but cognate and accidental fragments generated by explosions may acquire a certain degree of roundness by successive phases of projection and falling back into the vent before final ejection (Fisher and Schminke, 1984).

Accretionary lapilli (Moore and Peck, 1962) are unusual types of lapilli-sized pellets of fine ash commonly showing a concentric internal structure formed by a discontinuous process of accretion around water droplets or coarser pyroclastic fragments. Some accretionary lapilli have an internal structure that is not layered, but shows a radial grading from coarse glass shards in the core to finer-grained and platy shards forming the external rim. The formation of accretionary lapilli has been observed in pyroclastic falls as well as in base- and ground-surge deposits and appears to result from random to regular moist aggregation of ash in eruption clouds undergoing strong turbulence or by the action of rain falling through dry eruption clouds (Fisher and Schminke, 1984).

Consolidated aggregates in which lapilli predominate (more than 75% by volume), with or without an interstitial matrix of ashes, are called *lapillistones, lapilli-tuffs* (Schmid, 1981), or more commonly *lithic tuffs* (**Plate 15.A, B**). Welded varieties of lapillistones are essentially devoid of interstitial matrix, but display internal flow structures within individual lapilli due to their reciprocal deformation upon welding.

Vitric Particles.   Vitric constituents are particles of a size smaller than 2 mm released when expanding gases in various types of magmas form a froth, which shatters upon instant release of the gases when reaching the ground surface. These vitric particles have been studied in detail petrographically for many years (Ross et al., 1928; Swineford and Frye, 1946) and more recently by SEM in order to attempt to establish relationships between type of eruption, magma composition, and morphology of the particles (Heiken, 1972, 1974; Heiken and Wohletz, 1985). These recent studies revealed the spectacular variety of shape of vitric particles in three dimensions. Under the petrographic microscope, it is possible to distinguish three end members among vitric constituents released by the vesiculation of silicic magma. *Platy shards* (**Plate 15.C**) represent glass walls separating two adjacent and large, flattened vesicles. *Cuspate or lunate shards* (**Plate 15.D**) are fragments of broken bubble walls, which are commonly Y-shaped in cross section and represent the remnants of three bubble junctions, or double-concave shards, which correspond to the wall between two adjoining bubbles. *Pumice shards* (**Plate 15.D**) are minute shreds of highly vesiculated glass that display a fibrous or cellular texture consisting of numerous minute, elongate, or circular cavities enclosed by their thin glass walls.

The shape of silicic shards unquestionably results from a number of factors, many of which are not yet understood; but Izett (1981) showed that pumice shards tend to form in relatively high-viscosity rhyolitic magmas with temperatures < 850°C, whereas cuspate and platy shards tend to develop from lower-viscosity rhyolitic magmas at temperatures > 850°C. SEM pictures of silicic glass shards from ancient marine and nonmarine environments show patterns of etchings and borings similar to those occurring on carbonate and silicate grains (Ross and Fisher, 1986). On the basis of morphology, and in particular systematic angles and annulations, the grooves were interpreted as the result of dissolution of the silica substrate by acid released by microorganisms (fungi) during their growth or while searching for organic matter in interstitial areas. The early formation of such groove patterns could be an important initial step in the development of soils in volcaniclastic terranes.

Glass shards of intermediate magmatic composition display generally irregular lumpy forms containing ovoid to spherical vesicles. Shards from mafic eruptions of Strombolian or Hawaiian type form when the rapid expansion of gas disrupts low-viscosity magma, producing lava fountains or coarse-ash-size droplets, lapilli, and block-size lava clots. Expansion of gas within individual glassy particles generates a variable degree of internal vesicularity, but generally does not cause its rupturing into smaller pieces. Shards consist of slightly vesicular spheres, ovoids, teardrops, dumbbells, and other rounded shapes, known as Pele's tears, as well as irregular fragments of these droplets. Highly drawn glass filaments, called Pele's hair, can reach up to 1 m in length, with diameters ranging from 15 to 500 μ.

When silicic pumice occurs in fragments coarser than the above-mentioned tiny shreds called pumice shards, they display even better the two textural end members. The fragments either show subspherical to spherical vesicles, essentially undeformed, or have a fibrous aspect, which results from highly elongate, tubular, and subparallel vesicles, indicating intense distortion and stretching, although the tiniest vesicles may remain spherical. Ewart (1963) related these two textural end members of silicic pumice to conditions of vapor pressure during eruption: high vapor pressure favoring the formation of pumice with spherical vesicles, and lower vapor pressure favoring that of fibrous pumice.

Pumice of intermediate magmatic composition shows ovoid to tubular vesicles, whereas mafic (basaltic) pumice generally contains small spheroidal vesicles. Most of these vesicles, except in the case of highly vesiculated froth, appear isolated within the glass; wherever vesicles touch adjacent ones, there is no indentation of the walls, but they connect with each other with little distortion. Consequently, the junctions between these vesicle intersections are cuspate, with points ranging from sharp to highly rounded. The roundness is apparently caused by liquid withdrawal due to surface tension (Fisher and Schminke, 1984).

Pyrogenic Minerals. These minerals consist of magmatic phenocrysts and microlites deposited as intact or broken crystals separated from the volcanic glass or partially enclosed in it. Size of crystals is not specified, but it is usually smaller than 2 mm. The fast cooling rate of pyroclastic material is a more favorable condition to observe arrested stages of crystallization, growth, and resorption of these minerals than their study in the lavas themselves. However, since pyrogenic minerals are sorted to a variable extent during eruptions and dispersal processes, their abundance and

size distribution are not, in most cases, a direct expression of the composition of the erupting magma (Fisher and Schminke, 1984).

Pyrogenic minerals undergo a relatively free growth in the magma; therefore, characteristic perfect crystals are the rule (**Plate 15.E**), such as bipyramidal quartz, pseudohexagonal biotite, high-temperature (disordered) subrectangular crystals of plagioclase, and potassic feldspars (sanidine)— often with spectacular zoning, overgrowths, and resorption figures—euhedral olivine, pyroxene, and amphibole. Rapid pressure decrease during explosive eruptions is responsible for widespread breakage of the various phenocrysts upon ejection or subsequent impact.

Minute features of the surface texture of shards and pyrogenic minerals as seen under the SEM can be used to discriminate the various processes of deposition of tuffs. For instance, Sheridan and Marshall (1983), in a study of glass shards, were able to distinguish between deposition from fallout processes, pyroclastic flow, and pyroclastic base surge.

Volcanic ash is an aggregate consisting of more than 75% by volume of various proportions of vitric particles, entire or broken pyrogenic minerals, and lithic particles of juvenile, cognate, or accidental origin. The average size is below 2 mm (with the exception of some pyrogenic minerals); coarse ash is defined as formed of particles ranging between 2 and $\frac{1}{16}$ mm, and fine ash (dust ash) consists of particles smaller than $\frac{1}{16}$ mm (Schmid, 1981). Consolidated equivalents of volcanic ash are designated as tuffs, with equivalent coarse and fine tuffs (dust tuffs).

Tuffs

Tuffs are subdivided in terms of composition (Cook, 1965) and according to a triangular diagram (**Fig. 4.1**) whose poles are glass-pumice (vitric tuffs), crystals and crystal fragments (crystal tuffs), and lithic fragments (lithic tuffs), giving the various possible combinations and related terms. The explosive generation of tuffs covers the entire spectrum of magmatic compositions, but with a stronger development in silicic and intermediate magmas. Genetic classification of tuffs is based on the inferred transport process (fallout, flow, surge, and others) or on the environment of deposition (subaerial, subaqueous), both combined with grain-size distributions expressed by various statistical parameters (Lirer and Vinci, 1991).

Subaerial Fallout Tuffs. These tuffs represent the portion of the subaerial fallout tephra recently discussed in detail by Fisher and Schminke (1984) that can be effectively investigated petrographically in thin sections. Subaerial fallout tuffs of silicic and intermediate composition are more common than mafic ones because of the high viscosity of the corresponding magmas, which generate more powerful ex-

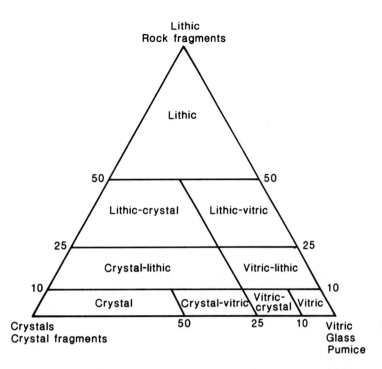

FIGURE 4.1 Textural classification of volcanic tuffs. From Cook (1965). Reprinted by permission of the Nevada Bureau of Mines and Geology.

plosions and greater discharge rates of particles. Furthermore, these tuffs are distributed over large areas across all depositional environments, and their widespread and thin layers are recognized as important time markers in stratigraphy because they represent extremely short periods of time. Tephrochronology dealing with the dating of such tuffs and tephrostratigraphy, which is concerned with their correlation, are new fields in active progress (Fisher and Schminke, 1984).

Transport processes involved in the formation of subaerial fallout tuffs are ballistic trajectory and turbulent suspension. The energy supplied to the fragments in provided initially by the eruption and its expanding cloud and later by low-altitude winds or jet streams. Particles with large settling velocities follow ballistic trajectories and fall close to the vent because they are weakly affected by the eruption expanding cloud or winds. Particles suspended by turbulence in the eruption cloud are next to fall according to their settling velocities as the energy of the cloud dissipates. Finally, particles with small settling velocities compared to wind force can remain in the air for long distances as "volcanic dust," which in some cases may circle the earth several times before final settling (Lamb, 1970). This eolian fractionation within an ash cloud, due to differences in settling velocities, is expressed by changes in the relative amounts of the major constituents, following an order that is frequently as follows: lapilli, denser pyrogenic crystals or fragments of crystals, less dense pumice fragments and lithics, and eventually the lightest pumice and glass shards. This fractionation also has a bulk chemical expression that is a tendency of the fallout tuff to become more silicic away from the source as glass shards become predominant over crystals and lithics (Fisher and Schminke, 1984).

However, the fall of particles according to settling velocities is far from being a simple process as outlined above. Increasing evidence shows that patterns of distribution are complicated by the premature fall of large amounts of very small particles as a result of several processes of agglutination by moisture in eruption clouds forming accretionary lapilli, by the flushing effect of raindrops, by the trapping of fine particles between coarser ash and lapilli, and even, as recently pointed out in the 1980 eruption of Mt. St. Helens (Sorem, 1982), by the formation of clusters of small particles through the combined effect of mechanical interlocking and electrostatic attraction. But, in spite of the above-mentioned complications, it is possible on a large scale within the circular or fan-shaped distribution areas of subaerial fallout tuffs to use, under the microscope, a number of parameters as tools for establishing dispersal directions. The maximum size of constituents, or apparent maximum size in thin sections, similar to the clasticity index of Carozzi (1989), when applied in particular to pumice fragments and lithic debris, yields data on inferred vent location, relative volcanic energy, and wind direction. All studies (Fisher and Schminke, 1984) show an exponential decrease in maximum diameter of both types of constituents away from the volcanic source, with some complications introduced if shapes of particles are more platy than equant. Median diameters have also been used for the same purpose as maximum diameters; they also decrease exponentially away from the volcanic source, but in detail the decrease is rarely systematic. For median diameters, complication is introduced when layers are relatively thick and display a vertical variation of grain size and composition. Grain-size distribution expressed by various sorting parameters is an important criterion for distinguishing fallout ash deposits from pyroclastic surge and flow deposits (Walker, 1971, 1984) and consequently equivalent tuffs. The degree of general sorting becomes better with increasing distance from the vent, but again in detail the pattern is complex (Fisher, 1964) for the same reasons discussed above, which control the settling of particles. Another factor influencing sorting is the relative increase in density of pumice with decreasing grain size due to the decrease of the ratio of vesicles to glass volume; therefore, at small sizes the densities of glass, lithic, and pyrogenic crystal fragments may become very similar (Fisher and Schminke, 1984).

Subaerial fallout tuffs display megascopically a well-developed bedding and consist often of alternating coarse- and fine-grained layers with gradational or sharp contacts. The characteristics of contacts depend on numerous factors, such as fluctuations in the energy of the eruption, its duration, variations in the discharge rate of the ash, variations in wind direction and strength, and finally the existence of intervals of weathering and winnowing between successive eruptions. Individual beds show generally few or no internal structures and simply differ from each other by color, which is related to differences in composition, grain size, and sorting of juvenile constituents. Graded bedding would be expected to be common in subaerial fallout tuffs, but many factors tend to reduce it, prevent it, or even generate reverse graded bedding, mainly in the proximal portions of the deposits. These factors are again fluctuations in eruption energy, turbulence within the eruptive cloud, effect of rain flushing, and changes of wind force and direction. However, in more distal parts of distribution areas where wind action predominates, settling velocities take over and normal graded bedding is common (Fisher and Schminke, 1984).

The major petrographic features of subaerial fallout tuffs can be appropriately discussed as a function of magmatic composition since they are essentially those of corresponding ashes (Heiken, 1972, 1974; Heiken and Wohletz, 1985).

Rhyolitic (**Plate 15.C, D**) and dacitic fallout tuffs are characterized by abundant vitroclastic material of colorless glass associated with intact and broken phenocrysts and microlites of quartz and sanidine, often embayed by resorption, biotite, and small amounts of augite and hornblende. Lithic grains of rhyolite, dacite, or xenolithic material, usually equant, are extremely rare. The vitric fragments are typically elongate to equant pumice fragments with thin vesicle walls and characteristic glass shards, curved, Y-shaped, or flat, representing the various types of vesicle junctions. General texture ranges from random in tuffs of proximal distribution areas, where small and large particles are often deposited together, either to isotropic, when relatively equant pumice fragments and shards predominate, or to anisotropic in distal

distribution areas, where elongate and platy glass shards tend to settle with their longest dimensions parallel to bedding.

SEM pictures and optical petrography of glassy fragments (Heiken, 1972, 1974; Heiken and Wohletz, 1985) show that vesicle walls are very smooth and that pumice fragments have very rough surfaces with broken vesicles. The petrographic study of rhyolitic and dacitic fallout tuffs raises the question that pertains actually to all fallout tuffs, that is, the origin of the finest interstitial material whose grain size eventually reaches the resolution power of the petrographic microscope. SEM studies indicate that this material, whenever unaltered by subsequent diagenesis, consists of extremely fine glass dust, the smallest products of the explosive process. This material often displays a light brown color, which is apparently due to the unequal refraction and internal reflection of light passing through finely aggregated fibers and particles. This brownish color of the interstitial finest groundmass is unrelated to brown staining by diagenetic iron oxides, which occurs in many rhyolitic and dacitic fallout tuffs.

Andesitic subaerial fallout tuffs tend to associate variable proportions of vitric, crystal, and lithic constituents. The vitric material consists of pale brown glass shards and pumice fragments with elongate microlites. Crystal fragments have shapes controlled by cleavage planes; they include plagioclase and pyroxene with abundant grains of opaque minerals. Lithic fragments, equant to rounded, are predominantly andesites of varying texture and andesitic tuffs at different stages of alteration to clay minerals and zeolites, together with some igneous and sedimentary xenoliths. SEM pictures and optical petrography of the glassy fragments (Heiken, 1972, 1974; Heiken and Wohletz, 1985) reveal equant to elongate pumice fragments whose shape is directly related to that of the vesicles they enclose; for instance, elongate pumice fragments contain elongate, ovoid, and tubular vesicles. Flat, pointed shards with smooth or conchoidal fracture surfaces represent broken vesicle walls. The surface of lithic fragments depends directly on the texture and fracture of the original rock type.

Basaltic fallout tuffs consist mostly of droplets and broken droplets of clear-brown basaltic glass (sideromelane) or black, submicrocrystalline basalt (tachylite) mixed with a variety of phenocrysts of plagioclase, amphibole, pyroxene, olivine, and opaque minerals (**Plate 15.A**). SEM pictures and optical petrography of the glassy fragments (Heiken, 1972, 1974; Heiken and Wohletz, 1985) show typical forms, such as irregular droplets with fluidal shapes (spheres, ovoids, dumbbells, teardrops), broken droplets, thin strands of glass, polygonal, latticelike networks of glass rods, and highly vesiculated glass froth. All droplets are highly vesicular; those with spheroidal and ovoidal shapes characterize very low viscosity lava, whereas irregularly elongate and broken ones indicate higher viscosities. The droplets often show smooth fluidal surfaces broken by smooth-lipped vesicle cavities because these vesicles broke open when the surface was still fluid, whereas cavities with angular rims indicate rupturing of vesicles after the surface of the grain was chilled.

Subaqueous Fallout Tuffs. Most of the emphasis on these tuffs resulted from early deep-sea exploration, which focused on Pleistocene to Pliocene oceanic sediments, followed by the Deep Sea Drilling Project, which yielded numerous results from as far back as the Jurassic. These new data were used to reconstruct the cyclic history of volcanism and its relationship with plate tectonics and climatic changes (Fisher and Schminke, 1984; Cas and Wright, 1987).

Marine and to a certain extent more restricted lacustrine fallout tuffs are derived from eruptions on land, initially dispersed by the wind and settled as fallout sheets in bodies of water and then modified to a variable extent by the action of subaqueous currents and by mixing with pelagic and hemipelagic sediments in the marine environment. The shape of the distribution areas is regular to irregular fan-shaped away from the vent. These sheets of tuffs have a general tendency to thin away from the source, but variations of thickness are frequent and probably result from subaqueous current action. They usually form single beds less than 50 cm thick, but thicker ones with multiple laminae represent successive fall units.

The absence of pervasive secondary alteration products in Cenozoic to Recent subaqueous fallout tuffs (Scheidegger et al., 1978) reveals a certain number of characteristic features. The lower contacts of the plane parallel beds are generally sharp and the upper ones gradational, mostly due to reworking by burrowers. The abundance of fresh pyroclastic material versus nonvolcanic components decreases upward, either gradually or irregularly, together with a well-developed normal graded bedding from crystal- and lithic-rich lower parts to shard-rich upper parts. This grading tends to be better than that of fallout tuffs on land (Ledbetter and Sparks, 1979). Grading results from the size fractionation due to the various settling velocities of the particles in bodies of water (Bramlette and Bradley, 1942); another aspect of this texture is an increase upward in SiO_2 content due to the predominance of shards over crystals. Inverse graded bedding occurs often when large pumice fragments are present because they contain many intact vesicles. Textural complications are introduced if the pyroclastic material is bioturbated, reworked, or involved in slumping processes and turbidity currents after its first settling episode (Fisher and Schminke, 1984). In general, sorting is good to poor depending on bioturbation and possible accidents of the depositional history. Size and sorting parameters display irregular variations from the source, but general grain size tends to decrease with distance of transport.

The composition of subaqueous fallout tuffs ranges from mafic to silicic, but silicic ash is more widespread.

Therefore, attempting to determine the chemical composition of these glass shards is most important for correlating marine and lacustrine ash beds across a basin and to their volcanic sources on land, as well as for establishing magmatic evolution through time.

A broad correlation has been known for a long time between color of glass shards and chemical composition, with colorless and more vesicular shards being silicic, whereas deep brown less vesicular to nonvesicular shards are mafic (Horn et al., 1969). Measurements of the refractive index of glass shards have been used to characterize them and to estimate their chemical composition, since an inverse relationship exists between refractive index and SiO_2 content (Mathews, 1951; Huber and Rinehart, 1966). However, glass shards hydrate with increasing age before altering to clay minerals and zeolites, and the refractive index decreases with increasing water content (Ross and Smith, 1955); moreover, other chemical parameters and the oxidation of iron also influence the refractive index. Modern techniques effectively used for characterizing shards rely mainly on microprobe analysis of individual shards (Scheidegger et al., 1978).

Investigations of pre-Cenozoic subaqueous fallout tuffs that are now bentonites (to be discussed petrographically in the section on diagenesis) display numerous similarities with their more recent equivalents. For instance, the bentonites cyclically interbedded among the various members of the Lower Cretaceous Mowry Formation in Wyoming are a perfect example of widespread thin time-line units, which also indicate an explosive volcanism-tectonic control of the associated sedimentation (Slaughter and Earley, 1965). These bentonites were dacite, latite, and quartz–latite ash beds ranging in thickness from a few centimeters to several meters. The thick beds consist of graded bedded multiple units that have sharp bottom contacts and show at the base sand-size pyrogenic crystals and glass shards grading upward into predominantly finer clay-size glassy material. Some multiple units contain thin individual beds, which have a remarkable lateral persistence. Slaughter and Earley (1965) described a bed, 3 cm thick, that was traced for a distance of 160 km and covers an area of 2,540 km^2. Each bentonite layer consists of elongate coalescing lobes or tongue-shaped bodies with a distribution pattern indicating multiple volcanic sources to the west. Decrease in percentage of phenocrysts, corresponding to a general decrease in grain size in the direction of transport, is characteristic of each bentonite layer. Transport was by high-velocity jet streams to the final site of deposition in predominantly quiet marine waters, followed in places by reworking and local transport by currents in the upper part of some layers.

In other instances of subaqueous fallout tuffs from the shallow marine Ordovician of Wales, loadcasts, pseudonodules, and slump features were described in the vicinity of the inferred source (Schiener, 1970) and attributed to seismic activity related to volcanism, which was responsible for vertical foundering of the ash into underlying muds by thixotropic reaction. Elsewhere, widespread ripple marks and cross-laminations indicate a polymodal distribution of relatively weak currents ascribed to wave action and tides in subtidal conditions. Similar occurrences of low-angle cross-stratification and channeling were also reported by Brenchley (1972) in Ordovician graded-bedded tuffs, indicating a definite marine influence.

In subaqueous fallout tuffs deposited in lacustrine environments (Hansen et al., 1963), wide lateral distribution and normal graded bedding are characteristic features, and inverse graded bedding occurs when pumice fragments are present. Depending upon the energy of the particular lacustrine environment, a variety of traction bedforms occur, such as laminations, scour and fill, and cross-bedding.

Pyroclastic Flow Tuffs. The mode of origin and transport of pyroclastic flows as well as their unusual and widespread deposits (sometimes called ignimbrites), which combine the features of pyroclastic rocks and lava flows, have been the subject of intense debate since their discovery a century ago (Fisher and Schminke, 1984).

Pyroclastic flow tuffs were thoroughly described and interpreted by many authors since the work of Gilbert (1938), who gave the first detailed petrographic description of these rocks and introduced the term of "welded tuffs" for welded parts of the Bishop Tuff of California. Major summaries indicating progress in the knowledge of these tuffs were given by Smith (1960a, b), Ross and Smith (1960), Peterson (1970), Chapin and Elston (1979), Heiken (1979), and Fisher and Heiken (1982).

Pyroclastic flow deposits consist of a variable proportion of glass shards, pumice fragments, pyrogenic crystals, and juvenile and xenolithic lithic fragments, depending on the composition of the magma and the origin of the flows. Pyroclastic flow tuffs, particularly emphasized in this discussion, rather than pyroclastic surge tuffs, consist by definition of more than 50% by volume of components within ash-size range (< 2 mm), which in some flows can make up the entire rock with a spectacular vitroclastic texture. Pumice fragments are always present. In some cases, they can be dominant and reach the size of lapilli. Commonly, pumice fragments display subparallel tubular vesicles, which give them a fibrous structure. More rarely, they show equidimensional spherical bubbles. Phenocrysts are the second major constituents, ranging from 1% to 60% by volume, and are essentially present in all flow tuffs. They are predominantly broken during eruption, transport, and subsequent compaction. This is shown, under the microscope, by incipiently separated quartz crystal fragments with glass-filled fractures, feldspar crystals dislocated along twinning planes, and crumpled pseudohexagonal biotite flakes. The abundance of crystals is often greater than in associated lava flows of the same composition. Sorting processes, which occur during eruption and emplacement of pyroclastic flow tuffs and associated fallout, result in concentration of crystals and lithics, depletion of fine-grained vitric shards, and reduction in size of pumice fragments by abrasion, which frequently makes them rounded (Fisher and Schminke, 1984).

Large-volume pyroclastic flow tuffs are dacitic to rhyolitic in composition; therefore, most of the phenocrysts consist of quartz, sanidine, and plagioclase, with minor amounts of amphibole, pyroxene, biotite, ilmenite, with zircon and sphene as accessory minerals. Quartz phenocrysts may be sharply euhedral with typical high-temperature hexagonal symmetry. More often they show rounding and embayments due to partial resorption by the melt. Sanidine and plagioclase crystals are also sharply euhedral, but commonly subhedral. Some show one side with a typical crystal face and irregular or fractured edges elsewhere; other crystals are rounded or irregularly embayed. In case of extreme fracturation, feldspar fragments can be sharply angular. Lithic fragments represent rarely more than 5% by volume; their origin is juvenile and from magmatic chamber and conduit walls; others are accidental and picked up by pyroclastic flows from their substratum, which can be of any type of igneous, sedimentary, metamorphic, or volcanic rocks. However, accidental lithics are commonly andesitic rocks, which express the well-known fact that andesitic eruptions normally precede explosive rhyolitic ones (Ross and Smith, 1960). In general, andesitic rock fragments are unaltered, with perfectly preserved augite, biotite, and hornblende crystals.

Pyroclastic flow tuffs are well stratified and have graded basal zones, discontinuous streaks of large fragments, and alternating coarse- to fine-grained layers. A single flow unit can show normal, inverse, symmetrical, or multiple grading with, simultaneously, pumice fragments inversely graded and lithic fragments normally graded as a result of their great respective differences in density. Maximum sizes of both pumice and lithics tend to decrease with distance from the vent.

The frequent superposition of pyroclastic flow tuff sheets is of great interest because their changes of mineralogical and chemical composition indicate zoned magmatic chambers and provide an understanding of the evolution in time of a given volcanic vent (Fisher and Schminke, 1984). Discrimination and correlation of pyroclastic flow tuff sheets is important in volcanic areas, which are structurally complex and lasted for a long period of time, because such sheets are the best markers of the regional stratigraphy. Hildreth and Mahood (1985) evaluated the problems and pitfalls encountered in the lithologic, magnetic, petrographic, chemical, and isotopic criteria used for correlating pyroclastic flow tuff sheets. Distinctive phenocrysts, pumice clasts, and lithic fragments are among the most reliable criteria, as are high-precision K–Ar ages and thermal remanent magnetization (TRM) directions in unaltered welded tuffs. Chemical correlation methods should rely on welded or unwelded pumice fragments, rather than on the vitroclastic matrix, which can be contaminated during emplacement. Compositional variations, which are characteristic of the numerous cooling units forming large sheets, require the analysis of many samples before trends based on phenocrysts, glass, or whole-rock chemical data can be used with confidence for correlations.

In pyroclastic flow tuffs, a fundamental stratigraphic and field distinction is to be made between flow units and cooling units (Smith, 1960b). A flow unit is depositional and represents a single pyroclastic flow deposited in one lobe; individual flow units vary in thickness and may follow each other in time very rapidly. Boundaries between flow units are shown by changes in grain size, composition, internal texture, concentration of pumice lapilli or blocks, and so on. When several hot flow units are rapidly superposed on top of each other, they may cool as a single *cooling unit*, which indicates that superposed flows cool as a unit with no sharp changes in temperature gradient. Cooling from depositional to ambient temperature may take years. Since the cooling process has fundamental implications on the texture of the ash-flow deposit, many of them are mapped as cooling units, although consisting of several flow units (Smith, 1960b).

Pyroclastic flow tuffs deposited at high temperature, although being particulate masses, are unusual heat-preserving systems. One of their most characteristic features is the plastic deformation and welding together of glass shards and pumice fragments, which represent a welding compaction preceded sometimes by a minor mechanical compaction in which elongate particles tend to become oriented horizontally (Sheridan and Ragan, 1977). Smith (1960b) defined within a cooling unit a systematic pattern of zones that differ in degree of welding, and therefore of density, produced by different cooling conditions (**Fig. 4.2**). They are the zones of no welding, partial welding, and dense welding. If dense welding is developed in the lower half of the cooling unit because it remained for the longest period of time at the maximum emplacement temperature, the zones of partial or no welding have upper and lower counterparts above and below the zone of dense welding. The degree of welding is controlled primarily by the viscosity of the particles, which in turn depends on the temperature, gas content, and rates of cooling and compaction, and not so much on the overlying load pressure, as suggested by the existence of welded fallout tuffs (Sparks and Wright, 1979; Wright, 1980).

The basal and upper parts of a cooling unit are zones of no welding, which are commonly the least spectacular parts of a pyroclastic flow tuff, but probably the most important because they are the only ones showing the original character of the erupted material. The absence of welding in the upper part is due to rapid cooling by heat convection and radiation in the atmosphere and in the lower part also to rapid cooling upon contact with the cold substratum.

The zones of partial welding (**Plate 15.F**) show all textures from incipient to advanced welding, and therefore a greater variety of textures than any other zones because of their wide range of porosity and degree of deformation of glassy particles. The transition from partial to dense welding is best shown by changes in the pumice fragments. Indeed, pumice fragments change by decreasing porosity and a general darkening of color until they become black and obsidianlike. The darkening of pumice fragments precedes that of the shard matrix. Ideally, the zone of dense welding is defined as that zone in which complete coalescence of glassy shards has eliminated all pore space. The end stage of welding (**Plate 16.A**) is a dense black glass called *vitrophyre* in which pumice fragments and glassy shards are megascopically indistinguishable. However, petrographically, it is still possible to see extremely elongate pumice fragments with their tubular vesicles reduced to undulose lines, which crinkle and crenulate around phenocrysts within the glassy mass, forming a streaky foliate structure called *eutaxitic structure.* Mass flowage may also occur with complete welding, as shown by the fact that flattened pumice fragments, which are normally disclike in the plane of flattening, may instead be elongate and show preferred orientation. However, this stretching indicates a mass movement that may reach from several inches to a few meters at the most.

Contemporaneous with welding processes or slightly later and under conditions of high initial temperatures and slow cooling rate, another series of processes develops in pyroclastic flow tuffs: partial to complete crystallization. Physically and chemically different conditions within the cooling unit give rise to various types and degrees of crystallization. Smith (1960a, b) defined the crystallization zones that are generally characteristic or dominant at different levels of a cooling zone and that are superimposed upon the various

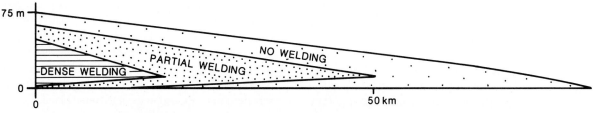

FIGURE 4.2 Welding zones in ashflow tuff cooling unit. Modified from Smith (1960b).

zones of welding. In order of frequency of occurrences, these types of crystallization are devitrification, vapor-phase crystallization, granophyric crystallization, and late fumarolic effects.

Devitrification, the most common crystallization process at relatively high temperature, tends to be more prevalent in welded portions, but can be present everywhere, even in the most porous zones. In rhyolitic welded tuffs, it consists of the simultaneous crystallization of cristobalite and alkali feldspars forming spherulitic and axiolitic intergrowths, with general preservation of the shape of pumice fragments and glass shards within which it occurs. Pumice is more susceptible than shards and develops coarser aggregates of crystals than shards in the partially welded to unwelded zones.

Vapor-phase crystallization can occur only in the presence of porosity, which allows movement of vapors and crystal growth in pore spaces. Its products are generally coarser grained than those of devitrification. In rhyolitic welded tuffs, it consists also of alkali feldspars, tridymite, and cristobalite in fibrous or granular intergrowths or in individual euhedral crystals. Vapor-phase crystallization typically makes welded tuffs light in color, and in certain types of pyroclastic flow tuffs called *sillar,* originally very porous, it is believed to be the major induration process rather than welding. Indeed, welding was not active because of the absence of loading pressure, although the temperature of pyroclastic flow tuffs could have been similar to that of welded tuffs. Vapor-phase crystallization destroys mafic phenocrysts such as biotite, hornblende, and orthopyroxenes; their former presence is shown by concentrations of opaque relicts and confirmed by their existence in the vitric welded zones. In some cases, on the contrary, there is generation of secondary mafic minerals, such as biotite, amphibole, fayalite, scapolite, and others, including magnetite.

Granophyric crystallization occurs only in fresh, unaltered rhyolitic welded tuffs thicker than 180 m. It is characterized by groundmass quartz, forming here directly rather than as a conversion of cristobalite, intergrown or associated with alkali feldspars and minor accessory minerals. These irregular and small aggregates have granophyric or micrographic textures similar to those shown by many slowly cooled rhyolitic lavas. However, many older pyroclastic flow tuffs, and eventually all of them, will probably contain a groundmass quartz, generally of secondary origin, resulting from the subsequent conversion of cristobalite and tridymite.

Late fumarolic effects are the fourth type of crystallization process. They are not easy to recognize in ancient pyroclastic flow tuffs. Although not fundamentally different from vapor-phase crystallization, their products are of another kind because of lower temperature and pressure. Fumarolic effects can cut across entire cooling units, as well as concentrate in upper, partially welded to nonwelded porous zones. They consist of metallic sublimates (realgar, orpiment) and metallic oxides. They are also responsible for the

local conversion of cristobalite and tridymite to quartz, chalcedony, and opal.

The final and characteristic textures, visible under the petrographic microscope, produced by welding combined with the various types of crystallization described above are very well displayed in rhyolitic welded tuffs of the Pleistocene Bishop Tuff in eastern California. They are briefly summarized here from the classical study of Gilbert (1938).

In the zone of no welding, near the surface of the cooling unit, the dustlike groundmass consists of the smallest glass particles, among which are scattered shards of clear and colorless glass, reaching 1 mm in length, usually free of microlites. Larger vitric fragments, which are pumiceous, display many tubeline vesicles, in some of which are concentrated minute grains of magnetite and hematite probably produced by fumarolic action. Many euhedral crystals of sanidine show faint soda-rich rims, 0.02 mm wide, interfingering irregularly with the cores and easily distinguished from them by their optical orientation, which is not quite parallel. The material forming the rims extends also along fractures within the crystals, indicating its late fumarolic origin. In some parts of the nonwelded zone, the vitroclastic texture can be partly obscured by crystallization of the fragments. The small angular shards are rimmed by a fringe of fibrous sanidine and tridymite growing inward normal to the edges of the shards. The center of each small shard remains a cavity, whereas in the larger shards and pumice lapilli, the fibrous fringe encloses a porous mass of tiny, irregular spherulites composed of radiating sanidine fibers and small blebs of tridymite. The original pumice structure is largely destroyed by crystallization, and only the general shape of the fragments is diagnostic.

Samples from the partial welding zone are similar to those of the no welding zone, except that pumice lapilli are slightly collapsed, and each shard and fragment of pumice has a thin, faintly birefringent rim. Because of the collapse, the tubular vesicles in the pumice are not so completely open, and there is a tendency for the major dimensions of each fragment to lie in the horizontal plane. The birefringent rims are mostly smaller than 0.01 to 0.02 mm in thickness, but they are obviously fibrous, with fibers growing inward normal to the edges of the shards. A faint crack may separate the glassy core of each fragment from the marginal fibrous rim and may be due to contraction during the change from glassy to crystalline state. Incipient crystallization may become complete and crystalline fragments appear, scattered at random through a weakly birefringent matrix of very fine constituents. The crystalline fragments can be recognized as original glass shards and pumice lapilli because their shapes and mutual relations are similar to those of the overlying zone of no welding, but they are slightly more flattened and more definitely aligned in the horizontal plane. The products of crystallization are fibers of potassic feldspar, probably sanidine, and crystals of tridymite. Clear rims occur on euhedral sanidine crystals, but are not so well developed as in the zone of no welding; some of these rims do not encase the sanidine completely; their outer margins are irregular and seem to fade out into the matrix. Oligoclase crystals may also be surrounded by rims of clear feldspar of considerably lower refractive index than the core. Finally, in larger lapilli of collapsed pumice, two zones of inward-growing fibers may occur, although the dark lines, representing closed and welded vesicles, are still visible.

In samples located just above the dense welded zone, crystallization features are the most advanced. Each fragment has crystallized as a distinct unit. In place of the central cavity present in shards and lapilli in overlying levels, there is commonly a xenomorphic aggregate of sanidine and tridymite. The larger fragments of collapsed pumice become finely fibrous along their margins; toward the centers the fibers become coarser and grade into aggregates that destroy completely the fine curving lines representing collapsed vesicles. Within some of the largest pumice fragments, fibrous crystals form a series of hemispherulites that radiate inward from a number of points spaced along the margins. Eventually, complete spherulites develop, with central portions consisting of granular aggregates of sanidine and tridymite. In the small shards, crystallization spreads from only one or two marginal centers, changing each shard into a plume of branching fibers.

In the zone of dense welding, the largest fragments, such as lapilli of collapsed pumice, are the first to flatten into dense glass, whereas the finest constituents are the last. This indicates that the larger fragments crystallized more readily in spite of consisting of bundles of welded glass threads, whereas the smallest particles are structureless. This process was probably induced by rising fumarolic gases acting on the surfaces of lapilli. In the dense welded zone, all pumice lapilli are collapsed and aligned in the horizontal plane; the shards are so flattened and molded to each other that they form similarly oriented linear elements.

Vitric constituents are not only flattened and reciprocally molded, but also tightly squeezed into the narrow spaces between adjacent phenocrysts and into embayments of resorbed quartz crystals. They eventually fan out when more space is available. The eutaxitic texture is generalized at all levels of observation.

HYDROCLASTIC ROCKS

The designation of hydroclastic rocks was generalized by Fisher and Schminke (1984) and set at the same level as "pyroclastic," so as to include all the products of hydro-explosions due to steam released from interaction between ascending magma and any kind of water. These phreatomagmatic eruptions occur in submarine or sublacustrine environments, as well as upon contact with groundwater, ice, and water-saturated sediments. The released steam generates the explosion, rather than the internal magmatic gases as in purely magmatic eruptions.

Major Constituents

Deposits from hydroclastic eruptions can be designated as hyaloclastites, using the term in a general sense as proposed by Fisher and Schminke (1984) for vitroclastic material produced by the interaction between water and hot magma or lava, with or without associated venting. Therefore, this generalization includes under the terms of hyaloclastites and hyalotuffs the products of conditions under which clastic products are formed by weakly explosive to nonexplosive spalling and granulation of magma in contact with water (Honnorez and Kirst, 1975). Most hyaloclastites and hyalotuffs are basaltic in composition, although intermediate to silicic types also occur (Heiken, 1972, 1974; Furnes et al., 1980; Stix, 1991).

Deposits from hydroclastic eruptions are generally fine grained and poorly sorted due to the abundant water (vapor) in the system, but lapillistones and breccias occur in the immediate vicinity of the vents. Blocks of breccias consist of irregularly shaped clasts of highly shattered sideromelane glass, with rare microlites and smooth fracture surfaces when the glass contracts upon rapid quenching, and granulates. Some blocks consist of tachylite and aphanitic basalt. Lapilli have a composition similar to blocks and are often of accretionary type. This particular structure expresses the abundance of water and steam in the eruption column, which favors aggregation of shards, but under completely different conditions than those regulating their formation in pyroclastic fallout deposits. Fisher and Schminke (1984) pointed out that hydroclastic lapilli may display vesicles in their outside layers and in their cores that do not occur in the other types. Armored lapilli consisting of a crystal or rock fragment as core, coated by concentric rings of fine to coarse ash shards, have been reported only in hydroclastic deposits (Waters and Fisher, 1971). Their origin is attributed again to the abundance of water, which generates widespread cohesive ash that adheres to coarser particles.

Under the SEM (Heiken, 1972, 1974; Heiken and Wohletz, 1985), vitric shards are typically blocky or crescent-shaped fragments of massive sideromelane glass with phenocrysts and microlites, but only few vesicles. They may show networks of incipient internal cracks due to rapid quenching. However, some shards may be variably vesicular when vesiculating basalt and quenching by water and steam are combined in the eruptive process.

Accidental clasts of all types, derived from the substratum, occur in hydroclastic deposits. They show no signs of thermal metamorphism, indicating the generally low temperature ($< 100°C$) of most hydroclastic eruptive systems, a situation also shown by the absence of welding of the ash deposits.

Basaltic Hyaloclastites

Numerous examples of deposits from phreatomagmatic eruptions have been described (Fisher and Schminke, 1984). A little-known, but very characteristic case is presented here from the Lagoa Feia Formation (Lower Cretaceous) of the Campos Basin, offshore Brazil, deposited in saline alkaline lakes oscillating between playa and pluvial conditions (Bertani and Carozzi, 1984, 1985). The sediments are an association of carbonates and clastics that filled a rift valley basin, with sublacustrine basaltic eruptions located along the major fault lines. These eruptions formed flat mound structures consisting of hyaloclastic breccias and several types of hyaloclastites grading upward and laterally into hyaloclastites with carbonate matrix, and eventually into very fine hyalotuffs, which occur as discontinuous distal layers and may contain shards also representing pyroclastic fallout constituents (**Fig. 4.3**).

Hyaloclastic breccias formed close to the vent and at the base of the volcanic system during each episode of subaqueous extrusion of basaltic magma. They consist of chunks of highly fractured sideromelane glass, sometimes plastically deformed and larger than 2 mm in size, set in a matrix of fine basaltic glass particles (**Plate 16.B**). These breccias grade laterally into lapillistones and finer hyaloclastites consisting of small globules, ranging between 1 and 2 mm in size, of sideromelane glass, often with ellipsoidal and elongate shapes due to plastic deformation while hot (**Plate 16.C, D**). Many of these globules are in fact accretionary lapilli and display internal concentric structures; they are set in an interstitial matrix of fine basaltic glass particles. An inverse correlation can be observed between size of glass globules and intensity of glass fracturation. The lack of fracturation in basaltic glass globules is favored by low glass viscosity and small globule size. Under these conditions, the internal stresses produced by different rates of cooling and contraction in the inner and outer parts of the globule are not sufficient to cause shattering (Carlisle, 1963; Bonatti, 1969). These two types of hyaloclastites grade upward and laterally into hyaloclastites with carbonate matrix (**Plate 16.D**). This matrix consists of micrite with ostracods and pelecypod bioclasts that, in contact with the plastically deformed glass

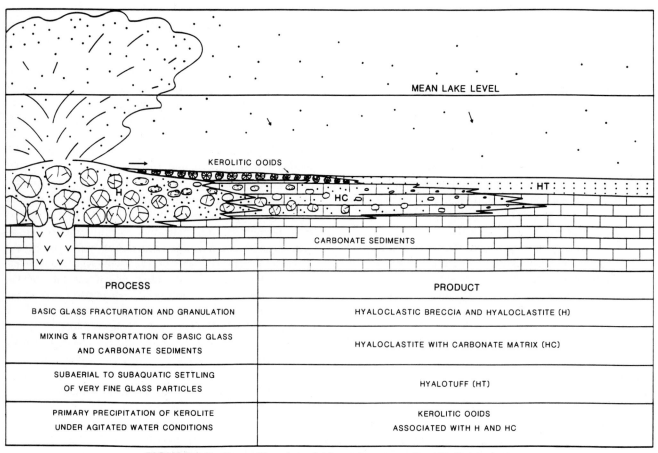

FIGURE 4.3 Depositional model for subaqueous basaltic hyaloclastites, Lagoa Feia Formation (Lower Cretaceous), Campos Basin, offshore Brazil. From Bertani and Carozzi (1985). Reprinted by permission of the *Journal of Petroleum Geology* and Scientific Press Ltd.

globules, was often thermally metamorphosed into a rim of short prismatic calcite crystals.

During the late stage of an episode of sublacustrine extrusion of basaltic magma, perfectly developed kerolitic ooids (kerolite is a fine-grained, hydrous, and disordered variety of talc) were formed on top of the flat hydroclastitic mound (**Plate 16.E**). Apparently, the presence of a heat source generated convective currents of heated lake water circulating through and over the mound. In this agitated environment, basaltic glass particles were reworked, altered to and surrounded by a thick coating of layers of tangentially oriented kerolite crystals. The following features indicate that the cortex of the kerolitic ooids was formed by direct chemical precipitation in a Mg-rich and agitated environment: constant tangential orientation of kerolite crystals; sharp contact with nuclei, which frequently display geometric forms indicating reworking of internally fractured glass globules; occurrence of broken and recoated ooids; and lack of impurities within the cortex such as glass and terrigenous and carbonate particles. Good sorting of the ooids and the presence of cross-stratification are additional evidence of an agitated environment.

Very fine basaltic glass particles ejected into the water column and into the air settled as hyalotuffs, parallel laminated with rare fragments of sideromelane glass and very rare grains of quartz and feldspars. They were deposited from near the vent, where they overlaid hyaloclastites, to more distant areas, where they covered lacustrine carbonates undisturbed by volcanic action.

TUFFITES

The general term tuffite (Schmid, 1981) encompasses rocks consisting of a mixture of pyroclasts (defined as individual crystals, crystal fragments, glass shards, and volcanic rock fragments generated by disruption due to direct volcanic action) and epiclasts in the proportion of < 75% pyroclasts and > 25% epiclasts by volume. Epiclasts, in turn, are defined as crystals, crystal fragments, and rock fragments released from any type of preexisting rocks (volcanic or nonvolcanic) by weathering or erosion and transported from their places of origin by gravity, air, water, or ice (Schmid, 1981).

This definition includes rocks generally called tuffaceous conglomerates, breccias, sandstones (**Plate 16.F**), siltstones, mudstones, and shales. It is obviously biased in favor of siliciclastic rocks and should be enlarged to include also the entire spectrum of carbonate rocks (**Plate 17.A**). In fact, tuffites express the direct contribution of fallout pyroclasts to a variety of marine and lacustrine sedimentary environments. Recognition of the various types of pyroclasts is relatively easy in sandstones and calcarenites, where bombs, lapilli, glass globules, pumice fragments, and large shards indicating contemporary silicic to mafic volcanism are scat-

tered respectively among quartz and silicate grains or bioclasts. Identification of the direct volcanic contribution becomes increasingly difficult with decreasing grain size of the pyroclasts, particularly when they occur in an argillaceous or micritic groundmass, to the extent that it often escapes routine petrographic examination. Assuming that minute debris of glass shards and small fragments of phenocrysts of quartz, sanidine, and other feldspars have not been highly modified by diagenetic alteration to clay minerals, calcite, or zeolites, it is still possible to recognize the pyroclastic contribution by the abundance of minute straight slivers of glass (extinguished under crossed nicols) and minute crystal fragments scattered at random as they fell into the fine argillaceous or carbonate matrix (**Plate 17.B, C**). The combination of random distribution of the pyroclastic particles and their identification as glass and crystals is the best tool for defining tuffaceous mudstones, shales, and fine-grained carbonates. The abundance of these types of rocks is another demonstration of the widespread and generally underestimated direct pyroclastic contribution to siliciclastic rocks and carbonates.

DIAGENESIS OF VOLCANICLASTIC ROCKS

Volcaniclastic rocks, characterized by their high content of glassy material, are highly susceptible to diagenetic modifications because high-temperature volcanic glass is thermodynamically unstable at temperature and pressure conditions of the earth's surface.

The process of alteration to clay minerals begins when glassy material is involved in processes of soil formation, where soluble organic compounds, derived from vegetation, control the release and transport of solutes by complexing Al and Fe and by generating low pH values (Antweiler and Drever, 1983).

In shallow marine conditions, glassy material is subject to hydration and hydrolysis into more stable polymorphs of silica and to replacement by clay minerals and zeolites. The simultaneous or successive precipitation of all these authigenic minerals into original interstitial spaces leads to rapid loss of initial porosity and very rapid induration.

Further diagenetic changes take place during burial, with increasing temperature and changes of composition of circulating fluids. Zeolites of the burial diagenetic type are formed and then gradually replaced by chlorite and epidote when metamorphic conditions prevail.

This analysis is mainly concerned with alteration at low temperatures and submarine conditions, where the nature of the alteration is mainly controlled by the type of volcanic glass. Consequently, the logical approach is to discuss first the alteration of basaltic glass and, second, the alteration of silicic glass with some of its diagenetic derivatives (bentonites), although both types of glass may share some processes and generate similar alteration products.

Alteration of Basaltic Glass

One of the most common products of submarine pore fluid alteration of sideromelane glass is a yellowish, bright orange to brownish material called palagonite (**Plate 17.D**). Its precise structural and chemical composition and its mode of formation are still controversial and have been debated for many years (Fisher and Schminke, 1984). Indeed, palagonite is not a specific mineral, but a mixture of minerals whose composition is a function of a certain number of physical and chemical conditions. Furthermore, palagonite undergoes a mineralogical evolution through geologic time. Recent studies by X-ray diffraction, microprobe, and experimental simulation have provided significant new data on the important redistribution of major and trace elements during this complex microsolution and precipitation process (Noack, 1981; Fisher and Schminke, 1984).

Microscopically, the formation of palagonite is clearly shown for individual glass fragments by a sharp boundary between the yellow to brown sideromelane glass and a light-yellow-orange red outside irregular rim, often color banded and variably birefringent. Palagonitization penetrates cracks in the fresh glass and may develop internally along vesicle walls. In some instances, several zones of palagonite may be observed from the sideromelane internal residue toward the outside. A zone of isotropic, gellike, often banded palagonite is followed by several zones of birefringent fibrous palagonite, more intensely colored and with segregation of zones of finely divided iron and titanium oxides (Fisher and Schminke, 1984).

According to Noack (1981), palagonitization of sideromelane is a complex hydrolysis process in contact with seawater, accompanied by the oxidation of Fe^{2+} to Fe^{3+} and relative gains and losses of other elements. A glass of tholeitic composition gradually changes into trioctahedral smectite and then into K- and Fe-rich dioctahedral smectite associated with the formation of phillipsite. It is impossible to identify the precise type of smectite because of the heterogeneity of the product and its low degree of crystallinity. However, in a few cases, nontronite and saponite were identified. In a simplified way, sideromelane evolves into thermodynamically more stable smectitic crystalline phases. This submarine alteration is a function of three major factors: original composition of the glass (tholeitic or alkaline basalt), temperature, and residence time of contact between glass and seawater. If the temperature is below 70°C, the glass–palagonite system gains K_2O and loses SiO_2, Al_2O_3, MgO, CaO, and Na_2O; a part of the K and Ca goes into the formation of phillipsite. If the temperature is above 70°C, there is loss of potassium and enrichment in magnesium, and rates of reaction are also increased. Generally, calcite and phillipsite fill fissures and pores in the palagonite, but they

do not belong to palagonite since they do not replace the sideromelane glass, although they derive a portion of their constituents from it. Besides the widespread occurrence of phillipsite, other zeolites have been reported: chabazite, natrolite, and analcite. Whenever several zeolites occur, the order of crystallization begins with potassic zeolite (phillipsite) and terminates with sodic zeolite (analcite) through intermediate stages of calco–sodic zeolites. The generation of these associated minerals tends to block the circulation of interstitial water and hence palagonitization itself, allowing the preservation of fresh sideromelane glass in very old lavas (Noack, 1981).

Therefore, palagonitization can be an intermediate stage of the widespread submarine alteration of sideromelane glass to a variety of zeolites. Zeolitization of basaltic glass is a rapid process in terms of marine diagenesis. This process apparently increases as a function of time, but in a highly irregular fashion. Eventually, zeolites disappear in older Mesozoic and Paleozoic sequences, when they in turn alter to authigenic feldspars (Hay, 1978).

Another alteration path of sideromelane glass under shallow burial conditions is into a complex mixture of smectite, mixed-layer Ca–smectite/chlorite, and chlorite as a final product (Surdam, 1973). Further burial, corresponding to low-grade metamorphism, results in the development of zeolites of the burial diagenetic type (laumontite, analcite, heulandite, clinoptilolite, and others) associated with prehnite, pumpellyite, epidote, and albite. These minerals occur often in distinct zones with increasing metamorphism (Hay, 1978).

In the above-described basaltic hyaloclastites formed in the saline alkaline lakes of the rift valley phase in the Lagoa Feia Formation of the Campos Basin (Bertani and Carozzi, 1984, 1985), several complex diagenetic stages were recognized. In the initial lava–water–sediment interaction, controlled by high temperatures, the sideromelane glass underwent intense globulization and fracturation, followed by palagonitization of glass globules along boundaries and fractures. Thermal neomorphism of the micrite matrix was shown by the formation of a rim of fibrous to prismatic calcite, with crystals oriented normally to the surface of the glass globules. Aquathermal diagenesis followed under the influence of heated lake water circulation. The basaltic glass globules were extensively replaced by kerolite (which also formed ooids by itself), zeolites, dolomite, calcite, and quartz. Spherulites of zeolites were then replaced by calcite, and dolomitization occurred along the outer rims of the glass globules, followed by calcitization of the unreplaced glass and feldspar phenocrysts. Coarse sparitic calcite cement filled fractures within glass globules as well as interparticle spaces between kerolitic ooids. A final aquathermal diagenetic event consisted of extensive development of microcrystalline quartz replacing calcite, which had previously replaced glass globules and kerolitic ooids.

Lacustrine phreatic diagenesis by normal lake water

consisted of intense alteration of basaltic glass to trioctahedral smectite and sepiolite. Finally, under freshwater vadose conditions, secondary porosity developed through partial or complete dissolution of glass globules and pelecypod bioclasts. But this type of porosity was subsequently occluded by sparitic calcite cement in the freshwater phreatic environment.

Alteration of Silicic Glass

The alteration of silicic glass seems to involve also, as in the case of sideromelane, an initial stage of hydration and alkali exchange controlled by diffusion processes and with minor chemical changes. However, the initial phase is followed by destruction of the silicic glass and precipitation of secondary minerals in the pores created by dissolution of the glass. It appears as if the initial high Na + K content of silicic glass and the early dissolution of these elements in the circulating pore fluids during hydrolysis reactions generates high pH conditions that promote rapid silica dissolution.

The initial stage of hydration changes obsidian into high-water-content perlite, with loss of Na and enrichment in K. Perlite (**Plate 17.E**) is characterized by curved concentric "perlitic" sets of cracks that give an onion-skin appearance to the silicic glass and are produced by an expansion of the glass during hydration (Friedman and Smith, 1958). The boundary between obsidian and perlite is generally sharp, with a strain birefringence developing in the latter. The processes by which perlite is changed in turn into material consisting mainly of smectites and zeolites are complex (Noh and Boles, 1989). Smectite formation seems to begin along the outer leached skins of glass shards or the fractures of perlite as detected by electron microprobe studies (Jesek and

Noble, 1978); then partial to complete dissolution of the remaining glass occurs, followed by precipitation of zeolites and other authigenic minerals, both inside the original porosity and the newly formed dissolution cavities.

Although zeolites formed from alteration of silicic glass in a saline alkaline environment are economically very important and those found in marine conditions have been extensively studied (Fisher and Schminke, 1984), this petrographic presentation focuses on the extensive layers of relatively pure silicic ash, entirely altered to smectite, which are called *bentonites,* and on those altered to kaolinite called *tonsteins.* Both types of deposits have also been the subject of many studies, in particular bentonites because of their economic importance (Grim and Güven, 1978).

Bentonites

The terms bentonite and tonstein have been used through the years in a variety of ways, and it seems reasonable to follow the suggestion of Fisher and Schminke (1984) to eliminate the term tonstein and to use the word bentonite in a broad sense, prefixed by its dominant clay mineral species, such as smectite, (s)-bentonite; illite, (i)-bentonite; or kaolinite, (k)-bentonite. It should be pointed out, to avoid nomenclature confusion, that the terms K-bentonite and metabentonite have been used previously and are still used for illite-rich bentonites (Weaver, 1953; Grim and Güven, 1978).

There is overwhelming evidence that most bentonites and kaolinite-rich types (tonsteins) were generated by diagenetic alteration of silicic vitric fallout ash. Bentonites form thin layers (generally less than 10 cm) consisting of several tongue- or lobe-shaped bodies typical of subaerial fallout ash (**Fig. 4.4**), with a lateral extent of hundreds of kilometers,

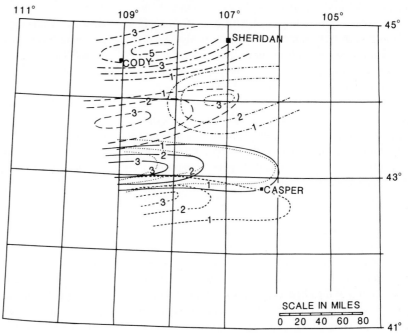

FIGURE 4.4 Schematic isopach maps of single bentonite units in a portion of the Lower Cretaceous Mowry Shale bentonites in Wyoming. From Slaughter and Earley (1965).

and with sharp upper and lower contacts with enclosing shallow-water marine and nonmarine sedimentary rocks. Consequently, they are isochronous and datable marker horizons of great importance for long-distance correlation (Person, 1982; Baadsgaard and Lerbekmo, 1982). Petrographic and SEM studies (Heiken and Wohletz, 1985) reveal relicts of pumice fragments and shards forming the typical residual vitroclastic texture of silicic tuffs (**Plates 17.F, 18.A**). Some bentonites are interlayered with fresh silicic vitric tuffs and also grade laterally into them.

The mineralogy of the sand-size fraction of bentonites is characterized by euhedral and high-temperature phases of pyrogenic minerals (Slaughter and Earley, 1965). Using the lower Cretaceous Mowry bentonites of Wyoming as an example, the sand-size fraction consists of orthoclase, sanidine, and microcline as representative species of alkali feldspars, sanidine being the most abundant, generally euhedral and often with resorption embayments. Among plagioclases, oligoclase and andesine predominate with columnar to equant shapes; zoned crystals are common with internal parts more calcic than external ones. Quartz is usually of two types: (1) anhedral and clear with inclusions of apatite and rutile, and (2) euhedral and clear with high-temperature forms, devoid of strain shadows, at times subrounded to rounded. The high-temperature form is always rounded as a result of in-melt solution. Biotite is very common and invariably euhedral or subhedral, showing well-developed pseudohexagonal flakes. It is deep brown with inclusions of rutile and apatite. Other minerals are magnetite, zircon, apatite, sphene, hornblende, titanite, and augite (Weaver, 1963). Clearly, all these petrographic and mineralogical features are different from those of the detrital mineral suites of enclosing rocks.

Smectite belonging to the montmorillonite–beidellite series is predominant in geologically young bentonites. With increasing diagenesis, mixed-layer illite/smectite forms and becomes more illite rich; sometimes chlorite is also generated. Cristobalite and tridymite occur in some bentonites, as well as analcite, which presents a peculiar aspect of microcrystalline to macrocrystalline masses of anhedral crystals, silt to sand sized and displaying perfectly the shape of replaced glass shards (Slaughter and Earley, 1965). The original composition of the altered glass, whenever it survives diagenesis, indicates a rhyolitic to dacitic composition, in agreement with the observed residual vitroclastic texture.

Modern chemical techniques that utilize chemical discriminants, that is, elements that remain immobile during the alteration process, have been used with success to infer the original composition of volcanic ashes, and hence to characterize individual bentonites, in particular, plots of Zr to TiO_2 ratios versus Nb to Y ratios (Huff and Türkmenoglu, 1981; Spears and Duff, 1984; Teale and Spears, 1986). Possible temporal evolution of the chemical composition of the original magmas can likewise be detected. Further sophistication in correlating and characterizing bentonite beds can be done by trace element fingerprinting. Huff (1983) analyzed six K-bentonites in the middle Ordovician of southeastern Ohio and northern Kentucky for 3 major and 22 trace elements by instrumental neutron activation analysis and X-ray fluorescence. The results were submitted to stepwise discriminant analysis. Eighteen elements were sufficient to establish 100% classification in a group matrix of the six bentonites with the principal assignment contributions coming from Th, Ta, Rb, Cr, Fe, Ga, Yb, and K. Trace element fingerprinting is an excellent technique for K-bentonite identification and correlation on a regional scale, and discriminant analysis provides a powerful tool for achieving these goals. A similar approach was used by Kolata et al. (1987) for middle Ordovician K-bentonites of the upper Mississippi Valley and of the Southern Appalachians (Huff and Kolata, 1990). These authors estimated that such bentonites covered a minimum area of 600,000 km^2 and represented at least 1,122 km^3 of precompaction volcanic ash, ranking them among the largest fallout tuffs in the geological record.

The origin of K-bentonites, which differ from other bentonites by containing mixed-layer illite/smectite as the dominant clay mineral instead of smectite, has been attributed to various mechanisms, but increasing evidence indicates that the introduction of potassium is related to burial metamorphic conditions (Nadeau and Reynolds, 1981). Altaner et al. (1984) added further support along these lines in a study of K-bentonites from the Marias River Formation (Upper Cretaceous) in the Montana disturbed belt. A 2.5-m-thick K-bentonite layer was found to be zoned mineralogically and chemically, the upper and lower contacts of the bed being more illitic and K-rich than the middle of the bentonite. K–Ar dates of illite/smectite extracted from the bed were also zoned, yielding older ages at the contacts than at the center. These authors presented a model for the formation of K-bentonites enclosed by K-rich shales in which K-bentonites were formed from smectite bentonites during a thermal event (T = 100° to 200°C) when buried under Laramide thrust sheets. Potassium was assumed to have been derived from long-range mineralogic breakdown, probably of mica and K-feldspars, in the shale host rock and to have migrated inward from the margins of the bentonite by diffusion through pore fluids.

Kaolinite-rich bentonites (tonsteins) are typically developed in the European Carboniferous as thin, widespread time markers among coal beds. They have been extensively studied and considered as bentonites having undergone unusual conditions of diagenesis in the acid and reducing environment of coal swamps, where low salinity, abundant organic matter, and the presence of sulfides led to large-scale crystallization of authigenic kaolinite (Price and Duff, 1969; Spears, 1970; Burger, 1980; Fisher and Schminke, 1984).

Petrographically, kaolinite has at least four major habits: a cryptocrystalline groundmass intimately associated with organic matter, an irregular pelletoidal texture (*Grau-*

pentonstein) consisting of ovoid to flattened pellets indicating intraformational reworking (**Plate 18.B**), and a spectacular development of reciprocally deformed large vermicular crystals of kaolinite (*Kristalltonstein*) set in the same dark cryptocrystalline matrix (**Plate 18.C**). Where kaolinite crystals are abundant, they are closely packed against each other and very frequently cleaved in place as if the sediment has been somewhat reworked just after the growth of the crystals. This condition is confirmed by the general orientation of the vermicular individuals, parallel or slightly imbricated with respect to bedding. The same observation can be made for the fourth habit, which consists of pseudomorphs of kaolinite after muscovite, biotite, feldspars, and particularly sanidine. Among the interstitial matrix of the four types of habits are relics of vitroclastic texture, biotite flakes, angular fragments of quartz and feldspars phenocrysts, and even relics of accretionary lapilli (Bohor and Triplehorn, 1984).

Attempts have also been made to chemically identify the parent material of kaolinite-rich bentonites using mineralogical composition and trace elements that remain immobile during diagenesis (Zielinski, 1985). Spears and Kanaris-Sotiriou (1979), in a study of British and European kaolinite-rich bentonites, applied discriminant function analysis using the variables Ti to Al, Cr to Al, Zr to Al, and Ni to Al and were able to classify the kaolinite-rich bentonites according to original composition. Although most of these kaolinite-rich bentonites were derived from silicic volcanic ash and shared a common distant source with French and German ones, a few British ones were derived from mafic volcanic ash mixed with variable amounts of sediments and interpreted as having originated from local eruptions.

RESERVOIR PROPERTIES OF VOLCANICLASTIC ROCKS

Theoretically, unaltered agglomerates, pyroclastic breccias, lapillistones, and subaerial fallout tuffs would have a primary intergranular porosity as high as 30% to 40%. However, widespread diagenesis related to the instability of quenched silicic and mafic glass, expressed by numerous processes such as hydration, devitrification, extensive replacement by clay minerals, silica, calcite, and zeolites, leads to rapid loss of initial porosity. Furthermore, this loss is irregularly distributed, destroying permeability as well as the possible connection between areas that might have been spared. However, intense hydrothermal alteration of crystal–lithic andesitic tuffs can take place, as described in eastern Georgia, Soviet Union (Vernik, 1990), in which laumontite occurs in intergranular spaces as a replacement product of plagioclase crystals and lithic fragments and as vein filling. This metasomatic process results in completely reorganized rocks, with laumontite content of 50% to 80% of the matrix,

and displaying appreciable effective porosity and permeability.

Secondary porosity in well-indurated tuffs by burial dissolution of the diagenetic replacement minerals, particularly zeolites and calcite, combined with dissolution of fracture fillings can also be considered. Interesting examples of burial diagenesis are known in the upper Cretaceous of the Rio Grande do Norte Basin, offshore Brazil: amygdaloidal basaltic lava flows and breccias interbedded, with baked contacts, in deep-water pelagic micrite, which were entirely calcified to the extent of simulating high intertidal to supratidal fenestral micrites. Subsequent burial dissolution of feldspar crystals, calcite, and zeolite filling of the amygdules, accompanied by intense fracturation, led to unusual reservoirs characterized by a combination of vuggy and fracture porosity. Such conditions could occur also with calcified and zeolitized hyaloclastites.

TYPICAL EXAMPLES

Plate 15.A. Vesicular basaltic lithic tuff consisting essentially of angular fragments (lapilli) of fresh tan basaltic glass with a variable amount of vesicles and phenocrysts of labradorite, amphiboles, and pyroxenes, many of which were replaced by calcite mosaic, associated with smaller lithoclasts of deeply altered mafic volcanic tuffs, tuffaceous siltstones, and porphyritic lavas. The dark and partially devitrified vitroclastic matrix contains small debris of phenocrysts of feldspars and abundant zeolitic alteration products. Cenozoic, Rocher Corneille, Puy-en-Velais, France.

Plate 15.B. Trachytic lithic tuff consisting of rounded dark fragments (lapilli) of trachytic pumiceous material with empty vesicles and cavities corresponding to dissolved feldspar phenocrysts. The vitroclastic matrix, deeply altered to clay minerals, contains smaller debris of trachytic material and broken phenocrysts of hornblende. Cenozoic, Nemi, Latina, Italy.

Plate 15.C. Rhyolitic vitric tuff consisting entirely of subrectangular to irregularly shaped fragments of transparent, incipiently devitrified silicic glass produced by the explosive fragmentation of flattened vesicular pumice. Debris of phenocrysts of quartz and sanidine, weathered pyroxene and amphibole, and biotite flakes are scattered between the glass fragments, which are poorly consolidated by incipient pressure welding. Oligocene, Black Hills, South Dakota, U.S.A.

Plate 15.D. Rhyolitic vitric tuff with typical vitroclastic texture in which large shards made of interbubble walls of vesicular pumice and pumice shards are set in a brown groundmass of finer-grained glassy fragments and abundant opaque iron oxide flecks. Scattered in the groundmass are large or broken phenocrysts of resorbed quartz and sanidine, biotite books, and rare lithoclasts of other types of dark altered vitric tuffs (upper-left and lower-right corners of picture). Bishop Tuff, Pleistocene, Morrison Quadrangle, Eastern California, U.S.A.

Plate 15.E.   Crystal tuff consisting of an assemblage of large and resorbed phenocrysts of fresh zoned plagioclases, fresh and deeply altered potassic feldspars, chloritized biotite and hornblende associated with rare lithoclasts of altered porphyritic volcanics. Interstitial devitrified vitroclastic groundmass is heavily altered to opaque iron oxides, chlorite, and sericite and contains smaller debris of phenocrysts of quartz and feldspars. Pinde Formation, Pliocene, Lafayette, California, U.S.A.

Plate 15.F.   Rhyolitic welded tuff, from middle part of cooling unit, consisting of appreciably welded devitrified glass shards and fragments of collapsed pumice with small amount of interstitial brown devitrified glass dust. Large phenocrysts of resorbed quartz and sanidine, deeply indented, are scattered along the welded glassy material. Bishop Tuff, Morrison Quadrangle, Eastern California, U.S.A.

Plate 16.A.   Highly welded rhyolitic tuff, from lower part of cooling unit, consisting of completely flattened, distorted, and strongly welded glass shards and fragments of collapsed pumice. The well-developed generalized fluidal texture is combined with a small amount of brown, interstitial, devitrified glass dust and rare, small sanidine phenocrysts. Bishop Tuff, Pleistocene, Morrison Quadrangle, Eastern California, U.S.A.

Plate 16.B.   Basaltic blocky hyaloclastite consisting of subangular fragments of basaltic glass altered to trioctahedral smectites and separated by calcite-filled cracks. Interstitial matrix if formed by fine glass material altered to clay minerals and partially replaced by calcite. Lagoa Feia Formation, Lower Cretaceous, Campos Basin, offshore Brazil.

Plate 16.C.   Basaltic globular hyaloclastite composed of small basaltic glass globules displaying internal concentric structure. The glass particles have ellipsoidal to elongated shapes due to plastic deformation when still hot. Silicification is extensive, but some glass globules still remain unaltered. Interstitial matrix consists of fine basaltic glass replaced by microcrystalline quartz. Lagoa Feia Formation, Lower Cretaceous, Campos Basin, offshore Brazil.

Plate 16.D.   Basaltic globular hyaloclastite composed of large basaltic glass globules plastically deformed and altered to sepiolite and kerolite, whereas enclosed plagioclase phenocrysts are calcitized (lower-right corner of picture). Carbonate matrix consists of argillaceous calcite, which, in places, has thermally neomorphosed along the margins of glass globules into rims of short, prismatic, clear calcite crystals. Lagoa Feia Formation, Lower Cretaceous, Campos Basin, offshore Brazil.

Plate 16.E.   Basaltic hyaloclastite consisting entirely of kerolitic ooids cemented by interstitial microcrystalline

quartz. Thick cortex of ooids is composed of tangentially oriented fibrous kerolite, and small nuclei consist of basaltic glass particles altered to clay minerals. partial silicification of some ooids also occurs. Lagoa Feia Formation, Lower Cretaceous, Campos Basin, offshore Brazil.

Plate 16.F.   Tuffaceous sandstone consisting of a poorly sorted and grain-supported framework of lithoclasts of porphyritic rhyolites, coarse rhyolitic tuffs with broken phenocrysts of quartz and sanidine (lower-left corner and right of picture), and fine vitroclastic rhyolitic tuffs. Interstitial glassy matrix, incipiently devitrified and zeolitized, contains smaller fragments of quartz and sanidine phenocrysts and concentrations of opaque iron oxides. Oligo-Miocene, Caliente, Nevada, U.S.A.

Plate 17.A.   Tuffaceous calcisiltite with bioclasts of echinoids, gastropods, and pelecypods neomorphosed to pseudomicrosparite, associated with abundant lithic fragments and globules of rhyolitic glass, which is fresh or altered to celadonite. Broken phenocrysts of quartz and pyrite flecks are scattered throughout the matrix. Chachao Formation, Lower Cretaceous, Malargüe, Mendoza Basin, Argentina.

Plate 17.B.   Tuffaceous and argillaceous calcilutite in which are randomly scattered entire or broken phenocrysts of quartz and potassic feldspars and rodlike, cuspate, or triangular shards of fresh silicic volcanic glass. Patches of coarse, irregular calcite mosaic represent filled molds of plant stems. Rosebud Formation, Lower Miocene, South Dakota, U.S.A.

Plate 17.C.   Clotted tuffaceous and argillaceous calcisiltite with abundant scattered minute and unsorted fragments of phenocrysts of quartz and potassic feldspars, rare chloritized biotite, and grains of opaque iron oxides. Morrison Formation, Upper Jurassic, Spearfish, South Dakota, U.S.A.

Plate 17.D.   Palagonitized basaltic hyaloclastite consisting of a framework of irregularly shaped fragments of various types of sideromelane glass with olivine phenocrysts. All fragments display a wide range of orange to brown colors characteristic of submarine palagonite alteration. Interstitial cement is coarse calcite mosaic. Holocene, Diamond Head, Oahu, Hawaii, U.S.A.

Plate 17.E.   Perlitic obsidian consisting of a dark tan silicic glass groundmass with typical interfering centers of multiple concentric cracks with onion-skin appearance. Large phenocrysts of sanidine with marginal and internal resorbed zones are scattered throughout the groundmass. Cenozoic, Lipari Islands, Italy.

Plate 17.F.   Coarse tuffaceous bentonite consisting of a vitroclastic groundmass entirely replaced by smectite in which are scattered many shards of fresh to highly replaced silicic glass surrounded by rims of matrix with pectinate structure. Abundant phenocrysts of quartz and sanidine, frequent irregularly shaped lithoclasts of dark fine tuffs, and aggregates of opaque iron oxides are scattered throughout the groundmass. Chira Formation, Upper Eocene, Vichayal, Sechura, Peru.

Plate 18.A. Fine tuffaceous bentonite consisting of a vitroclastic groundmass with abundant relicts of silicic glass shards and small pumice fragments entirely replaced by smectite. Scattered minute debris of quartz and sanidine phenocrysts, irregularly shaped hazy lithoclasts of dark fine tuffs, flecks of opaque iron oxides, and rare isolated filled vesicles are scattered throughout the smectite groundmass. Chira Formation, Upper Eocene, Vichayal, Sechura, Peru.

Plate 18.B. *Graupentonstein* consisting of clear microcrystalline to slightly vermicular kaolinite pellets of variable size and incipiently flattened parallel to bedding, associated with a few larger vermicular crystals of kaolinite and broken phenocrysts of quartz and sanidine. Interstitial

groundmass is microcrystalline kaolinite, stained orange-brown by organic material and displaying scattered minute grains of quartz and sanidine, muscovite flakes, and streaks of carbonaceous material. Wesphalian C, Couche Hanas, Charbonnages Unis, Mons, France.

Plate 18.C. *Krystalltonstein* consisting of an irregular framework of large vermicular crystals of kaolinite, often distorted and cleaved by appreciable pressure solution against adjacent large and angular fragments of phenocrysts of quartz and sanidine. Interstitial groundmass is dark-brown intimate association of microcrystalline kaolinite and organic material, with scattered small debris of quartz and sanidine phenocrysts, rare flakes of biotite and muscovite, and irregular concentrations of carbonaceous material. Wesphalian C, Veine à Terre, Charbonnages Unis, Mons, France.

REFERENCES

ALTANER, S.P., HOWER, J., WHITNEY, G., and ARONSON, J. L., 1984. Model for K-bentonite formation: evidence from zoned K-bentonites in the disturbed belt, Montana. *Geology,* 12, 412–415.

ANTWEILER, R. C., and DREVER, J. I., 1983. The weathering of a late Tertiary volcanic ash: importance of organic solutes. *Geochim. Cosmochim. Acta,* 47, 623–629.

BAADSGAARD, H., and LERBEKMO, J. F., 1982. The dating of bentonite beds. In G.S. Odin (ed.), *Numerical Dating in Stratigraphy,* Part One. John Wiley & Sons, New York, pp. 423–440.

BERTANI, R.T., and CAROZZI, A. V., 1984. Microfacies, depositional models, and diagenesis of Lagoa Feia Formation (Lower Cretaceous), Campos Basin, offshore Brazil. *Petrobrás, Cenpes, Ciência Técnica Petróleo,* 14, 104 pp.

———, and ———, 1985. Lagoa Feia Formation (Lower Cretaceous), Campos Basin, offshore Brazil: rift-valley stage lacustrine carbonate reservoirs. *J. Petroleum Geol.,* part I, 8, 37–58, part II, 8, 199–220.

BOHOR, B. F., and TRIPLEHORN, D. M., 1984. Accretionary lapilli in altered tuffs associated with coal beds. *J. Sed. Petrology,* 54, 317–325.

BONATTI, E., 1969. Mechanisms of deep-sea volcanism in the South Pacific. In P. Abelson (ed.), *Researches in Geochemistry,* J. Wiley & Sons, New York, pp. 453–491.

BRAMLETTE, M. N., and BRADLEY, W. H., 1942. Geology and biology of North Atlantic deep-sea cores between Newfoundland and Ireland. Part 1. Lithology and geologic interpretations. *U.S. Geol. Survey Prof. Paper 196 A,* pp. 1–34.

BRENCHLEY, P. J., 1972. The Cwm Clwyd Tuff, North Wales: a paleogeographical interpretation of some Ordovician ash-shower deposits. *Yorkshire Geol. Soc. Proc.,* 39, 199–224.

BURGER, K., 1980. Kaolin-Kohlentonstein im flötzführenden Oberkarbon des Niederrheinisch-Westfälischen Steinkohlenreviers. *Geol. Rundschau,* 69, 488–531.

CARLISLE, D., 1963. Pillow breccias and their aquagene tuffs, Quadra Island, British Columbia. *J. Geol.,* 71, 48–71.

CAROZZI, A. V., 1989. *Carbonate Rock Depositional Models: A Microfacies Approach.* Prentice Hall, Englewood Cliffs, N.J., Advanced Reference Series, 604 pp.

CAS, R. A. F., and WRIGHT, J. V., 1987. *Volcanic Successions Modern and Ancient. A Geological Approach to Processes, Products, and Successions.* Allen and Unwin, London, 528 pp.

CHAPIN, C. E., and ELSTON, W. E. (eds.), 1979. *Ash-flow tuffs. Geol. Soc. Amer.,* Special Paper 180, 211 pp.

COOK, E. F., 1965. Stratigraphy of Tertiary volcanic rocks in eastern Nevada. *Nevada Bureau of Mines,* Report No. 11, 66 pp.

EWART, A., 1963. Petrology and petrogenesis of the Quaternary pumice ash in the Taupo area, New Zealand. *J. Petrology,* 4, 392–431.

FISHER, R. V., 1961. Proposed classification of volcaniclastic sediments and rocks. *Geol. Soc. Amer. Bull.,* 72, 1409–1414.

———, 1964. Maximum size, median diameter, and sorting of tephra. *J. Geophys. Res.,* 69, 341–355.

———, 1966. Rocks composed of volcanic fragments. *Earth Science Rev.,* 1, 287–298.

———, and HEIKEN, G., 1982. Mt. Pelée, Martinique: May 8 and 20, 1902 pyroclastic flows and surges. *J. Volcanol. Geotherm. Res.,* 13, 339–371.

———, and SCHMINKE, H. -U., 1984. *Pyroclastic Rocks.* Springer-Verlag, New York, 472 pp.

———, and SMITH, G. A. (eds.), 1991. *Sedimentation in Volcanic Settings.* Soc. Econ. Paleontologists and Mineralogists, Special Publ. No. 45, 257 pp.

FRIEDMAN, I., and SMITH, R. L., 1958. The deuterium content of water in some volcanic glasses. *Geochim. Cosmochim. Acta,* 15, 218–228.

FURNES, H., FRIDLEIFSSON, I. B., and ATKINS, F. B., 1980. Subglacial volcanics—on the formation of acid hyaloclastites. *J. Volcanol. Geotherm. Res.,* 8, 95–110.

GILBERT, C. M., 1938. Welded tuffs in eastern California. *Geol. Soc. Amer. Bull.,* 49, 1829–1862.

GRIM, R. E., and GÜVEN, N., 1978. *Bentonites. Geology, Mineralogy, Properties, and Uses.* Developments in Sedimentology 24, Elsevier Scientific Publishing Co. New York, 256 pp.

HANSEN, W. R., LEMKE, R. W., CATTERMOLE, J. M., and GIBBONS, A. B., 1963. Stratigraphy and structure of the Rainier and U.S.G.S. tunnel areas, Nevada Test Site. *U.S. Geol. Survey Prof. Paper 382-A,* 49 pp.

HAY, R. L., 1978. Geologic occurrences of zeolites. In L. B. Sand and F. A. Mumpton (eds.), *Natural Zeolites, Occurrence, Properties, Use.* Pergamon Press, Elmsford, N.Y., pp. 135–143.

HEIKEN, G., 1972. Morphology and petrography of volcanic ashes. *Geol. Soc. Amer. Bull.,* 83, 1961–1988.

————, 1974. *An Atlas of Volcanic Ash.* Smithsonian Contributions to the Earth Sciences, No. 12, 101 pp.

————, 1979. Pyroclastic flow deposits. *Amer. Scientist,* 67, 564–571.

————, and WOHLETZ, K., 1985. *Volcanic Ash.* University of California Press, Berkeley, Calif., 246 pp.

HILDRETH, W., and MAHOOD, G., 1985. Correlation of ash-flow tuffs. *Geol. Soc. Amer. Bull.,* 96, 968–974.

HONNOREZ, J., and KIRST, P., 1975. Submarine basaltic volcanism: morphometric parameters for discriminating hyaloclastites from hyalotuffs. *Bull. Volcanol.,* 39, 441–465.

HORN, D. R., DELACH, M. N., and HORN, B. M., 1969. Distribution of volcanic ash layers and turbidites in the north Pacific. *Geol. Soc. Amer. Bull.,* 80, 1715–1723.

HUBER, N. K., and RINEHART, D. D., 1966. Some relationship between the refractive index of fused glass beads and the petrologic affinity of volcanic rock suites. *Geol. Soc. Amer. Bull.,* 77, 101–110.

HUFF, W. D., 1983. Correlation of Middle Ordovician K-bentonites based on chemical finger printing. *J. Geol.,* 91, 657–669.

————, and KOLATA, D. R., 1990. Correlation of the Ordovician Deicke and Millbrig K-bentonites between the Mississippi Valley and the Southern Appalachians. *Amer. Assoc. Petroleum Geologists Bull.,* 74, 1736–1747.

————, and TÜRKMENOGLU, A. G., 1981. Chemical characteristics and origin of Ordovician K-bentonites along the Cincinnati Arch. *Clays and Clay Minerals,* 29, 113–123.

IZETT, G. A., 1981. Volcanic ash beds: recorders of Upper Cretaceous silicic pyroclastic volcanism in the western United States. *J. Geophys. Res.,* 86, B11, 10200–10222.

JESEK, P. A., and NOBLE, D. C., 1978. Natural hydration and ion exchange of obsidian: an electron microscope study. *Amer. Mineralogist,* 63, 266–273.

KOLATA, D. R., FROST, J. K., and HUFF, W. D., 1987. Chemical correlation of K-bentonite beds in the Middle Ordovician Decorah Subgroup, upper Mississippi Valley. *Geology,* 15, 208–211.

LAMB, H. H., 1970. Volcanic dust in the atmosphere, with a chronology and assessment of its meteorological significance. *Royal Soc. London Phil. Trans.,* A266, 425–533.

LEDBETTER, M. T., and SPARKS, R. S. J., 1979. Duration of large-magnitude explosion eruptions deduced from graded bedding in deep-sea ash layers. *Geology,* 7, 240–244.

LIRER, L., and VINCI, A., 1991. Grain-size distributions of pyroclastic deposits. *Sedimentology,* 38, 1075–1083.

MATHEWS, W. H., 1951. A useful method for determining approximate compositions of fine-grained igneous rocks. *Amer. Mineralogist,* 36, 92–101.

MOORE, J. G., and PECK, D. L., 1962. Accretionary lapilli in volcanic rocks of the western continental United States. *J. Geol.,* 70, 182–193.

NADEAU, P. H., and REYNOLDS, R. C., 1981. Burial and contact metamorphism in the Mancos Shale. *Clays and Clay Minerals,* 29, 249–259.

NOACK, Y., 1981. La palagonite: caractéristiques, facteurs d'évolution et mode de formation. *Bull. Minéralogie,* 104, 36–46.

NOH, J. H., and BOLES, J. R., 1989. Diagenetic alteration of perlite in the Guryongpo area, Republic of Korea. *Clays and Clay Minerals,* 37, 47–58.

PERSON, A., 1982. The genesis of bentonites. In G. S. Odin (ed.), *Numerical Dating in Stratigraphy,* Part One, John Wiley & Sons, New York, pp. 407–421.

PETERSON, D. W., 1970. Ash-flow deposits—their character, origin, and significance. *J. Geol. Education,* 18, 66–76.

PRICE, N. B., and DUFF, P. McL. D., 1969. Mineralogy and chemistry of tonsteins from Carboniferous sequences in Great Britain. *Sedimentology,* 13, 45–69.

ROSS, C. S., and SMITH, R. L., 1955. Water and other volatiles in volcanic glass. *Amer. Mineralogist,* 40, 1071–1089.

————, and ————, 1960. Ash-flow tuffs: their origin, geologic relations, and identification. *U.S. Geol. Survey Prof. Paper 366,* 81 pp.

————, MISER, H. D., and STEPHENSON, L. W., 1928. Water-laid volcanic rocks of early upper Cretaceous age in southwestern Arkansas, southeastern Oklahoma, and northeastern Texas. *U.S. Geol. Survey Prof. Paper 154-F,* pp. 175–202.

ROSS, K. A., and FISHER, R. V., 1986. Biogenic grooving on glass shards. *Geology,* 14, 571–573.

SCHEIDEGGER, K. F., JESEK, P. A., and NINKOVICH, D., 1978. Chemical and optical studies of glass shards in Pleistocene and Pliocene ash layers from DSDP Site 192, Northwest Pacific Ocean. *J. Volcanol. Geotherm. Res.,* 4, 99–116.

SCHIENER, E. J., 1970. Sedimentology and petrography of three tuff horizons in the Caradocian sequence of the Bala area (North Wales). *Geol. J.,* 7, 25–46.

SCHMID, R., 1981. Descriptive nomenclature and classification of pyroclastic deposits and fragments: recommendations of the I.U.G.S. Subcommission on the Systematics of Igneous Rocks. *Geology,* 9, 41–43.

SHERIDAN, M. F., and MARSHALL, J. R., 1983. Interpretation of pyroclast surface features using SEM images. *J. Volcanol. Geotherm. Res.,* 16, 153–159.

————, and RAGAN, D. M., 1977. Compaction in ash-flow tuffs. In G. V. Chilingarian and K. H. Wolf (eds.), *Compaction of Coarse-*

grained Sediments II. Developments in Sedimentology 18B, Elsevier Scientific Publishing Co. New York, pp. 677–717.

SLAUGHTER, M., and EARLEY, J. W., 1965. Mineralogy and geological significance of the Mowry bentonites, Wyoming. *Geol. Soc. Amer. Special Paper 83,* 116 pp.

SMITH, R. L., 1960a. Ash flows. *Geol. Soc. Amer. Bull.,* 71, 795–842.

———, 1960b. Zones and zonal variations in welded ash flows. *U.S. Geol. Survey Prof. paper 354F,* pp. 149–159.

SOREM, R. K., 1982. Volcanic ash clusters: tephra rafts and scavengers. *J. Volcanol. Geotherm. Res.,* 13, 63–71.

SPARKS, R. S. J., and WRIGHT, J. V., 1979. Welded air-fall tuffs. In C. E. Chapin and W. E. Elston (eds.), Ash-flow tuffs. *Geol. Soc. Amer. Special Paper 180,* pp. 155–166.

SPEARS, D. A., 1970. A kaolinite mudstone (tonstein) in the British Coal Measures. *J. Sed. Petrology,* 40, 386–394.

———, and DUFF, P. McL. D., 1984. Kaolinite and mixed-layer illite–smectite in Lower Cretaceous bentonites from the Peace River Coalfield, British Columbia. *Can. J. Earth Sci.,* 21, 465–476.

———, and KANARIS-SOTIRIOU, R., 1979. A geochemical and mineralogical investigation of some British and other European tonsteins. *Sedimentology,* 26, 407–425.

STIX, J., 1991. Subaqueous intermediate to silicic-composition explosive volcanism: a review. *Earth-Science Rev.,* 31, 21–53.

SURDAM, R. C., 1973. Low-grade metamorphism of tuffaceous rocks in the Karmutsen Group, Vancouver Island, British Columbia. *Geol. Soc. Amer. Bull.,* 84, 1911–1922.

SWINEFORD, A., and FRYE, J. C., 1946. Petrographic comparison of Pliocene and Pleistocene volcanic ash from eastern Kansas. *Kansas Geol. Survey Bull.,* 84, Pt. I, pp. 1–32.

TEALE, C. T., and SPEARS, D. A., 1986. The mineralogy and origin of some Silurian bentonites, Welsh Borderland, U.K. *Sedimentology,* 33, 757–765.

VERNIK, L., 1990. A new type of reservoir rock in volcaniclastic sequences. *Amer. Assoc. Petroleum Geologists Bull.,* 74, 830–836.

WALKER, G. P. L., 1971. Grain-size characteristics of pyroclastic deposits. *J. Geol.,* 79, 696–714.

———, 1984. Characteristics of dune-bedded pyroclastic surge bedsets. *J. Volcanol. Geotherm. Res.,* 20, 281–296.

WATERS, A. C., and FISHER, R. V., 1971. Base surges and their deposits: Capelinhos and Taal volcanoes. *J. Geophys. Res.,* 76, 5596–5614.

WEAVER, C. E., 1953. Mineralogy and petrology of some Ordovician K-bentonites and related limestones. *Geol. Soc. Amer. Bull.,* 64, 921–943.

———, 1963. Interpretation value of heavy minerals from bentonites. *J. Sed. Petrology,* 33, 343–349.

WRIGHT, J. V., 1980. Stratigraphy and geology of the welded air fall tuffs of Pantelleria, Italy. *Geol. Rundschau,* 69, 263–291.

———, SMITH, A. L., and SELF, S., 1980. A working terminology of pyroclastic deposits. *J. Volcanol. Geotherm. Res.,* 8, 315–336.

ZIELINSKI, R. A., 1985. Element mobility during alteration of silicic ash to kaolinite—a study of tonstein. *Sedimentology,* 32, 567–579.

CHAPTER 5

LIMESTONES

INTRODUCTION

The formation of carbonate sediments results mainly from biochemical and biological processes in warm shallow marine and lacustrine environments. The only exceptions are the less frequent cold-water carbonates deposited in temperate, subpolar, and glaciomarine conditions characterized by particular petrographic and geochemical signatures (Rao, 1981, 1983, 1988). In contrast to siliciclastic sediments, warm-water marine carbonates, to which this presentation is limited, are intrabasinal in origin. Transport of carbonate sediments by waves and tidal currents is limited to local and small-scale redistribution of their constituents. The only exception consists of possible transport from carbonate platforms to adjacent basins by turbidity currents.

Limestones, which represent the final indurated product of carbonate sediments, range from the Precambrian to the Present. Their petrographic study reveals the evolution and extinction of numerous groups of marine organisms, which either precipitated calcium carbonate from seawater or secreted carbonate skeletons, which, upon death of their respective organisms, released skeletal particles, also called *bioclasts*, of greatly variable shape and size. These types of particles are associated with nonskeletal grains such as ooids, peloids, and intraclasts. Together they may form a grain-to-grain framework whose interstitial spaces are filled by carbonate mud matrix and/or precipitated calcite cement called *sparite*.

The environmental interpretation of limestones relies to a large extent on the identification of the above-mentioned types of grains, in particular bioclasts, which represent the remains of organic communities living essentially in place and therefore provide critical data on depth, salinity, clarity, and temperature of water.

Although the grain size of bioclasts reflects mainly the original size of the released skeletal constituents and the biological factors controlling their breakdown, the mechanical action of waves and currents eventually shapes skeletal and nonskeletal grains into particles whose grain size expresses to a large extent the energy level (bedshear) of the environment of deposition. Naturally, carbonate particles differ in their hydrodynamic behavior from quartz grains because of their more complicated original shape, lower density due to internal microporosity, and variable content in organic matter. Consequently, sorting and rounding coefficients of carbonate particles, particularly skeletal grains, are not reliable criteria as in siliciclastic rocks. Nevertheless, the general application of the index of clasticity or maximum apparent grain size, together with the frequency index of resistant and almost equidimensional bioclasts, such as crinoid columnals, visible under the microscope, has been applied in numerous studies and provided fundamental data for the final reconstruction of carbonate rock depositional environments (Carozzi, 1989).

Two other approaches are used for evaluating the energy level of a carbonate environment: (1) the absence or presence of a grain-to-grain depositional framework (mud-supported versus grain-supported texture), and (2) the mud to cement ratio in the interstitial spaces of the grain-supported framework. Since carbonate mud deposition occurs only under shallow or deep quiet conditions, increasing energy of the environment leads to a decrease in the amount of carbonate mud and to the generation of a grain-supported framework, with open interstitial pores subsequently filled by a variety of precipitated aragonite or calcite cements. This generalization of the significance of carbonate mud can suffer some exceptions when it is trapped under high-energy conditions in spaces within organically constructed structures such as reefs, or stabilized by algal mats or possibly precipitated as a form of microcrystalline cement during early submarine diagenetic stages.

The unusual variety of skeletal and nonskeletal constituents of limestones is combined with a relatively simple mineralogy. In Recent carbonate skeletal and nonskeletal particles, the following calcium carbonate minerals predominate: low-Mg calcite with less than 4 mol % $MgCO_3$; high-Mg calcite with more than 4 mol % $MgCO_3$ (usually 11 to 19 mol % $MgCo_3$); and aragonite, which has a very low magnesium content but may show up to 1% Sr substituting for calcium. Aragonite is a metastable form that alters to calcite, and high-Mg calcite loses its magnesium so that eventually all calcium carbonate minerals are converted to low-Mg calcite, the stable form of calcium carbonate under ordinary temperature and pressure conditions. This general conversion is obviously highly destructive of original structures in bioclasts and original textures in carbonate mud and cements. However, in any given carbonate sequence, sufficient relicts are preserved to allow a reasonable reconstruction of depositional features and their subsequent diagenetic history.

The intensity and even the very existence of the conversion to low-Mg calcite in older limestones is based on the uniformitarian concept that carbonate deposition in oceans of the past was in the form of aragonite and high-Mg calcite as today. However, numerous recent geochemical and petrographic studies, using the most refined techniques, revealed that such was not the case, and that periods of low-Mg calcite seas alternated with periods of high-Mg calcite and aragonite seas (**Fig. 5.1**). This cyclic pattern was closely related to worldwide eustatic rise and fall of sea level and hence to global tectonics (Fischer, 1981; Wilkinson, 1982; Sandberg, 1983; Wilkinson et al., 1985; Wilkinson and Given, 1986).

During the existence of calcite seas, the atmosphere was supposed to be under conditions of "greenhouse," with high CO_2 concentration resulting from peaks of volcanic and tectonic activity. Sea level was high, and shallow epicontinental seas were widespread as a consequence of active growth of the midoceanic ridge system. Conversely, during the existence of aragonite seas, as at present, the atmosphere was under conditions of "icehouse," with lower CO_2 concentration and reduced volcanic and tectonic activity. Sea level was low, and shallow seas were reduced to narrow encroachments along continental margins as an effect of subsidence of the midoceanic ridge system.

This cyclic pattern of calcite and aragonite seas, which appears to be amply documented, clearly indicates that data obtained from present-day carbonate deposition in the oceans can only be applied to certain intervals of the geological record: Precambrian to Cambrian, Pennsylvanian to Triassic, and Cenozoic to Present (**Fig. 5.1**). Under these conditions, many aspects of the diagenetic evolution of limestones, and in particular the mineralogical composition of their cements, need to be carefully reevaluated in each specific case. For instance, the low-Mg interstitial sparite cement of many high-energy calcarenites, generally attributed to the diagenetic phreatic environment, can no longer be used as a criterion by itself (Given and Wilkinson, 1985) since it could well be a marine precipitate during calcite seas (Wilkinson et al., 1982).

Further mineralogical complication is introduced among limestones by the formation of diagenetic carbonates such as ferroan calcite, ferroan dolomite (ankerite), and dolomite (Chapter 6). Because of the similarity in optical properties between the various carbonate minerals, reliable distinction of these minerals generally requires the use of well-established staining techniques, such as alizarin red-S and potassium ferricyanide (Dickson, 1966; Hutchinson, 1974). Although the stain absorption is related to grain size and crystallographic orientation, staining techniques emphasize many textural features otherwise invisible under the microscope and represent an extremely valuable tool for characterizing diagenetic stages. Even more spectacular results are obtained in this domain by the use of cathodoluminescence petrography (Amieux, 1982; Machel, 1985). This technique is able to distinguish delicate manganese-rich and manganese-poor zones in calcite cements that express minute chemical changes in the chemistry of circulating pore fluids. Recognition and correlation of these zones over wide areas in a carbonate basin leads to the establishment of a cement stratigraphy that is critical for the understanding of the diagenetic evolution of carbonates (Meyers, 1974, 1978; Frank et al., 1982; Fairchild, 1983; Kaufman et al., 1988; Goldstein, 1988).

The general intracratonic depositional setting of most limestones tends to make them relatively uncontaminated by extrabasinal detrital influxes, such as detrital quartz and clay minerals. However, a great variety of early to late authigenic minerals occurs, such as glaucony (see Chapter 1), pyrite, hematite, phosphate, silica (chert), barite, feldspars (albite), and various evaporitic minerals, among which anhydrite and gypsum predominate. Analyses and classifications for mixed siliciclastic–carbonate rocks have been proposed (Mount, 1985; Doyle and Roberts, 1988).

The above-mentioned general conversion of calcium carbonate minerals to stable low-Mg calcite is but one aspect, albeit a fundamental one, of the very complex diagenetic processes that change carbonate sediments deposited in seawater into limestones submitted to shallow and deep

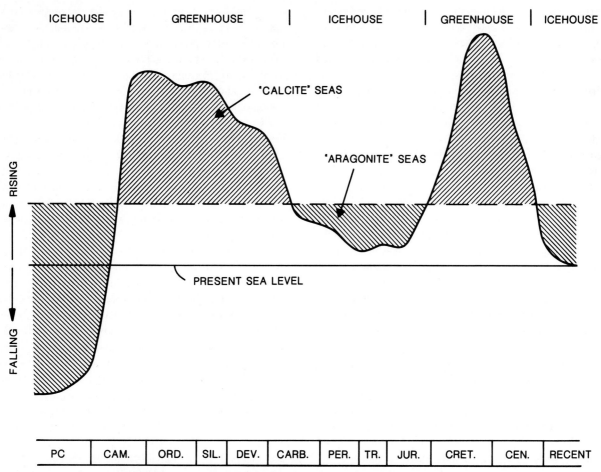

FIGURE 5.1 Possible distribution of calcite and aragonite seas and associated climatic conditions through geological time as related to absolute positions of global sea levels. The horizontal dashed line is arbitrarily fixed at a position that best separates primary aragonite and primary calcite components. Modified from Wilkinson (1982). Reproduced by permission of the author and the *Journal of Geological Education.*

burial conditions. Compaction, dissolution with pore generation, cementation, and neomorphism occur in successive and interfering generations. They all tend to be highly destructive of depositional and earlier diagenetic fabrics, although in many cases they have an economically interesting effect in generating reservoir conditions for mineral-depositing solutions and liquid or gaseous hydrocarbons (Roehl and Choquette, 1985). Again, detailed petrographic studies reveal sufficient relict textures, with relative geometrical relationships, that have escaped these processes and allow sketching basinwide step-by-step diagenetic changes forming the diagenetic sequences (Carozzi, 1989). Nevertheless, the pervasive effect of diagenesis dominates the entire understanding of limestones, a factor to be kept in mind constantly during any investigation. Furthermore, since mineralogy of carbonate deposits has always been highly controlled by marine organic communities, and a number of them have become extinct, it is also critical to realize that the depositional environments of many ancient limestones can only be

interpreted on the basis of their own intrinsic paleoecological, paleotectonic, and physiographic features (Carozzi, 1989). Any comparison with Recent carbonates, although tempting and therefore widely abused, is impaired by the fact that they are being deposited in aragonite seas, under restricted shallow marine conditions, which may not be representative of past conditions. At any rate, a full review of the numerous types of carbonate rock depositional environments is beyond the scope of this presentation and has been given in many publications (Wilson, 1975; Scholle et al., 1983; James, 1984a, b, c; McIlreath and James, 1984; Carozzi, 1989; Tucker et al., 1990; Tucker and Wright, 1990).

MAJOR CONSTITUENTS

The major constituents of limestones can be divided for the purpose of general classification into four groups: skeletal grains, nonskeletal grains, carbonate mud, and microsparite and sparite cements. Well-illustrated and systematic treatment of these constituents was given by Cayeux (1916, 1935), Horowitz and Potter (1971), Majewske (1974), and Scholle (1978).

Skeletal Grains

Identification of skeletal grains represents a particular aspect of sedimentary petrography. Such a procedure is handicapped by the fact that shapes and sizes represent a two-dimensional view from which a geometric body is to be reconstructed in three dimensions. Consequently, any identification beyond the genus level, with a few exceptions, is rarely possible. Internal microstructures, unless altered by diagenesis, are sufficiently characteristic to allow identification of shell fragments of brachiopods, pelecypods, gastropods, and cephalopods; carapace debris of trilobites and ostracods; plates and columnals of pelmatozoans; zooecia of bryozoans; secreted frameworks of corals; structures of calcareous red and green algae, and of encrusting or trapping cyanobacteria; and tests of benthonic and pelagic foraminifers.

The shells of articulate brachiopods consist of low-Mg calcite, and their internal structure is therefore well preserved (**Plates 19.A, 20.E**). Their structure displays a very thin outer layer of calcite fibers, oriented perpendicular to the shell surface and overlying a thicker internal layer of oblique fibers. Crenulations of the tests are very common. Shells of punctate brachiopods display fine tubes perpendicular to the shell surface, but restricted to the inner layer. These tubes are generally filled by carbonate mud or sparite cement. Shells of pseudopunctate brachiopods lack tubes, but are crossed by fine rods of coarse calcite against which the oblique calcite fibers are bent.

The microfabric of pelecypod shells is more complex. It consists generally of two distinct layers, which can be aragonitic, aragonite–calcite, or entirely calcitic. The shell is made of an inner nacreous layer consisting of sheets of oblique aragonite tablets overlain by an external prismatic layer of aragonite or calcite prisms. The latter in oysters are often released as isolated fragments upon decay of the organic constituents of the shell. The aragonite portions of pelecypod shells are very susceptible to conversion into stable low-Mg calcite, with the original structure destroyed by the development of a sparite mosaic, which may occasionally preserve some relicts of internal layering (**Plates 19.E, 25.A**).

Gastropod shells have a similar complex structure as pelecypods, but with two to three distinct layers. The aragonitic layers are generally destroyed and entirely replaced by a sparite mosaic. However, the highly coiled shape of most gastropods, which is a characteristic feature for identification of microfabric, is preserved (**Plate 19.E**).

Cephalopod (nautiloids and ammonoids) shells are extremely thin and delicately curved with septae and siphuncles. Their aragonitic composition leads to a fine sparitic mosaic, with rare preservation of the original structure consisting generally of one- to three-layered and prismatic units.

Among arthropods, the minute ostracod shells appear as articulated or disarticulated thin valves, smooth or ornamented, consisting of massive or fibroradiated calcite (**Plate 18.E, F**). Trilobite remains are thick shells of calcite with characteristic sweeping wavy extinction, with or without canal system, often displaying a typical "shepherd-hook" aspect corresponding to the curved margin of the carapace (**Plate 19.A**).

Although modern bryozoans have skeletons of aragonite, high-Mg calcite or a combination of both, ancient forms appear to have had calcitic skeletons. Those of fenestrate types consist of strings of perfectly preserved zooecia or cells connected by a stem of dark laminated calcite (**Plate 20.F**).

Ancient echinoid and blastoid plates and crinoid columnals consisted of low-Mg calcite, although modern forms are generally of high-Mg calcite. Crinoid columnals are easily identifiable bioclasts because they are cylindrical bodies with a central canal. Each columnal consists of a large single calcite crystal with unit extinction (**Plates 19.C, F, 20.C**). Pelmatozoan fragments have a "dusty" appearance under the microscope produced by their typical porous microstructure, which may be partially infilled by carbonate mud or authigenic glaucony. In high-energy crinoidal and echinoidal calcarenites, very abundant in the Paleozoic and the Mesozoic, skeletal grains are frequently surrounded by individual syntaxial overgrowths of sparite cement in optical continuity with them (**Plates 19.F, 24.B**). The overgrowths fill the interstitial spaces of the framework and interfere among themselves, displaying compromise boundary faces. This texture is comparable to that of a quartzite. Transverse sections of echinoid spines, which are also single calcite crystals, are characterized by complex radial stellate patterns.

Paleozoic rugose and tabulate corals have skeletons of predominantly low-Mg calcite and hence well-preserved complex internal structures, which are characterized either by numerous types of internal septas radiating from a central cavity filled with sparite cement or by rectangular patterns of intersecting systems of plates outlining sparite-filled chambers. Scleratinian corals (Mesozoic to Recent) are aragonitic and therefore are often poorly preserved (**Plates 22.C, 24.E**).

Calcareous red algae have well-preserved skeletons of cryptocrystalline, dark-colored low-Mg calcite, forming various patterns of rectangular cells overlain by a dark massive external crust, but modern forms have high-Mg content in their calcite. Each genus has its own characteristic pattern of chambers (**Plates 22.B, D, 24.F**). Red algae, very widespread in the Cenozoic, display a very variable morphology closely related to the energy level of the environment. Encrustation and irregular nodules characterize high-energy conditions, whereas delicate branching forms develop in low-energy situations.

Among the green calcareous algae, the freshwater Characeae, which appeared in the Devonian, are only partially calcified by low-Mg calcite, but their characteristic stems and oogonia are well preserved (**Plate 18.E**). They are excellent environment indicators of lacustrine conditions, although some oogonia may be floated for long distances and finally deposited in marine environments.

In marine and brackish environments, Dasycladaceae are widespread but poorly preserved, as a result of their incomplete encrustation by aragonite. Among the Codiaceae, *Halimeda* has a characteristically segmented thallus that breaks down into plates with an internal irregularly tubular structure (**Plate 21.D**). Various types of encrusting or trapping cyanobacteria build in combination with other bacteria, filamentous mats that trap and bind fine carbonate particles into laminated structures called *stromatolites* (Walter, 1976). They were very important in the Precambrian, but gradually decreased during the Phanerozoic as an effect of grazing organisms, in particular gastropods. The general shapes of stromatolites range from planar through hemispheroids into columnar. The laminated structure, consisting of alternating couplets of dark, organic-matter-rich bacterial mats and lighter-colored layers of trapped carbonate mud, is believed to represent a diurnal to seasonal periodicity (**Plate 26.C**). The variable morphology of stromatolites is a direct function of environmental conditions such as water depth, wave and tidal energy and amplitude, frequency of subaerial exposure, and rate of carbonate mud supply.

Many bioclasts display dark micrite envelopes (**Plate 22.E**) produced by endolithic coccoid algae that riddle their margins with tiny perforations. Successive generations of perforations and infillings by carbonate cement, precipitated by biochemical processes related to the algae and associated bacteria, eventually build a thick dark envelope called a *micrite envelope*. Completion of the process may lead to total destruction of the original substrate. This micritization is a paleoindicator of the photic zone.

Foraminifers, either benthonic or planktonic, have a low- or high-Mg calcite skeleton, rarely of aragonite. Consequently, most are extremely well preserved in Phanerozoic limestones. Their tests show an extraordinary variety of shapes and sizes (**Plates 18.D, 21.B, C, E**). Their internal structures range from simple, microgranular thin walls to thick, highly complex multilayered walls, each type characteristic of a particular genus or even species.

Nonskeletal Grains

These types of grains consist of the following: stromatolitic bacterial mats forming a variety of bodies with irregular internal structures called oncoids (**Plates 19.B, 23.C, 25.F**) and generally designated under the name of coated grains (Peryt, 1983); ooids and pisoids with characteristic regular concentric and radial fabrics; pellets or peloids; and intraclasts. The origin of of some of these nonskeletal grains is still in part controversial, particularly the role played by bacteria, and they deserve some comments.

Ooids (Simone, 1981) are spherical to subspherical grains displaying one or more concentric envelopes of variable thickness precipitated around a nucleus consisting of a carbonate particle (bioclast, pellet, ooid fragment) or any detrital grain, generally quartz or glaucony (**Plates 19.E, 20.A to D, 22.F, 23.F**). Although the grain-size boundary between ooids and pisoids is usually set at 2 mm, most ooids range in diameter between 0.5 and 1 mm. Almost all marine ooids formed today consist of concentric layers of tangentially oriented acicular aragonite needles; however, early Holocene ooids display a combined radial and concentric structure and consist of high-Mg calcite (Marshall and Davies, 1975; Milliman and Barretto, 1975). Ooids formed in quiet hypersaline waters such as the Great Salt Lake, Utah (Sandberg, 1975), are aragonitic and show similar radial–concentric structures.

Calcitic ooids of ancient limestones, reaching as far back in time as the Precambrian, have a typical radial structure interrupted by concentric growth lines. The origin of ancient ooids remains disputed with respect to their mineralogy (precipitated calcite or aragonite) and with respect to their internal structure: originally radial–concentric or a diagenetic replacement product of an original tangential arrangement of the crystals. Very likely both origins existed in the past (Wilkinson et al., 1985) and alternated cyclically during the above-mentioned Phanerozoic intervals of calcitic and aragonitic seas (**Fig. 5.1**).

The general term of pellets, or peloids, is applied in a pure descriptive sense to spherical and ovoidal bodies consisting of cryptocrystalline to microcrystalline carbonate, generally devoid of internal structure (**Plate 19.D**). Although some pellets can reach a size of several millimeters, most range between 0.1 and 0.5 mm. Pellets can have several distinct origins, which are still debated (Macintyre, 1985). Most of them are fecal pellets of gastropods, crustaceans, and worms. These pellets, being relatively soft at the time of excretion, tend to be easily deformed and to merge eventually into an almost structureless mud. However, the differences in content of organic matter of the individual pellets give a clotted aspect to the deposit, which indicates its fecal origin. Other pellets appear to be zoned and to represent early stages of repeated nucleation during the precipitation of submarine high-Mg calcite cements (Macintyre, 1985). Still other pellets are interpreted as calcite precipitates within and around clumps of bacteria and induced by the vital activity of bacterial colonies (Chafetz, 1986). Finally, lithic pellets are small and variably rounded intraclasts of structureless carbonate mud reworked and transported by intraformational mechanical processes. They can be distinguished from fecal pellets by a greater variability in color and texture, expressing various types of reworked carbonate muds. Even more irregularly shaped pellets result from the complete micritization of small bioclasts by endolithic algae, and their irregular shape is the only distinguishing character. Incipient neomorphism in the form of irregularly distributed centers, where the original homogeneous carbonate mud underwent a slight aggrading process, may lead to a pseudopelletoidal texture in which the boundaries of the pseudopellets are hazy and complex in three dimensions.

Intraclasts are generally subangular to angular fragments of indurated to semiindurated carbonate mud reworked and redeposited by intraformational mechanical processes (**Plate 21.A**). These chips, flakes, or flat pebbles show very large size variations. They indicate either submarine syneresis processes or subaerial exposure and desiccation. Whenever their size is smaller than 0.5 mm and their shape subrounded by abrasion during transport, they fall into the above-mentioned category of lithic pellets.

Carbonate mud was originally defined as formed by grains of microcrystalline calcite with a grain size of less than 4 μ and designated by the term micrite (Folk, 1962). SEM studies showed that micrite is a heterogeneous association of crystals of variable sizes with numerous types of planar, curved, irregular, and even microstylolitic intercrystalline boundaries (Fischer et al., 1967; Flügel et al., 1968; Steinen, 1978). The proposed upper size limit of 4 μ for micrite is beyond the resolution power of the petrographic microscope. Consequently, for practical purposes, the limit has been raised in a variety of ways by numerous authors to the extent that a certain confusion has arisen. In this presentation, the fine carbonate muds are divided into calcilutites with an upper size limit of 30 μ and calcisiltites for those ranging in grain size from 30 to 60 μ. Calcilutites rarely reveal clues about their origin, whereas calcisiltites show that they consist of minute skeletal and nonskeletal particles that can be related to the abrasion or breakdown of the few sand-size constituents these calcisiltites may contain.

The origin of carbonate mud is complex and highly disputed because many processes are involved and can interfere on a permanent basis. The possibility of inorganic precipitation cannot be discarded, but a biochemical precipitation of aragonite or calcite by bacterial actions and algal photosynthesis appears more likely, although neither of the two possible origins has been convincingly demonstrated. Although some carbonate muds result from the accumulation of pelagic coccoliths and other marine phytoplankton, the temptation is strong at the present time to consider the disintegration of shallow-water calcareous green algae as the major process of carbonate mud generation. Neumann and Land (1975), in a remarkable paper on the carbonate muds of the Bahamian platform, demonstrated that the disintegration of these algae into a vast quantity of aragonite needles ac-

counts not only for the deposition of carbonate mud (< 60 μ in size) in lagoons and adjacent tidal flats, but that a large excess is also available for offshore export in suspension. These conditions of overproduction of carbonate mud, followed by transport and final deposition into open oceanic waters, change in a fundamental manner the traditional concept of "pelagic" carbonate sedimentation. Other processes of carbonate mud generation of lesser importance are bioerosion by perforating organisms (sponges and algae), that is, micritization of bioclasts and the mechanical abrasion of skeletal grains during transport by waves and currents. The above-mentioned speculations on the origin of carbonate mud arise from the question of whether this mud was deposited as aragonite or high-Mg calcite, because its conversion to stable low-Mg calcite is an aggrading and destructive process of original fabrics. This conversion is generally followed by pervasive aggrading neomorphism during burial, which further changes the low-Mg calcite into a coarser mosaic of pseudomicrosparite (< 30 μ) and pseudosparite (> 30 μ, no upper size given) with additional damaging effects. However, Lasemi and Sandberg (1984) described Pleistocene aragonite-dominated muds that indurated by inversion to both micrite and microsparite during a single event of neomorphism under freshwater vadose conditions. In this case, the aragonitic mud did not calcitize first to micrite and then alter to microsparite by aggrading neomorphism. Neomorphism is discussed below in more detail in the section on diagenesis.

Microsparite and sparite cements form the last major constituent of limestones. For purposes of simplification and general classification, before analyzing the effects of diagenetic processes, microsparite is defined as a chemically precipitated calcite of a crystal size smaller than 30 μ and sparite as a chemically precipitated, coarse-grained calcite of a crystal size larger than 30 μ, with no upper size given. Sparite is a cavity-filling cement that occupies the interstitial spaces of the framework of grain-supported calcarenites and calcirudites (**Plates 22.B, 24.A**). Typically, it shows a bladed texture at the contacts with framework grains and grades or abruptly changes into an equant mosaic toward the center of interstitial cavities. Pseudosparite developed by strongly aggrading neomorphism has specific characters, discussed below, that prevent its confusion with cavity-filling sparite cement.

On the basis of this simplified description of limestones as consisting of four major constituents, skeletal grains, nonskeletal grains, carbonate mud, and microsparite and sparite cements, general classifications can be attempted. These classifications can only be of preliminary scope, because a final understanding of limestones requires the simultaneous investigation of the effects of diagenesis on the above-mentioned four major constituents. The study of diagenetic features refines the subdivisions of any classification, while providing a history of diagenetic events or diagenetic sequence that traces the stages of evolution of the carbonate sediment into an indurated buried limestone.

CLASSIFICATION

In the early 1960s, sedimentary petrologists began to realize the necessity for an adequate, relatively simple, and practical classification of limestones that would codify the increasing research undertaken on these rocks, as well as its application to oil exploration.

As mentioned earlier for sandstones, attempts at classification tend to blend, in different manners, descriptive and genetic properties with the added and inevitable effect of regionalism and human vanity. It should be pointed out that most classifications presented in the 1960s did not take into account the effects of diagenesis, a field barely known at the time.

Many proposed classifications (Ham, 1962), emphasizing either composition or texture, shared the common genetic feature of expressing the energy level (bedshear) of the environment of deposition. Indeed, limestones were presented as a continuous spectrum ranging between end types in which the carbonate mud predominates in low-energy environments (mud-supported limestones) and end types that consist of a framework of grains (grain-supported limestones) deposited in high-energy environments and with its interstitial porosity eventually filled by a precipitated sparite cement.

The classification proposed by Folk (1959, 1962) em-

phasized mainly composition over texture and is based on three major constituents: allochems (skeletal and nonskeletal grains), carbonate mud matrix called micrite, and a cement of cavity-filling sparite (**Fig. 5.2**). The related terminology can be easily derived and defines two major sequences based on the predominant constituents. The first, with a micrite matrix, contains skeletal grains (biomicrite), ooids (oomicrite), pellets (pelmicrite), and intraclasts (intramicrite). The second sequence has a sparite interstitial cement between skeletal grains (biosparite), ooids (oosparite), pellets (pelsparite), and intraclasts (intrasparite). Terms can be modified if two types of allochems occur together; the designation would be, for instance, biopelsparite. The term of biolithite pertains to limestones consisting of a framework constructed by colonial organisms.

The classification of Dunham (1962) emphasized texture and introduced the valuable concept of mud-supported versus grain-supported limestones, that is, the distinction between sand-size grains distributed within a carbonate mud and not in reciprocal contact as opposed to sand-size grains forming a self-supporting framework (**Fig. 5.3**). The boundary between these two major textures can vary widely because of the irregular shape of bioclasts, but 30% to 35% grains generally can build a framework, and the distinction is not as highly subjective as it may appear at first glance. Extreme cases are represented by unusually platy bioclasts such as phylloid algal blades and fenestrate bryozoan fronds, which build self-supporting frameworks when reaching 15%, leaving between themselves primary depositional porosity on the order of 75%, the so-called "potato chip poros-

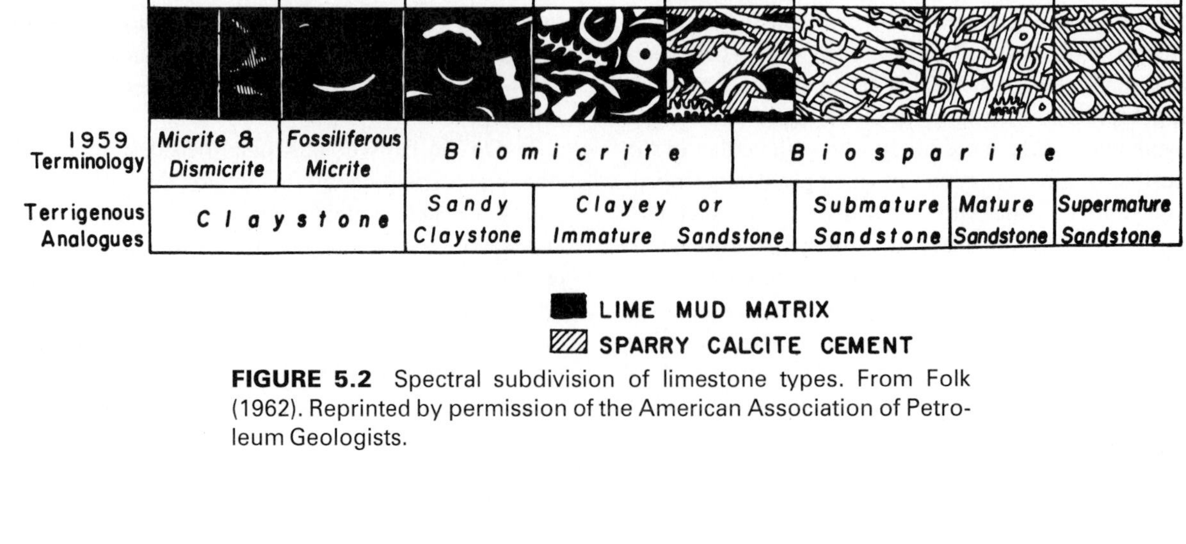

Percent Allochems	OVER 2/3 LIME MUD MATRIX				SUBEQUAL SPAR & LIME MUD	OVER 2/3 SPAR CEMENT		
	0-1%	1-10%	10-50%	OVER 50%		SORTING POOR	SORTING GOOD	ROUNDED & ABRADED
Representative Rock Terms	MICRITE & DISMICRITE	FOSSILI-FEROUS MICRITE	SPARSE BIOMICRITE	PACKED BIOMICRITE	POORLY WASHED BIOSPARITE	UNSORTED BIOSPARITE	SORTED BIOSPARITE	ROUNDED BIOSPARITE
1959 Terminology	Micrite & Dismicrite	Fossiliferous Micrite	Biomicrite			Biosparite		
Terrigenous Analogues	Claystone		Sandy Claystone	Clayey or Immature Sandstone		Submature Sandstone	Mature Sandstone	Supermature Sandstone

■ LIME MUD MATRIX
▨ SPARRY CALCITE CEMENT

FIGURE 5.2 Spectral subdivision of limestone types. From Folk (1962). Reprinted by permission of the American Association of Petroleum Geologists.

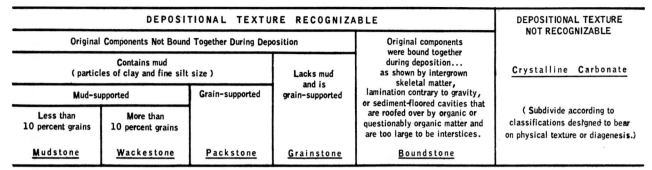

DEPOSITIONAL TEXTURE RECOGNIZABLE					DEPOSITIONAL TEXTURE NOT RECOGNIZABLE
Original Components Not Bound Together During Deposition				Original components were bound together during deposition... as shown by intergrown skeletal matter, lamination contrary to gravity, or sediment-floored cavities that are roofed over by organic or questionably organic matter and are too large to be interstices.	Crystalline Carbonate
Contains mud (particles of clay and fine silt size)		Grain-supported	Lacks mud and is grain-supported		
Mud-supported					(Subdivide according to classifications designed to bear on physical texture or diagenesis.)
Less than 10 percent grains	More than 10 percent grains				
Mudstone	Wackestone	Packstone	Grainstone	Boundstone	

FIGURE 5.3 Classification of carbonate rocks according to depositional texture. From Dunham (1962). Reprinted by permission of the American Association of Petroleum Geologists.

ity." The terminology of Dunham's classification ranges also from low- to high-energy conditions. A first group includes limestones containing carbonate mud. The term mudstone is used when mud contains less than 10% sand-size grains and the term wackestone when it contains more than 10%. When a grain framework is present, but mud still exists, the term packstone is used. Finally, a grain framework with interstitial porosity filled with precipitated sparite cement is called grainstone. The term boundstone refers to limestones bounded together during deposition by constructing organisms. The designation of crystalline carbonate is proposed for limestones in which depositional textures are not recognizable.

Further refinements were introduced by Embry and Klovan (1971) into Dunham's classification to characterize limestones consisting of grains coarser than sand size (> 2 mm). The new terms are floatstone, when the larger particles are not in reciprocal contact and are therefore matrix supported (the matrix being sand-size grains and carbonate mud), and rudstone, when the larger particles build themselves a grain-supported framework with the interstitial material consisting of sand-size grains and precipitated calcite cement. The term boundstone was subdivided according to the inferred type of organic binding, that is, framestone when the organisms built a rigid framework, bindstone when tabular or lamellar organisms encrusted or bound together the sediment during deposition, and finally bafflestone when stalked organisms growing in place trapped carbonate sediments by baffling effects.

Under the influence of major oil companies interested only in a fast pigeonholing approach, Dunham's classification became the most popular one in spite of its shortcomings, which consist mainly of the choice of ambiguous terms. A few examples should suffice: mudstone leads to confusion with argillaceous rocks; therefore, alternating argillaceous mudstones and fine-grained limestones need to be designated as alternating argillaceous mudstones and carbonate mudstones. Similarly, the term wackestone leads to confusion with immature sandstones called wackes. Another aspect of misleading terminology has been a fast method of studying limestones devised by Wilson (1975) and Flügel

(1982), which consists of comparing them with standard microfacies types (SMF). This system recognizes 24 standard microfacies considered diagnostic of particular environments, which in turn are classified into nine standard facies belts. Unfortunately, the use of Dunham's classification, combined with very generalized definitions of these standard microfacies, makes their environmental interpretation highly questionable, if not misleading. In reality, carbonate deposition is far more complex than this approach would suggest, mainly as the result of its high control by the evolution of the contributing organisms. Consequently, the number of microfacies and their combinations are almost innumerable (Carozzi, 1989), and oversimplifications cannot replace detailed and complete descriptions and related interpretations.

Since their initial presentation, the classifications of Folk and Dunham were endlessly modified and combined by researchers who adapted them to their particular purposes or regional settings. A general confusion developed gradually, and at present the situation is such that each author has to clearly define in which particular sense, both in terms of grain size and texture, he or she uses a given petrographic term borrowed from either classification. An additional problem was introduced by the fact that both classifications are in fact oversimplifications and no longer precise enough to account for many petrographic features, such as the relative role played by various constituents, the variability of depositional textures, and the fundamental effects of diagenetic processes. A simple term like biosparite or packstone, even with qualifiers, simply does not express the complex image of a limestone as seen under the petrographic microscope. This is the reason why the author of this book developed a more elaborate classification of carbonate rocks, which was informally and successfully tested for more than 30 years on graduate students and co-workers all over the world and only recently published (Carozzi, 1983, 1989). This classification, like the above-mentioned ones, is descriptive–genetic and based on a modified Wentworth grain-size scale combined with the textural concept of mud supported versus grain supported. Instead of using a single general term, this classification (**Fig. 5.4**) requires a full sentence or paragraph

PRACTICAL PETROGRAPHIC CLASSIFICATION OF CARBONATE ROCKS

INCREASING ENERGY OF DEPOSITIONAL ENVIRONMENT ↓

1. CALCILUTITE
   (micrite, 10 to 30μ) — with up to 10% of sand-size components (bio-litho)

2. CALCISILTITE
   (30 to 60μ) — with up to 10% of sand-size components (bio-litho)

3. MUD-SUPPORTED CALCARENITE
   (floating grains, bio-pel-litho) — 10 to 20 (30) % sand-size components (60μ to 2.5-5mm), bio-litho in:
   A. Calcilutite matrix
   B. Calcisiltite matrix
   C. Bioclastic matrix

4. GRAIN-SUPPORTED CALCARENITE
   — more than 20(30)% of sand-size components (60μ to 2.5-5mm), bio-pel-litho-ool in interstitial material of:
   (grain framework, bio-pel-litho-ool)
   A. Calcilutite matrix
   B. Calcisiltite matrix
   C. Bioclastic matrix
   D. Cement, drusy to sparite, overgrowth, 20μ to no upper crystal size limit
   E. Pressure-welded

5. MATRIX-SUPPORTED CALCIRUDITE
   (floating pebbles, bio-litho) — 10 to 20 (30)% pebble-size components (2.5-5mm to no upper size limit) with interstitial material of types 1 through 4 (except 4D)

6. CLAST-SUPPORTED CALCIRUDITE
   (pebble framework, bio-litho) — more than 30% of pebble-size components (2.5-5mm to no upper size limit) with interstitial material of types 1 through 4

7. BIOACCUMULATED LIMESTONE
   (loose accumulation) — accumulation by sessile non-framework builder organisms with interstitial material of types 1 through 3 (low energy)

8. BIOCONSTRUCTED LIMESTONE
   (constructed framework) — colonial framework builder organisms with interstitial material of types 1 through 6 (entire low to high energy spectrum)

FIGURE 5.4 Practical classification of carbonate rocks. From Carozzi (1989). Reprinted by permission of Prentice Hall, Englewood Cliffs, N.J.

describing all constituents and textural–diagenetic relationships, for instance, "coarse, grain-supported, crinoid–bryozoan calcarenite with submarine isopachous fibrous calcite cement and freshwater vadose cavity-filling mosaic of equant sparite cement." This approach may appear at first sight rigid and cumbersome, but it forces the observer to integrate all his or her microscopic observations into a single package that includes depositional and diagenetic features. The study of limestones by means of microfacies techniques (Carozzi, 1989) could not have been implemented without such a detailed classification.

MAJOR PETROGRAPHIC TYPES

Short descriptions are presented below of the major types of limestones in order of increasing energy level of the depositional environment.

Among calcilutites, those resulting from the accumulation of pelagic foraminifers are the most characteristic (**Plate 18.D**). They display a random distribution of delicate undistorted tests, often with internal microsparite filling, within a groundmass of fine carbonate mud, usually undisturbed by bioturbation. This texture indicates a quiet environment of deposition of at least a few hundred meters deep. Aggrading neomorphism in these pelagic calcilutites is generally minimal, and a relatively rapid submarine cementation destroyed their original microporosity and prevented appreciable compaction.

Chalks are petrographically similar to these pelagic calcilutites, although they consist of 80% coccolith plates and fragments associated with thin-shelled pelecypods, calcispheres, and echinoderm bioclasts. Chalks are essentially nonindurated and characterized by a high microporosity (36% to 60%), interpreted as having largely preserved their original composition and texture by the fact that coccolith plates consist of stable low-Mg calcite and hence did not undergo any mineralogical conversion. SEM studies of chalks revealed that a rigid framework was achieved soon after deposition by means of three delicate and incipient processes that took place at the contact points between coccolith fragments: meniscus-shaped cementation, pressure solution, and slight selective overgrowth (Mapstone, 1975; Scholle, 1977).

Lacustrine calcilutites are among the finest in grain size and are characterized by Characeae oogonia and stem debris associated with thin-shelled ostracods (**Plate 18.E**). Shallow-water calcilutites and calcisiltites can display in a striking manner the effects of numerous types of intense bioturbation by infaunas. Bioturbation has a profound effect on depositional fabrics by increasing the degree of comminution of the small skeletal debris and by redistributing them into elongate or spiral-shaped concentrations representing burrow fillings (**Plate 18.F**).

Mud-supported calcarenites are relatively rare transitional lithologies to grain-supported calcarenites with calcilutite or calcisiltite matrix (**Plate 19.A, B, C**). Crinoid columnals are among the most frequent contributors to these limestones. The interstitial matrix is usually of coarse silt size and designated as bioclastic matrix because it consists of the most comminuted but still recognizable products of the mechanical abrasion of the columnals themselves. Crinoids, particularly during the Paleozoic, were also important contributors to carbonate turbidites, which originated from the carbonate platforms and deposited their load on frontal slopes and basinal environments (Davies, 1977).

Generally, the finest grain-supported calcarenites with interstitial microcrystalline cement are pelletoidal, hence of fecal or lithic origin, and the pellets themselves may have undergone a slight mechanical abrasion during the moderate transport they underwent (**Plate 19.D**). Higher-energy, grain-supported calcarenites with sparite cement are extremely widespread over carbonate platforms. They display a vast spectrum of bioclasts, ooids, and intraclasts (**Plate 19.E**). When consisting again of abundant crinoid columnals, they are cemented by syntaxial overgrowths in optical continuity around individual columnals, showing compromise crystal faces due to their interfering growth patterns (**Plate 19.F**). Oolitic calcarenites are among the most spectacular limestones in thin section, and their occurrence indicates shallow, agitated conditions (**Plate 20.A**). In many instances, the interstitial sparite cement of oolitic calcarenites consists of a mosaic of very large anhedral calcite crystal forming the poikilotopic texture in which ooids and bioclasts seem to "float" in very large cement crystals (**Plate 20.B**). However, this texture still reflects a grain-supported framework and is a good example of the problem encountered when reconstructing three-dimensional objects from thin-section observation.

A frequent feature of grain-supported biocalcarenites and oolitic calcarenites is to display the effects of early to late diagenetic pressure solution (**Plate 20.C, D, E**). This process, in addition to leading to reciprocal deformation and interpenetration of the various types of grains, eliminates in most cases any initial matrix or cement, thus depriving the

observer of a major criterion for determining the energy level of the depositional environment.

Matrix-supported calcirudites are relatively rare (**Plate 20.F**) and are produced by debris flows of carbonate mud and its constituents, usually along the frontal slopes of carbonate platforms (Cook and Enos, 1977), or over platforms submitted to submarine reworking and redeposition by storm processes (Einsele and Seilacher, 1982). Clast-supported calcirudites consist either of a single type of intraformational lithic clasts (see Chapter 2) or, in more rare instances, of a variety of types of well-indurated and older carbonate clasts that were apparently exposed along the margins of cratonic basins (**Plate 21.A**). These types of calcirudites, either intraformational or extraformational in origin, are varieties of the rudaceous rocks described in Chapter 2.

The accumulation in place of sessile nonframework builder organisms expresses the occurrence of banks of small and large foraminifers, pelecypods, brachiopods, crinoids, codiacean algae, ostracods, and so on, which developed in shallow, quiet, and protected areas of carbonate platforms (**Plate 21.B to E**). Consequently, these originally loose accumulations of tests have interstitial materials ranging from calcilutites and calcisilites to mud-supported calcarenites. These rocks, designated as bioaccumulated limestones, indicate a general low-energy environment in which bioturbation processes are very widespread; they also have a great paleoecological significance because of their formation in place.

Colonial framework builder organisms were numerous throughout the geological record; among them are scleratinian corals, bryozoans, stromatoporoids, red calcareous algae, encrusting foraminifers, and annelids (**Plates 21.F, 22.A, B, C**). These organisms played an important role in carbonate depositional systems by constructing buildups or "reefs," often located along the frontal margin of carbonate platforms, thus controlling to a great extent their sedimentological evolution (Laporte, 1974; Toomey, 1981; James, 1984a, b, c; Crevello et al., 1989). The solid carbonate frameworks resulting from baffling, encrusting, binding, and secreting processes are designated as bioconstructed limestones. The materials filling the interstitial spaces of the framework can cover the entire spectrum of carbonate sediments from low-energy calcilutites to high-energy calcarenites and calcirudites and to pure calcite cements. They are excellent indicators of the local conditions under which the framework was constructed. Frequently, silt- to sand-size grains settle at the bottom of framework cavities as geopetal internal sediments, often graded bedded and overlain by a variety of calcite cements.

DIAGENETIC EVOLUTION

The diagenetic evolution of limestones consists of a continuous sequence of processes that modify their depositional texture and composition through time from deposition to deep burial. Innumerable workers have investigated the petrographic and geochemical aspects of limestone diagenesis. A number of fully documented overviews were published in recent years on this subject (Bricker, 1971; Longman, 1980; Schneidermann and Harris, 1985; James and Choquette, 1990a, b; Choquette and James, 1990; Tucker and Wright, 1990).

The fundamental feature of the diagenesis of limestones is their interaction with marine and meteoric waters and with subsurface solutions. Each of these fluids leaves a unique diagenetic imprint at the time of its reaction with the limestone. This imprint can be considered as a distinct diagenetic environment and qualified in time as a diagenetic phase, although this imprint may be largely or totally destroyed by subsequent diagenetic phases. In a simplified way, it is convenient to distinguish four diagenetic environments when dealing, for instance, with an exposed carbonate platform that has a well-developed and active freshwater lense (**Fig. 5.5**). For the sake of demonstration, the carbonate sediment is assumed to begin its evolution in the marine phreatic, then pass through the mixed marine vadose–freshwater vadose on the strandline, continue into the freshwater vadose (with subaerial exposure), descent into the freshwater phreatic, the mixed freshwater phreatic–marine phreatic (mixing zone), and finally reach deep burial conditions.

In reality, from the time of deposition as a sediment in the marine phreatic environment, a given limestone can follow different pathways until final burial, which may or may not involve passing through all the above-described diagenetic zones. Reconstruction of the resulting diagenetic sequence by the study of surviving textures and geochemical properties of the limestones indicates that many factors regulate their diagenetic evolution, such as tectonic framework, climate, composition and flow rate of the waters, and the length of residence time within a particular diagenetic environment.

The Marine Phreatic Environment

The marine phreatic environment, in which pores are entirely filled with seawater, is characterized by boring (**Plate 22.D**) and micritization (**Plate 22.E**) of bioclasts and constructed frameworks by endolithic algae and fungi, by deposition of aragonite or calcite muds, and by several types of precipitated cements (James and Choquette, 1990a). Microcrystalline cement, originally either high-Mg calcite, low-Mg calcite, or aragonite, was reported, but its interpretation is difficult because of the effects of aggrading neomorphism (Mountjoy and Riding, 1981).

Lasemi and Sandberg in a study of Pleistocene micrites from south Florida and the Bahamas lithified under early vadose conditions (1984) showed, by means of polished and etched surfaces, that microcrystalline muds with aragonite precursors display neomorphic calcite crystals with pitted surfaces and relict aragonite inclusions, both absent in coexisting void-filling cements. These authors concluded from the similarities between micrite and microspar crystals that they have a common origin through a one-step neomorphic process of calcitization; that is, the aragonitic mud did not first calcitize to micrite and then alter to microsparite by aggrading neomorphism.

Isopachous rims of finely fibrous calcite cement (originally aragonite or high-Mg calcite) can be extremely widespread and of variable thickness. (**Plates 21.F, 22.B, F, 23.A**) The interference of their growth around grains produces polygonal boundaries. Two other types of fibrous cements are often associated in a given cement fringe. The first type is fascicular–optic calcite, which consists of nested cones of fibrous calcite with divergent optical axes and diverging subcrystals (Kendall, 1977); the second type is radiaxial fibrous calcite in which the crystals have consertal boundaries, curved twins, and convergent optical axes and

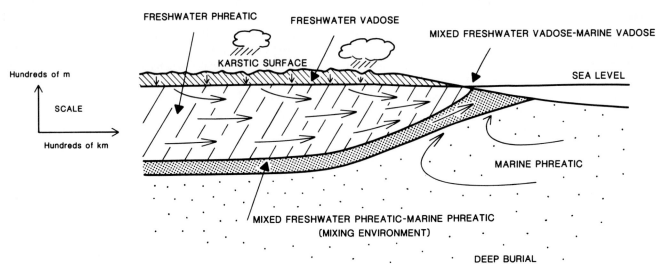

FIGURE 5.5 Main diagenetic environments in an exposed carbonate platform.

also display divergent subcrystals (Bathurst, 1959; Kendall and Tucker, 1973; Kendall, 1985). These two particular types of fibrous calcite cements are not forming in present-day oceans. It appears that both were precipitated in the past as low-Mg and high-Mg calcite (Kendall, 1985; Saller, 1986); the latter has undergone conversion as shown by its microdolomite rhombic inclusions (Lohmann and Meyers, 1977).

Stromatactis structures are irregular masses of fibrous radiaxial calcite cement often displaying a flat base and digitated to flamelike upper boundaries. These structures occur mainly in calcilutite mounds. Bathurst (1959, 1980, 1982) and Wallace (1987) interpreted them as the product of a complex submarine diagenesis during which cavity development and sea-floor incrustations alternated and interfered. Botryoidal masses of fibrous calcite cement occur frequently in older limestones (**Plate 23.B**) and closely resemble forms of aragonite cement found in Holocene reefs. Their fabric consists of mosaics of minute crystals or fan-shaped pseudosparite, both indicating replacement of original aragonite (Assereto and Folk, 1980; Mazzullo, 1980; Aissaoui, 1985). However, Upper Cretaceous botryoidal cements from widespread Mediterranean localities were interpreted as former high-Mg calcite (Ross, 1991). The petrographic and geochemical features indicating this original mineralogy include rhombic terminations to precursor crystallites, microdolomite inclusions, and low strontium concentrations. The former calcitic mineralogy of Upper Cretaceous botryoids is in agreement with the concept that ocean chemistry at that time (calcite seas and greenhouse conditions) inhibited inorganic aragonite precipitation.

Scalenohedral calcite cement, consisting of clear columnar crystals with unit extinction and rare inclusions, has been reported and interpreted as marine in origin (Kerans et al., 1986). This type of cement can be distinguished from a similar one formed during deep burial, which has a typically dull cathodoluminescence.

Marine epitaxial calcite cements developed as isopachous crusts on a variety of substrates of skeletal and nonskeletal nature, such as brachiopods, corals, stromatoporoids, tentaculites, calcispheres, and ooids, most of which have a primary fibrous calcite microstructure (Kerans et al., 1986).

Longman (1980) found little obvious evidence of syntaxial overgrowths on pelmatozoan debris in present marine phreatic environments. However, such overgrowths were reported by Kerans et al. (1986) in the Devonian of western Australia, and Walker et al. (1990) recently recognized them throughout the Paleozoic. They noticed that the marine phreatic syntaxial overgrowths are very thin (tens of microns). Therefore, they are difficult to microsample for geochemical studies, and an association of petrographic criteria appears a more reliable approach for establishing the marine origin of the initial overgrowth stage. Walker et al. (1990) described

the following criteria: overgrowths truncated by marine fibrous or pseudofibrous cements (radiaxial, radial fibrous, and fascicular optic); synchronous growth of fibrous marine cements and overgrowths that share growth zones that can be traced laterally in a continuous manner from one type of cement to the other, indicating coeval development; marine fibrous cements underlying and overlying syntaxial overgrowths expressing a criterion of superposition; several types of compromise boundaries between unzoned crystals, indicating interference during growth of adjacent fibrous cements and overgrowths; oolitic and oncolitic accretionary rings and coralline algae encrustations coating syntaxial overgrowths; various types of marine borings in overgrowths, reaching or not their pelmatozoan substrate; micrite matrix completely surrounding overgrowths; and turbid or cloudy aspect in plane-polarized light of early parts of overgrowths in contrast with the limpid later parts, showing interference with meteoric and burial cements. This turbidity is partly attributed to microdolomite inclusions similar to those described by Lohmann and Meyers (1977) as a criterion for high-Mg calcite deposited in marine conditions and partly to fluid inclusions, minute organic debris, and other suspended solids in seawater. These early cloudy overgrowths are nonluminescent, indicating the low Mn to Fe ratio of typical marine precipitation; they contrast with the multiple-zoned and highly luminescent later phases of overgrowths (Grover and Read, 1983; Kaufman et al., 1988).

An attempt by Walker et al. (1990) to draw a curve of variation of the abundance of syntaxial marine overgrowths, which certainly consisted of high-Mg calcite, during the Phanerozoic appears to show peaks corresponding with low-Mg calcite seas instead of aragonite, high-Mg calcite seas. This strange situation may result from insufficient or unreliable data for post-Paleozoic times and points out the necessity for further studies. Clearly, syntaxial overgrowths appear to develop rapidly in the freshwater phreatic environment where they display typical zonations in cathodoluminescence, which were used for establishing cement stratigraphy (Meyers, 1974, 1978). They may continue to grow in the deep burial environment.

The Mixed Marine–Freshwater Vadose Environment

The transition between the marine phreatic environment and the freshwater vadose environment corresponds to the strandline, which can be as narrow as a beach and as wide as a tidal flat. The system consists here of sediment–water–air, with the water being commonly a mixture of marine and freshwater.

A first aspect of the mixed marine–freshwater vadose environment is represented by the formation of beachrock, which is an extremely rapid geological process. The precipitated cements are aragonite and high-Mg calcite; they display the same spectrum of fabrics as those in the marine phreatic environment, but microcrystalline aragonite cement is widespread (Bricker, 1971). However, characteristic features produced by the alternating presence of air and water in the pores are meniscus and pendant structures of these cements (**Plates 19.B, 23.C**). Open cavities often contain, at their bottom, geopetal vadose silts, at times graded bedded (**Plate 23.D**), which are concentrations of minute bioclasts, pellets, and crystals that escaped dissolution processes, and are overlain by various types of cements (Dunham, 1969). There is general agreement that the precipitation of beachrock cement is from seawater or from a variable mixture of seawater–meteoric water by evaporation and CO_2 degassing as an intertidal process taking place during the low-tide phase (Stoddart and Cann, 1965; Bricker, 1971; Moore, 1973; Meyers, 1987). It is possible that some of this precipitation may be biologically induced by algae and bacteria (Hanor, 1978). Ancient beachrocks have not received the attention they deserve (James and Choquette, 1990 a), although their incomplete cementation, associated with dissolution processes, makes them potential petroleum reservoirs of commercial interest (Carozzi et al., 1983).

A second aspect of the marine–freshwater vadose environment is represented by carbonate tidal flats. Pelletoidal calcilutites deposited in intertidal to supratidal conditions display several types of fenestral cavities, ranging in size from a few millimeters to several centimeters, which are either irregular in shape, or laminoid, that is, elongated parallel to bedding, or tubular, vertical to subvertical. Their formation is attributed to the entrapment in the carbonate mud of gas originating from organic matter decay combined with shrinkage processes (Shinn, 1968; Shinn et al., 1969). Fenestral cavities are considered reliable criteria for the intertidal environment. Recent fenestral structures are often filled with acicular to bladed aragonite cement, ancient ones by cavity-filling sparite cement. In both cases, geopetal marine vadose silts frequently occur at the bottom of the cavities and are overlain by various types of cements. Fenestral limestones have been the subject of numerous descriptions (Fischer, 1964; Grover and Read, 1978; Shinn, 1983).

The Freshwater Vadose Environment

The freshwater vadose environment corresponds to the subaerial zone that lies between the land surface and the zone of water saturation, or water table (Longman, 1980). Meteoric freshwater and air alternatively occur in the pores during phases of wetting and drying. Consequently, water is generally held between grains by capillary forces as meniscus drops or under grains as pendant drops. The freshwater vadose environment is under direct climatic control, which regulates precipitation, evaporation, and vegetation.

At present, two end-member products of this environment are recognized (Esteban and Klappa, 1983; James and Choquette, 1990b): the surface karstic facies and the calcrete or caliche facies. The limestone surface, either directly exposed or located beneath a soil cover, comes in contact with aggressive meteoric waters, which predominantly dissolve but can also lead to precipitation by evaporation and biological action. Therefore, limestone reacts essentially in two ways: formation of a surface karst due to dissolution, or generation of calcrete by a combination of solution and precipitation. These two features coexist and overlap, but karst develops under all climatic conditions, whereas calcrete indicates semiarid conditions. Paleokarstic phenomena are beyond the scope of this presentation; they have been extensively investigated (Wright, 1982a and b; James and Choquette, 1988; Wright et al., 1991). Calcretes, paleocalcretes (**Plate 23.E, F**), and paleosols have equally been studied in detail in recent years (Wright, 1986). Calcretes, which range from <0.5 to 3 m in thickness, are characterized by a vertical zoning that can be described as follows from top to bottom. At the surface is a hardpan consisting of a solid crust of cryptocrystalline to microcrystalline calcite, generally well laminated and brecciated by dissolution and precipitation processes grading downward into irregular plates and crusts. The underlying nodular and crumbly zone is very complex. It shows organic-rich diagenetic pellets and coated grains (**Plate 23.E**) often surrounded by circular shrinkage–expansion cracks; skeletal and lithic grains at all stages of micritization including the generation of pockets of mud-supported fabrics; multiple brecciation fabrics with intersecting and anastomosing systems of calcite-filled cracks; fenestral and alveolar cavities partially filled with precipitated cements of microcrystalline, acicular, and bladed calcite overlying geopetal vadose silts; and well-formed vadose ooids and pisoids (**Plate 23.F**). Finally, the massive chalky zone, which represents the transition to the underlying unaltered limestone, consists of uncemented silt-size calcite grains with scattered coated grains. Since roots of plants can penetrate any of the above-mentioned zones, rhizoconcretions that result from the precipitation of calcite around rootlets, or from filling of root molds, appear as cylindrical, straight, wavy or branching calcite molds or calcite-filled tubules, all of which are extremely diagnostic of calcretes (**Plate 23.E**). All the above-mentioned features indicate the combined effect of solution, precipitation, micritization, expansion, contraction, and slow rotation of constituents under the strong influence of plants, lichens, fungi, and bacteria (Walls et al., 1975; Esteban, 1976; Harrison and Steinen, 1978; Adams, 1980; Peryt, 1983; Wright, 1982a and b, 1987). In the complex domain of calcrete petrography, cathodoluminescence is a very useful technique to unravel their evolution (Solomon and Walkden, 1985). In the solution zone of the freshwater vadose environment, aragonitic bioclasts are preferentially dissolved and, as mentioned above (Lasemi and Sandberg, 1984), some aragonitic muds may already invert to both micrite and microsparite during a one-step neomorphic process, although the main diagenetic environment for neomorphism is the freshwater phreatic one described below.

When meteoric waters continue to move downward into the lower part of the freshwater vadose zone, saturation in calcium carbonate is eventually reached, and typical vadose meniscus and pendant calcite cements (**Plates 19.B, 23.C**) are precipitated, which essentially reflect the distribution of pore waters (Longman, 1980). Debris of these cements may fall to the bottom of cavities forming geopetal accumulations of vadose silts, often graded bedded (**Plate 23.D**). These vadose cements consist of very fine, equant calcite crystals, but initial crystals may not be preserved within the final filling of the pores by coarser calcite crystals increasing in size toward the center of the cavities, thus establishing the transition toward the types of cements developed in the underlying freshwater phreatic environment.

The Freshwater Phreatic Environment

The freshwater phreatic environment is located between the vadose freshwater and the mixed freshwater–marine phreatic environment, the mixing zone (Longman, 1980). Pores of the phreatic zone are always filled with water, and precipitated calcite crystals can develop without obstacles, except intercrystalline interferences. Thick isopachous layers of bladed-prismatic calcite crystals form around grains and grade toward the center of the pores into coarsening equant mosaics of blocky calcite (**Plates 21.F, 22.B, 24.A**). However, these cement morphologies develop also under deep burial conditions, making crystallographic shapes by themselves unreliable criteria, but the dull cathodoluminescence of deep burial cements provides the answer.

In the freshwater phreatic environment, all aragonite is theoretically replaced by stable, equant low-Mg calcite, and this environment represents the beginning of the final lithification of limestones. For instance, syntaxial overgrowths around pelmatozoan fragments develop rapidly under these conditions (**Plates 19.F, 24. B**), in some instances with a displacive mode (Maliva, 1989) that under the SEM shows either an unimpeded displacive precipitation in which adjacent grains and matrix offer little resistance to crystal growth, or an impeded displacive precipitation in which resistance decreases calcite precipitation at overgrowth–obstacle contacts. Other types of crystallographic anomalies, such as inhibition, distortion, and dissolution of overgrowths on pelmatozoan fragments, were described by Braithwaite and Heath (1989). At any rate, and as mentioned above, these overgrowths may continue to grow in the deep burial environment. At this stage, very little porosity is left, which had been depositional primary or developed earlier in beachrock conditions or in the freshwater vadose zone.

The freshwater phreatic environment is also the major location of the process called *neomorphism,* a term created by Folk (1965) to include all transformations, by solution–reprecipitation, between one mineral and itself or a polymorph. This definition includes therefore the inversion of aragonite to calcite (which can already occur in the freshwater vadose environment), but mainly the calcite to calcite "recrystallization" (Bathurst, 1975, 1983). Most neomorphism in limestones is of the aggrading type, which, when completed, generates a coarse mosaic of crystals highly destructive of previous depositional and diagenetic textures. Some of the most instructive examples of aggrading neomorphism occur in oolitic calcarenites (**Plate 24.C**), in calcilutites and calcisiltites (**Plate 24.D**), and in large framework-building colonial organisms (**Plates 22.C, 24.E**). Two major textural aspects are generated according to Folk (1965). One, called coalescive, corresponds to a gradual enlargement of all crystals maintaining a uniform crystal size at all given times. In the second, called porphyroid, a few crystals grow scattered within an unchanged background. Upon completion of the neomorphic process, the result of both types is indistinguishable (**Fig. 5.6**). Neomorphic calcite crystals are designated as pseudomicrosparite when crystal

NEOMORPHISM

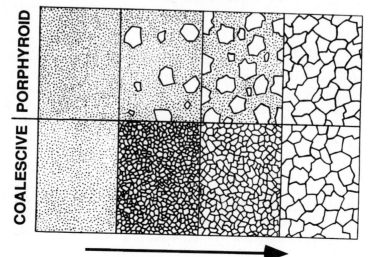

FIGURE 5.6 Neomorphic fabrics in limestones. The conversion of small crystals to large ones may take place either (1) by growth of a few large crystals in a static groundmass (porphyroid neomorphism) or (2) by gradual enlargement maintaining a uniform crystal size at all given times (coalescive neomorphism). The end points are indistinguishable. From Folk (1965). Reprinted by permission of the Society of Economic Paleontologists and Mineralogists.

size is smaller than 30 μ and pseudosparite when it is greater than 30 μ, with no upper size given; size limits are purely descriptive and vary from author to author. It is important to recognize the textural features of aggrading neomorphism so as to avoid confusing it with diagenetic precipitated cements. These features are (**Plate 24.C to E**) irregular or curved intercrystalline boundaries; very irregular crystal size distribution; patchy and complex three-dimensional shapes of the coarse mosaic; gradational boundaries between neomorphic calcite and the surrounding matrix; and inclusions of unaffected patches of matrix or of bioclasts in the neomorphic mosaic, which have the appearance of "floating," with a distribution pattern that is incompatible with any grain to grain depositional framework. In calcilutites, calcisiltites, and mud-supported calcarenites, neomorphism can display still other aspects, such as radiating clusters of calcite needles, fine mosaics within bioclasts, bladed fringes on various bioclasts, and even irregular overgrowths around crinoid columnals extending into the matrix. However, cathodoluminescence showed that, in some cases, these overgrowths were not formed by neomorphism but represent a passive cement filling solution voids that developed around pelmatozoan debris in diagenetic environments ranging from near surface to deep burial with pressure solution (Walkden and Berry, 1984).

Degrading neomorphism, that is, a decrease of the size of calcite crystals or "crystal diminution," is relatively rare in limestones and appears to occur under conditions of tectonic deformation or very low grade metamorphism (Wardlaw, 1962; Tucker and Kendall, 1973; Bathurst, 1975).

The Mixing Environment

The boundary between the lower part of the freshwater phreatic water lens and the underlying marine phreatic zone corresponds to a layer of variable thickness of brackish water formed by the mixing of freshwater and seawater (Longman, 1980). Diagenetic processes taking place in this mixing zone are still poorly understood, although hydrologic and geochemical aspects are being actively investigated (Back et al., 1986). The most interesting diagenetic process that occurs in the case of very active mixing when the freshwater phreatic lens is abundantly fed by meteoric waters is dolomitization (see Chapter 6). Badiozamani (1973) showed that mixing of about 10% seawater with 90% freshwater produces a solution slightly undersaturated with calcite and oversaturated with dolomite. This type of dolomitization by slow replacement of calcite has been called the "Dorag dolomitization model." It was recognized as operating today in Yucatan, Bermuda, and Florida and extended to many cases of older dolostones (Hanshaw et al., 1971; Badiozamani, 1973; Land, 1973; Land et al., 1975; Carozzi, 1989). Similar conditions were assumed by Knauth (1979) for the generation of replacement cherts in limestones (see Chapter 7).

The Deep-Burial Environment

The deep-burial environment extends essentially from the mixing zone downward to a variable and unspecified depth where low-grade metamorphism begins. It is directly influenced by the lithostatic pressure of the overlying column of rocks, which in turn controls temperature and pressure, circulation of fluids, and pore-water chemistry, which are factors reacting in time with the mineralogy, grain size, and textures of a given limestone. The predominant processes are dewatering, decrease of porosity, mechanical compaction, microfracturation, chemical compaction (represented by pressure solution), cementation, various types of replacement by dolomite and other minerals, and finally alteration and thermal maturation of organic matter into hydrocarbons (Scholle and Halley, 1985; Choquette and James, 1990).

Although the general result of deep-burial diagenesis is a gradual decrease of porosity by the combined effect of compaction and cementation, pore fluids may often change composition to the extent of becoming aggressive solutions capable of generating secondary porosity, an aspect discussed below in more detail.

The results of physical compaction when the thickness of a given limestone bed decreases and constituents are reorganized in a tighter fabric are mainly a function of the depositional mud-supported or grain-supported texture. In mud-supported limestones, the results are very striking, including strong crushing and deformation of bioturbation textures; flattening of burrow fillings; collapse of fenestral cavities; flattening and reciprocal merging of pellets; rotation toward the horizontal plane of elongate grains, which may be crushed when brought into reciprocal contact; draping of organic-matter-rich laminae around more rigid grains; and fracturation and rotation of apparently more resistant laminae (Shinn and Robbin, 1983). The general textural result is a drastic change of calcilutites and calcisiltites, which had originally less than 10% sand-size constituents, to mud-supported calcarenites (with 20% to 30% sand-size constituents), and even to grain-supported calcarenites with a carbonate mud matrix and more than 30% sand-size constituents. This situation, if it remains undetected, affects the environmental interpretation of these rocks.

In grain-supported limestones with interstitial mud matrix or sparite cement, the plastic deformation of pellets, coated grains, and skeletal fragments is accompanied by brittle fracturation and crushing. Very spectacular are the effects on oolitic calcarenites with sparite cement, in which ooids are distorted, with spalling of their external cortical layers (**Plates 20.D, 25.B**). Similar situations are displayed by micrite envelopes, often overlain by isopachous fibrous rim cement, that are spalled from their substrates and broken into fragments dispersed in the sparite cement. In some instances, broad and flat or weakly curved skeletal fragments such as phylloid algal blades or brachiopod and pelecypod shells or even the inside of articulated shells show a sheltering effect. This effect results in the preservation of the original texture, which, upon comparison with the outside compacted texture, provides a relative measure of the amount of physical compaction. Several attempts have been made to determine more precisely the amount of physical compaction, mostly by using originally cylindrical bodies such as bioturbation tubes or trace fossils (Byers and Stasko, 1978; Gaillard and Jautee, 1987; Ricken, 1987).

It seems reasonable to increase the scale of observation of the effects of physical compaction and to infer that entire limestone formations could be thinned by the cumulative results of the process. Beach and Schumacher (1982) demonstrated that the final result could be a thinning reaching 31%; this thinning was shown under the microscope to be expressed by crushed bioclasts and other textural features, indicating that only physical compaction was the responsible agent.

Compaction of fine-grained carbonate sediments is generally assumed to be related to the amount of clay minerals and organic matter they contain. Ricken (1987) suggested that the inverse relationship between the percentage of carbonate and the amount of compaction has the value of a general law that can be used for calculating all aspects of compaction and decompaction.

Chemical compaction represented by pressure solution is an important diagenetic process in limestones (**Plates 20.C to E, 24.F**). It produces reductions of the thickness of individual beds and entire formations on the order of 20% to 35%, which are added to the effects of physical compaction. This cumulative effect of the two types of compaction (**Fig. 5.7**) is particularly well shown by the study of crinoid–bryozoan calcarenites (Meyers, 1980; Meyers and Hill, 1983). Pressure solution by its very essence is also a producer of carbonate cement in deep-burial conditions, which completes the process of porosity destruction that began in the initial stages of limestone diagenesis (Wong and Oldershaw, 1981; Scholle and Halley, 1985; Simpson, 1985). The most spectacular products of pressure solution are stylolites, which can be at times sufficiently abundant to generate a pseudobedding appearance. Stylolites (**Plate 24.F**) display great variations in shape and amplitude, and their pattern depends upon numerous factors, such as lithostatic pressure, calcium carbonate mineralogy, presence of clay minerals and organic matter, porosity and permeability, and chemistry and flow rate of circulating solutions. Numerous workers have dealt with stylolites, but Wanless (1979, 1983) stressed the important distinction between the sutured seams or common stylolites and the nonsutured seams or microstylolites or clay seams that are wispy to planar surfaces ranging laterally from a few centimeters to hundreds of meters. Like sutured seams, they show concentrations of clay minerals, organic matter, and pyrite. Nonsutured seams appear to characterize limestones relatively rich in clay minerals, in silt-size detrital quartz, and in organic matter. The pressure-

solution origin of nonsutured seams is difficult to demonstrate, particularly in the light of similar features produced experimentally by physical compaction called pseudostylolites (Shinn and Robbin, 1983). Nonsutured seams can also be closely spaced to generate a pseudobedding appearance. If they were indeed produced by chemical compaction, they would be, like stylolites, producers of burial calcium carbonate cements, in particular for argillaceous limestones (Bathurst, 1987).

Deep-burial cements are extremely important aspects of the diagenetic evolution of limestones (Scholle and Halley, 1985; Choquette and James, 1990). They represent by far most of the cementation observed under the microscope, which has practically eliminated all porosity and permeability except the irreducible types along crystal boundaries. The major types of deep-burial cements include calcite, dolomite, anhydrite, and barite, all of which may occur in successive generations, filling voids and microfractures, replacing the limestone themselves as well as being mutually replacive. Deep-burial cements contain various types of fluid and hydrocarbon inclusions, which can be used as geothermometers. Geochemical studies showed that these various cements can form under a relatively large range of conditions, which include temperatures between 40° to 50°C and 200° C or higher; formational fluids range from brackish to highly saline, with acidic to strongly alkaline pH and moderately to highly negative Eh; fluid flow rates range from nearly stagnant to flow of normal groundwaters; and, finally, the time of formation extends from thousands to millions of years (Choquette and James, 1990). Coarse-grained, deep-burial calcite cements are generally ferroan and enriched in Mn^{2+} and impoverished in Sr^{2+} compared with earlier types. Their cathodoluminescence is typically dull with or without compositional zonation (Grover and Read, 1983). Besides the geochemical characterization of deep-burial cements, the most diagnostic features of their origin are the observable cross-cutting relationships under the microscope, which relate them to earlier diagenetic cements, physical and chemical compaction, and episodes of tectonic microfracturation (**Plate 24.F**). Among these time relationships are deep-burial cement crystals healing fractured ooids or spalled ooids, postdating reciprocal interpenetration of grains by physical compaction, growing across stylolites, or filling systems of nonfabric selective fractures.

Deep-burial calcite cement is the most frequent; it displays a bladed-prismatic habit showing elongate scalenohedral crystals growing directly on grain surfaces or over earlier diagenetic or marine cements. It is usually the oldest burial cement and is overlain by a coarse, equant sparite mosaic. As mentioned earlier, these fabrics occur also in calcite precipitated in the freshwater phreatic environment, but here cathodoluminescence is different. In the bladed-prismatic, it is typically dark with a few bright zones, whereas in the coarse sparite mosaic, which may or may not be appreciably

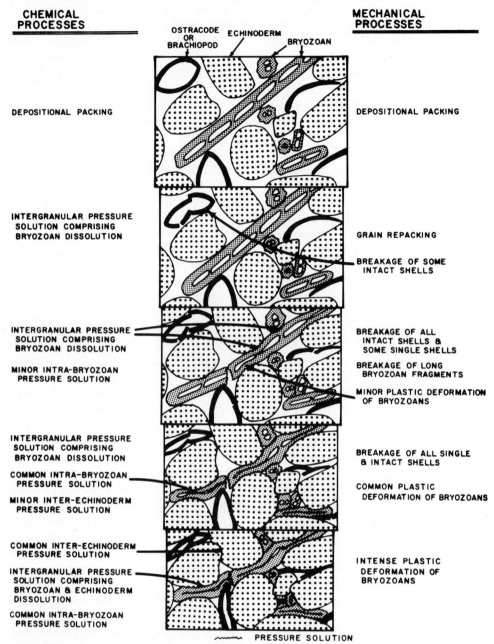

CHEMICAL
PROCESSES

MECHANICAL
PROCESSES

OSTRACODE
OR
BRACHIOPOD   ECHINODERM
BRYOZOAN

DEPOSITIONAL PACKING

DEPOSITIONAL PACKING

INTERGRANULAR PRESSURE
SOLUTION COMPRISING
BRYOZOAN DISSOLUTION

GRAIN REPACKING

BREAKAGE OF SOME
INTACT SHELLS

INTERGRANULAR PRESSURE
SOLUTION COMPRISING
BRYOZOAN DISSOLUTION

MINOR INTRA-BRYOZOAN
PRESSURE SOLUTION

BREAKAGE OF ALL
INTACT SHELLS &
SOME SINGLE SHELLS

BREAKAGE OF LONG
BRYOZOAN FRAGMENTS

MINOR PLASTIC DEFORMATION
OF BRYOZOANS

INTERGRANULAR PRESSURE
SOLUTION COMPRISING
BRYOZOAN DISSOLUTION

COMMON INTRA-BRYOZOAN
PRESSURE SOLUTION

MINOR INTER-ECHINODERM
PRESSURE SOLUTION

BREAKAGE OF ALL SINGLE
& INTACT SHELLS

COMMON PLASTIC
DEFORMATION OF BRYOZOANS

COMMON INTER-ECHINODERM
PRESSURE SOLUTION

INTERGRANULAR PRESSURE
SOLUTION COMPRISING
BRYOZOAN & ECHINODERM
DISSOLUTION

COMMON INTRA-BRYOZOAN
PRESSURE SOLUTION

INTENSE PLASTIC
DEFORMATION OF
BRYOZOANS

PRESSURE SOLUTION

FIGURE 5.7   Schematic diagram showing the interpretation of processes that occurred during progressively more intense compaction in Lake Valley Formation (Mississippian) biocalcarenites. Calcite cements and cherts are omitted for simplification. From Meyers (1980). Reprinted by permission of the Society of Economic Paleontologists and Mineralogists.

ferroan, cathodoluminescence is generally dull, zoned, or unzoned. Poikilotopic calcite, which in places takes the place of the equant mosaic, behaves in the same way. Calcite-filling successive generations of tectonic microfractures show frequently a tooth-comb texture resulting from the interfering growth of bladed-prismatic crystals from both walls of the fracture (**Plate 24.F**).

Other deep-burial cements are represented by patches of vug-filling or replacive saddle dolomite (Radke and Mathis, 1980; Machel, 1987; Kretz, 1992) and of xenotopic (nonplanar) dolomite cement (Gregg and Sibley, 1984), which appears as a mosaic of anhedral crystals of variable size with irregular and curved boundaries and undulatory extinction. The formation of this type of dolomite cement, often replacive, falls into the temperature range of saddle dolomite, which is between 50° and 150°C.

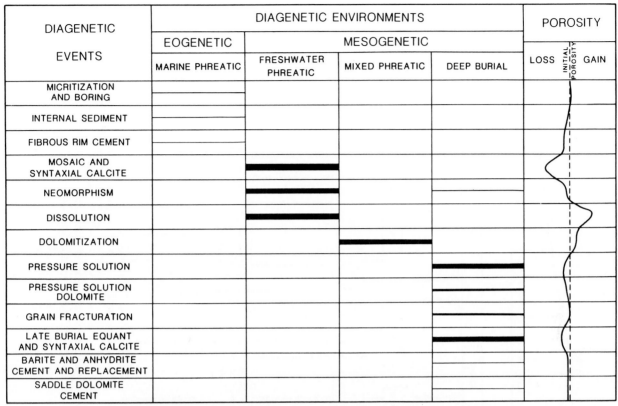

DIAGENETIC EVENTS	DIAGENETIC ENVIRONMENTS				POROSITY		
	EOGENETIC	MESOGENETIC			LOSS	INITIAL POROSITY	GAIN
	MARINE PHREATIC	FRESHWATER PHREATIC	MIXED PHREATIC	DEEP BURIAL			
MICRITIZATION AND BORING							
INTERNAL SEDIMENT							
FIBROUS RIM CEMENT							
MOSAIC AND SYNTAXIAL CALCITE		▬▬▬					
NEOMORPHISM		▬▬▬					
DISSOLUTION		▬▬▬					
DOLOMITIZATION			▬▬▬				
PRESSURE SOLUTION				▬▬▬			
PRESSURE SOLUTION DOLOMITE				▬▬▬			
GRAIN FRACTURATION				▬▬▬			
LATE BURIAL EQUANT AND SYNTAXIAL CALCITE				▬▬▬			
BARITE AND ANHYDRITE CEMENT AND REPLACEMENT				▬▬▬			
SADDLE DOLOMITE CEMENT				▬▬▬			

FIGURE 5.8 Composite diagenetic sequence for carbonate depositional model 2, Amapá Formation (Paleocene–Middle Miocene), Foz do Amazonas Basin, offshore NE Brazil. From Wolff and Carozzi (1984).

Mosaics of coarse anhydrite, reaching in many cases poikilotopic texture, occur also as burial cement and replacement, generally postdating calcite and dolomite. Only anhydrite exists in deep burial because it is the stable form, gypsum converting to anhydrite at about 1,000-m burial (Murray, 1964; Kendall, 1984). Barite is relatively rare as a fracture-filling mosaic of anhedral crystals.

The sequential study of diagenetic environments of limestones leads to the preparation of a diagenetic sequence to which the relative gains and losses of porosity may be added (**Fig. 5.8**). In the case of a well-documented depositional model based on detailed microfacies studies, it is possible to go one step further and unravel the diagenetic pathways for each individual microfacies (**Fig. 5.9**) within their respective depositional belts (Wolff and Carozzi, 1984). Numerous examples of diagenetic sequences for limestone of all ages were presented by Carozzi (1989).

The final aspect of deep-burial diagenesis is the controversial question about the origin of its various cements. In the most common case of calcite, it is obvious that calcium carbonate needs to be dissolved in some zones of the carbonate–water system and precipitated elsewhere after a certain distance of transfer. The most debated aspect is precisely the distance of transport, the apparently large volumes of fluids required for the process, and their flow rates. All these aspects of cementation of limestones are similar to those previously discussed for quartz arenites (see Chapter 1). However, for limestones, an additional process has to be considered, that is, the increasing evidence that pressure solution, in a partially closed system, may be responsible for their autocementation under deep burial diagenesis (Hudson, 1975, 1977; Scholle and Halley, 1985). Indeed, many observers, assuming that stylolitization liberates carbonate for burial cements, have searched for relationships between the distribution of stylolites and zones of abnormally low porosity (Nelson, 1981). Wong and Oldershaw (1981) were able to demonstrate a space and time relationship, that is, development of stylolites and simultaneous precipitation of coarse-zoned calcite cements produced by pressure solution, even to the extent that stylolites in their later stages of development grew at the expense of some of the earlier cements. Similarly, in the case of the Salem Limestone of west-central Indiana, Finkel and Wilkinson (1990) demonstrated a genetic relationship between stylolitization and significant contribution (47% to 90%) to the intergranular sparite cement in adjacent calcarenites. Such studies might tip the scale in favor of a generalized autocementation of limestones in deep-burial conditions.

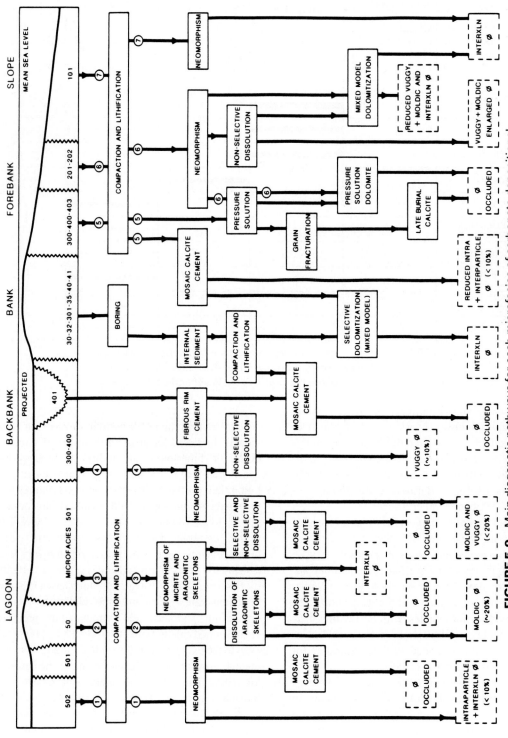

FIGURE 5.9 Main diagenetic pathways for each microfacies of carbonate depositional model 2, Amapá Formation (Paleocene–Middle Miocene), Foz do Amazonas Basin, offshore NE Brazil. From Wolff and Carozzi (1984).

RESERVOIR PROPERTIES

The above-described sequence of diagenetic environments, regardless of its various possible pathways, ends with deep-burial conditions and the gradual destruction of depositional porosity and early acquired secondary porosity (in the fresh-water vadose environment where Wright et al. (1991) described paleokarstic reservoirs), which terminates with irreducible porosity along crystal boundaries. However, porosity can be generated in the deep-burial environment by dissolution processes on a much larger scale than assumed for buried sandstones and following the same scenario (see Chapter 1). Solution porosity in sandstone bodies is assumed to be produced by the action of aggressive solutions rich in CO_2 originating from the thermal decarboxylation of maturing organic matter in adjacent compacting basinal shales. Such aggressive fluids would rise from the middle part of basins toward sandstone masses along the basin margins. Replacing the sandstone bodies with carbonate platforms would make the process applicable to limestones. Indeed, secondary deep-burial porosity has been documented in many instances at depths of more than 2 to 3 km (Moore and Druckman, 1981; Elliott, 1982; Druckman and Moore, 1985; Moore, 1985, 1989). The occurrence of this secondary porosity brought the same general realization as for sandstones of the great importance of deep-seated reservoirs generated by burial dissolution (Mazzullo and Harris, 1992).

The terminology and classification of porosity in limestones and dolostones was codified by Choquette and Pray (1970) as either fabric selective: interparticle (**Plate 25.A**), intraparticle (**Plate 25.B**), intercrystal, moldic (**Plate 25.C, D**), fenestral, shelter, and growth framework; or nonfabric selective: fracture, channel, vug, and cavern; and fabric selective or not: breccia, boring, burrow, and shrinkage (**Fig. 5.10**). An attempt at a genetic classification of natural porosity in oolitic limestones was presented by Rich and Carozzi (1981), and the importance of natural stylolitic porosity as a generator of deep-burial reservoirs was stressed by Carozzi and Von Bergen (1987).

Deep-burial porosity ranges from fabric selective to nonfabric selective and includes molds (**Plate 25.C, D**), enlarged molds (**Plate 25.E**), vugs (**Plate 25.E**), solution-enlarged interstitial voids, channels (**Plate 25.F**), stylolites (**Plate 26.A**), and fractures (**Plate 26.B**). The discovery of deep-burial porosity in sandstones prompted experimental studies on developing burial porosity in limestones (Donath et al., 1980). Samples of oolitic calcarenites of the Ste. Genevieve Limestone, Mississippian, of the Illinois basin were placed in a specially designed triaxial apparatus that permitted circulation of pore fluid (CO_2-charged water) under constant pressure, while subjecting the rock specimen to constant vertical pressure and lateral confining pressure simulating burial while temperature was kept at 24°C. The results (**Fig. 5.11**) indicated that, under conditions of stable low-Mg calcite, the process of oomoldic porosity begins with the selective dissolution of the highly porous outermost cortical layers of the ooids. Where individual ooids were in contact, permeability develops along pathways formed from one ooid margin to another. As solution progresses further and after having removed much of the highly vulnerable cortical layers, it begins to attack the nuclei of ooids. Pelletoidal nuclei seem most vulnerable and disappear first because of their intercrystalline microporosity; but subsequently skeletal nuclei (crinoid columnals) also dissolve. Before disappearance, the nuclei may fall to the bottom of the cavities in a geopetal fashion. The interstitial sparite cement remains essentially unaffected to form the framework of the oomoldic carbonate and to appear often as delicate hourglasslike islands between adjacent ooids. Etching observed along intracrystalline glide planes and along the irreducible porosity of crystal boundaries of the sparite indicates that, if the tests had been carried farther, nonfabric, channellike platy cavities would have been generated. These lamellar pores, combined with those originating from adjacent enlarged oolitic molds, would have eventually collapsed the general framework of the rock.

Closed stylolite systems investigated under similar experimental conditions in Atokan (Pennsylvanian) limestones of the Delaware basin in Texas (Von Bergen and Carozzi, 1990) developed stylolitic porosity similar to that reported from similar limestones buried at depths in excess of 4 km. The dissolution sequence in the experimental development of stylolitic porosity is as follows (**Fig. 5.12**): incipient pores follow the traces of stylolites and grade into stylolitic lamellar pores; these enlarge eventually into nonfabric selective channels. These stylolite-controlled conduits are surrounded by adjacent incipient-to-complete oomolds, enlarged oomolds, incipient-to-complete algal bioclast molds, enlarged biomolds, and halos of intercrystalline microporosity to porosity.

In summary, natural and experimental data indicate that for all practical purposes secondary deep-burial porosity is the major type of porosity occurring in limestones.

TYPICAL EXAMPLES

Plate 18.D. Pelagic calcilutite consisting of a cryptocrystalline groundmass of slightly argillaceous micrite with hematite flecks in which are randomly scattered tests of lagenids, globigerinids, and abundant *Globotruncana,* with chambers filled by microcrystalline calcite mosaic or calcilutite matrix. Minute angular grains of detrital quartz are distributed throughout the groundmass. Scaglia Rossa, Upper Cretaceous, Preggio, Tuscany, Italy.

Plate 18.E. Calcilutite with *Chara* oogonia consisting of a cryptocrystalline groundmass of dark calcilutite displaying longitudinal, tangential, and apical sections of oogonia with central cavities filled with sparite mosaic and

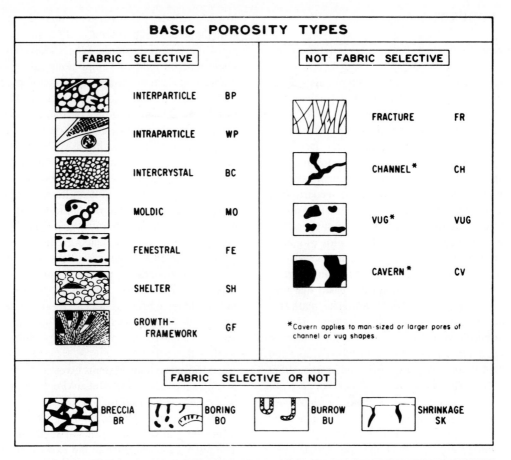

FIGURE 5.10 Geological classification of pores and pore systems in carbonate rocks. From Choquette and Pray (1970). Reprinted by permission of the American Association of Petroleum Geologists.

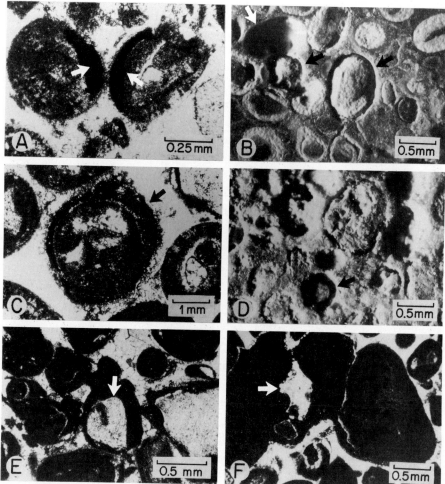

FIGURE 5.11 Oomoldic porosity experimentally developed in Mississippian oolitic limestone. **A.** Thin section of tested oolitic calcarenite displaying two adjacent ooids with partially dissolved cortical layers (arrows). Notice preservation of delicate bridge of sparite cement. Nicols crossed. **B.** Ground internal surface of tested oolitic calcarenite revealing several stages of fabric selective dissolution: removal of outermost cortical layers (center arrow), isolation of core (left arrow), total removal of ooid (upper left arrow). Natural reflected light. **C.** Thin section of tested oolitic calcarenite showing selective dissolution of all cortical layers of a normal ooid (arrow) and preservation of pelletoidal core. Nicols crossed. **D.** Ground bottom surface of tested oolitic calcarenite revealing advanced stages of selective dissolution with isolated nuclei (arrow). General aspect of the surface indicates undersaturation of fluids exiting the sample. Natural reflected light. **E.** Thin section of tested oolitic calcarenite showing several interconnected dissolved ooids, with the largest ooid displaying dissolved cortical layers and geopetal settling of crinoidal core at bottom of cavity (arrow). Nicols crossed. **F.** Thin section of tested oolitic calcarenite showing oomoldic porosity consisting of several juxtaposed and totally dissolved ooids separated by a delicate island of sparite cement (arrow). Nicols crossed. From Donath et al. (1980). Reprinted by permission of the Society of Economic Paleontologists and Mineralogists.

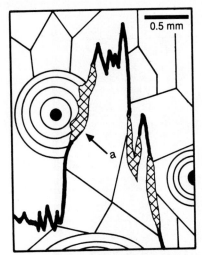

Porosity

Cement

Stylolite   Seam

Stage 1:   Incipient dissolution along seam (a)
and incipient oomoldic porosity.

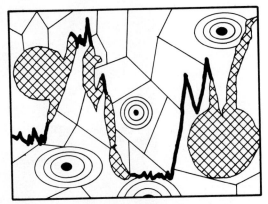

Stage 3:   Expanded lamellar stylolitic pores
with some nonfabric selective
and oomoldic porosity.

Ooid

1 mm

Stage 2:   Lamellar stylolitic porosity and incomplete
oomoldic porosity.

Stage 4:   Stylolite controlled channel porosity (a),
nonfabric selective porosity (b),
and enlarged oomoldic porosity (c).

FIGURE 5.12 Simplified sketch of experimental stylolitic porosity
developed in Atokan (Middle Pennsylvanian) limestone, Delaware
Basin, Texas. From Von Bergen and Carozzi (1990). Reprinted by permission of the *Journal of Petroleum Geology,* and Scientific Press Ltd.

rare internal sediment. Scattered in the matrix are smaller debris of *Chara* stems and oogonia, smooth ostracod shells,
rare minute grains of detrital quartz, and pyrite flecks. Upper
Portlandian (Purbeckian), Salève, France.

Plate 18.F. Bioturbated calcisiltite showing an irregularly textured matrix due to concentrations of tests of brachiopods, thin tests of gastropods and pelecypods, smooth
ostracods, monaxonic sponge spicules, and abundant
calcispheres. Irregular patches of neomorphic pseudosparite
represent unidentified bioclasts. Kinkaid Formation, Upper
Mississippian, Bunscombe, Illinois, U.S.A.

Plate 19.A. Mud-supported biocalcarenite consisting of subparallel sand-size bioclasts of various types of
brachiopods, bryozoans, large leperditid ostracods, hook-

shaped trilobites, and rare crinoids. Interstitial matrix is a
slightly but variably argillaceous calcisiltite with minute
bioclasts, phosphatic debris, and pyrite flecks. Galena
Group, Middle Ordovician, Guttenberg, Iowa, U.S.A.

Plate 19.B. Oncolitic calcarenite with calcisiltite
matrix consisting of a framework of large single or double
algal oncoids with nuclei formed by smaller reworked
oncoids or other bioclastic debris. The oncoids are surrounded by a thin rim of fibroradiated beachrock (mixed
freshwater–marine vadose) calcite cement locally with pendant texture (upper-left corner of picture). Interstitial subsequent matrix is a finely pelletoidal calcisiltite with minute
bioclasts. Macaé Formation, Albian–Cenomanian, Campos
Basin, offshore Brazil.

Plate 19.C. Crinoidal calcarenite with bioclastic matrix consisting of a bioturbated framework of crinoid columnals, bioclasts of brachiopods, neomorphosed gastropods, fenestrate bryozoans, and ostracods. Appreciable pressure solution is visible between all grain constituents. Dark interstitial matrix consists of argillaceous calcilutite with abundant pyrite flecks. Brereton Limestone, Middle Pennsylvanian, Middle Copperas Creek, southern Illinois, U.S.A.

Plate 19.D. Pelletoidal calcarenite with microsparite cement consisting of a framework of subrounded fecal pellets and irregularly shaped lithic pellets of unfossiliferous calcisiltite. Many pellets are reciprocally merged and all of them display hazy boundaries with the interstitial cement. Wapsipinicon Formation, Middle Devonian, Iowa City, Iowa, U.S.A.

Plate 19.E. Biocalcarenite with sparite cement consisting of a framework of subrounded bioclasts of bryozoans, brachiopods, neomorphosed pelecypods, partially micritized crinoid columnals, and *Lenticulina* associated with reworked normal ooids and rare lithoclasts of oolitic calcarenite. Interstitial cement is bladed to irregular mosaic of freshwater phreatic cavity-filling calcite with scattered patches of calcilutite and pyrite flecks. Calcaire Roux, Valanginian, Noirmont, Jura Mountains, Switzerland.

Plate 19.F. Crinoidal calcarenite with overgrowth calcite cement. The grain-supported framework consists of coarse, subrounded, abraded crinoid columnals cemented by extensive syntaxial overgrowths developed in freshwater phreatic to deep-burial conditions around each columnal and with compromise crystal faces. Atokan, Middle Pennsylvanian, Reeves County, Delaware Basin, Texas, U.S.A.

Plate 20.A. Oolitic calcarenite with sparite cement. The grain-supported framework consists of large normal ooids with well-developed concentric structure in places interfering with fibroradiated structure, and showing pelletoidal or bioclastic nuclei. Many broken ooids occur with shapes controlled by concentric or fibroradiated internal structures, together with rare subrounded intraclasts of oolitic calcarenite (upper-right corner of picture), bioclasts of punctate brachiopods, and crinoid columnals with superficial oolitic coating. Interstitial cement is freshwater phreatic, coarsely crystalline mosaic of sparite. St. Louis Limestone, Middle Mississippian, Cave-in-Rock, southern Illinois, U.S.A.

Plate 20.B. Oolitic calcarenite with poikilotopic calcite cement. The grain-supported framework consists of dark, mature and highly micritized ooids with small crinoidal or pelletoidal nuclei and showing only relicts of fine concentric structure. Rare crinoid bioclasts are associated with the ooids. The well-developed interstitial poikilotopic sparite cement, deposited in freshwater phreatic environment, contains minute concentrations of opaque iron oxides. Gilmore City Limestone, Lower Mississippian, Humboldt, Iowa, U.S.A.

Plate 20.C. Oolitic–crinoidal calcarenite with pressure solution. The highly interlocked framework consists of predominant crinoid bioclasts penetrating deeply by pressure solution into well-developed normal ooids with crinoidal and pelletoidal nuclei. The irregularly shaped interstitial spaces are filled with single crystals or a mosaic of coarsely crystalline calcite of freshwater phreatic origin. Bajocian, Neuchâtel, Jura Mountains, Switzerland.

Plate 20.D. Oolitic calcarenite with intense pressure solution. The framework consists of highly deformed normal ooids with small to large crinoidal or pelletoidal nuclei. The strong compaction has led to interlocking polygonal shapes with local triangular spalling of external concentric layers, and even complete transverse rupturing of elongated ooids. Ooid in upper-left corner of picture shows euhedral crystal of authigenic quartz. The interstitial cement shows remains of an isopachous bladed submarine calcite cement followed, in remaining larger spaces, by an irregular mosaic of freshwater phreatic sparite. Ste. Genevieve Limestone, Middle Mississippian, Dongola, Illinois, U.S.A.

Plate 20.E. Biocalcarenite with intense pressure solution. The framework consists of highly interpenetrated bioclasts of various types of brachiopods predominantly oriented parallel to bedding. The extremely advanced compaction has generated microstylolitic contacts between the bioclasts emphasized by linings of clay minerals and pyrite flecks that represent the insoluble residues of an original interstitial matrix. Helderberg Limestone, Lower Devonian, Bald Hill, Pennsylvania, U.S.A.

Plate 20.F. Mud-supported biocalcirudite (probably a tempestite) displaying large bioclasts of fenestrate bryozoans associated with randomly scattered sand-size bioclasts of crinoids, echinoids, and ostracods. Groundmass is a dark bioturbated calcisiltite with minute bioclasts of the above-mentioned organisms. Platteville Formation, Middle Ordovician, Dixon, Illinois, U.S.A.

Plate 21.A. Lithocalcirudite with sparite cement consisting of a framework of subrounded intraclasts of various types of dark calcilutites, pelletoidal calcilutites, and arenaceous calcisiltites, associated with rare ooids with fibroradiated internal structure. The intense pressure solution has reciprocally deformed the intraclasts along microstylolitic contacts. Interstitial cement is clear poikilotopic sparite deposited in freshwater phreatic conditions and subsequently twinned by tectonic stresses, which are also responsible for generalized oblique microfacturing. Lyell Formation, Upper Cambrian, Sunwapta Pass, Alberta, Canada.

Plate 21.B. Foraminifer bioaccumulated limestone consisting of a highly pressure welded framework of often deformed tests of textularids, *Orbitolites, Alveolina,* and *Quinqueloculina,* aligned subparallel to bedding and stained brown by bituminous matter. The very reduced interstitial material consists of argillaceous microsparite with small debris of the above-mentioned organisms and minute grains of detrital quartz. Paleocene, Lerida, Spain.

Plate 21.C. *Spiroclypeus* bioaccumulated limestone. Tests of this large benthic foraminifer are aligned subparallel to bedding, but dislocated and interpenetrated by the effect of appreciable pressure solution. Interstitial argillaceous calcisiltite, partially bioturbated, shows sand-size bioclasts of echinoids, miliolids, globigerinids, and scattered flecks of pyrite. Middle Miocene, Tabuelan, northern Cebu, Philippines.

Plate 21.D. *Halimeda* bioaccumulated limestone. Most of the fragments of fronds of this codiacean alga are aligned subparallel to bedding, but others are oriented obliquely by effect of bioturbation. Interstitial, highly argillaceous calcilutite matrix displays smaller algal bioclasts. Middle Miocene, Tabuelan, northern Cebu, Philippines.

Plate 21.E. *Amphistegina* bioaccumulated limestone. Accumulation of complete tests of this benthic foraminifer randomly oriented by effect of bioturbation. Interstitial argillaceous calcisiltite matrix displays smaller bioclasts of lepidocyclinids and echinoids. Middle Miocene, Iloilo Basin, Panay, Philippines.

Plate 21.F. *Donezella* bioconstructed limestone. The uncompacted branching network of tubes of dark cryptocrystalline to microcrystalline calcite typical of this alga is cemented by fibrous marine phreatic rim cement of calcite, followed by cavity-filling freshwater phreatic sparite cement. Atokan, Middle Pennsylvanian, Reeves County, Delaware Basin, Texas, U.S.A.

Plate 22.A. Annelid bioconstructed limestone. Transverse to oblique sections of annelid tubes, associated in a colony, show walls with fibroradiated and concentric structures of organic-rich calcite. Pointed internal structures are reinforcements of the tubes, which are empty or partially filled with dark calcilutite. Pleistocene, Salvador, Bahia, Brazil.

Plate 22.B. Red algae bioconstructed limestone. Open branched nodular colony of *Lithophyllum* displays geopetal pelletoidal calcisiltite internal sediment overlain by thin rim of fibrous, isopachous marine phreatic calcite cement, which is also the first lining of the internal cavities. This first cement is in turn followed by cavity-filling freshwater phreatic sparite mosaic in central parts of cavities. Amapá Formation, Paleocene–Middle Miocene, Foz do Amazonas Basin, offshore Brazil.

Plate 22.C. Coral bioconstructed limestone. Scleratinian coral framework is entirely neomorphosed into a mosaic of pseudomicrosparite, which becomes coarser toward the central parts of the septas. Interseptal spaces are filled with massive calcisiltite, with scattered minute grains of detrital quartz and feldspar. Middle Miocene, Iloilo Basin, Panay, Philippines.

Plate 22.D. Perforations, marine phreatic environment. Massive nodular colony of red algae appears heavily perforated. Borings were filled subsequently by an internal

sediment consisting of calcisiltite, with minute bioclasts of red algae, benthic and pelagic foraminifers, or by patches of marine phreatic calcite combined with the calcilutite, or by the cement alone. Amapá Formation, Paleocene–Middle Miocene, Foz do Amazonas Basin, offshore Brazil.

Plate 22.E. Micritization, marine phreatic environment. Dark micrite envelope of irregular thickness was formed as the result of endolithic algae boring into a pelecypod aragonite shell elongated parallel to bedding (center of picture). The shell was subsequently dissolved and refilled by cavity-filling sparite in the freshwater phreatic environment. Galena Group, Middle Ordovician, Guttenberg, Iowa, U.S.A.

Plate 22.F. Isopachous rim calcite cement, marine phreatic environment. Normal ooids with numerous concentric rings around bioclastic or pelletoidal nuclei are cemented by well-developed isopachous rim of fibrous calcite, which was, in places, preceded by early compaction of the ooids. Massive to finely pelletoidal calcilutite internal sediment occupies the remaining pore spaces. Atokan, Middle Pennsylvanian, Reeves County, Delaware Basin, Texas, U.S.A.

Plate 23.A. Fibrous calcite cement, marine phreatic environment. Large cavity in scleratinian coral reef is partially filled with thick fringes of fibrous submarine calcite cement. Central portion of cavity displays geopetal accumulation of dark pelletoidal calcisiltite overlain by freshwater phreatic sparite mosaic. Middle Miocene, Negros, Philippines.

Plate 23.B. Botryoidal calcite cement, marine phreatic environment. Hemispherical development of botryoidal, tan, pseudosparite, enclosed relicts of original fan-shaped acicular aragonite crystals with scattered patches of clear pseudosparite. The botryoidal cement grew on top of a phylloid algal blade neomorphosed to coarse pseudosparite. Interstitial groundmass is neomorphosed calcisiltite. Raytown Limestone, Iola Formation, Upper Pennsylvanian, Iola, Kansas, U.S.A.

Plate 23.C. Pendant calcite cement, freshwater vadose environment. Interstitial pores within a framework of grain-supported large oncoids are lined with isopachous rim of fibrous to prismatic calcite, with pendant structure underneath the largest oncoid (center of picture). This beachrock cement may involve mixed marine–freshwater vadose conditions. Macaé Formation, Albian–Cenomanian, Campos Basin, offshore Brazil.

Plate 23.D. Graded-bedded vadose silt, freshwater vadose environment. Accumulation of graded-bedded vadose silt consists mostly of minute bioclasts in the lower half and fine pellets and crystal debris in the upper half. Deposition took place between two neomorphosed and highly micritized phylloid algal fronds in subvertical position. The vadose silt is overlain by subsequent freshwater phreatic sparite mosaic. Atokan, Middle Pennsylvanian, Reeves County, Delaware Basin, Texas, U.S.A.

Plate 23.E. Paleocalcrete with coated grains, freshwater vadose environment. Irregularly shaped and concretionary grains at various stages of development and of destruction by neomorphic pseudomicrosparite developing within interstitial cavities and network of fine cracks. Annular coated grain with partially porous center (black) and containing a grain of detrital quartz (center left of picture) is interpreted as a rhizoconcretion. Baurú Formation, Cretaceous, Ponte Alta, Minas Gerais, Brazil.

Plate 23.F. Paleocalcrete with vadose ooids, freshwater vadose environment. Perfectly developed normal ooids with nuclei of broken ooid (left) and angular grain of detrital quartz (right) are surrounded by several layers of isopachous fibroradiated calcite. Interstitial spaces are filled by dark, fine vadose silt (in geopetal position in upper part of picture), followed by cavity-filling freshwater phreatic sparite cement. Cenozoic, San José de Itaborai, Rio de Janeiro, Brazil.

Plate 24.A. Bladed and equant calcite cement, freshwater phreatic to deep-burial environments. Interstitial space of a framework of grain-supported opaque oncoids shows cavity-filling calcite cement displaying a rim of irregularly bladed crystals, followed by mosaic of equant, coarser sparite. Macaé Formation, Albian–Cenomanian, Campos Basin, offshore Brazil.

Plate 24.B. Syntaxial overgrowth calcite cement, freshwater phreatic to deep-burial environments. Large, curved bioclast of echinoid test is surrounded by a well-developed syntaxial overgrowth, which, in addition to filling the adjacent interstitial pores, also partially replaced surrounding bioclasts and calcisiltite matrix, showing euhedral crystal faces. Amapá Formation, Paleocene–Middle Miocene, Foz do Amazonas Basin, offshore Brazil.

Plate 24.C. Neomorphism of ooids, freshwater phreatic environment. Oolitic biocalcarenite with interstitial calcisiltite matrix shows normal and composite ooids in which the numerous original fine concentric rings were differentially neomorphosed into coarser annular layers of pseudomicrosparite, with concentration of impurities in intervening zones. Glen Dean Formation, Upper Mississippian, Branchville, Perry County, Indiana, U.S.A.

Plate 24.D. Neomorphism of calcisiltite, freshwater phreatic environment. Mud-supported biocalcarenite consists of abundant, thin brachiopod shells oriented parallel to bedding. Interstitial calcisiltite matrix was submitted to intense aggrading neomorphism and changed into predominantly subrectangular crystals of clear pseudosparite, often elongated perpendicular to individual brachiopod shells and separated by remaining unaffected matrix. Galena Group, Middle Ordovician, Dickeyville, Wisconsin, U.S.A.

Plate 24.E. Neomorphism of coral framework, freshwater phreatic environment. The polygonal to circular sections of the septas of this scleratinian coral were transformed into a fine and pure mosaic of pseudomicrosparite,

with tendency of the crystals to be oriented perpendicularly to the darker margins of the septas. The interseptal spaces, originally occupied by a calcisiltite matrix, were also neomorphosed into an irregular mosaic of coarse pseudosparite, with numerous areas showing relicts of the original matrix. Middle Miocene, Negros, Philippines.

Plate 24.F. Fracture-filling calcite, deep-burial environment. Grain-supported biocalcarenite consists of a framework of irregularly shaped to subrounded bioclasts of several types of red algae displaying complex reciprocal sutures due to appreciable pressure solution. Interstitial cement is coarse bladed to irregular mosaic of freshwater phreatic sparite. Under burial conditions, vertical fracturing occurred (right side of picture), and late generations of calcite cement filled the tension gash cracks, often with crystals perpendicular to edges, which cut across both bioclasts and earlier freshwater phreatic sparite. Amapá Formation, Paleocene–Middle Miocene, Foz do Amazonas Basin, offshore Brazil.

Plate 25.A. Interparticle porosity. This type of primary porosity occurs in a grain-supported and well-sorted pelecypod biocalcarenite in which elongated bioclasts are neomorphosed to coarse pseudosparite and coated with an isopachous rim of lacustrine phreatic, bladed calcite cement, which slightly decreased the original depositional porosity. Lagoa Feia Formation, Lower Cretaceous, Campos Basin, offshore Brazil.

Plate 25.B. Intraparticle porosity. This type of secondary porosity occurs in a grain-supported biocalcarenite that consists of a framework of large bryozoan and crinoid bioclasts with superficial oolitic coatings; associated are normal ooids highly deformed and spalled. Interstitial cement is freshwater phreatic sparite mosaic. Intraparticle porosity developed within the ooids by differential dissolution of certain cortical layers, by enlargement of ruptured and spalled areas, and also by dissolution of contacts between bioclasts and superficial oolitic coatings (center left of picture). Great Oolite, Bathonian, Bath, Great Britain.

Plate 25.C. Oomoldic porosity. This type of secondary porosity occurs in a grain-supported oolitic calcarenite that consists of a framework of normal ooids with crinoidal and pelletoidal cores, associated with bioclasts of crinoids, echinoids, neomorphosed pelecypods, and bryozoans. Interstitial cement is freshwater phreatic microsparite with patches of calcisiltite matrix. Ooids were either completely dissolved, or only their concentric rings were dissolved and the nuclei dropped to the bottom of cavities in a nongeopetal position, indicating subsequent reworking. Secondary dissolution also developed as irregular intercrystalline patches in the interstitial microsparite cement. Some ooids were later filled by single crystals of late sparite. Quintuco-Loma Montosa Formation, Lower Cretaceous, Loma La Lata, Neuquén Basin, Argentina.

Plate 25.D. Biomoldic porosity. This type of secondary porosity occurs in a grain-supported biocalcarenite consisting of a framework of bioclasts of scleratinian corals, red algae, *Halimeda,* and large benthic foraminifers. Interstitial matrix is dark argillaceous calcisiltite. All bioclasts were completely dissolved into perfect, fabric-selective molds except the test of *Lenticulina* at center of picture. Middle Miocene, Iloilo Basin, Panay, Philippines.

Plate 25.E. Moldic enlarged, vuggy porosity. This type of secondary porosity occurs in a grain-supported biocalcarenite consisting of a framework of bioclasts of scleratinian corals, red algae, *Halimeda,* and large benthic foraminifers. Interstitial matrix is an argillaceous calcisiltite. The nonfabric selective dissolution affected all bioclasts and matrix, creating large irregular vugs. Middle Miocene, Iloilo Basin, Panay, Philippines.

Plate 25.F. Channel porosity. This type of secondary porosity occurs on a grain-supported biocalcarenite consisting of a framework of oncoids with an interstitial matrix of calcisiltite. The nonfabric selective dissolution generated a channel cutting across both oncoids and original matrix. Its margins are coated with a thin rim of deep-burial secondary calcite cement with subhedral crystals. The same occurred during dissolution of some interstitial calcisiltite spaces (upper center of picture), where larger rhombohedral crystals of calcite developed. Macaé Formation, Albian–Cenomanian, Campos Basin, offshore Brazil.

Plate 26.A. Stylolitic porosity. This type of deep-burial secondary porosity occurs in a grain-supported biocalcarenite consisting of a framework of predominant phylloid algal fronds neomorphosed to pseudomicrosparite, associated with bioclasts of brachiopods, pelecypods, *Endothyra,* and small calcareous foraminifers. Interstitial material is cavity-filling freshwater phreatic sparite with scattered concentrations of calcisiltite matrix. The double set of vertical, suture-shaped stylolitic pores (center) are associated with incomplete biomolds of phylloid algal fronds (left and right of stylolitic pores). Atokan, Middle Pennsylvanian, Reeves County, Delaware Basin, Texas, U.S.A.

Plate 26.B. Fracture porosity. This type of deep-burial secondary porosity occurs in a grain-supported biocalcarenite consisting of a framework of oncoids of variable sizes and rectangular bioclasts of pelecypods neomorphosed to pseudosparite. Interstitial material is a calcisiltite with smaller debris of oncoids. The open and irregular fractures developed in a roughly radiating pattern inside the larger oncoid and are branching away from a strongly developed stylolitic seam with abundant insoluble residue (lower-left corner to upper-right corner of picture). Macaé Formation, Albian–Cenomanian, Campos Basin, offshore Brazil.

REFERENCES

ADAMS, A. E., 1980. Calcrete profiles in the Eyam Limestone (Carboniferous) of Derbyshire: petrology and regional significance. *Sedimentology,* 27, 651–660.

AISSAOUI, D. M., 1985. Botryoidal aragonite and its genesis. *Sedimentology,* 32, 345–361.

AMIEUX, P., 1982. La cathodoluminescence: méthode d'étude sédimentologique des carbonates. *Bull. Centre Recherches S.N.E.A.P., Pau,* 6, 457–483.

ASSERETO, R., and FOLK, R. L., 1980. Diagenetic fabrics of aragonite, calcite, and dolomite in an ancient peritidal-spelean environment: Triassic Calcare Rosso, Lombardia. *J. Sed. Petrology,* 50, 371–395.

BACK, W., HANSHAW, B. B., HERMANN, J. S., and VAN DRIEL, J. N., 1986. Differential dissolution of a Pleistocene reef in the groundwater mixing zone of coastal Yucatan, Mexico. *Geology,* 14, 137–140.

BADIOZAMANI, K., 1973. The Dorag dolomitization model–application to the Middle Ordovician of Wisconsin. *J. Sed. Petrology,* 43, 965–984.

BATHURST, R. G. C., 1959. The cavernous structure of some Mississippian stromatactis reefs in Lancashire, England. *J. Geol.,* 67, 506–521.

———, 1975. *Carbonate Sediments and Their Diagenesis.* 2nd ed. Developments in Sedimentology 12. Elsevier Publishing Co., New York, 658 pp.

———, 1980. Stromatactis—origin related to submarine-cemented crusts in Paleozoic mudmounds. *Geology,* 8, 131–134.

———, 1982. Genesis of stromatactis cavities between submarine crusts in Paleozoic carbonate mud mounds. *J. Geol. Soc. London,* 139, 165–181.

———, 1983. Neomorphic spar versus cement in some Jurassic grainstones: significance for evaluation of porosity evolution and compaction. *J. Geol. Soc. London,* 140, 229–237.

———, 1987. Diagenetically enhanced bedding in argillaceous platform limestones: stratified cementation and selective compaction. *Sedimentology,* 34, 749–778.

BEACH, D. K., and SCHUMACHER, A., 1982. Stanley Field, North Dakota—a new model for a new exploration play. In J. E. Christopher and J. Kaldi (eds.), *4th International Williston Basin Symposium.* Saskatchewan Geological Society, Special Publ. No. 6, 235–243.

BRAITHWAITE, C. J. R., and HEATH, R. A., 1989. Inhibition, distortion, and dissolution of overgrowth cements on pelmatozoan fragments. *J. Sed. Petrology,* 59, 267–271.

BRICKER, O. P. (ed.), 1971. *Carbonate Cements.* Johns Hopkins University Studies in Geology No. 19, Johns Hopkins University Press, Baltimore, 376 pp.

BYERS, C. W., and STASKO, L. E., 1978. Trace fossils and sedimentologic interpretation—McGregor Member of Platteville Formation (Ordovician) of Wisconsin. *J. Sed. Petrology,* 48, 1303–1310.

CAROZZI, A. V., 1983. Modelos deposicionales carbonaticos. *Asociación Geológica Argentina, Buenos Aires, Series B. Didáctica y complementaria,* No. 11, vol. 1, 106 pp., vol. 2, 197 pp.

——, 1989. *Carbonate Rock Depositional Models. A Microfacies Approach.* Prentice Hall, Englewood Cliffs, N.J., Advanced Reference Series, 604 pp.

——, and VON BERGEN, D., 1987. Stylolitic porosity in carbonates: a critical factor for deep hydrocarbon production. *J. Petroleum Geol.,* 10, 267–282.

——, FALKENHEIN, F. U. H., and FRANKE, M. R., 1983. Depositional environment, diagenesis, and reservoir properties of oncolitic packstones, Macaé Formation (Albian–Cenomanian), Campos Basin, offshore Rio de Janeiro, Brazil. In T. M. Peryt (ed.), *Coated Grains.* Springer-Verlag, New York, 330–343.

CAYEUX, L., 1916. *Introduction à l'étude pétrographique des roches sédimentaires.* Mém. Carte géol. dét. France, Imprimerie nationale, Paris, vol. 1, text, 524 pp., vol. 2, atlas, 56 plates (reprinted in 1931).

——, 1935. *Les roches sédimentaires de France—Roches carbonatées (calcaires et dolomies).* Masson, Paris, 436 pp. See also L. Cayeux, 1970, *Sedimentary Rocks of France—Carbonate Rocks (Limestones and Dolomites),* translated and updated by A. V. Carozzi, Hafner Publishing Co., Darien, Connecticut, 438 pp.

CHAFETZ, H. S., 1986. Marine peloids: a product of bacterially induced precipitation of calcite. *J. Sed. Petrology,* 56, 812–817.

CHOQUETTE, P. W., and JAMES, N. P., 1990. Limestones: the burial diagenetic environment. In I. A. McIlreath and D. W. Morrow (eds.), *Diagenesis.* Geological Society of Canada, Geoscience Canada, Reprint Series 4, 75–112.

——, and PRAY, L. C., 1970. Geologic nomenclature and classification of porosity in sedimentary carbonates. *Amer. Assoc. Petroleum Geologists Bull.,* 54, 207–250.

COOK, H. E., and ENOS, P. (eds.), 1977. *Deep-water Carbonate Environments.* Soc. Econ. Paleontologists and Mineralogists, Special Publ. 25, 336 pp.

CREVELLO, P. D., WILSON, J. L., SARG, J. F., and READ, J. E. (eds.), 1989. *Controls on Carbonate Platform and Basin Development.* Soc. Econ. Paleontologists and Mineralogists, Special Publ. 44, 405 pp.

DAVIES, G. R., 1977. Turbidite debris sheets, and truncation structures in upper Paleozoic deep-water carbonates of the Sverdrup Basin, Arctic Archipelago. In H. E. Cook and P. Enos (eds.), *Deep-water Carbonate Environments.* Soc. Econ. Paleontologists and Mineralogists, Special Publ. 25, 221–247.

DICKSON, J. A., 1966. Carbonate identification and genesis as revealed by staining. *J. Sed. Petrology,* 36, 491–505.

DONATH, F. A., CAROZZI, A. V., FRUTH, L. S., JR., and RICH, D. W., 1980. Oomoldic porosity experimentally developed in Mississippian oolitic limestones. *J. Sed. Petrology,* 50, 1249–1260.

DOYLE, L. J., and ROBERTS, H. H. (eds.), 1988. *Carbonate–clastic Transitions.* Developments in Sedimentology 42, Elsevier Scientific Publishing Co., New York, 304 pp.

DRUCKMAN, Y., and MOORE, C. H., 1985. Late subsurface secondary porosity in a Jurassic grainstone reservoir, Smackover Formation, Mt. Vernon field, southern Arkansas. In P. O. Roehl and P. W.

Choquette (eds.), *Carbonate Petroleum Reservoirs.* Springer-Verlag, New York, 371–383.

DUNHAM, R. J., 1962. Classification of carbonate rocks according to depositional texture. In W. E. Ham (ed.), *Classification of Carbonate Rocks.* Amer. Assoc. Petroleum Geologists, Memoir 1, 108–121.

——, 1969. Early vadose silt in Townsend Mound (reef), New Mexico. In G. M. Friedman (ed.), *Depositional Environments in Carbonate Rocks: A Symposium.* Soc. Econ. Paleontologists and Mineralogists, Special Publ. 14, 182–192.

EINSELE, G., and SEILACHER, A. (eds.), 1982. *Cyclic Event Stratification.* Springer-Verlag, New York, 536 pp.

ELLIOTT, T. L. 1982. Carbonate facies, depositional cycles and development of secondary porosity during burial diagenesis. In J. E. Christopher and J. Kaldi (eds.), *4th International Williston Basin Symposium.* Saskatchewan Geological Society, Special Publ. 6, 131–151.

EMBRY, A. F., and KLOVAN, J. E., 1971. A late Devonian reef tract on northeastern Banks Island, N.W.T. *Canadian Petroleum Geol. Bull.,* 19, 730–781.

ESTEBAN, M., 1976. Vadose pisolite and caliche. *Amer. Assoc. Petroleum Geologists Bull.,* 60, 2048–2057.

——, and KLAPPA, C. F., 1983. Subaerial exposure. In P. A. Scholle and C. H. Moore (eds.), *Carbonate Depositional Environments.* Amer. Assoc. Petroleum Geologists, Memoir 33, 1–54.

FAIRCHILD, I. J., 1983. Chemical controls of cathodoluminescence of natural dolomites and calcites: new data and review. *Sedimentology,* 30, 579–583.

FINKEL, E. A., and WILKINSON, B. H., 1990. Stylolitization as source of cement in Mississippian Salem Limestone, west-central Indiana. *Amer. Assoc. Petroleum Geologists Bull.,* 74, 174–186.

FISCHER, A. G., 1964. The Lofer cyclothems of the Alpine Triassic. In D. F. Merriam (ed.), Symposium on cyclic sedimentation. *Geol. Survey Kansas Bull.,* 169, 107–149.

——, 1981. Climatic oscillations in the biosphere. In M. H. Nitecki (ed.), *Biotic Crises in Ecological and Evolutionary Time.* Academic Press, New York, 103–133.

——, HONJO, S., and GARRISON, R. E., 1967. Electron micrographs of limestones and their nannofossils. *Monogr. Geol. Paleont. 1* (A. G. Fischer, ed.), Princeton University Press, Princeton, N.J., 141 pp.

FLÜGEL, E., 1982. *Microfacies Analysis of Limestones.* Springer-Verlag, New York, 633 pp.

——, FRANZ, H. E., and OTT, W. F., 1968. Review of electron microscope studies of limestones. In G. Müller and G. M. Friedman (eds.), *Recent Developments in Carbonate Sedimentology in Central Europe.* Springer-Verlag, New York, 85–97.

FOLK, R. L., 1959. Practical classification of limestones. *Amer. Assoc. Petroleum Geologists Bull.,* 43, 1–38.

——, 1962. Spectral subdivision of limestone types. In W. E. Ham (ed.), *Classification of Carbonate Rocks.* Amer. Assoc. Petroleum Geologists, Memoir 1, 62–84.

——, 1965. Some aspects of recrystallization of ancient limestones. In L. C. Pray and R. C. Murray (eds.), *Dolomitization and*

Limestone Diagenesis. Soc. Econ Paleontologists and Mineralogists, Special Publ. 13, 14–48.

FRANK, J. R., CARPENTER, A. B., and OGLESBY, T. W., 1982. Cathodoluminescence and composition of calcite cement in the Taum Sauk Limestone (Upper Cambrian), southeast Missouri. *J. Sed. Petrology,* 52, 631–638.

GAILLARD, C., and JAUTEE, E., 1987. The use of burrows to detect compaction and sliding in fine-grained sediments: an example from the Cretaceous of S.E. France. *Sedimentology,* 34, 585–593.

GIVEN, R. K., and WILKINSON, B. H., 1985. Kinetic control of morphology, composition, and mineralogy of abiotic sedimentary carbonates. *J. Sed. Petrology,* 55, 109–119.

GOLDSTEIN, R. H., 1988. Cement stratigraphy of Pennsylvanian Holder Formation, Sacramento Mountains, New Mexico. *Amer. Assoc. Petroleum Geologists Bull.,* 72, 425–438.

GREGG, J. M., and SIBLEY, D. F., 1984. Epigenetic dolomitization and the origin of xenotopic dolomite texture. *J. Sed. Petrology,* 54, 908–931.

GROVER, G., and READ, J. F., 1978. Fenestral and associated vadose diagenetic fabrics of tidal flat carbonates, Middle Ordovician New Market Limestone, southwestern Virginia. *J. Sed. Petrology,* 48, 453–473.

———, and ———, 1983. Paleoaquifer and deep burial related cements defined by regional cathodoluminescent patterns, Middle Ordovician carbonates, Virginia. *Amer. Assoc. Petroleum Geologists Bull.,* 67, 1275–1303.

HAM, W. E. (ed.), 1962. *Classification of Carbonate Rocks.* Amer. Assoc. Petroleum Geologists, Memoir 1, 279 pp.

HANOR, J. S., 1978. Precipitation of beachrock cements: mixing of marine and meteoric waters vs. CO_2-degassing. *J. Sed. Petrology,* 48, 489–501.

HANSHAW, B. B., BACK, W., and DEIKE, R. G., 1971. A geochemical hypothesis for dolomitization by groundwater. *Economic Geol.,* 66, 710–724.

HARRISON, R. S., and STEINEN, R. P., 1978. Subaerial crusts, caliche profiles and breccia horizons: comparison of some Holocene and Mississippian exposure surfaces: Barbados and Kentucky. *Geol. Soc. Amer. Bull.,* 89, 385–396.

HOROWITZ, A. S., and POTTER, P. E., 1971. *Introductory Petrography of Fossils.* Springer-Verlag, New York, 302 pp., 100 plates.

HUDSON, J. D., 1975. Carbon isotopes and limestone cement. *Geology,* 3, 19–22.

———, 1977. Stable isotopes and limestone lithification. *J. Geol. Society London,* 133, 637–660.

HUTCHINSON, C. S., 1974. *Laboratory Handbook of Petrographic Techniques.* John Wiley & Sons, New York, 527 pp.

JAMES, N. P., 1984a. Introduction to carbonate facies models. In R. G. Walker (ed.), *Facies Models,* 2nd ed. Geological Association of Canada, Geoscience Canada, Reprint Series 1, 209–211.

———, 1984b. Shallowing-upward sequences in carbonates. In R. G. Walker (ed.), *Facies Models,* 2nd ed. Geological Association of Canada, Geoscience Canada, Reprint Series 1, 213–228.

———, 1984c. Reefs. In R. G. Walker (ed.), *Facies Models,* 2nd ed. Geological Association of Canada, Geoscience Canada, Reprint Series 1, 229–244.

———, and CHOQUETTE, P. W. (eds.), 1988. *Paleokarst.* Springer-Verlag, New York, 416 pp.

———, and ———, 1990a. Limestones: the sea floor diagenetic environment. In I. A. McIlreath and D. W. Morrow (eds.), *Diagenesis.* Geological Association of Canada, Geoscience Canada, Reprint Series 4, 13–34.

———, and ———, 1990b. Limestones: the meteoric diagenetic environment. In I. A. McIlreath and D. W. Morrow (eds.), *Diagenesis.* Geological Association of Canada, Geoscience Canada, Reprint Series 4, 35–74.

KAUFMAN, J., CANDER, H. S., DANIELS, L. D., and MEYERS, W. J., 1988. Calcite cement stratigraphy and cementation history of Burlington-Keokuk Formation (Mississippian), Illinois and Missouri. *J. Sed. Petrology,* 58, 312–326.

KENDALL, A. C., 1977. Fascicular-optic calcite: a replacement of bundled acicular carbonate cements. *J. Sed. Petrology,* 47, 1056–1062.

———, 1984. Evaporites. In R. G. Walker (ed.), *Facies Models,* 2nd ed. Geological Association of Canada, Geoscience Canada, Reprint Series 1, 259–298.

———, 1985. Radiaxial fibrous calcite: a reappraisal. In N. Schneidermann and P. M. Harris (eds.), *Carbonate Cements.* Soc. Econ. Paleontologists and Mineralogists, Special Publ. 36, 59–77.

———, and TUCKER, M. E., 1973. Radiaxial fibrous calcite: a replacement after acicular carbonate. *Sedimentology,* 20, 365–389.

KERANS, C., HURLEY, N. F., and PLAYFORD, P. E., 1986. Marine diagenesis in Devonian reef complexes of the Canning Basin, western Australia. In J. H. Schroeder and B. H. Purser (eds.), *Reef Diagenesis.* Springer-Verlag, New York, 357–380.

KNAUTH, L. P., 1979. A model for the origin of chert in limestone. *Geology,* 7, 274–277.

KRETZ, R., 1992. Carousel model for the crystallization of saddle dolomite. *J. Sed. Petrology,* 62, 190–195.

LAND, L. S., 1973. Holocene meteoric dolomitization of Pleistocene limestones, north Jamaica. *Sedimentology,* 20, 411–422.

———, SALEM, M. R. I., and MORROW, D. W., 1975. Paleohydrology of ancient dolomites; geochemical evidence. *Amer. Assoc. Petroleum Geologists Bull.,* 49, 1602–1625.

LAPORTE, L. F. (ed.), 1974. *Reefs in Time and Space. Selected Examples from the Recent and the Ancient.* Soc. Econ. Paleontologists and Mineralogists, Special Publ. 18, 356 pp.

LASEMI, Z., and SANDBERG, P. A., 1984. Transformation of aragonite-dominated lime muds to microcrystalline limestones. *Geology,* 12, 420–423.

LOHMANN, K. C., and MEYERS, W. J., 1977. Microdolomite inclusions in cloudy prismatic calcites—a proposed criterion for former high magnesium calcites. *J. Sed. Petrology,* 47, 1078–1088.

LONGMAN, M. W., 1980. Carbonate diagenetic textures from nearsurface diagenetic environments. *Amer. Assoc. Petroleum Geologists Bull.,* 64, 461–487.

MACHEL, H. G., 1985. Cathodoluminescence in calcite and dolomite and its chemical interpretation. *Geoscience Canada,* 12, 139–147.

———, 1987. Saddle dolomite as a by-product of chemical compaction and thermochemical sulfate reduction. *Geology,* 15, 936–940.

MACINTYRE, I. G., 1985. Submarine cements—The peloidal question. In N. Schneidermann and P. M. Harris (eds.), *Carbonate Cements.* Soc. Econ. Paleontologists and Mineralogists, Special Publ. 36, 109–116.

MAJEWSKE, O. P., 1974. *Recognition of Invertebrate Fossil Fragments in Rocks and Thin Sections.* E. J. Brill, Leiden, 101 pp., 106 plates.

MALIVA, R. G., 1989. Displacive calcite syntaxial overgrowths in open marine limestones. *J. Sed. Petrology,* 59, 397–403.

MAPSTONE, N. B., 1975. Diagenetic history of a North Sea chalk. *Sedimentology,* 22, 601–613.

MARSHALL, J. F., and DAVIES, P. J., 1975. High-magnesium calcite ooids from the Great Barrier Reef. *J. Sed. Petrology,* 45, 285–291.

MAZZULLO, S. J., 1980. Calcite pseudospar replacive of marine acicular aragonite and implications for aragonite cement diagenesis. *J. Sed. Petrology,* 50, 409–423.

———, and HARRIS, P. M., 1992. Mesogenetic dissolution: its role in porosity development in carbonate reservoirs. *Amer. Assoc. Petroleum Geologists Bull.,* 76, 607–620.

MCILREATH, I. A., and JAMES, N. P., 1984. Carbonate slopes. In R. G. Walker (ed.), *Facies Models,* 2nd ed. Geological Association of Canada, Geoscience Canada, Reprint Series 1, 245–257.

MEYERS, J. H., 1987. Marine vadose beachrock cementation by cryptocrystalline magnesian calcite—Maui, Hawaii. *J. Sed. Petrology,* 57, 558–570.

MEYERS, W. J., 1974. Carbonate cement stratigraphy of the Lake Valley Formation (Mississippian), Sacramento Mountains, New Mexico. *J. Sed. Petrology,* 44, 837–861.

———, 1978. Carbonate cements: their regional distribution and interpretation in Mississippian limestones of southwestern New Mexico. *Sedimentology,* 25, 371–400.

———, 1980. Compaction in Mississippian skeletal limestones, southwestern New Mexico. *J. Sed. Petrology,* 50, 457–474.

———, and HILL, B. E., 1983. Quantitative studies of compaction in Mississippian skeletal limestones, New Mexico. *J. Sed. Petrology,* 53, 231–242.

MILLIMAN, J. D., and BARRETTO, H. T., 1975. Relict magnesian calcite oolite and subsidence of the Amazon shelf. *Sedimentology,* 22, 137–145.

MOORE, C. H., 1973. Intertidal carbonate cementation, Grand Cayman, West Indies. *J. Sed. Petrology,* 43, 591–602.

———, 1985. Upper Jurassic subsurface cements: a case history. In N. Schneidermann and P. M. Harris (eds.), *Carbonate Cements.* Soc. Econ. Paleontologists and Mineralogists, Special Publ. 36, 291–308.

———, 1989. *Carbonate Diagenesis and Porosity.* Developments in Sedimentology 46, Elsevier Scientific Publishing Co., New York, 338 pp.

———, and DRUCKMAN, Y., 1981. Burial diagenesis and porosity evolution, Upper Jurassic Smackover, Arkansas and Louisiana. *Amer. Assoc. Petroleum Geologists Bull.,* 65, 597–628.

MOUNT, J., 1985. Mixed siliciclastic and carbonate sediments: a proposed first-order textural and compositional classification. *Sedimentology,* 32, 435–442.

MOUNTJOY, E. W., and RIDING, R., 1981. Foreslope stromatoporoid-renalcid bioherm with evidence of early cementation, Devonian Ancient Wall reef complex, Rocky Mountains. *Sedimentology,* 28, 299–321.

MURRAY, R. C., 1964. Origin and diagenesis of gypsum and anhydrite. *J. Sed. Petrology,* 34, 512–523.

NELSON, R. A., 1981. Significance of fracture sets associated with stylolite zones. *Amer. Assoc. Petroleum Geologists Bull.,* 65, 2417–2425.

NEUMANN, A. C., and LAND, L. S., 1975. Lime mud deposition and calcareous algae in the Bight of Abaco, Bahamas: a budget. *J. Sed. Petrology,* 45, 763–786.

PERYT, T. M. (ed.), 1983. *Coated Grains.* Springer-Verlag, New York, 655 pp.

RADKE, R. M., and MATHIS, R. L., 1980. On the formation of saddle dolomite. *J. Sed. Petrology,* 50, 1149–1168.

RAO, C. P., 1981. Criteria for recognition of cold-water carbonate sedimentation: Berriedale Limestone (Lower Permian), Tasmania, Australia *J. Sed. Petrology,* 51, 491–506.

———, 1983. Geochemistry of Early Permian cold-water carbonates (Tasmania, Australia). *Chem. Geol.* 38 (3–4), 307–319.

———, 1988. Oxygen and carbon isotope composition of cold-water Berriedale Limestone (Lower Permian), Tasmania, Australia. In C. S. Nelson (ed.), Non-tropical shelf carbonates: modern and ancient. *Sed. Geol.,* special issue, 60 (1–4), 221–231.

RICH, D. W., and CAROZZI, A. V., 1981. Natural porosity in oolitic limestones: an attempt at a genetic classification. *Actas VIII Congreso Geol. Argentino, II,* 593–635.

RICKEN, W., 1987. The carbonate compaction law: a new tool. *Sedimentology,* 34, 571–584.

ROEHL, P. O., and CHOQUETTE, P. W. (eds.), 1985. *Carbonate Petroleum Reservoirs.* Springer-Verlag, New York, 622 pp.

ROSS, D J., 1991. Botryoidal high-magnesium calcite marine cements from the Upper Cretaceous of the Mediterranean region. *J. Sed. Petrology,* 61, 349–353.

SALLER, A. H., 1986. Radiaxial calcite in lower Miocene strata, subsurface Enewetak Atoll. *J. Sed. Petrology,* 56, 743–762.

SANDBERG, P. A., 1975. New interpretations of Great Salt Lake ooids and of ancient non-skeletal mineralogy. *Sedimentology,* 22, 497–537.

———, 1983. An oscillatory trend in Phanerozoic non-skeletal carbonate mineralogy. *Nature,* 305, 19–22.

SCHNEIDERMANN, N., and HARRIS, P. M., 1985. *Carbonate Cements.* Soc. Econ. Paleontologists and Mineralogists, Special Publ. 36, 379 pp.

SCHOLLE, P. A., 1977. Chalk diagenesis and its relation to petroleum exploration: oil from chalk, a modern miracle? *Amer. Assoc. Petroleum Geologists Bull.,* 61, 982–1009.

———, 1978. *A Color Illustrated Guide to Carbonate Rock Constituents, Textures, Cements, and Porosites.* Amer. Assoc. Petroleum Geologists, Memoir 27, 241 pp.

———, and HALLEY, R. B., 1985. Burial diagenesis—out of sight, out of mind! In N. Schneidermann and P. M. Harris (eds.), *Carbon-*

ate Cements. Soc. Econ. Paleontologists and Mineralogists, Special Publ. 36, 309–334.

——, BEBOUT, D. G., and MOORE, C. H. (eds.), 1983. *Carbonate Depositional Environments.* Amer. Assoc. Petroleum Geologists, Memoir 33, 708 pp.

SHINN, E. A., 1968. Practical significance of birdseye structures in carbonate rocks. *J. Sed. Petrology,* 38, 215–223.

——, 1983. Birdseyes, fenestrae, shrinkage pores, and loferites: a reevaluation. *J. Sed. Petrology,* 53, 619–628.

——, and ROBBIN, D. M., 1983. Mechanical and chemical compaction in fine-grained shallow-water limestones. *J. Sed. Petrology,* 53, 595–618.

——, LLOYD, R. M., and GINSBURG, R. N., 1969. Anatomy of a modern carbonate tidal flat, Andros Island, Bahamas. *J. Sed. Petrology,* 39, 1202–1228.

SIMONE, L., 1981. Ooids: a review. *Earth-Science Rev.* 16, 319–355.

SIMPSON, J., 1985. Stylolite-controlled layering in a homogeneous limestone: pseudo-bedding by burial diagenesis. *Sedimentology,* 32, 405–505.

SOLOMON, S. T., and WALKDEN, G. M., 1985. The application of cathodoluminescence to interpreting the diagenesis of an ancient calcrete profile. *Sedimentology,* 32, 877–896.

STEINEN, R. P., 1978. On the diagenesis of lime-mud: scanning electron microscopic observations of subsurface material from Barbados, W. I. *J. Sed. Petrology,* 48, 1139–1148.

STODDART, D. R., and CANN, J. R., 1965. The nature and origin of beachrock. *J. Sed. Petrology,* 35, 243–247.

TOOMEY, D. F. (ed.), 1981. *European Fossil Reef Models.* Soc. Econ. Paleontologists and Mineralogists, Special Publ. 30, 546 pp.

TUCKER, M. E., and KENDALL, A. C., 1973. The diagenesis and low-grade metamorphism of Devonian styliolinid-rich pelagic carbonates from West Germany: possible analogues of Recent pteropod oozes. *J. Sed. Petrology,* 43, 672–687.

——, and WRIGHT, V. P., 1990. *Carbonate Sedimentology.* Blackwell Scientific Publications, Oxford, 482 pp.

——, WILSON, J. L., CREVELLO, P. D., SARG, J. F., and READ, J. F. (eds.), 1990. *Carbonate Platforms: Facies, Sequences, and Evolution.* Internat. Assoc. Sedimentologists, Special Publ., 9, Blackwell Scientific Publications, Oxford, 328 pp.

VON BERGEN, D., and CAROZZI, A. V., 1990. Experimentally-simulated stylolitic porosity in carbonate rocks. *J. Petroleum Geology,* 13, 179–192.

WALKDEN, G. M., and BERRY, J. R., 1984. Syntaxial overgrowths in muddy crinoidal limestones: cathodoluminescence sheds new light on an old problem. *Sedimentology,* 31, 251–267.

WALKER, K. R., JERNIGAN, D. G., and WEBER, L. J., 1990. Petrographic criteria for the recognition of marine syntaxial overgrowths, and their distribution in geologic time. *Carbonates and Evaporites,* 5, 141–151.

WALLACE, M. W., 1987. The role of internal erosion and sedimentation in the formation of stromatactis mudstones and associated lithologies. *J. Sed. Petrology,* 57, 695–700.

WALLS, R. A., HARRIS, W. B., and NUNAM, W. E., 1975. Calcareous crust (caliche) profiles and early subaerial exposure of Carboniferous carbonates, northeastern Kentucky. *Sedimentology,* 22, 417–440.

WALTER, M. R. (ed.), 1976. *Stromatolites.* Developments in Sedimentology 20, Elsevier Scientific Publishing Co., New York, 790 pp.

WANLESS, H. R., 1979. Limestone response to stress-pressure solution and dolomitization. *J. Sed. Petrology,* 49, 437–462.

——, 1983. Burial diagenesis in limestones. In A. Parker and B. W. Sellwood (eds.), *Sediment Diagenesis.* NATO ASI Series C: Mathematical and Physical Sciences, vol. 115, J. Reidel and Company, Lancaster, 379–417.

WARDLAW, N. C., 1962. Aspects of diagenesis in some Irish Carboniferous limestones. *J. Sed. Petrology,* 32, 776–780.

WILKINSON, B. H., 1982. Cyclic cratonic carbonates and Phanerozoic calcite seas. *J. Geological Education,* 30, 189–203.

——, and GIVEN, R. K., 1986. Secular variations in abiotic marine carbonates: constraints on Phanerozoic atmospheric carbon dioxide contents and oceanic Mg/Ca ratios. *J. Geology,* 94, 321–333.

——, JANECKE, S. V., and BRETT, C. E., 1982. Low-magnesian calcite marine cement in Middle Ordovician hardgrounds from Kirkfield, Ontario. *J. Sed. Petrology,* 52, 47–59.

——, OWEN, R. M., and CARROLL, A. R., 1985. Submarine hydrothermal weathering, global eustasy, and carbonate polymorphism in Phanerozoic marine oolites. *J. Sed. Petrology,* 55, 171–183.

WILSON, J. L., 1975. *Carbonate Facies in Geologic History.* Springer-Verlag, New York, 471 pp.

WOLFF, B., and CAROZZI, A. V., 1984. Microfacies, depositional environments, and diagenesis of the Amapá carbonates (Paleocene–Middle Miocene), Foz do Amazonas Basin, offshore NE Brazil. *Petrobrás, Cenpes, Ciência Técnica Petróleo,* No. 13, 103 pp.

WONG, P. K., and OLDERSHAW, A., 1981. Burial cementation in the Kaybob reef complex, Alberta, Canada. *J. Sed. Petrology,* 51, 507–520.

WRIGHT, V. P., 1982a. The recognition and interpretation of paleokarsts: two examples from the lower Carboniferous of south Wales. *J. Sed. Petrology,* 52, 83–94.

——, 1982b. Calcrete palaeosols from the lower Carboniferous Llanelly Formation, South Wales. *Sed. Geol.,* 33, 1–33.

——, (ed.), 1986. *Paleosols: Their Recognition and Interpretation.* Princeton University Press, Princeton, N.J., Blackwell Scientific Publications, London, 315 pp.

——, 1987. The role of fungal biomineralization in the formation of Early Carboniferous soil fabrics. *Sedimentology,* 33, 831–838.

——, ESTEBAN, M., and SMART, P. L. (eds.), 1991. *Palaeokarsts and Palaeokarstic Reservoirs.* A course book prepared for a University of Reading short course, May 1, 1991. University of Reading, Postgraduate Research Institute for Sedimentology, Contribution No. 152, 158 pp.

CHAPTER 6

DOLOSTONES

INTRODUCTION

The origin of dolomite and of its secular variations in amount and texture remains a major unsolved problem of sedimentary petrography (Sibley, 1991). In spite of a great amount of experimental and theoretical data concerning the chemical behavior of the mineral dolomite $CaMg(CO_3)_2$, it remains difficult to understand because it is highly ordered and kinetic constraints hinder its precipitation. To obtain results, experiments were conducted at temperatures above 100°C and with artificial fluids; therefore, comparisons with precipitation of dolomite under natural conditions remain very hypothetical (Shukla and Baker, 1988; Morrow, 1990a; Tucker and Wright, 1990). Nevertheless, numerous attempts were made at devising models of dolomitization in order to account for the bewildering settings under which dolostones of the present and the past appear to have formed predominantly by replacement of a calcium carbonate precursor. These attempts make by themselves a fascinating history of the evolution of ideas about scientists trying to solve a multifaceted problem with insufficient available data. Of comparable interest is the recently unraveled complicated history of the term *dolomite* (Carozzi and Zenger, 1991).

DOLOMITIZATION MODELS

For the past 30 years at least, the interpretation of ancient dolostones, representing widespread and thick units, was attempted on the basis of uniformitarian principles by comparison with various types of modern occurrences of dolomitization. These were documented at the surface, but on a relatively small scale in marine intertidal, supratidal, and evaporitic environments, and at depth on a relatively large scale in the freshwater phreatic–marine phreatic mixing zone (see Chapter 5). During this time interval, five general types of dolomitization models were proposed for the genesis of ancient dolostones. They were exhaustively described petrographically and geochemically and periodically updated and reviewed in numerous comprehensive papers (Pray and Murray, 1965; Zenger et al., 1980; Machel and Mountjoy, 1986; Morrow, 1990b; Tucker and Wright, 1990). All these models, which attempted at all costs to connect the past with the present, were highly debated because none of them was able to account for some of the fundamental geochemical and kinetic requirements. Explanations were lacking for the following major factors: an enormous source of magnesium; an effective large-scale pumping process capable of moving large quantities of fluids through carbonate sediments; the elimination of calcium; the nature of the fluids; the sites of replacement by solution–precipitation processes; the timing; and, finally, the amount of time required to complete the process.

Seawater, a magnesium-rich fluid, is the obvious source, but kinetic factors hinder dolomite precipitation in a normal marine environment. Therefore, most of the proposed models relied heavily on some chemical modification of seawater, either by evaporative concentration or by freshwater dilution. The major general models of dolomitization are the seepage-reflux model, which explains that magnesium-rich hypersaline fluids, generated by evaporation in a lagoon protected by a barrier or a reef, flow downward through the porous sedimentary body, dolomitizing it on their way to the adjacent ocean; the evaporative drawdown model, active during falling sea level, which is similar to the preceding one and explains dolomitized intertidal to subtidal facies underlying evaporites; the sabkha model, which refers to seawater, invading supratidal flats through storms or very high tides, that partly evaporates and dolomitizes its substrate; the Coorong lagoon or lacustrine model, which implies that recharged freshwater evaporates in ephemeral lakes into magnesium-rich fluids that precipitate primary dolomite or dolomitize their substrate as in numerous small-scale occurrences in Quaternary lakes (Last, 1990; Rosen and Coshell, 1992); the meteoric–marine mixing zone or Dorag model, which explains that, under active hydrological circulation, both in confined and unconfined aquifers, mixing of seawater and freshwater creates chemical conditions favorable to dolomite precipitation; the burial dolomitization model, in which carbonate platform margins are dolomitized by the action of magnesium-rich fluids expelled by the compaction of adjacent basinal shales; and the hydrothermal-convection model, which proposes that dense hypersaline brines, migrating from deep in the crust, recirculate in shallower areas by thermal convection and dolomitize the porous limestones they encounter in their path.

Land (1985) stressed that three major aspects of massive dolomitization of ancient limestone sequences have not been adequately accounted for in the above-mentioned models: (1) dolomitization requires a massive addition of magnesium, (2) an equally massive export of calcium must occur, and (3) an active long-term fluid circulation, or pumping, is needed. Seawater is the only natural and almost inexhaustible solution that contains enough magnesium to cause massive dolomitization. Consequently, most dolostones must have formed relatively early in the depositional–diagenetic history of carbonate sediments when seawater or seawater-derived fluids were involved in an active pumping mechanism, at a time when these sediments were as yet appreciably porous and permeable and almost still in their depositional context, or barely buried. The awareness that normal or slightly modified seawater can be an important mechanism for large-scale dolomitization of carbonate platforms generated a number of seawater dolomitization models. The presentation of these models also coincided with the discovery of small amounts of apparently marine dolomite in a variety of environments, such as hardgrounds associated with fringing reefs (Mitchell et al., 1987), carbonate cool-water platforms (Bone et al., 1992), carbonate platform slopes (Mullins et al., 1984a and b), and several instances of deep-water pelagic environment in organic-matter-rich muds (Kelts and McKenzie, 1982, 1984; Shimmield and Price, 1984; Kulm et al., 1984; Baker and Burns, 1985). Conditions of precipitation in deep-water muds are unclear, but seem related to the decrease in sulfate content and increased alkalinity of pore waters due to microbial activity.

A first model of seawater dolomitization was proposed by Saller (1984) in his study of cores from the Enewetak Atoll in the Pacific. He assumed that thermally circulated seawater caused dolomitization of porous late Eocene limestones at the base of the atoll, which are today at depths between 1,250 and 1,400 m below sea level and rest on the underlying volcanic basement. The calcite saturation depth in this region of the Pacific is about 1,000 m; therefore, Saller (1984) proposed that seawater would still be supersaturated with respect to dolomite at this depth and below. Under these conditions, only seawater is involved, and the driving mechanism for large-scale circulation of cold, deep-ocean water, undersaturated with calcite and supersaturated with respect to dolomite, is thermal convection due to heat flow from the volcanic basement. Tucker and Wright (1990) seem to imply that because ocean water can freely flow deep within the Enewetak Atoll, there could be another type of dolomitization model, called ocean-current pumping, in which oceanic currents and tides impinge on the margins of a carbonate platform and dolomitize them in the absence of any geothermal convection system. This could be the case of the Bahama Platform, where the Gulf Stream waters encroach the Bahama Escarpment and generate an internal circulation of seawater that could dolomitize Pleistocene carbonates beneath the meteoric mixing zone (Smart et al., 1988). A dynamic circulation of interstitial seawater in a north Jamaican fringing reef was also documented by Land et al. (1989). This circulation appears unrelated to astronomical tides, local wind-driven surges, or meteoric water discharge, and results from as yet unexplained forces. It is associated with the previously reported occurrence of marine dolomite cement (Mitchell et al., 1987).

Large-scale circulation of seawater into carbonate platform margins, but under a well-established geothermal convection system, was proposed as early as 1967 by Kohout and applied to dolomitization by Simms (1984). The thermal convection system proposed by Kohout is generated by the horizontal density gradient between cold marine water adjacent to a carbonate platform margin and geothermally heated groundwater inside the platform. Ocean waters are drawn inside the platform margin where they displace lighter groundwater, which eventually emerges as springs on the platform or along its edge (Kohout, 1967; Fanning et al., 1981). According to Simms (1984), this open cell convection operates when the depth of the adjacent ocean water is about 2 to 3 km and provides a mechanism in which cold ocean water, undersaturated with respect to calcium carbonate and saturated with respect to dolomite, is pumped through platform margin carbonates where dolomitization would occur. Mullins et al. (1984a, b), considered this mechanism responsible for the diagenetic changes they observed in the periplatform oozes off the Little Bahama Bank, where high-Mg calcite and aragonite are eliminated, low-Mg calcite is increased, and small crystals of authigenic dolomite develop. Such a circulation taking place for long periods of time could naturally interact with meteoric mixing pumps, although it appears now that subsurface dolomitization in Florida, previously interpreted as resulting from a Dorag model, might involve only seawater (Fanning et al., 1981).

Simms (1984) suggested and demonstrated on the Bahama Platform yet another large-scale dolomitization process called *reflux*. It results from the fact that seawater at the surface of a carbonate platform, as a consequence of several types of circulation restrictions, can become slightly to highly hypersaline. If the pore waters within the platform consist of normal seawater, a downward large-scale reflux of even the slightly more saline waters takes place and could generate subsurface dolomitization.

The concept of normal marine water dolomitization led also to the reinterpretation of tidal flat carbonate sediments in Florida (Carballo et al., 1987) and in Belize (Mazzullo et al., 1987) as resulting from active tidal pumping during spring tides of near-normal seawater, instead of the previously assumed evaporation process.

In summary, recent studies and reevaluations seem to conclude that large-scale dolomitization occurs only in association with active pumping mechanisms of normal or slightly modified seawater. This is not to say that the other previously proposed models are invalid, but they probably account for only very limited processes of dolomitization under very specific environmental and climatic conditions. Land (1985) pointed out that limestones in contact with static or slowly moving hypersaline, normal marine, or mixed meteoric–marine fluids do not seem to be sites for dolomite nucleation and hence are not able to generate significant quantities of dolostones. On the other hand, active dynamic systems promote dolomite nucleation and crystal growth possibly by modifying substrates and reducing the kinetic constraints of dolomite formation by their continuous flushing of fluids.

It was mentioned above that, as long as the modern occurrences of environments of dolomitization were limited to geographically restricted and small-scale intertidal, supratidal, and evaporitic conditions, they could not be successfully used for understanding large-scale ancient dolostones by applying uniformitarian principles. Now that dolomitization on a large scale is visualized as produced by the circulation of normal seawater through carbonate platforms, the process becomes directly dependent upon the chemical and eustatic changes of the oceans through geologic time in the same manner as the alternating aragonite and calcite seas that directed limestone deposition and diagenesis. The question can be raised whether seawater dolomitization, as assumed today, still remains beyond the application of uniformitarian principles, but this time for other reasons. Before dealing with these reasons, it is necessary to dispose of the old and persistent belief that the distribution of dolomite from the beginning of geologic time to the present is a gradual decrease in abundance. This assumed evolution implied that dolomitization is a cumulative process taking place

slowly during the entire burial history of carbonate sequences. Available data do not support the assumption of any direct correlation between dolomite content and age. However, significant fluctuations of relative dolomite abundance occur throughout Phanerozoic carbonate sequences, suggesting that dolomite formation was favored by factors related to global eustasy (Lumsden, 1985; Given and Wilkinson, 1987; Sibley, 1991), that is, the intervals of high global sea-level stands of the calcite seas (greenhouse intervals) during which epicontinental areas were flooded by marine transgressions (see **Fig. 5.1**). These conditions certainly favored the above-mentioned models of deep circulation of seawater through carbonate platforms. However, the present is a time of low global sea-level stand of the aragonite sea (icehouse interval), and hence the circulation of seawater beneath carbonate platforms may be attenuated or different from what took place during the intervals of calcite seas. Therefore, today's circulation might not be fully representative of the past.

Furthermore, the two other models of potentially large scale dolomitization, which have attracted many supporters, burial dolomitization, in which fluids are derived from adjacent compacting basinal shales, and mixed meteoric–marine dolomitization, seem to encounter serious difficulties. Global mass calculations by Given and Wilkinson (1987) indicate that marine-derived connate waters in basinal sediments cannot be a major source of magnesium for dolomitization. Moreover, if dolomitization were a burial process, it would be cumulative, and the older a carbonate unit the greater its probability of having undergone burial. This would lead to a return to the possible relationship between dolomite content and geological age, which is not substantiated by observation of facts. With respect to the mixed meteoric–marine dolomitization model, the very high hydrologic requirements are a major obstacle (Hardie, 1987).

In summary, present-day views favor the formation of large-scale dolomitized sequences by using models based on the circulation of normal or slightly modified seawater. This approach thus makes dolomitization an aspect of secular variations and of geotectonic control on carbonate sedimentology and diagenetic processes. Therefore, conditions observed today may only be applicable to certain intervals of the past, but not to the entire history of dolomitization.

CLASSIFICATION

Earlier classifications of dolostones (Cayeux, 1935) were based only on the relative percentage of the mineral dolomite. The terminology was as follows: 0% to 5% dolomite, limestones; 5% to 10% dolomite, magnesian limestone; 10% to 50% dolomite, dolomitic limestone; 50% to 90% dolomite, calcitic dolostone; and 90% to 100% dolomite, dolostone. At present, more elaborate techniques such as X-ray diffraction (Hardy and Tucker, 1988), electron microprobe, and back-scattered electron microscopy are used to establish chemical composition in dolostones. Because dolomitization processes generally reach completion, the need for such determinations concentrates on the detailed study of the chemical composition of individual crystals, rather than on whole rock. Transmission electron microscopy is used to study the complex structure of dolomite crystals and its frequent lattice defects (Wenk et al., 1983).

The various types of petrographic classifications of limestones discussed in Chapter 5 can be applied to dolostones resulting from pseudomorphic replacement and having therefore a high degree of preservation of original depositional fabrics. In the classification proposed by Carozzi (1989), either the prefix dolo- is added, generating terms such as dololutite, dolosiltite, dolarenite, and dolorudite, or the term dolostone is added, as in bioaccumulated dolostone and bioconstructed dolostone; finally, the term dolostone is applied alone when all original textures have been destroyed. The term is further qualified by the size of the crystals of the mosaic: finely crystalline dolostone, coarsely crystalline dolostone, and so on.

PETROGRAPHY

This section pertains mainly to replacement dolostones, which occur most commonly. Undisturbed depositional fabrics indicate that dolomitization is obviously a volume by volume dissolution–precipitation process in which a metastable phase (calcium rich or disordered) is first precipitated and subsequently changed by complex geochemical and neomorphic processes into more stable, ordered, and coarser-grained phases (Land, 1985; Shukla and Baker, 1988; Tucker and Wright, 1990).

Dolostones of all ages display a variety of common textural features, which are as follows: pseudomorphic (mimic) replacement; nonpseudomorphic (nonmimic) replacement of constituents; biomoldic porosity; cloudy centered and clear-rimmed rhombs; mosaic texture, called sucrosic, when displaying intercrystalline porosity; and dolomite cement. The variety of dolomitization environments mentioned above naturally has a time implication, which is clearly expressed petrographically. The timing ranges from early syndepositional dolomitization, as shown by stromatolites (**Plate 26.C**) and some types of burrows (**Plate 26.D**), to diagenetic incipient dolomitization as scattered zoned rhombs (**Plate 26.E, F**), and complete dolomitization as fully developed mosaics of anhedral to subhedral crystals in which pseudomorphic and nonpseudomorphic textures can be associated (**Plate 27.A**). These textural features are controlled by the mineralogy of the precursor (low- or high-Mg calcite or aragonite), the availability of dolomite nucleation sites, and the degree of saturation of the circulating solution with respect to the carbonate being replaced (Sibley, 1982).

However, the relative influence of mineralogy versus crystal size remains a highly debated question. In general, bioclasts of aragonite and low-Mg calcite (corals, gastropods, pelecypods) tend to have their original skeletal fabric destroyed by nonmimic replacement dolomite mosaics. High-Mg bioclasts (coralline algae, some foraminifiers, echinoids, and crinoids) tend to have their original skeletal fabric perfectly preserved by pseudomorphic replacement, particularly when they consist originally of microporous crystals with single optical orientation (echinoid plates, crinoid columnals). Bullen and Sibley (1984) experimentally

---

FIGURE 6.1   Examples of well-preserved depositional textures in replacement dolostones, Burnt Bluff Group (Silurian), Wisconsin. **A.** Coarsely crystalline unzoned dolomite exhibiting curved cleavage and large, unbroken leperditid ostracod; originally a calcilutite. **B.** Medium crystalline subhedral dolomite; originally a calcisiltite resulting from winnowing of a calcilutite. **C.** Alternating irregular layers of lighter-colored coarsely crystalline dolomite and darker medium-crystalline dolomite; originally interbedded calcilutite and calcisiltite. **D.** Finely crystalline angular fragments of dark dolomite set in coarsely crystalline dolomite matrix; originally calcisiltite clasts in calcilutite. **E.** Medium crystalline, unzoned dolomite matrix with large, partially dissolved ooid; originally calcisiltite with scattered ooids. **F.** Finely crystalline dolomite with whole leperditid ostracods, crenulated brachiopods, and gastropod; originally a fossiliferous calcisiltite. **G.** Medium to finely crystalline dolomite containing microcrystalline, dark, irregular layers; originally calcisiltite with cyanobacterial mats. **H.** Dolomitized ooids, some of which exhibit coarse microcrystalline rings, with interstitial clear dolomite cement; originally oolitic calcarenite with clear sparite cement. All photomicrographs: plane–polarized light. From Soderman and Carozzi (1963). Reprinted by permission of the American Association of Petroleum Geologists.

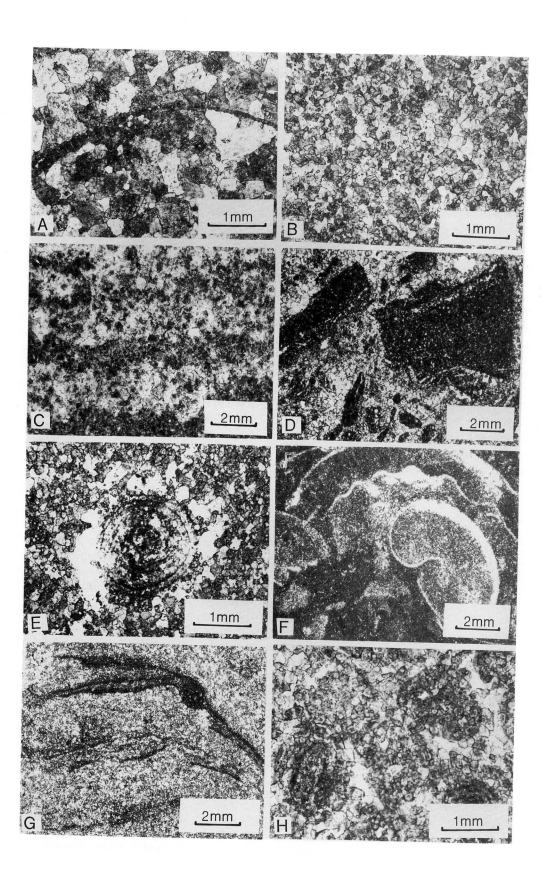

reproduced dolomitization of skeletal particles and found that coralline algae, echinoids, Halimedae, and foraminifers dolomitized faster than the other fossils. They observed that echinoids and foraminifers were mimically replaced, whereas pelecypods, gastropods, and corals were replaced by dolomite with destruction of their original skeletal fabric. However, artificial conversion of high-Mg coralline red algae and echinoid fragments to low-Mg calcite before dolomitization (simulating freshwater diagenesis) had no apparent effect on the rate of dolomitization or on the fabric of the dolomitized bioclasts; furthermore, echinoids and foraminifers exhibited mimic replacement whether they were high-Mg calcite or low-Mg calcite. These observations raised again the question of mineralogy versus crystal size in dolomitization. Bullen and Sibley (1984) considered that a general correspondence exists between rate of dolomitization and fabric of dolomite; that is, rapidly dolomitized cryptocrystalline fossils show mimic replacement, whereas more slowly dolomitized microcrystalline fossils show nonmimic replacement. These authors interpreted this situation to be a function of the number of nucleation sites available: the greater the number of sites, the more rapid the replacement and hence the more likely mimic fabric would result. In summary, according to Bullen and Sibley (1984), mimic replacement and selective dolomitization are more likely affected by crystal size than by mineralogy; but the question remains open, particularly in the light of mimic dolomitization of bioclasts with such large differences in original crystallinity as between coralline algae and echinoid or crinoid bioclasts.

The importance of pseudomorphic or mimic dolomitization has been downplayed for a long time, and dolostones have suffered from the old and persistent belief that most of them are mosaics of anhedral to subhedral rhombs that have essentially destroyed the original limestone depositional fabrics. Consequently, the petrographic study of dolostones is still commonly considered of little value and not usable for inferring depositional environments. Earlier studies (Murray, 1964; Murray and Lucia, 1967) and our own petrographic studies, which were published over a span of more than 35 years and include thousands of thin sections of dolostones of all ages (Carozzi, 1989), contradict the above assumption. Most dolostones, naturally with the exception of the coarsest mosaic or sucrosic types, when properly examined under the microscope with diffused, plane-polarized light to preserve contrast, reveal, to an astonishing degree, depositional textures (Soderman and Carozzi, 1963) to the extent that complete and detailed microfacies determinations can be obtained (**Figs. 6.1, 6.2**). Similar results are obtained by using fluorescence microscopy (Dravis and Yurewicz, 1985). This preservation of depositional textures results from the generalized pseudomorphic replacement by dolomite with slight crystal size increase of the original calcite. An extreme case is shown by the dolomitization of oolitic calcarenites with sparite cement in which the original fine concentric structure of the ooids becomes a coarser annular structure consisting of a few rings of rhombs of almost the same size as those developed at the expense of the interstitial sparite cement. However, the higher content in organic matter of ooids outlines their shape in contrast with the clearer aspect of the replaced cement. Organic matter plays the same role in completely dolomitized oncoids, cyanobacterial mats, and pellets, which also stand out within a clearer cement.

FIGURE 6.2 Examples of well-preserved depositional textures in replacement dolostones, Burnt Bluff Group (Silurian), Wisconsin. **A.** Finely crystalline, organic-rich dark oncoids with vague concentric structure in a cement of clear dolomite; originally oncoidal calcarenite with sparite cement. **B.** Coarsely crystalline oncoids in a dark cryptocrystalline, organic-rich dolomite containing a small, smooth ostracod; originally oncoidal calcarenite with interstitial calcisiltite matrix. **C.** Coarsely crystalline unzoned dolomite mosaic with large leperditid ostracod carapaces; originally an algal bioconstructed limestone with intraframework concentrations of smooth ostracods. **D.** Uneven-textured, coarsely crystalline dolomite with a bimodal association of large, unzoned crystals and medium-sized aggregates; originally an algal bioconstructed limestone. **E.** Medium-crystalline dolomite with poorly preserved fragmented leperditids, large, well-rounded quartz grain (lower left), and dark, subrounded lithoclasts; originally a biocalcarenite with calcisiltite matrix. **F.** Irregular layers of dark, cryptocrystalline dolomite interbedded with light-colored, coarsely crystalline dolomite; originally alternating cyanobacterial mats and calcilutite. **G.** Band of cryptocrystalline dark dolomite with finely crystalline organic network; originally a cyanobacterial mat (*Spongiostromata*). **H.** Dolomitized and silicified tabulate coral with surrounding coarsely crystalline dolomite; originally a calcilutite matrix. All photomicrographs: plane–polarized light. From Soderman and Carozzi (1963). Reprinted by permission of the American Association of Petroleum Geologists.

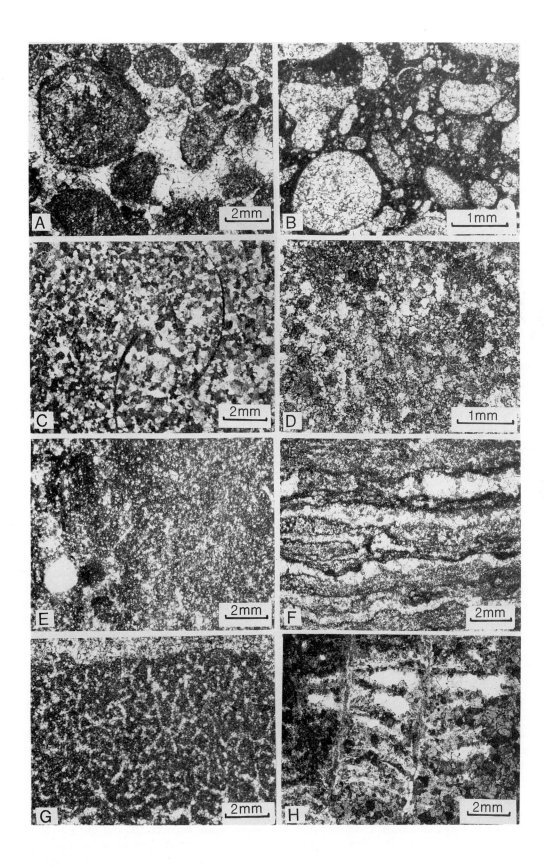

Even some calcisiltite matrixes rich in organic matter appear as dark-colored dolosiltite groundmasses.

Biomoldic porosity is an extremely common feature of dolostones, but still of disputed origin. Sibley (1982) showed that it can develop concurrently with pseudomorphic replacement, but it may also result from the subsequent dissolution of constituents consisting originally of low-Mg calcite or that were converted to this type of calcite before dolomitization and thus were resistant to it.

Cloudy centered, but clear-rimmed dolomite rhombs are also frequently encountered in dolostones (**Plate 26.E, F**). These rhombs developed in dolostones that consisted originally of low-Mg calcite, because cores are made cloudy either by inclusions of this type of calcite, by voids resulting from its subsequent dissolution, or by fluid-filled microcavities. Sibley (1980) suggested that these cloudy centers develop because initial dolomitizing fluids were saturated with respect to low-Mg calcite and therefore minute crystals of it were included in early rhomb centers. Later, as diagenetic fluids became undersaturated with low-Mg calcite, dolomite zones were free of such inclusions. Since the low-Mg precursor rock was relatively resistant to dolomitization and not entirely dissolved, nucleation sites are widely scattered, and the size of these particular rhombs is usually much larger than that of rhombs without cloudy centers. Chemical differences occur frequently between the inner and outer parts of these rhombs. These differences, due mainly to variations in iron and manganese content, are well displayed by cathodoluminescence and back-scattered electron microscopy (Choquette and Steinen, 1980; Fairchild, 1980; Gawthorpe, 1987). Land et al. (1975) confirmed that trace element differences between cloudy centers and clear rims could be interpreted as indicating that clear rims precipitated in more dilute solutions, which could be undersaturated with respect to low-Mg calcite, thus explaining why inclusions are missing (Sibley, 1980).

Mosaics of anhedral to subhedral dolomite crystals apparently represent the final product of dolomitization, with total destruction of original depositional fabrics. Some of the crystals of these mosaics show the previously described cloudy cores and clear rims, indicating that the precursor carbonate was low-Mg calcite or had been converted to it before dolomitization. The generation of these dolomite mosaics certainly involves a long diagenetic history, which can be surmised from textural features such as successive episodes of fracturing and rehealing of crystals, various types of overgrowths and neomorphism, internal dissolution surfaces, and other crystallographic anomalies that have time significance and developed through the diagenetic sequence.

Dolomite mosaics have been extensively studied petrographically and experimentally, and terminologies have varied accordingly from complex (Gregg and Sibley, 1984) to much simpler ones (Sibley and Gregg, 1987) in which crystal size distribution (unimodal and polymodal) and crystal boundary shapes (planar and nonplanar, previously

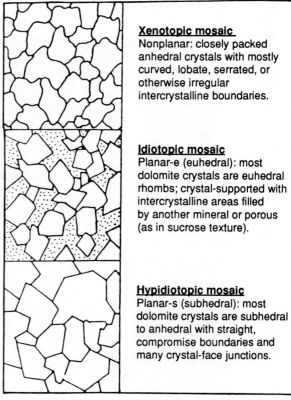

Xenotopic mosaic
Nonplanar: closely packed anhedral crystals with mostly curved, lobate, serrated, or otherwise irregular intercrystalline boundaries.

Idiotopic mosaic
Planar-e (euhedral): most dolomite crystals are euhedral rhombs; crystal-supported with intercrystalline areas filled by another mineral or porous (as in sucrose texture).

Hypidiotopic mosaic
Planar-s (subhedral): most dolomite crystals are subhedral to anhedral with straight, compromise boundaries and many crystal-face junctions.

FIGURE 6.3 Common types of dolomite textures. From Sibley and Gregg (1987). Reprinted by permission of the Society of Economic Paleontologists and Mineralogists.

called, respectively, idiotopic and xenotopic) were emphasized (**Fig. 6.3**). The simplified classification is descriptive, but carries genetic implications, because size distribution is controlled by both nucleation and growth kinetics and crystal boundary shapes by growth kinetics. The interesting part of this kind of investigation is to determine the temperature of diagenetic formation of these mosaics, which, according to Gregg and Sibley (1984), is the major factor of their growth. Apparently, mosaics with planar crystal boundary shapes (idiotopic) indicate growth temperatures below 50°C, and those with nonplanar boundary shapes (xenotopic) result from elevated temperatures greater than 50°C. In the same range of temperatures is saddle dolomite (Radke and Mathis, 1980; Kretz, 1992), also called baroque dolomite (**Plate 27.B**), clear to turbid, with characteristic curved cleavage traces and crystal faces, and a peculiar sweeping extinction under crossed nicols caused by a distorted crystal lattice. It often displays zonations due to iron and manganese concentration, which are well expressed in cathodoluminescence. Saddle dolomite, which may contain oil inclusions, is assumed to be an excellent geothermometer, indicating burial temperatures between 50° and 150°C. These values, which also apply to calcite, are within the window of hydrocarbon thermal maturation. Saddle dolomite appears to be a by-product of chemical compaction and thermochemical sulfate reduction, thus confirming its high-temperature formation from hypersaline brines (Machel, 1987).

Precipitated dolomite cement is of common occurrence in late Cenozoic dolostones of the Caribbean and Bahama regions, where it occurs between dolomitized grains or within biomoldic porosity either as clear cavity-filling dolosparite, which does not show any crystal size increase toward the center of the cavity, or as cavity-lining of clear euhedral rhombs (Land, 1973; Supko, 1977; Sibley, 1980, 1982; Kaldi and Gidman, 1982). Recognition of the real cement nature of dolomite is often difficult because it depends upon the petrographic evidence that dolomite did not replace a precursor calcite cement (Sibley, 1982). Cements in older dolostones present similar problems, but if zoned or internally discontinuous by dissolution and overgrowths processes, their real origin can be demonstrated by cathodoluminescence (Fairchild, 1980; Tucker, 1983).

Detrital dolomite is a product of reworking of penecontemporaneous or, more commonly, older dolostones followed by redeposition after transport in a variety of freshwater and marine sandstones (Amsbury, 1962; Sabins, 1962; Lindholm, 1969; Freeman et al., 1983).

Detrital dolomite appears either as moderately to well-rounded grains of microcrystalline dolomite or individual rhombohedral crystals partly rounded, but with often straight sides due to rupture along cleavage planes during transport. Detrital rhombs can be surrounded by subsequent overgrowths of clear dolomite cement in optical continuity, at times extending into available pore spaces of the final host rock. Confusion with the well-zoned, cloudy centered and clear-rimmed rhombs described above is impossible because of the internal discordance between abraded rhombic cores and subsequent overgrowths.

In summary, interpreting the complex diagenetic history of dolostones requires careful petrographic studies, first, on the manner by which skeletal and nonskeletal grains were replaced pseudomorphically or not, partially replaced, or spared by the process and subsequently dissolved to generate biomoldic porosity; second, on how precursor matrixes, cements, and original porosity were affected in ways similar to those of the grains; third, on how and where dolomite cements were precipitated; and, finally, on whether detrital dolomite is possibly present. To establish diagenetic sequences in dolostones, which can often be a pure matter of successive crystallographic states, is a far more complex task than to apply the same approach to limestones.

DEDOLOMITIZATION

Petrographic evidence for dedolomitization or calcitization of dolomite consists generally of two major fabrics: first, single crystals of rhombohedral calcite (**Plate 27.C, D**), which is a pseudomorphic replacement (calcite rarely occurs in isolated rhombohedral form), and, second, an equicrystalline mosaic of small anhedral calcite crystals within the rhombohedral outline of the precursor dolomite crystal (Evamy, 1967). These two fabrics indicate that calcite is indeed a dolomite replacement. Other instances correspond to a calcite cement filling the rhombohedral cavity with a marginal bladed structure grading toward the center into a blocky mosaic. This is obviously a cavity-filling cement, indicating that the process involved the intermediate stage of a rhombohedral pore. If cementation does not proceed beyond the bladed rim cement, an incomplete crystal moldic porosity is left. Dedolomitization may also take the shape of large patches of replacive calcite mosaic (Budai et al., 1984) that contains dolomite rhombs with their edges corroded to a variable extent. This dissolution can penetrate deeply into the rhombs as irregular embayments and may destroy their cores. The final result may be irregular remnants of dolomite rims and cores enclosed between reciprocal contacts of large crystals in the replacive calcite mosaic. This nonfabric-selective replacement can be associated with fabric-selective calcitization, where only dolomite rhomb cores, or specific zones if present, are replaced by calcite as if the penetration of the replacive fluids took place very selectively along the dolomite cleavage planes. In all instances, the replacive calcite is in optical continuity with the dolomite substrates; its crystals are anhedral, of variable size, but always larger than those of the precursor dolomite.

Interpretations of processes responsible for dedolomitization have grown in number with increasing knowledge. Today, calcitization of dolomite seems to cover the entire spectrum of diagenetic environments from subaerial exposure and near-surface conditions to deep burial and even metamorphism.

Dedolomitization was long interpreted as a product of near-surface diagenesis, generally associated with evaporite minerals (Lucia, 1961; Warrak, 1974). The suggested mechanism was the reaction of dolomite with calcium-sulfate-rich solutions, producing calcium carbonate and magnesium sulfate, which is soluble and rarely preserved (Shearman et al., 1961; Evamy, 1963; Goldberg, 1967; Folkman, 1969). However, it was not clearly established if calcite replaced dolomite directly or through an intermediate stage of replacive anhydrite. It was also assumed that under these near-surface oxidizing conditions, ferroan dolomite (Al-Hashimi and Hemingway, 1973) and calcian dolomite (Katz, 1968) were preferentially replaced, and that iron oxides and hydroxides, which occur with many calcitized ferroan dolomites, were by-products of the replacement process (Shearman et al., 1961; Evamy, 1963; Katz, 1968, 1971; Wolfe, 1970; Wood and Amstrong, 1975; Franck, 1981).

Furthermore, near-surface dedolomitization was also interpreted as indicating an erosional unconformity within the involved sequence (Schmidt, 1965; Goldberg, 1967;

Braun and Friedman, 1970; Scholle, 1971) or an expression of late postburial weathering (Mossler, 1971; Chafetz, 1972).

Dedolomitization processes are also active in various subsurface conditions, some of them still related to the circulation of sulfate-rich groundwater. Mattavelli et al. (1969) described dedolomitization that took place in the Gela oilfield in Sicily by the action of deeply penetrating sulfate-rich waters during early tectonic movements in the Jurassic, followed by subsequent partial dissolution of calcite under CO_2-rich waters related to igneous activity. Other instances of dedolomitization were considered directly related to the circulation of meteoric freshwater in the past or at present. Indeed, Margaritz and Kafri (1981) reported in the Cenomanian of northern Israel a dedolomitization restricted to a narrow transition zone between shallow-marine hypersaline dolostones and basinal limestones and chalks, which they attributed to a diagenetic process through exposure of the transition zone to meteoric freshwater. Holail et al. (1988) described in the Upper Cretaceous of Egypt a dolomitization that occurred in two successive episodes. First was generation of zoned rhombs with cloudy iron-rich rhombohedral cores surrounded by euhedral clear rims with decreasing iron content. These cores have a mottled fabric under cathodoluminescence, indicating partial dissolution and replacement of an unstable precursor dolomite phase, probably formed in marine waters. Second were clear dolomite cement overgrowths that eventually may fill the interstitial pores between the zoned dolomite rhombs. In the vicinity of an unconformity, dolomite rhombs and cement underwent extensive dissolution and calcitization. Dedolomitization was either fabric selective with precipitated Mg-enriched calcite in dissolution voids of the iron-rich rhomb cores and in narrow zones within the rhombs, or nonfabric selective when entire rhombs and the overgrowths of dolomite cement were replaced. In all instances, the replacive calcite is in optical continuity with dolomite substrates, anhedral, with more variable and larger crystal size than the precursor dolomite. Geochemical data indicate that the dolomite–dolomite replacement, that is, the two successive phases of dolomitization, as well as subsequent dedolomitization, occurred within a single-fluid system that evolved in time from possibly marine to mixed meteoric–marine dolomitization to a purely meteoric environment of dedolomitization.

Longman and Mench (1978) described dedolomitization taking place within the fault-bounded freshwater portion of the Edwards aquifer in Texas and attributed it to the high Ca to Mg ratio of the circulating freshwater in a shallow subsurface environment. Back et al. (1983) suggested that dedolomitization in the Madison aquifer of North Dakota resulted from dolomite and gypsum dissolution during groundwater migration.

In the deep subsurface, Land and Prezbindowski (1981) reported dedolomitization by hot, calcium-rich brines moving up-dip into a dolomitic reservoir in the Edwards Group of the Gulf Coast. The high calcium content of these particular brines was attributed to albitization of plagioclase in sandstones located down-dip. In essence, any brine responsible for carbonatization can dedolomitize. Budai et al. (1984) described widespread dedolomitization in the Madison Group of the Western Overthrust Belt for which petrographic, isotopic, and trace element data indicate that at least four distinct phases of burial dedolomitization took place in a succession of diagenetic environments. The inferred chronological order is as follows: prestylolite, early fracture-related dedolomitization; calcitization of dolomite within anhydrite nodules; stylolite-related dedolomitization; and, finally, late fracture-related dedolomitization.

Freeman et al. (1983) reported dedolomitization in Miocene calcarenites of Minorca (Spain) containing abundant detrital dolomite as abraded rhombs and as dolomite overgrowths on them. Dedolomitization was either nonfabric selective and total, involving rhombs and overgrowths, or fabric selective, concentrating on the centers of the rhombs and the inner margin of the dolomite overgrowths. Movements of fluids probably took place along the dolomite cleavage planes (Franck, 1981).

To complete the spectrum of dedolomitization environments, the effects of contact metamorphism were assumed for dedolomitization of the Lisburne Group (Mississippian) of the Sadlerochit Mountains of Alaska (Wood and Amstrong, 1975).

RESERVOIR PROPERTIES

Dolostones represent extremely important reservoirs for hydrocarbons (Moore, 1989; Choquette et al., 1992) produced by associations of various types of porosity, resulting not only from replacement dolomitization processes, but also from dedolomitization and selective leaching of calcite under the action of CO_2-rich circulating waters. Detailed petrographic and geochemical studies by Prather (1992) revealed in the Smackover Formation (Oxfordian) of the northeastern Gulf Coast, in spite of the effects of recrystallization, the existence of distinct dolomitization phases among regionally extensive dolostones. These phases are seawater-seepage reflux; near-surface mixed water; shallow-burial mixed water; and, finally, deep burial. These phases of dolomitization, which overlapped in time and space, eventually built a regional platform body of dolostone reservoirs resulting from a complex combination of processes.

When replacement dolomitization develops a coarse mosaic texture, appreciable intercrystalline porosity is produced along the contacts between the various types of rhombohedra. Subsequent dissolution may involve particular zones of the rhombs or their cores, consisting of calcian dolomite or even of low-Mg calcite generating unusual types of

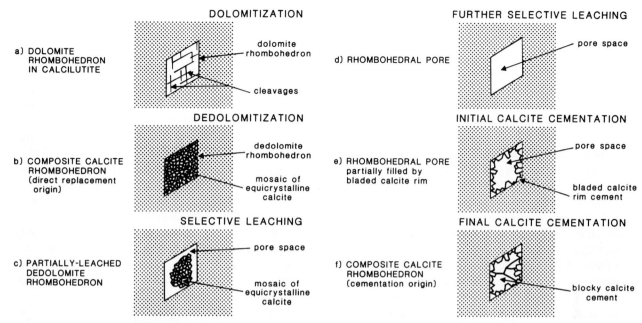

FIGURE 6.4 Schematic diagenetic history of certain dedolomitized limestones. From Evamy (1967). Reprinted by permission of the Society of Economic Paleontologists and Mineralogists.

fabric-selective intracrystalline porosity. Under the same conditions of diagenetic dissolution, low-Mg calcite skeletal and nonskeletal constituents that escaped dolomitization generate biomoldic porosity. This combination of intercrystalline, intracrystalline, and biomoldic porosities is responsible for the high quality of dolostone reservoirs (**Plate 27.E**).

Dedolomitization or calcitization of dolomite is also a factor in generating porosity in dolostones through dissolution of the replacive calcite. As mentioned above, this calcitization can be nonfabric selective, generating patches of calcite mosaic, or extremely fabric selective within certain rhomb zones and cores, generating crystal moldic and intracrystalline porosity. One of the most striking aspects of dedolomitization porosity involves the formation of rhombohedral pores (**Fig. 6.4**) when dolomite rhombs are scattered within partially dolomitized calcisiltites and mud-supported calcarenites (Evamy, 1967). The dolomite rhombohedra, either replaced by a single calcite crystal or by a mosaic of fine equicrystalline grains, are dissolved by selective leaching into rhombohedral pores. If these pores are close enough to each other, they may join into hybrid voids consisting of several crystal molds. Further dissolution of the surrounding calcite matrix can expand these molds into enlarged crystal molds and eventually into nonfabric-selective vugs. The rhombohedral porosity can be subsequently reduced by pore spaces becoming lined with precipitated rims of bladed calcite and even obliterated when the rim is followed by deposition of a coarser, blocky calcite cement filling the rest of the cavity. Actually, the obliteration of all the above-mentioned types of porosity in dolostones can occur not only by the late precipitation of calcite, but also by late generations of dolomite cements and of anhydrite during subsequent stages of the diagenetic sequence.

If a large amount of detrital dolomite is redeposited and mixed with a small amount of detrital quartz and feldspar, an unusual rock, which may be called "regenerated dolostone," is formed. Such was apparently the case for a shallow marine fan delta described in the Atokan (Pennsylvanian) of the deep Anadarko basin, with the detrital dolomite derived from the uplifting and erosion of a prominent high of Cambro-Ordovician Arbuckle Dolomite (Lyday, 1985). In such an unusual dolostone, the detrital depositional fabric was apparently almost destroyed by widespread aggrading neomorphism, accompanied by replacement of quartz by dolomite, under unspecified depth and temperature conditions, which developed a coarse-grained mosaic with sufficient intercrystalline porosity and permeability to generate a commercial hydrocarbon reservoir.

TYPICAL EXAMPLES

Plate 26.C. Stromatolitic dolostone. Bioconstructed dolostone consisting of thin, upward-curved, dark and irregular cyanobacterial mats that were highly pelleted in place by desiccation. They are separated by thicker and lighter laminae of dolosiltite to pseudodolomicrosparite due to aggrading neomorphism of the calcisiltite matrix originally trapped by the cyanobacterial mats. Small concentrations of opaque pyrite and rare fenestrae filled with dolosparite cement are scattered throughout. Shakopee Dolomite, Lower Ordovician, Wyalusing, Wisconsin, U.S.A.

Plate 26.D.   Partly dolomitized burrow. Horizontal burrow in a calcisiltite was partially and selectively replaced by early marine phreatic concentration of subhedral to perfectly euhedral rhombs of dolomite. Galena Group, Middle Ordovician, Dickeyville, Wisconsin, U.S.A.

Plate 26.E.   Incipient dolomitization. Rudistid-constructed limestone with dark interstitial matrix of bituminous calcisiltite. Incipient dolomitization in the mixed freshwater phreatic–marine phreatic environment occurred in the form of scattered, large-zoned rhombs with poorly defined cloudy centers and clear rims. Upper Cretaceous, Lenola, Latina, Italy.

Plate 26.F.   Dolomitization by zoned rhombs. This well-advanced type of dolomitization occurred in an argillaceous calcilutite in the mixed freshwater phreatic–saline lacustrine phreatic environment. The large rhombohedral crystals of dolomite show well-defined cloudy nuclei rich in calcite inclusions and clear rims. Lagoa Feia Formation, Lower Cretaceous, Campos Basin, offshore Brazil.

Plate 27.A.   Selective dolomitization. This type of dolomitization in the mixed freshwater phreatic–marine phreatic environment occurred in a grain-supported biocalcarenite consisting originally of a framework of bioclasts of several types of red algae, with an interstitial matrix of pelletoidal calcisiltite with smaller bioclasts of red algae and small calcareous foraminifers. Nonpseudomorphic dolomitization converted the original matrix into a mosaic of subhedral rhombs of dolomicrosparite with relicts of former small constituents. Pseudomorphic dolomitization replaced the red algae bioclasts, with excellent preservation of original skeletal structures. Amapá Formation, Paleocene–Middle Miocene, Foz do Amazonas Basin, offshore Brazil.

Plate 27.B.   Saddle dolomite. Irregular, large dissolution vug in a grain-supported biocalcarenite consisting of a framework of bioclasts of crinoids, neomorphosed pelecypods, *Endothyra,* gastropods, lithoclasts of fossiliferous calcisiltite, and superficial and normal ooids. Pressure solution is moderate among all constituents, and cement is cavity-filling microsparite. The saddle dolomite filling the vug developed in deep-burial environment and displays its peculiar type of sweeping extinction. Salem Limestone, Middle Mississippian, Texaco B-1, Francis Wente well, Cumberland County, Southern Illinois, U.S.A.

Plate 27.C.   Dedolomitization. This process is shown by fabric-selective dedolomitization in which dolomite rhombs reduced to their clear rims and filled in part by iron-rich calcite "float" in clear poikilotopic calcite cement. Thin section stained with Alizarin red-S. Galena Group, Rockford, Winnebago County, Illinois, U.S.A.

Plate 27.D.   Dedolomitization. This process is shown in a slightly pelletoidal, pyritic, and organic-rich calcisiltite in which dissolved rhombs of dolomite scattered in the matrix were dissolved and refilled with an irregular microsparite mosaic. Thin section stained with Alizarin red-S. Cedar Valley Limestone, Middle Devonian, New Jersey Zinc core DDH-6, Macon County, Missouri, U.S.A.

Plate 27.E.   Dolorudite with intercrystalline and moldic porosity. Originally, this rock was a biocalcirudite consisting of a framework of red algae bioclasts with a cavity-filling cement of sparite. Extensive solution, following dolomitization, developed moldic to moldic-enlarged porosity from the red algae bioclasts and intercrystalline porosity in the cement. Amapá Formation, Paleocene–Middle Miocene, Foz do Amazonas Basin, offshore Brazil.

REFERENCES

AL-HASHIMI, W. S., and HEMINGWAY, J. E., 1973. Recent dedolomitization and the origin of the rusty crusts of Northumberland. *J. Sed. Petrology,* 43, 82–91.

AMSBURY, D. L., 1962. Detrital dolomite in central Texas. *J. Sed. Petrology,* 32, 5–14.

BACK, W., HANSHAW, B. B., PLUMMER, L. N., RAHN, P. H., RIGHTMIRE, C. T., and RUBIN, M., 1983. Process and rate of dedolomitization: mass transfer and ^{14}C dating in a regional carbonate aquifer. *Geol. Soc. Amer. Bull.,* 94, 1415–1429.

BAKER, P. A., and BURNS, S. J., 1985. Occurrence and formation of dolomite in organic-rich continental margin sediments. *Amer. Assoc. Petroleum Geologists Bull.,* 69, 1917–1930.

BONE, Y., JAMES, N. P., and KYSER, T. K., 1992. Synsedimentary detrital dolomite in Quaternary cool-water carbonate sediments, Lacepede Shelf, South Australia. *Geology,* 20, 109–112.

BRAUN, M., and FRIEDMAN, G. M., 1970. Dedolomitization fabric in peels: a possible clue to unconformity surfaces. *J. Sed. Petrology,* 40, 417–419.

BUDAI, J. M., LOHMANN, K. C., and OWEN, R. M., 1984. Burial dedolomitization in the Mississippian Madison Limestone, Wyoming and Utah thrust belt. *J. Sed. Petrology,* 54, 276–288.

BULLEN, S. B., and SIBLEY, D. F., 1984. Dolomite selectivity and mimic replacement. *Geology,* 12, 655–658.

CARBALLO, J. D., LAND, L. S., and MISER, D. E., 1987. Holocene dolomitization of supratidal sediments by active tidal pumping, Sugarloaf Key, Florida. *J. Sed. Petrology,* 57, 153–165.

CAROZZI, A. V., 1989. *Carbonate Rock Depositional Models: A Microfacies Approach.* Prentice Hall, Englewood Cliffs, N.J., Advanced Reference Series, 604 pp.

———, and ZENGER, D. H., 1991. The original chemical analysis of dolomite by Nicolas-Théodore de Saussure (1792): a laboratory error and its historical consequences. *Archives Sciences, Genève,* 44, 163–196.

CAYEUX, L., 1935. *Les roches sédimentaires de France—Roches carbonatées (calcaires et dolomies).* Masson, Paris, 436 pp. See also L. Cayeux, *Sedimentary Rocks of France—Carbonate Rocks (Limestones and Dolomites),* translated and updated by A. V. Carozzi, Hafner Publishing Company, Darien, Conn., 438 pp.

CHAFETZ, H. S., 1972. Surface diagenesis of limestone. *J. Sed. Petrology,* 42, 325–329.

CHOQUETTE, P. W., and STEINEN, R. P., 1980. Mississippian non-supratidal dolomite, Ste. Genevieve Limestone, Illinois Basin: evidence for mixed-water dolomitization. In D. H. Zenger, J. B. Dunham, and R. L. Ethington (eds.), *Concepts and Models of Dolomitization.* Soc. Econ. Paleontologists and Mineralogists, Special Publ. 28, 168–196.

———, COX, A., and MEYERS, W. J., 1992. Characteristics, distribution and origin of porosity in shelf dolostones: Burlington-Keokuk Formation (Mississippian), U.S. Mid-Continent. *J. Sed. Petrology,* 62, 167–189.

DRAVIS, J., and YUREWICZ, D. A., 1985. Enhanced carbonate petrography using fluorescence microscopy. *J. Sed. Petrology,* 55, 795–804.

EVAMY, B. D., 1963. The application of a chemical staining technique to a study of dolomitisation. *Sedimentology,* 2, 164–170.

———, 1967. Dedolomitization and the development of rhombohedral pores in limestones. *J. Sed. Petrology,* 37, 1204–1215.

FAIRCHILD, I. J., 1980. Stages in a Precambrian dolomitization, Scotland: cementing versus replacement textures. *Sedimentology,* 27, 631–650.

FANNING, K. A., BYRNE, R. H., BRELAND, J. A., BETZER, P. R., MOORE, W. S., ELSINGER, R. J., and PYLE, T. E., 1981. Geothermal springs of the west Florida continental shelf: evidence for dolomitization and radionuclide enrichment. *Earth Planet. Sci. Letters,* 52, 345–354.

FOLKMAN, Y., 1969. Diagenetic dedolomitization in the Albian–Cenomanian Yagur Dolomite on Mount Carmel (Northern Israel). *J. Sed. Petrology,* 39, 380–385.

FRANCK, J. R., 1981. Dedolomitization in the Taum Sauk Limestone (Upper Cambrian), southeast Missouri. *J. Sed. Petrology,* 51, 7–18.

FREEMAN, T., ROTHBARD, D., and OBRADOR, A., 1983. Terrigenous dolomite in the Miocene of Menorca (Spain): provenance and diagenesis. *J. Sed. Petrology,* 53, 543–548.

GAWTHORPE, R. L., 1987. Burial dolomitization and porosity development in a mixed carbonate–clastic sequence: an example from the Bowland Basin, northern England. *Sedimentology,* 34, 533–558.

GIVEN, R. K., and WILKINSON, B. H., 1987. Dolomite abundance and stratigraphic age: constraints on rates and mechanisms of Phanerozoic dolostone formation. *J. Sed. Petrology,* 57, 1068–1078.

GOLDBERG, M., 1967. Supratidal dolomitization and dedolomitization in Jurassic rocks of Hamakhtesh Haqatan, Israel. *J. Sed. Petrology,* 37, 760–773.

GREGG, J. M., and SIBLEY, D. F., 1984. Epigenetic dolomitization and the origin of xenotopic dolomite texture. *J. Sed. Petrology,* 54, 908–931.

HARDIE, L. A., 1987. Dolomitization: a critical view of some current views. *J. Sed. Petrology,* 57, 166–183.

HARDY, R., and TUCKER, M. E., 1988. X-ray diffraction. In M. E. Tucker (ed.), *Techniques in Sedimentology.* Blackwell Scientific Publications, London, 191–228.

HOLAIL, H., LOHMANN, K. C., and SANDERSON, I. D., 1988. Dolomitization and dedolomitization of Upper Cretaceous carbonates: Bahariya Oasis, Egypt. In V. Shukla and P. A. Baker (eds.), *Sedimentology and Geochemistry of Dolostones.* Soc. Econ. Paleontologists and Mineralogists, Special Publ. 43, 191–207.

KALDI, J., and GIDMAN, J., 1982. Early diagenetic dolomite cements: examples from the Permian Lower Magnesian Limestone and the Pleistocene carbonates of the Bahamas. *J. Sed. Petrology,* 52, 1073–1085.

KATZ, A., 1968. Calcian dolomites and dedolomitization. *Nature,* 217, 439–440.

———, 1971. Zoned dolomite crystals. *J. of Geology,* 79, 38–51.

KELTS, K., and MCKENZIE, J. A., 1982. Diagenetic dolomite formation in Quaternary anoxic diatomaceous muds of Deep Sea Drilling Project Leg 64, Gulf of California. Init. Repts. D.S.D.P., U.S. Government Printing Office, Washington, D.C., 64, 553–570.

———, and ———, 1984. A comparison of anoxic dolomite from deep-sea sediments: Quaternary Gulf of California and Messinian Tripoli Formation of Sicily. In R. E. Garrison, M. Kastner, and D. H. Zenger (eds.), *Dolomites of the Monterey Formation and Other Organic-rich Units.* Soc. Econ. Paleontologists and Mineralogists, Pacific Section, Publ. 41, 119–140.

KOHOUT, F. A., 1967. Groundwater flow and the geothermal regime of the Floridan plateau. *Trans. Gulf Coast Assoc. Geol. Societies,* 17, 339–354.

KRETZ, R., 1992. Carousel model for the crystallization of saddle dolomite. *J. Sed. Petrology,* 62, 190–195.

KULM, L. D., SUESS, E., and THORNBURG, T. M., 1984. Dolomites in organic-rich muds of the Peru forearc basins: analogue to the Monterey Formation. In R. E. Garrison, M. Kastner, and D. H. Zenger (eds.), *Dolomites of the Monterey Formation and Other Organic-rich Units.* Soc. Econ. Paleontologists and Mineralogists, Pacific Section, Publ. 41, 29–47.

LAND, L. S., 1973. Contemporaneous dolomitization of middle Pleistocene reefs by meteoric water, north Jamaica. *Bull. Marine Sci.,* 23, 64–92.

———, 1985. The origin of massive dolomite. *J. Geol. Education,* 33, 112–125.

———, and PREZBINDOWSKI, D. R., 1981. The origin and evolution of saline formation water, lower Cretaceous carbonates, south-central Texas, U.S.A. *J. Hydrology,* 54, 51–74.

———, LUND, H. J., and MCCULLOUGH, M. L., 1989. Dynamic circulation on interstitial seawater in a Jamaican fringing reef. *Carbonates and Evaporites,* 4, 1–7.

———, SALEM, M. R. I., and MORROW, D. W., 1975. Paleohydrology of ancient dolomites: geochemical evidence. *Amer. Assoc. Petroleum Geologists Bull.,* 59, 1602–1625.

LAST, M. W., 1990. Lacustrine dolomite—an overview of modern,

Holocene, and Pleistocene occurrences. *Earth Sci. Rev.,* 27, 221–263.

LINDHOLM, R. C., 1969. Detrital dolomite in Onondaga Limestone (Middle Devonian) of New York: its implication for the "dolomite question." *Amer. Assoc. Petroleum Geologists Bull.,* 52, 1035–1042.

LONGMAN, M. W., and MENCH, P. A., 1978. Diagenesis of Cretaceous limestones in the Edwards aquifer system of south-central Texas: a scanning electron microscope study. *Sed. Geol.,* 21, 241–276.

LUCIA, F. J., 1961. Dedolomitization in the Tansill (Permian) Formation. *Geol. Soc. Amer. Bull.,* 72, 1107–1110.

LUMSDEN, D. N., 1985. Secular variations in dolomite abundance in deep marine sediments. *Geology,* 13, 766–769.

LYDAY, J. R., 1985. Atokan (Pennsylvanian) Berlin field: genesis of recycled dolomite reservoir, Deep Anadarko Basin, Oklahoma. *Amer. Assoc. Petroleum Geologists Bull.,* 69, 1931–1949.

MACHEL, H. G., 1987. Saddle dolomite as a by-product of chemical compaction and thermochemical sulfate reduction. *Geology,* 15, 936–940.

———, and MOUNTJOY, E. W., 1986. Chemistry and environments of dolomitization—a reappraisal. *Earth Sci. Rev.,* 23, 175–222.

MARGARITZ, M., and KAFRI, U., 1981. Stable isotope and Sr^{2+}/Ca^{2+} evidence of diagenetic dedolomitization in a schizohaline environment: Cenomanian of northern Israel. *Sed. Geol.,* 28, 29–41.

MATTAVELLI, L., CHILINGARIAN, G. V., and STORER, D., 1969. Petrography and diagenesis of the Taormina Formation, Gela oil field, Sicily (Italy). *Sed. Geol.,* 3, 59–86.

MAZZULLO, S. J., REID, A. M., and GREGG, J. M., 1987. Dolomitization of Holocene Mg-calcite supratidal deposits, Ambergris Cay, Belize. *Geol. Soc. Amer. Bull.,* 98, 224–231.

MITCHELL, J. T., LAND, L. S., and MISER, D. E., 1987. Modern marine dolomite cement in a north Jamaican fringing reef. *Geology,* 15, 557–560.

MOORE, C. H., 1989. *Carbonate Diagenesis and Porosity.* Developments in Sedimentology 46, Elsevier Scientific Publishing Co., New York, 338 pp.

MORROW, D. W., 1990a. Dolomite—Part 1: The chemistry of dolomitization and dolomite precipitation. In I. A. McIlreath and D. W. Morrow (eds.), *Diagenesis.* Geological Association of Canada, Geoscience Canada, Reprint Series 4, 113–123.

———, 1990b. Dolomite—Part 2: Dolomitization models and ancient dolostones. In I. A. McIlreath and D. W. Morrow (eds.), *Diagenesis.* Geological Association of Canada, Geoscience Canada, Reprint Series 4, 125–139.

MOSSLER, J. H., 1971. Diagenesis and dolomitization of Swope Formation (Upper Pennsylvanian), southeast Kansas. *J. Sed. Petrology,* 41, 962–970.

MULLINS, H. T., WISE, S. W., JR., GRADULSKI, A. F., HINCHEY, E. J., MASTERS, P. M., and SIEGEL, D. I., 1984a. Shallow subsurface diagenesis of Pleistocene periplatform ooze: northern Bahamas. *Sedimentology,* 32, 473–494.

———, ———, LAND, L. S., SIEGEL, D. I., MASTERS, P. M., HINCHEY, E. J., and PRICE, K. R., 1984b. Authigenic dolomite in Bahamian peri-platform slope sediment. *Geology,* 13, 292–295.

MURRAY, R. C., 1964. Preservation of primary structures and fabrics in dolomite. In J. Imbrie and N. D. Newell (eds.), *Approaches to Paleoecology.* John Wiley & Sons, New York, 388–403.

———, and LUCIA, F. J., 1967. Cause and control of dolomite distribution by rock selectivity. *Geol. Soc. Amer. Bull.,* 78, 21–35.

PRATHER, B. E., 1992. Origin of dolostone reservoir rocks, Smackover Formation (Oxfordian), northeastern Gulf Coast, U.S.A. *Amer. Assoc. Petroleum Geologists Bull.,* 76, 133–163.

PRAY, L. C., and MURRAY, R. C. (eds.), 1965. *Dolomitization and Limestone Diagenesis—A Symposium.* Soc. Econ. Paleontologists and Mineralogists, Special Publ. 13, 180 pp.

RADKE, R. M., and MATHIS, R. L., 1980. On the formation of saddle dolomite. *J. Sed. Petrology,* 50, 1149–1168.

ROSEN, M. R., and COSHELL, L., 1992. A new location of Holocene dolomite formation, Lake Hayward, Western Australia. *Sedimentology,* 39, 161–166.

SABINS, F. F., JR., 1962. Grains of detrital, secondary, and primary dolomite from Cretaceous strata of the Western Interior. *Geol. Soc. Amer. Bull.,* 73, 1183–1196.

SALLER, A. H., 1984. Petrologic and geochemical constraints on the origin of subsurface dolomite, Enewetak Atoll: an example of dolomitization by normal seawater. *Geology,* 12, 217–220.

SCHMIDT, V., 1965. Facies, diagenesis, and related reservoir properties in the Gigas beds (Upper Jurassic), northwestern Germany. In L. C. Pray and R. C. Murray (eds.), *Dolomitization and Limestone Diagenesis: A Symposium.* Soc. Econ. Paleontologists and Mineralogists, Special Publ. 13, 124–168.

SCHOLLE, P. A., 1971. Diagenesis of deep-water carbonate turbidites, Upper Cretaceous Monte Antola Flysch, northern Apennines, Italy. *J. Sed. Petrology,* 41, 233–250.

SHEARMAN, D. J., KHOURI, J., and TAHA, S., 1961. On the replacement of dolomite by calcite in some Mesozoic limestones from the French Jura. *Proc. Geol. Assoc. London,* 71, 1–12.

SHIMMIELD, G. B., and PRICE, N. B., 1984. Recent dolomite formation in hemipelagic sediments off Baja California, Mexico. In R. E. Garrison, M. Kastner, and D. H. Zenger (eds.), *Dolomites of the Monterey Formation and Other Organic-rich Units.* Soc. Econ. Paleontologists and Mineralogists, Pacific Section, Publ. 41, 5–18.

SHUKLA, V., and BAKER. P. A. (eds.), 1988. *Sedimentology and Geochemistry of Dolostones.* Soc. Econ. Palentologists and Mineralogists, Special Publ. 43, 266 pp.

SIBLEY, D. F., 1980. Climatic control of dolomitization, Seroe Domi Formation (Pliocene), Bonaire, N. A. In D. H. Zenger, J. B. Dunham, and R. L. Ethington (eds.), *Concepts and Models of Dolomitization.* Soc. Econ. Paleontologists and Mineralogists, Special Publ. 28, 247–258.

———, 1982. The origin of common dolomite fabrics: clues from the Pliocene. *J. Sed. Petrology,* 52, 1087–1100.

———, 1991. Secular changes in the amount and texture of dolomite. *Geology,* 19, 151–154.

———, and GREGG, J. M., 1987. Classification of dolomite rock textures. *J. Sed. Petrology,* 57, 967–975.

SIMMS, M., 1984. Dolomitization by groundwater-flow systems in carbonate platforms. *Trans. Gulf Coast Assoc. Geol. Societies,* 34, 411–420.

PLATE 1

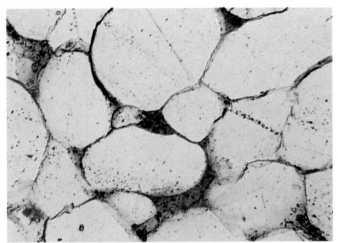

A. Pressure-solution quartz arenite

_____ 0.25 mm PL

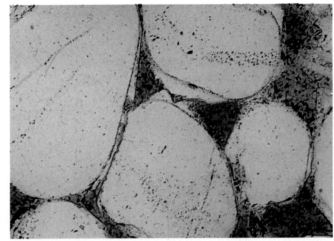

B. Overgrowth quartz arenite

_____ 0.25 mm PL

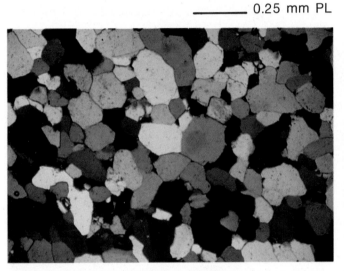

C. Quartzite by pressure solution

_____ 0.5 mm XN

D. Quartzite by overgrowth

_____ 0.5 mm XN

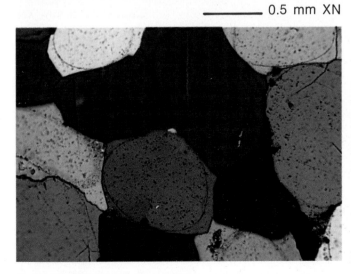

E. Quartzite by overgrowth

_____ 0.25 mm XN

F. Quartzite by overgrowth

_____ 0.25 mm CL

PLATE 2

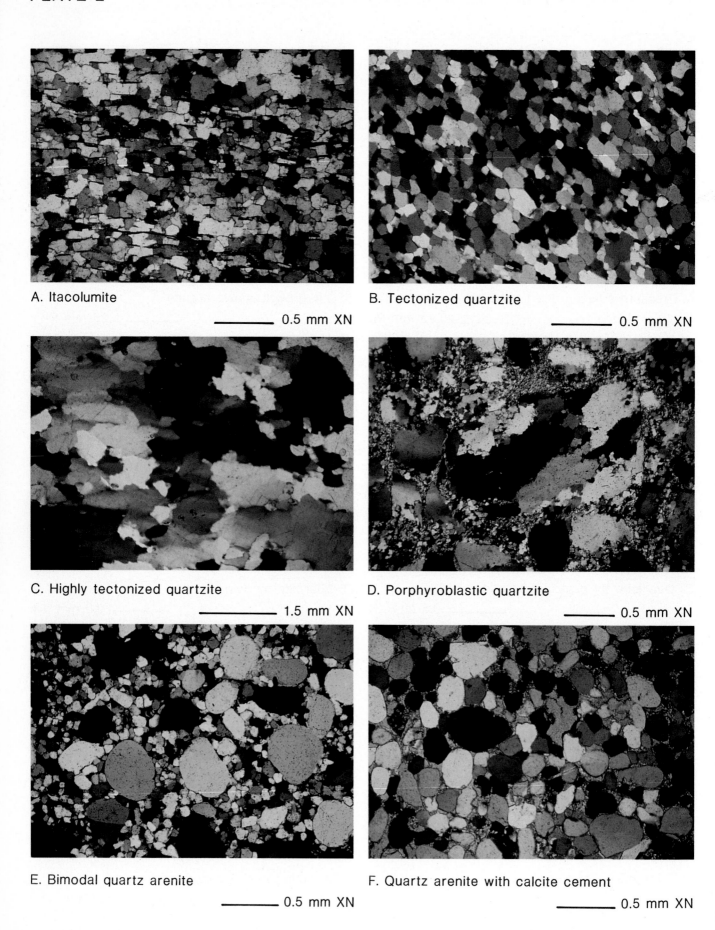

A. Itacolumite

——————— 0.5 mm XN

B. Tectonized quartzite

——————— 0.5 mm XN

C. Highly tectonized quartzite

——————— 1.5 mm XN

D. Porphyroblastic quartzite

——————— 0.5 mm XN

E. Bimodal quartz arenite

——————— 0.5 mm XN

F. Quartz arenite with calcite cement

——————— 0.5 mm XN

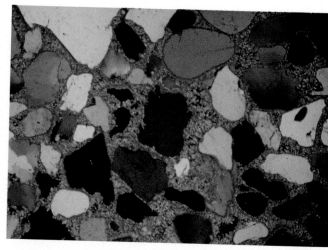

A. Quartz arenite, poikilotopic calcite cement
——— 0.5 mm XN

B. Quartz arenite, fibrous calcite cement
——— 0.5 mm XN

C. Quartz arenite, dolomite-anhydrite cement
——— 0.5 mm XN

D. Quartz arenite, concretionary opal cement
——— 0.5 mm PL

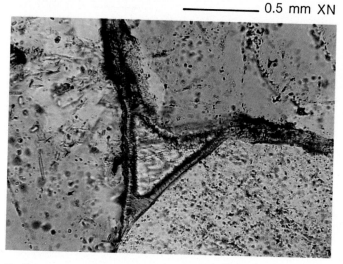

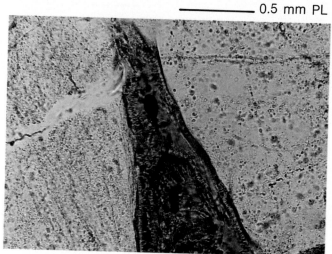

E. Quartz arenite, chlorite-calcite cement
——— 0.1 mm PL

F. Quartz arenite, chlorite cement
——— 0.1 mm PL

PLATE 3

PLATE 4

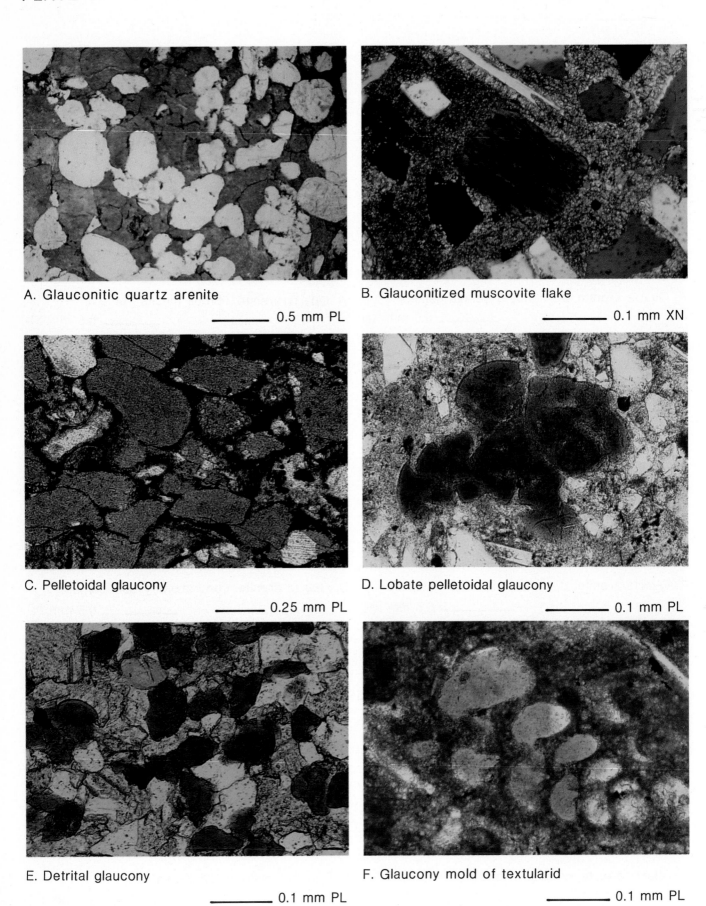

A. Glauconitic quartz arenite

—————— 0.5 mm PL

B. Glauconitized muscovite flake

—————— 0.1 mm XN

C. Pelletoidal glaucony

—————— 0.25 mm PL

D. Lobate pelletoidal glaucony

—————— 0.1 mm PL

E. Detrital glaucony

—————— 0.1 mm PL

F. Glaucony mold of textularid

—————— 0.1 mm PL

PLATE 5

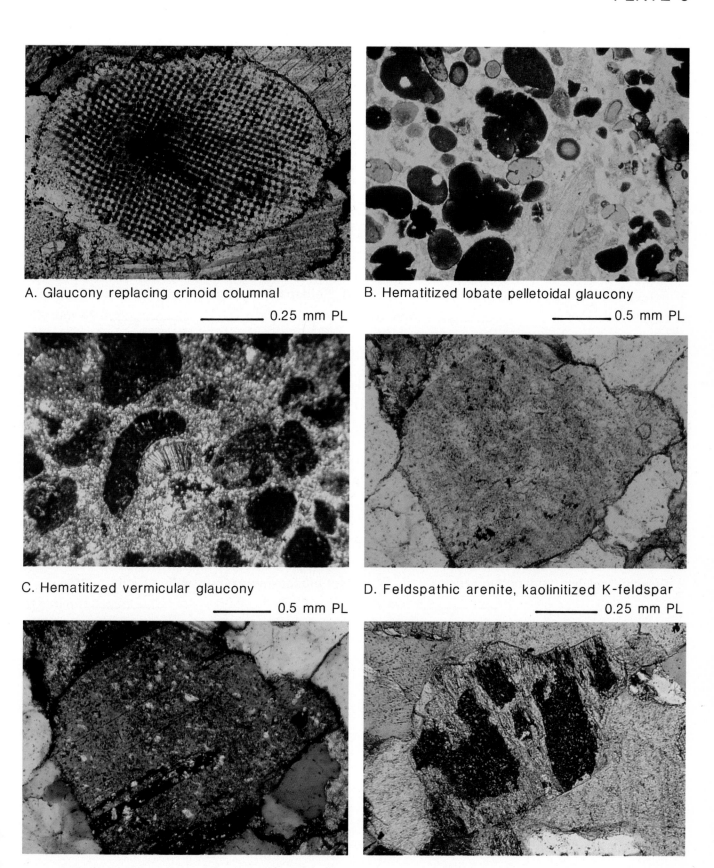

A. Glaucony replacing crinoid columnal

―――――― 0.25 mm PL

B. Hematitized lobate pelletoidal glaucony

―――――― 0.5 mm PL

C. Hematitized vermicular glaucony

―――――― 0.5 mm PL

D. Feldspathic arenite, kaolinitized K-feldspar

―――――― 0.25 mm PL

E. Feldspathic arenite, sericitized K-feldspar

―――――― 0.25 mm XN

F. Feldspathic arenite, K-feldspar altered to calcite-kaolinite

―――――― 0.25 mm XN

PLATE 6

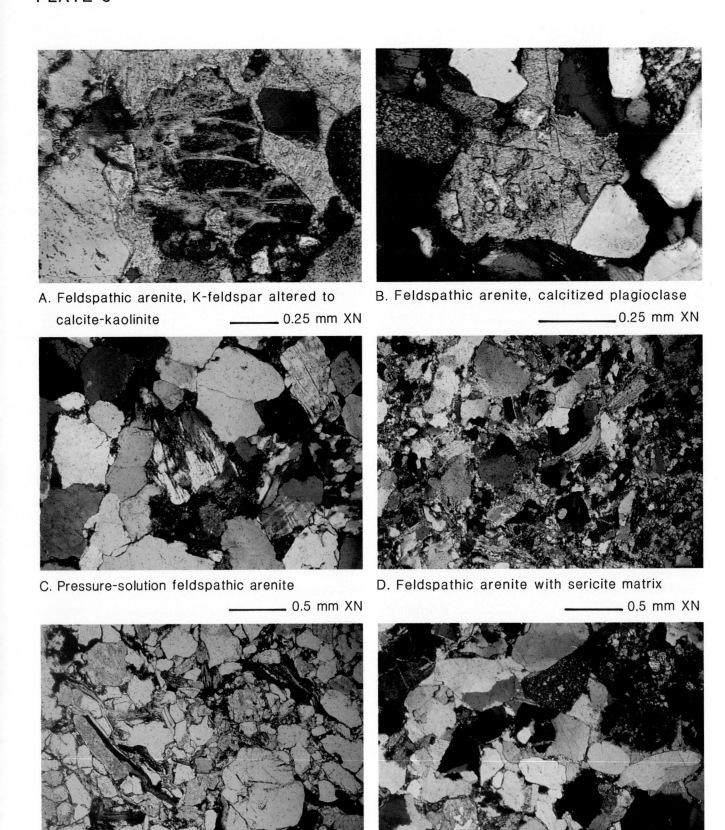

A. Feldspathic arenite, K-feldspar altered to calcite-kaolinite _____ 0.25 mm XN

B. Feldspathic arenite, calcitized plagioclase _____ 0.25 mm XN

C. Pressure-solution feldspathic arenite _____ 0.5 mm XN

D. Feldspathic arenite with sericite matrix _____ 0.5 mm XN

E. Feldspathic arenite with hematite cement _____ 0.5 mm PL

F. Feldspathic arenite with calcite cement _____ 0.5 mm XN

PLATE 7

A. Feldspathic arenite, anhydrite cement

——————— 0.5 mm XN

B. Feldspathic arenite, chlorite-kaolinite cement

——————— 0.25 mm PL

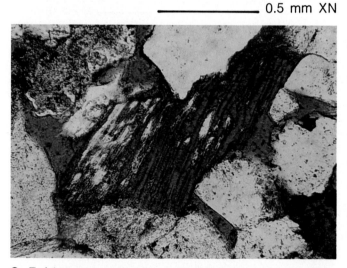

C. Feldspathic arenite, partly dissolved feldspar

——————— 0.25 mm PL

D. Lithic arenite, shale clast

——————— 0.25 mm PL

E. Lithic arenite, muscovite schist clast

——————— 0.25 mm XN

F. Lithic arenite, dolomite clasts

——————— 0.5 mm XN

PLATE 8

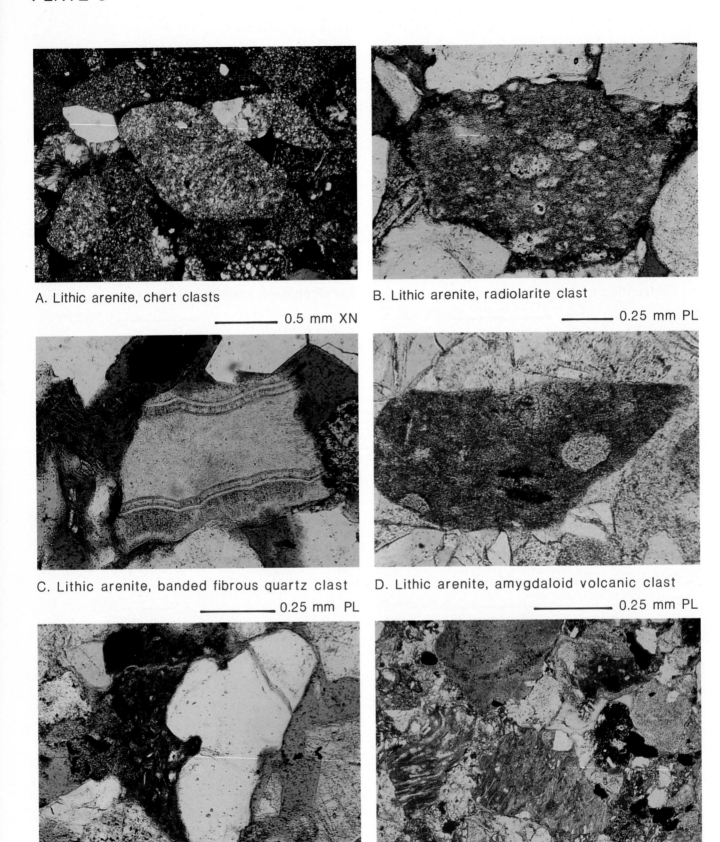

A. Lithic arenite, chert clasts

_____ 0.5 mm XN

B. Lithic arenite, radiolarite clast

_____ 0.25 mm PL

C. Lithic arenite, banded fibrous quartz clast

_____ 0.25 mm PL

D. Lithic arenite, amygdaloid volcanic clast

_____ 0.25 mm PL

E. Lithic arenite, volcaniclastic clast

_____ 0.25 mm PL

F. Lithic arenite, basic glass clasts

_____ 1 mm PL

PLATE 9

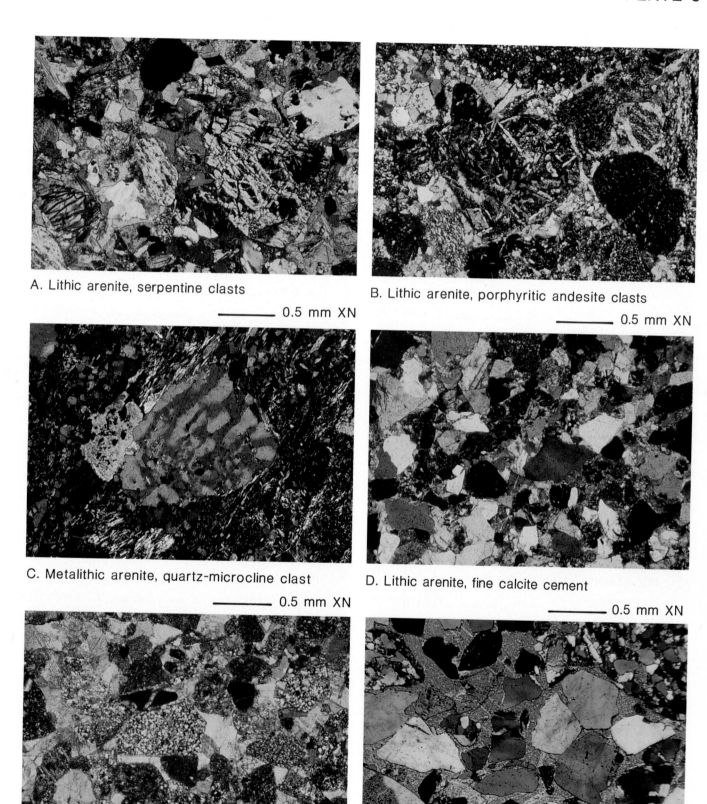

A. Lithic arenite, serpentine clasts

———— 0.5 mm XN

B. Lithic arenite, porphyritic andesite clasts

———— 0.5 mm XN

C. Metalithic arenite, quartz-microcline clast

———— 0.5 mm XN

D. Lithic arenite, fine calcite cement

———— 0.5 mm XN

E. Lithic arenite, coarse calcite cement

———— 0.5 mm XN

F. Lithic arenite, poikilotopic calcite cement

———— 0.5 mm XN

PLATE 10

A. Lithic arenite, chlorite-kaolinite cement

_____ 0.1 mm PL

B. Lithic arenite, chlorite-nontronite cement

_____ 0.1 mm PL

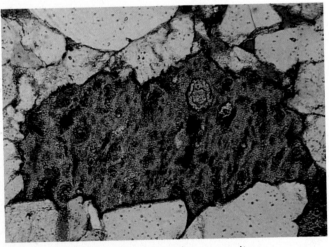

C. Lithic arenite, intragranular porosity,
   pumice clast

_____ 0.25 mm PL

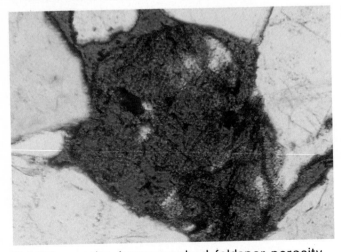

D. Lithic arenite, intragranular porosity,
   volcanic clast

_____ 0.25 mm PL

E. Lithic arenite, honeycombed feldspar porosity

_____ 0.25 mm PL

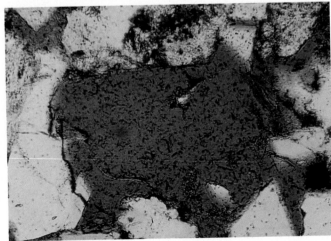

F. Lithic arenite, oversized pore

_____ 0.25 mm PL

PLATE 11

A. Lithic wacke

_____ 0.5 mm XN

B. Lithic wacke

_____ 0.5 mm XN

C. Volcanic-lithic wacke

_____ 0.5 mm PL

D. Volcanic-lithic wacke

_____ 0.5 mm XN

E. Feldspathic wacke

_____ 0.5 mm XN

F. Feldspathic-quartz wacke

_____ 0.5 mm XN

PLATE 12

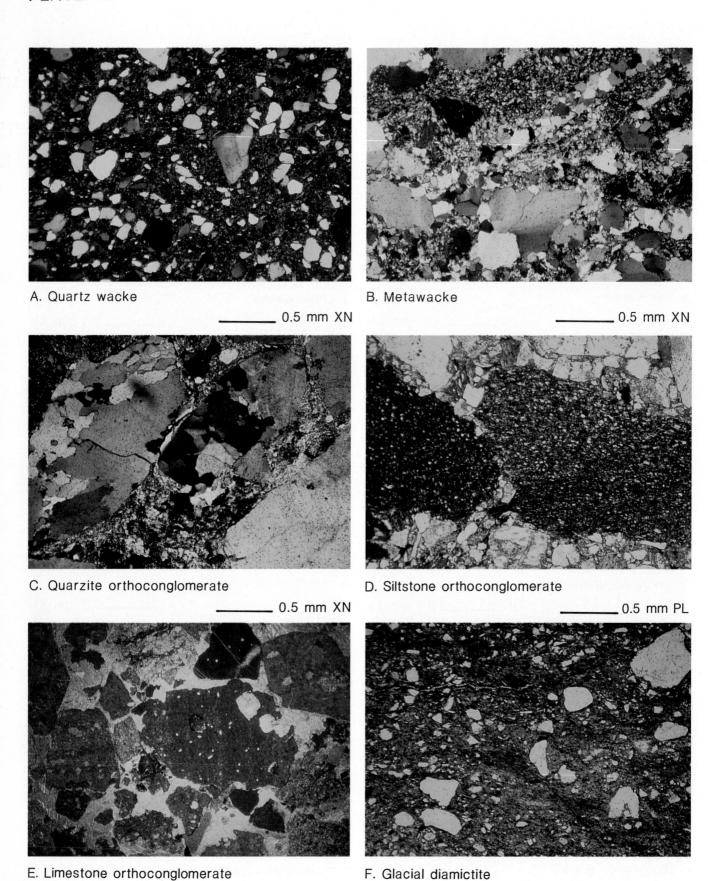

A. Quartz wacke

——— 0.5 mm XN

B. Metawacke

——— 0.5 mm XN

C. Quarzite orthoconglomerate

——— 0.5 mm XN

D. Siltstone orthoconglomerate

——— 0.5 mm PL

E. Limestone orthoconglomerate

——— 1.5 mm PL

F. Glacial diamictite

——— 0.5 mm PL

PLATE 13

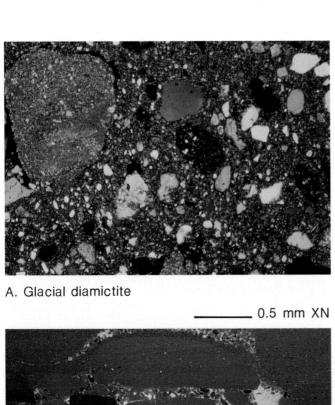

A. Glacial diamictite

_____ 0.5 mm XN

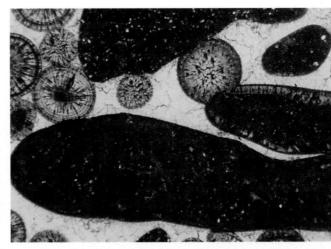

B. Intraformational limestone conglomerate

_____ 0.5 mm PL

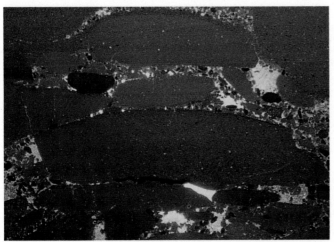

C. Intraformational limestone conglomerate

_____ 0.5 mm PL

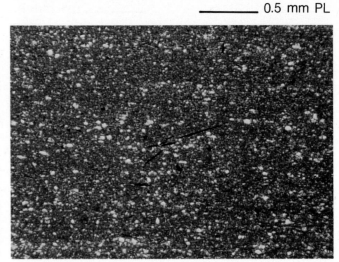

D. Limestone collapse breccia

_____ 0.5 mm PL

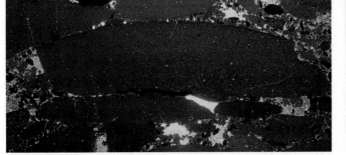

E. Pseudobreccia by dolomitization

_____ 0.5 mm PL

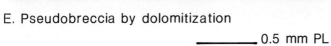

F. Arenaceous shale

_____ 0.5 mm PL

PLATE 14

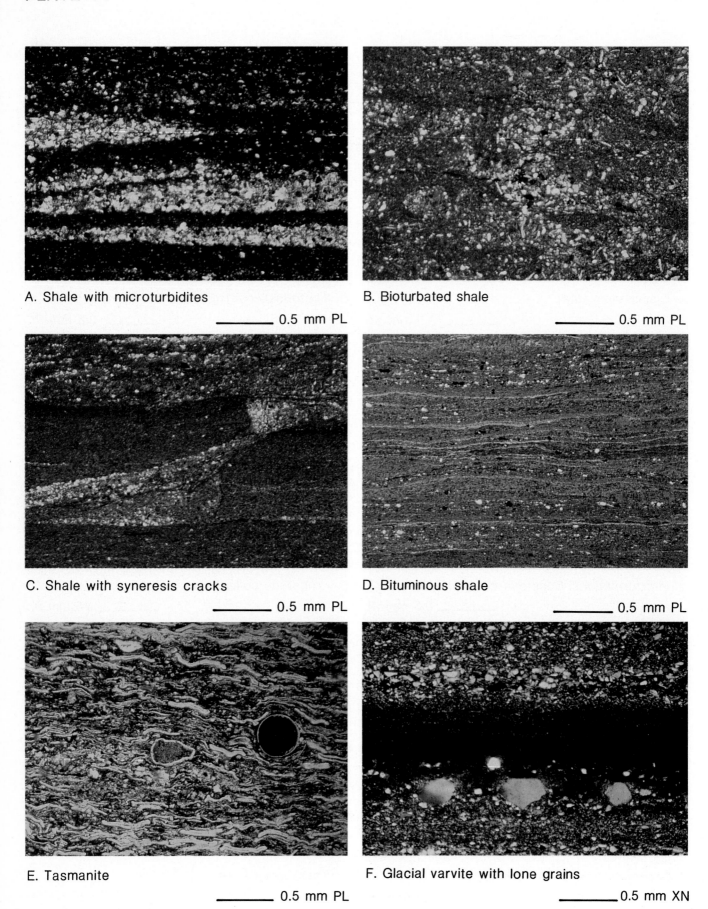

A. Shale with microturbidites

___ 0.5 mm PL

B. Bioturbated shale

___ 0.5 mm PL

C. Shale with syneresis cracks

___ 0.5 mm PL

D. Bituminous shale

___ 0.5 mm PL

E. Tasmanite

___ 0.5 mm PL

F. Glacial varvite with lone grains

___ 0.5 mm XN

PLATE 15

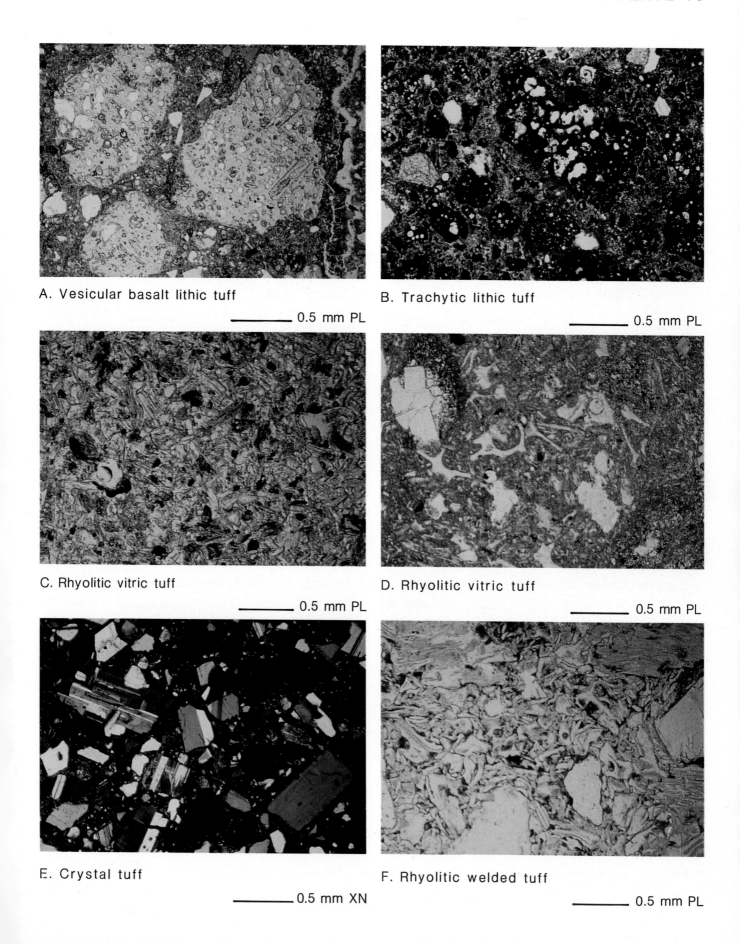

A. Vesicular basalt lithic tuff

B. Trachytic lithic tuff

_____ 0.5 mm PL

_____ 0.5 mm PL

C. Rhyolitic vitric tuff

D. Rhyolitic vitric tuff

_____ 0.5 mm PL

_____ 0.5 mm PL

E. Crystal tuff

F. Rhyolitic welded tuff

_____ 0.5 mm XN

_____ 0.5 mm PL

PLATE 16

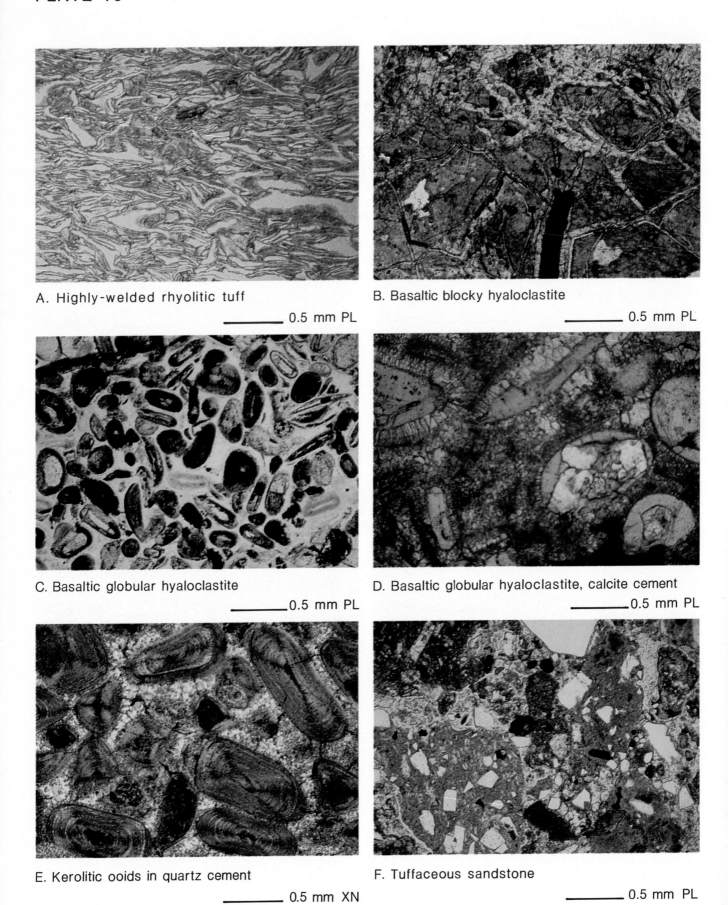

A. Highly-welded rhyolitic tuff

_____ 0.5 mm PL

B. Basaltic blocky hyaloclastite

_____ 0.5 mm PL

C. Basaltic globular hyaloclastite

_____0.5 mm PL

D. Basaltic globular hyaloclastite, calcite cement

_____0.5 mm PL

E. Kerolitic ooids in quartz cement

_____ 0.5 mm XN

F. Tuffaceous sandstone

_____ 0.5 mm PL

PLATE 17

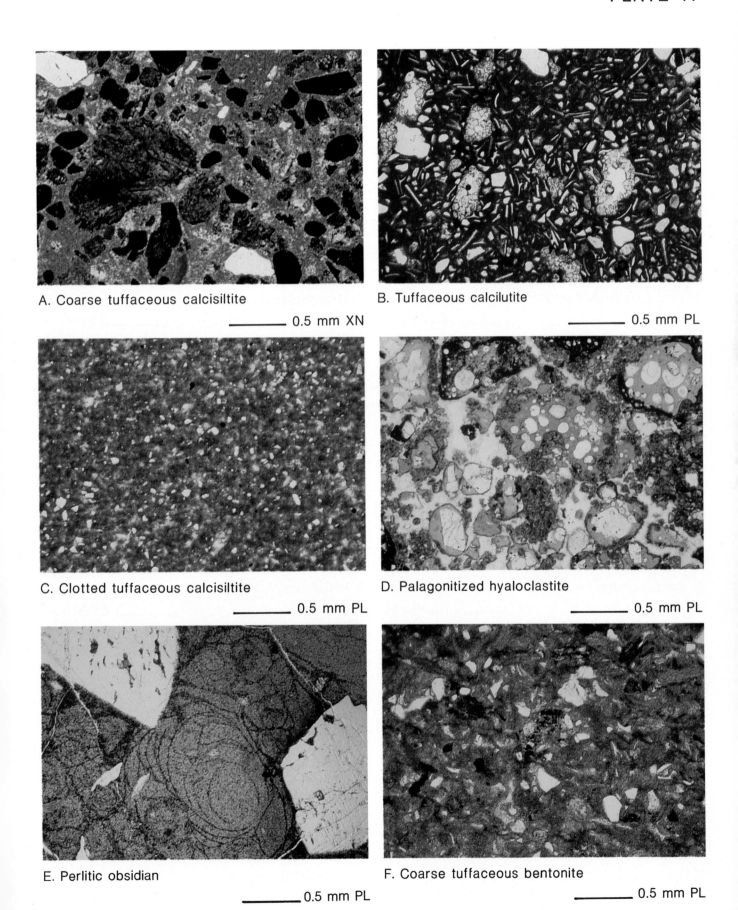

A. Coarse tuffaceous calcisiltite

_____ 0.5 mm XN

B. Tuffaceous calcilutite

_____ 0.5 mm PL

C. Clotted tuffaceous calcisiltite

_____ 0.5 mm PL

D. Palagonitized hyaloclastite

_____ 0.5 mm PL

E. Perlitic obsidian

_____ 0.5 mm PL

F. Coarse tuffaceous bentonite

_____ 0.5 mm PL

PLATE 18

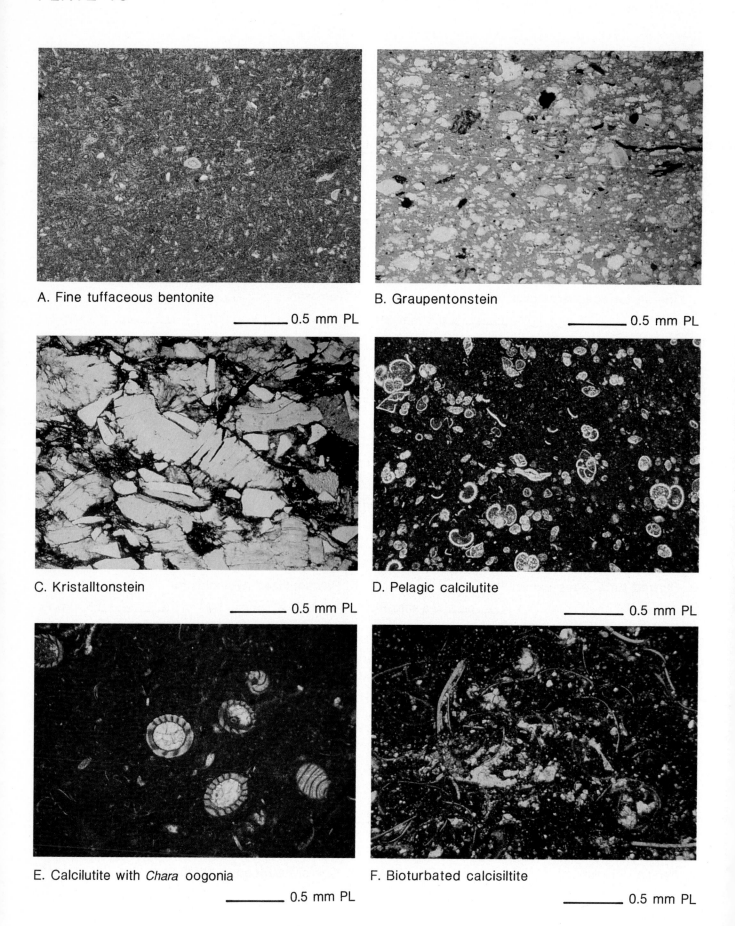

A. Fine tuffaceous bentonite

_____ 0.5 mm PL

B. Graupentonstein

_____ 0.5 mm PL

C. Kristalltonstein

_____ 0.5 mm PL

D. Pelagic calcilutite

_____ 0.5 mm PL

E. Calcilutite with *Chara* oogonia

_____ 0.5 mm PL

F. Bioturbated calcisiltite

_____ 0.5 mm PL

PLATE 19

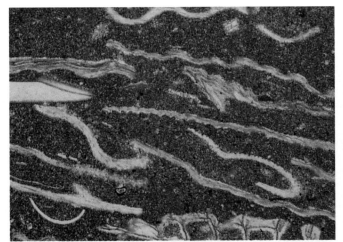

A. Mud-supported biocalcarenite

_____ 0.5 mm PL

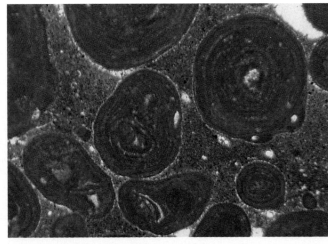

B. Oncolitic calcarenite, calcisiltite matrix

_____ 0.5 mm PL

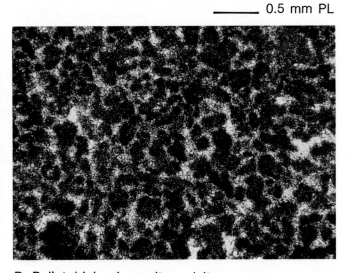

C. Crinoidal calcarenite, bioclastic matrix

_____ 0.5 mm PL

D. Pelletoidal calcarenite, calcite cement

_____ 0.5 mm PL

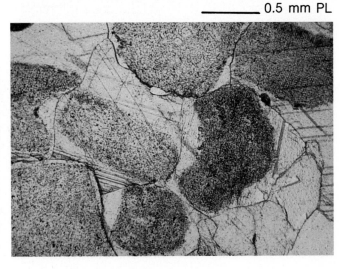

E. Biocalcarenite, sparite cement

_____ 0.5 mm PL

F. Crinoidal calcarenite, overgrowth cement

_____ 0.5 mm PL

PLATE 20

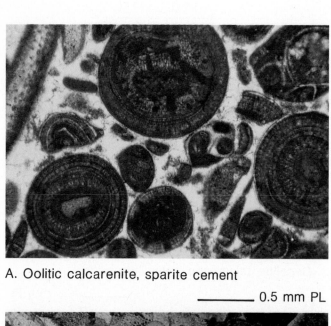

A. Oolitic calcarenite, sparite cement

—————— 0.5 mm PL

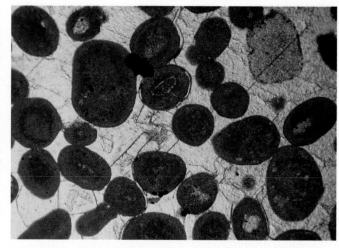

B. Oolitic calcarenite, poikilotopic calcite cement

—————— 0.5 mm PL

C. Oolitic-crinoidal calcarenite, pressure-solution

—————— 0.5 mm PL

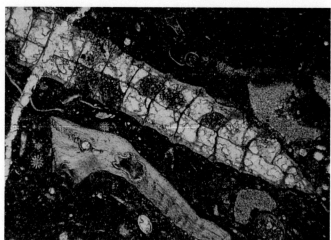

D. Oolitic calcarenite, pressure-solution

—————— 0.5 mm PL

E. Biocalcarenite, pressure-solution

—————— 0.5 mm PL

F. Mud-supported biocalcirudite

—————— 1.5 mm PL

PLATE 21

A. Lithocalcirudite, sparite cement

_____ 1.5 mm PL

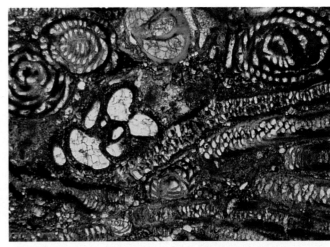

B. Foraminifer bioaccumulated limestone

_____ 0.5 mm PL

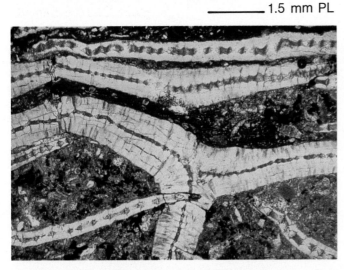

C. *Spiroclypeus* bioaccumulated limestone

_____ 0.5 mm PL

D. *Halimeda* bioaccumulated limestone

_____ 0.5 mm PL

E. *Amphistegina* bioaccumulated limestone

_____ 0.5 mm PL

F. *Donezella* bioconstructed limestone

_____ 0.5 mm PL

PLATE 22

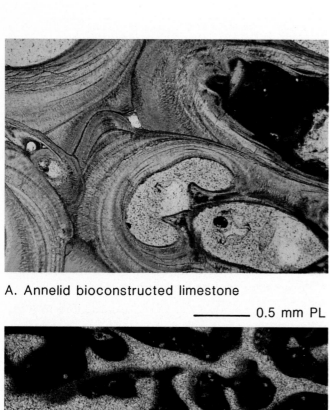

A. Annelid bioconstructed limestone

———— 0.5 mm PL

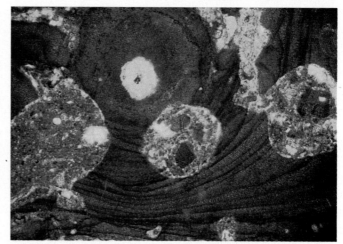

B. Red algae bioconstructed limestone

———— 0.5 mm PL

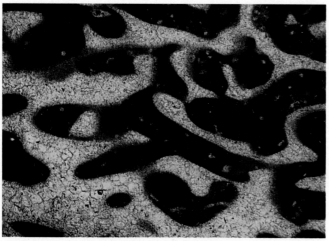

C. Coral bioconstructed limestone

———— 0.5 mm PL

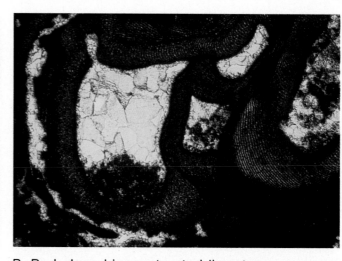

D. Perforations, marine phreatic

———— 0.5 mm PL

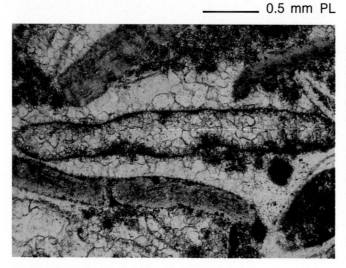

E. Micritization, marine phreatic

———— 0.25 mm PL

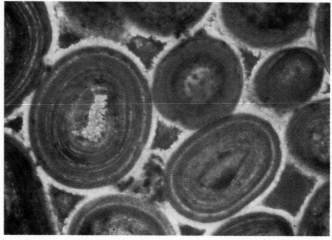

F. Isopachous fibrous calcite rim cement
   marine phreatic   ———— 0.5 mm PL

PLATE 23

A. Fibrous calcite cement, marine phreatic

———— 0.5 mm PL

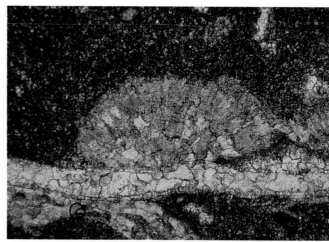

B. Botryoidal calcite cement, marine phreatic

———— 0.5 mm PL

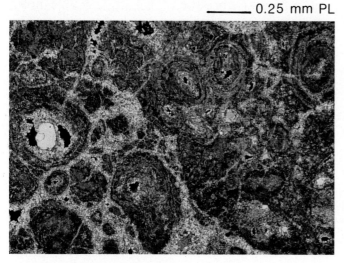

C. Pendant calcite cement, freshwater vadose

———— 0.25 mm PL

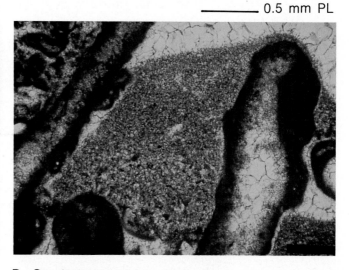

D. Graded bedded vadose silt,
   freshwater vadose ———— 0.5 mm PL

E. Paleocalcrete with coated grains,
   freshwater vadose ———— 0.5 mm XN

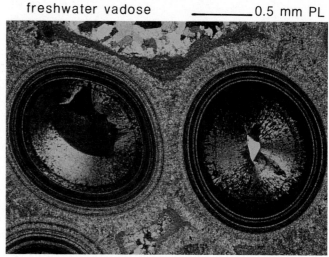

F. Paleocalcrete with vadose ooids,
   freshwater vadose ———— 0.25 mm XN

PLATE 24

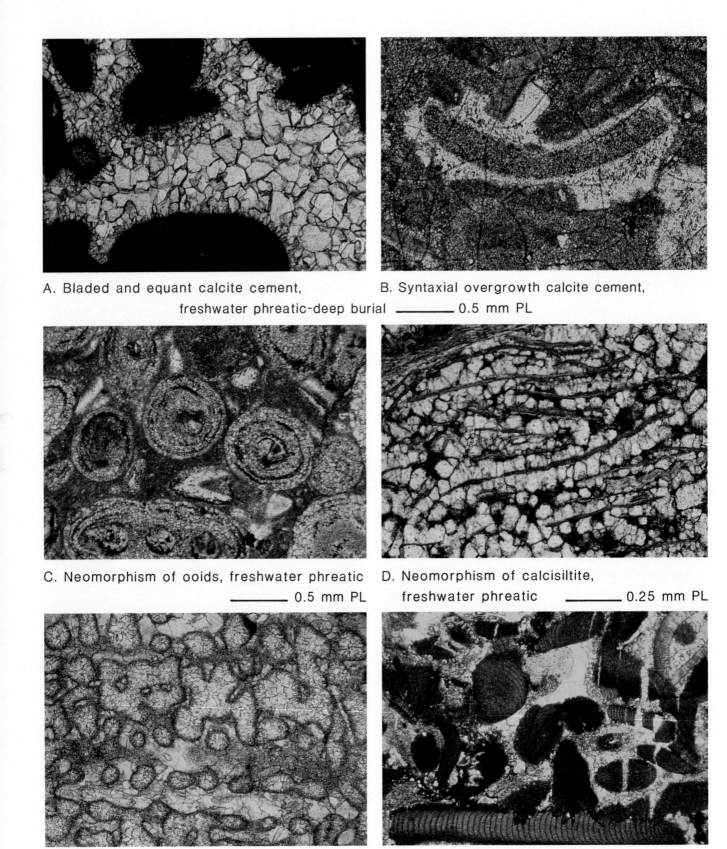

A. Bladed and equant calcite cement,
   freshwater phreatic-deep burial _____ 0.5 mm PL

B. Syntaxial overgrowth calcite cement,
   _____ 0.5 mm PL

C. Neomorphism of ooids, freshwater phreatic
   _____ 0.5 mm PL

D. Neomorphism of calcisiltite,
   freshwater phreatic _____ 0.25 mm PL

E. Neomorphism of coral framework,
   freshwater phreatic _____ 0.5 mm PL

F. Fracture-filling calcite, deep burial
   _____ 0.5 mm PL

PLATE 25

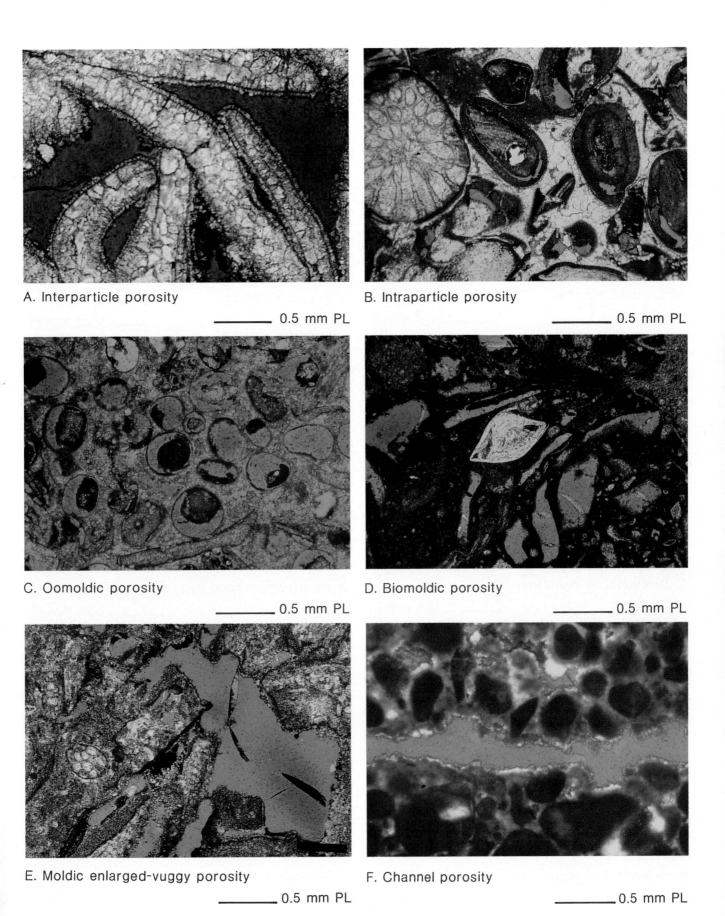

A. Interparticle porosity

_____ 0.5 mm PL

B. Intraparticle porosity

_____ 0.5 mm PL

C. Oomoldic porosity

_____ 0.5 mm PL

D. Biomoldic porosity

_____ 0.5 mm PL

E. Moldic enlarged-vuggy porosity

_____ 0.5 mm PL

F. Channel porosity

_____ 0.5 mm PL

PLATE 26

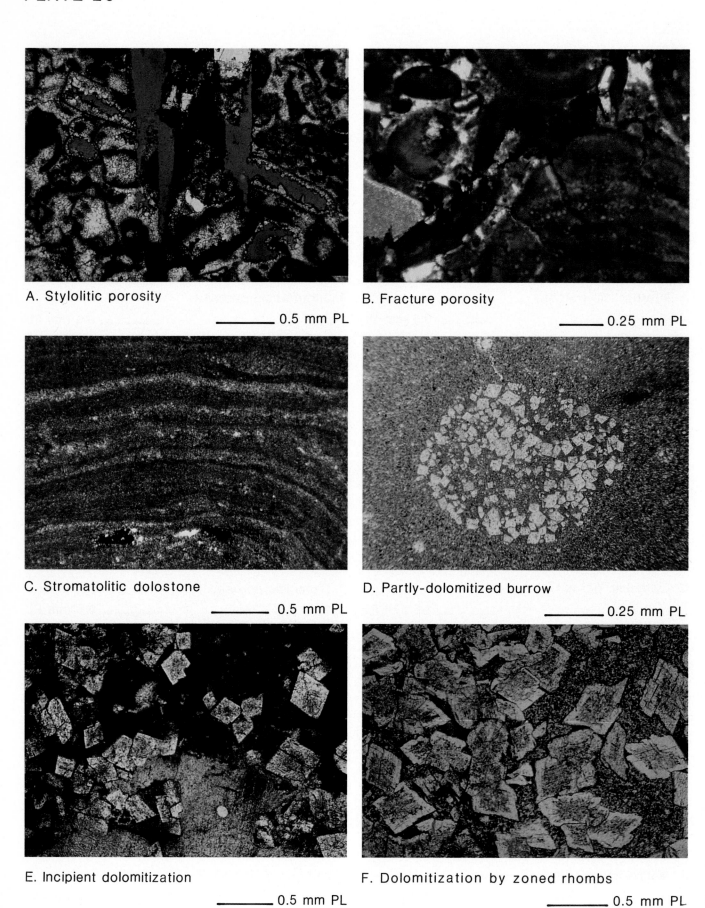

A. Stylolitic porosity

_____ 0.5 mm PL

B. Fracture porosity

_____ 0.25 mm PL

C. Stromatolitic dolostone

_____ 0.5 mm PL

D. Partly-dolomitized burrow

_____ 0.25 mm PL

E. Incipient dolomitization

_____ 0.5 mm PL

F. Dolomitization by zoned rhombs

_____ 0.5 mm PL

PLATE 27

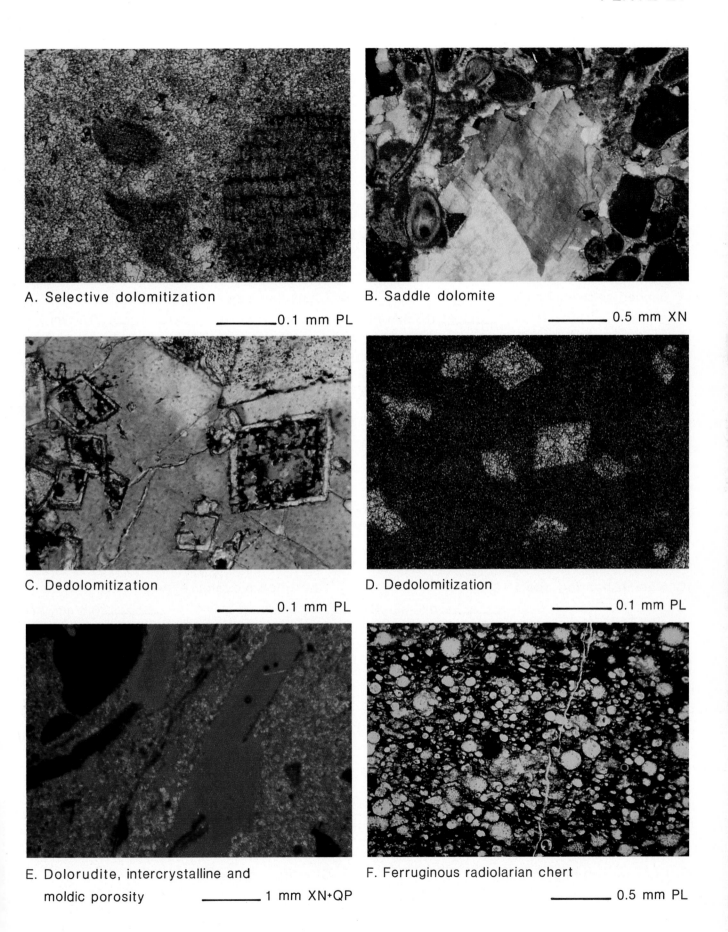

A. Selective dolomitization

_____0.1 mm PL

B. Saddle dolomite

_____ 0.5 mm XN

C. Dedolomitization

_____0.1 mm PL

D. Dedolomitization

_____0.1 mm PL

E. Dolorudite, intercrystalline and
moldic porosity _____ 1 mm XN+QP

F. Ferruginous radiolarian chert

_____ 0.5 mm PL

PLATE 28

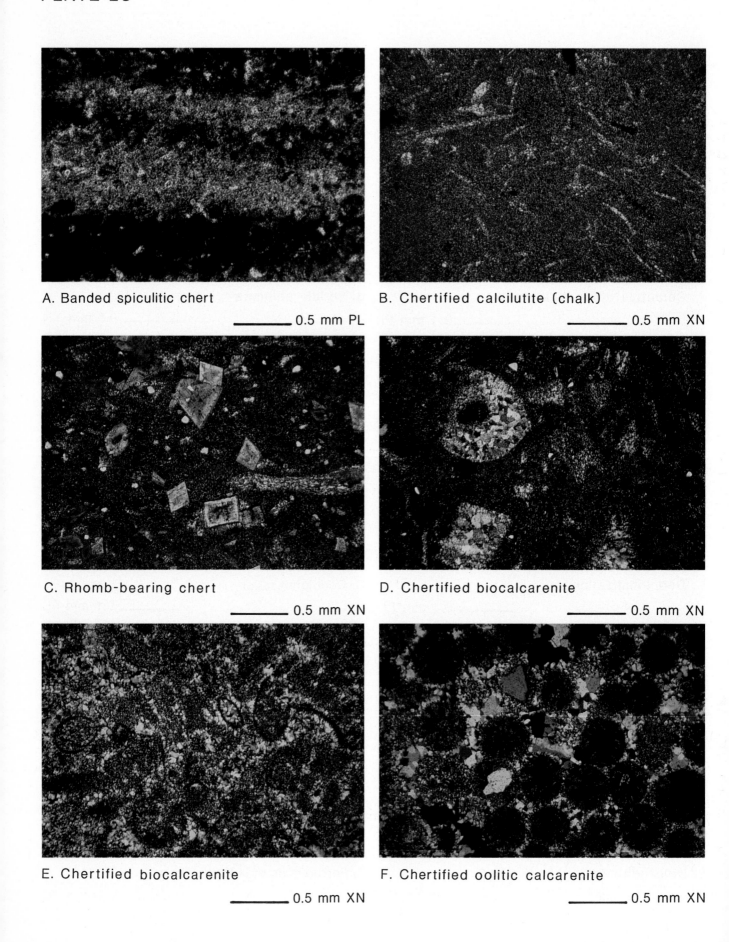

A. Banded spiculitic chert

_____ 0.5 mm PL

B. Chertified calcilutite (chalk)

_____ 0.5 mm XN

C. Rhomb-bearing chert

_____ 0.5 mm XN

D. Chertified biocalcarenite

_____ 0.5 mm XN

E. Chertified biocalcarenite

_____ 0.5 mm XN

F. Chertified oolitic calcarenite

_____ 0.5 mm XN

PLATE 29

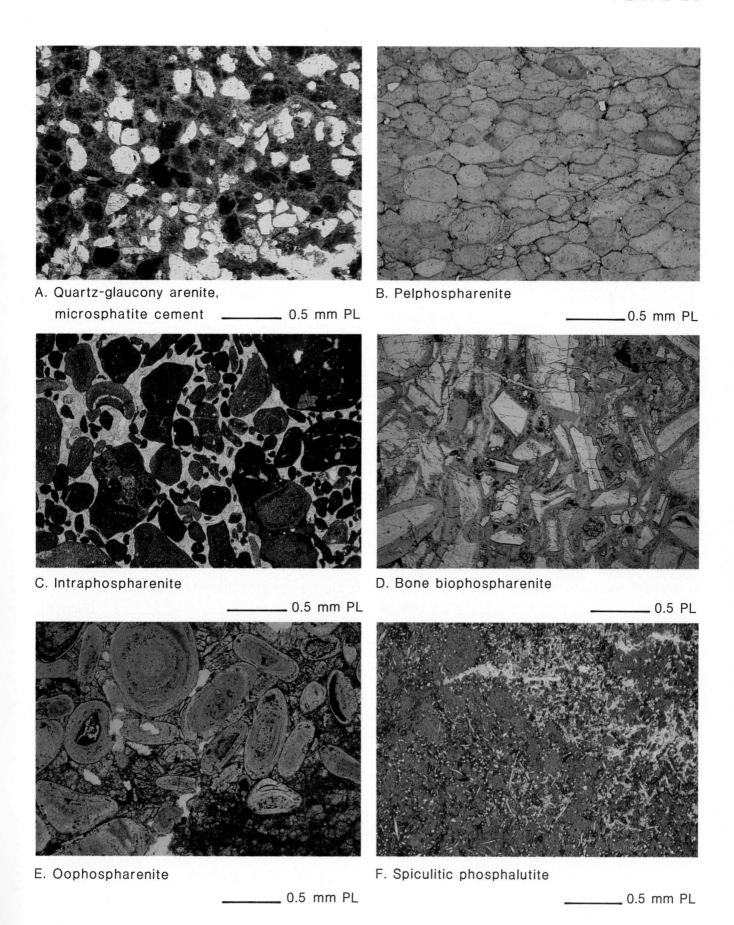

A. Quartz-glaucony arenite,

microsphatite cement ———— 0.5 mm PL

B. Pelphospharenite

———— 0.5 mm PL

C. Intraphospharenite

———— 0.5 mm PL

D. Bone biophospharenite

———— 0.5 PL

E. Oophospharenite

———— 0.5 mm PL

F. Spiculitic phosphalutite

———— 0.5 mm PL

PLATE 30

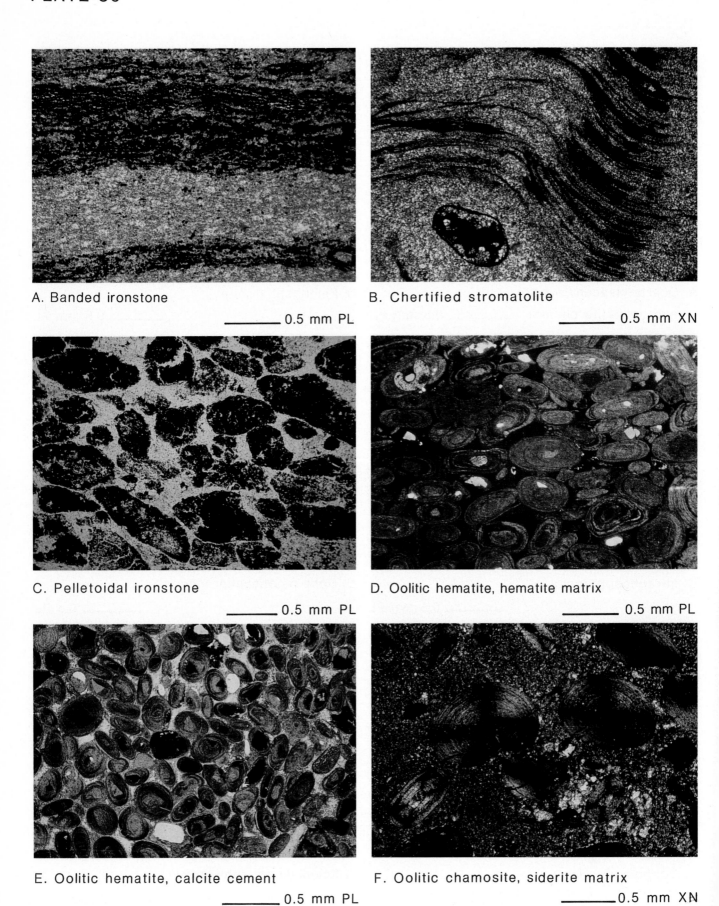

A. Banded ironstone

_____ 0.5 mm PL

B. Chertified stromatolite

_____ 0.5 mm XN

C. Pelletoidal ironstone

_____ 0.5 mm PL

D. Oolitic hematite, hematite matrix

_____ 0.5 mm PL

E. Oolitic hematite, calcite cement

_____ 0.5 mm PL

F. Oolitic chamosite, siderite matrix

_____ 0.5 mm XN

PLATE 31

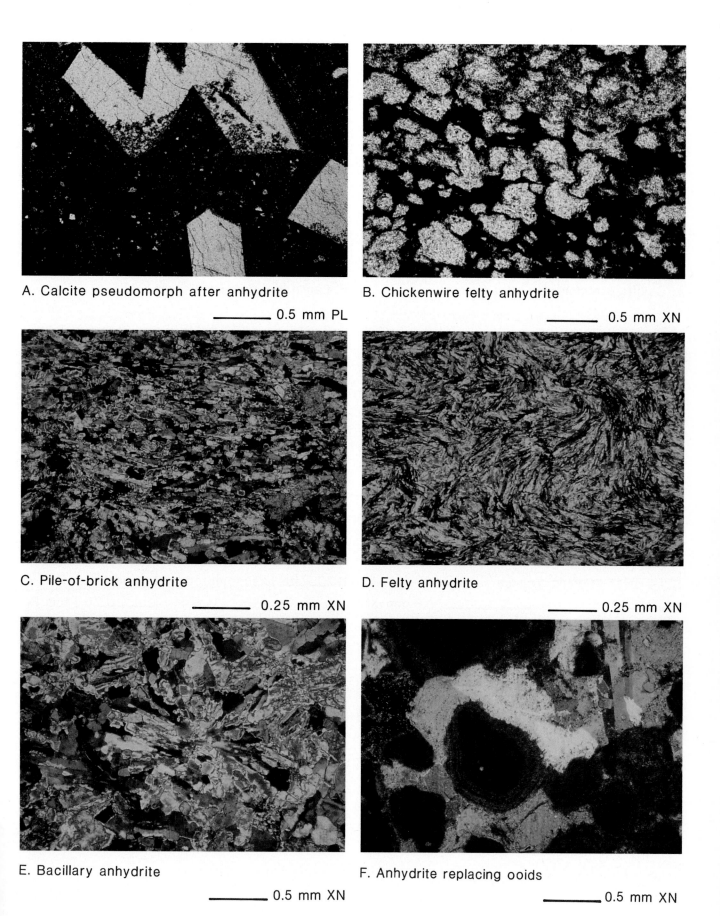

A. Calcite pseudomorph after anhydrite

_____ 0.5 mm PL

B. Chickenwire felty anhydrite

_____ 0.5 mm XN

C. Pile-of-brick anhydrite

_____ 0.25 mm XN

D. Felty anhydrite

_____ 0.25 mm XN

E. Bacillary anhydrite

_____ 0.5 mm XN

F. Anhydrite replacing ooids

_____ 0.5 mm XN

PLATE 32

A. Calcite pseudomorph after gypsum

—————— 0.5 mm PL

B. Laminated primary gypsum

—————— 0.5 mm XN

C. Incipient selenitic gypsum

—————— 0.5 mm XN

D. Calcite pseudomorph after selenite

—————— 0.5 mm PL

E. Calcitized anhydrite

—————— 1 mm PL

F. Replacive cubical halite

—————— 0.5 mm PL

SMART, P. L., DAWANS, J. M., and WHITAKER, F., 1988. Carbonate dissolution in a modern mixing zone. *Nature,* 335, 811–813.

SODERMAN, J. W., and CAROZZI, A. V., 1963. Petrography of algal bioherms in Burnt Bluff Group (Silurian), Wisconsin. *Amer. Assoc. Petroleum Geologists Bull.,* 47, 1682–1708.

SUPKO, P. R., 1977. Subsurface dolomites, San Salvador, Bahamas. *J. Sed. Petrology,* 47, 1063–1077.

TUCKER, M. E., 1983. Diagenesis, geochemistry, and origin of a Precambrian dolomite: the Beck Spring Dolomite of eastern California. *J. Sed. Petrology,* 53, 1097–1119.

————, and WRIGHT, V. P., 1990. *Carbonate Sedimentology.* Blackwell Scientific Publications, London, 482 pp.

WARRAK, M., 1974. The petrography and origin of dedolomitized, veined or brecciated carbonate rocks, the "cornieules" in the Fréjus region, French Alps. *J. Geol. Soc. London,* 130, 229–247.

WENK, H. R., BARBER, D. J., and REEDER, R. J., 1983. Microstructures in carbonates. *Mineral. Soc. Amer. Rev. Mineral.,* 11, 301–368.

WOLFE, M. J., 1970. Dolomitization and dedolomitization in the Senonian chalk of northern Ireland. *Geol. Magazine,* 107, 39–49.

WOOD, G. V., and AMSTRONG, A. K., 1975. Diagenesis and stratigraphy of the Lisburne Group limestones of the Sadlerochit Mountains and adjacent areas, northeastern Alaska. *U.S. Geol. Survey Prof. Paper 857,* 47 pp.

ZENGER, D. H., DUNHAM, J. B., and ETHINGTON, R. L., 1980. *Concepts and Models of Dolomitization.* Soc. Econ. Paleontologists and Mineralogists, Special Publ. 28, 320 pp.

CHAPTER 7

SILICEOUS ROCKS

INTRODUCTION

Investigations related to deep-sea drilling provided over the past 10 years data that significantly advanced the understanding of present-day genesis of biogenic siliceous sediments and their diagenetic changes into deep-water nodular and bedded cherts encountered in underlying Pliocene and older sections of the drillings (Wise and Weaver, 1974). These investigations also shed new light on problems debated for decades, such as the succession of complex diagenetic changes from siliceous oozes to chert.

Another important aspect, recently pointed out, is the secular change in chert environmental distribution through geological time, reflecting the evolving biological participation in the silica cycle (Maliva and Siever, 1989b; Maliva et al., 1989; Gao and Land, 1991). The abundance of cherts in upper Proterozoic peritidal carbonates suggests that silica was at that time removed from seawater by abiological processes. With the evolution of desmosponges in the Cambrian, subtidal shelf and platform biogenic cherts became more common, and with the Ordovician rise of Radiolaria, sedimented skeletons became increasingly important as sink for oceanic silica, although abiological removal processes may have continued during the early Paleozoic because of this still inefficient type of biogenic precipitation. At any rate, cherts of Silurian to Cretaceous age display many similarities in facies distribution and petrography. However, they differ from Cenozoic to Recent nodular and bedded cherts,

which are basinal to deep oceanic and reflect the mid-Cretaceous development of diatoms and their subsequent rise to domination of the silica cycle.

The dominant form of dissolved SiO_2 in natural water is the undissociated monomeric silicic acid H_4SiO_4, called also silicon. Silica-secreting organisms, mainly diatoms, Radiolaria, silicoflagellates, and siliceous sponges, extract silicon from seawater to build their tests, which consist of amorphous silica, or opal-A. This activity occurs mainly in nutrient-rich surface waters, in areas of oceanic upwelling, and therefore belongs to an oceanwide cyclic pattern. Silicon supplied in solution by contemporaneous volcanic hydrothermal systems, which are restricted to oceanic ridges and "hot spots," is volumetrically insignificant. Furthermore, it does not fit the widespread distribution of present and past oceanic siliceous oozes, thus ending a long-lasting controversy about basaltic submarine volcanism being a source of silica (Garrison, 1974; Hesse, 1990a). It is estimated that 90% to 99% of opal-A produced by siliceous organisms from surface seawater redissolves before burial and is endlessly reused by subsequent generations of these organisms (Hesse, 1990a). This dissolution begins after the death of the organisms while their tests sink through the water column and while they are exposed on the sea floor; it continues after burial within the unconsolidated sediment. A portion of the silica dissolved during burial returns to the ocean waters by diffusion at the sediment–water interface, but most of the dissolution within the sediment provides the silicon required

146

for diagenetic reprecipitation. This process generates deep-sea cherts and to a minor extent, in the presence of altered volcanic ash, authigenic zeolites and clay minerals such as palygorskite and sepiolite. The understanding of this large-scale internal oceanic recycling solves another long-debated problem about the biogenic output, which appears to be at least 25 times larger than the silicon input in the oceans from various sources, mainly rivers (Hesse, 1990a).

The complex diagenetic history of the changes undergone by biogenic siliceous oozes during their maturation into cherts was recently reviewed by Hesse (1990a). It consists essentially of two successive dissolution–reprecipitation events. The first one is the transition of opal-A to opal-CT (or association of cristobalite and tridymite); the second is the transition of opal-CT to low-temperature quartz, which is the final stage of the diagenetic evolution (Rad and Rösch, 1974; Wise and Weaver, 1974).

The deep-sea biogenic production of silica related to upwelling can also extend to shallow shelf conditions of continental margins, where it accounts for diatomites and related cherts of the Monterey Formation of California, and to the Cenozoic opaline claystones of the coastal plains of the southeastern United States. However, the spectrum of chert formation is much broader and includes the innumerable instances of syngenetic to early diagenetic replacement of semiconsolidated to incipiently cemented carbonate rocks and sequences, which require their own hydrological models, followed by late burial diagenetic replacement. Other particular conditions pertain to cherts derived in alkaline lakes from hydrous sodium silicate precursor minerals or from gels of hydrothermal–volcanic origin, to the low-temperature chemical silcrete deposits of pedological origin, and to the highly debated Precambrian cherty-banded iron ores to be discussed in Chapter 9.

It appears indispensable to present here a brief review of the various fabrics displayed by silica in siliceous rocks. Under the petrographic microscope, opal appears as an amorphous to weakly birefringent material and its varieties are determined by X-ray diffraction (Jones and Segnit, 1971) and SEM examination (Hesse, 1990a, b).

Biogenic opal-A is a near-amorphous substance; diagenetic opal-C (cristobalite) and opal-CT (cristobalite-tridymite) have their particular X-ray patterns and appear under SEM as tiny spheres of about 5 μ in diameter called lepispheres (Weaver and Wise, 1972), first described in Cenozoic deep-sea cherts. These lepispheres consist of either blades of cristobalite or of interpenetrative growth of cristobalite–tridymite blades; coalescence of individual lepispheres into composite ones is very common in deep-sea siliceous sediments; those described in older rocks (Meyers, 1977; Jones and Knauth, 1979; Carver, 1980; Maliva and Siever, 1988; Gao and Land, 1991) are of larger size and appear to be aggregates rather than simple individuals.

The diagenetic transformation by solution–reprecipi-tation of opal-A into opal-CT and finally into quartz is a complex and debated process (Hesse, 1990a, b). When diagenetic changes reached the mineralogical stage of quartz, whether precursor phases are not preserved or difficult to identify, at least seven different fabrics of quartz can be recognized petrographically. They belong to two major categories: equigranular and fibrous. Equigranular types consist of microcrystalline quartz or microquartz (Folk and Pittman, 1971) and megaquartz. Microquartz is formed by tiny crystals, usually ranging from less than 5 to 20 μ in size, associated in an equigranular mosaic that gives a pin-point extinction under crossed nicols. Megaquartz is formed by crystals that commonly range from 20 to 2,000 μ and above, with no designated upper size limit. Patches of megaquartz mosaic generally display an increase of crystal size from their margins toward their centers, which is similar to the typical cavity-filling texture of sparite cement. However, for reasons not yet clear, megaquartz shows this texture also when it is of replacive origin of carbonates or of a precursor opal-CT phase. Distinction of the two possible origins can only rely on the shape of the mosaics, on inclusions, and on a reasonable interpretation of the surrounding petrographic context.

The most abundant fibrous silica is chalcedonic quartz or, as it is commonly called, chalcedony, in fact length-fast chalcedony. It occurs in bundles of fibers radiating from particular points of the margins of cavities, eventually forming interfering spherulites. Chalcedony can also form successive crusts, called overlays, of parallel fibers growing perpendicular to cavity margins, with color banding due to a variety of inclusions. These isopachous rims can be followed toward the center of the cavities by megaquartz mosaic or spherulitic chalcedony. Fringing cements of parallel, nonradiating fibers occur in many tectonically deformed conglomerates and sandstones containing dolostone pebbles or grains in the Western Alps. These fringes are discontinuous and correspond to pressure shadows perpendicular to maximum tectonic stresses. In these fringes, length-fast and length-slow chalcedony are intimately intergrown and were subsequently followed or partially replaced by fibrous calcite cement (Hesse, 1987). As mentioned below, it is possible that length-slow chalcedony precipitated in the presence of magnesium-rich solutions derived from dissolution of dolostone pebbles and grains. However, chalcedony can also be of replacive origin (Wilson, 1966).

Both length-slow chalcedony, also called quartzine, and lutecite, another form of fibrous quartz, occur in chert nodules replacing evaporites. Quartzine in particular was originally proposed to be used as a criterion to identify these particular conditions (Folk and Pittman, 1971). Subsequently, Folk (1975) described quartzine-replaced bioclasts of pelmatozoans, brachiopods, and bryozoans in rocks showing no evidence of evaporitic environment. Folk suggested an alternative explanation consisting of the presence of Mg-rich diagenetic fluids. Keene (1983) showed that quartzine is the most abundant form of quartz in silicified pelagic sediments from the North Pacific, where it is associated with authigenic barite and dolomite. Keene concluded that the generation of quartzine indicates pore fluids rich in sulfate and magnesium, a situation supported by laboratory experiments in which quartzine was synthetized (Kastner, 1980). In summary, diagenetic fluids rich in sulfate and magnesium can occur in a variety of sedimentary environments ranging from lacustrine and intertidal–supratidal evaporitic to deep-sea pelagic. Therefore, quartzine reflects interstitial chemical conditions and is not an indicator of any particular depositional environment. However, the question of the particular composition of interstitial diagenetic fluids remains open because instances of replacement of carbonate bioclasts by spherulitic quartzine were recorded under conditions in which neither magnesium- nor sulfate-rich fluids were assumed (Brown et al., 1969; Meyers, 1977).

The remaining two fabrics of quartz are petrographically spectacular. The first, called zebraic chalcedony, is often fan shaped with beautiful, banded extinction under crossed nicols. The zebraic pattern results from a helicoidal twisting of the fiber axes around the c axis (McBride and Folk, 1977; Frondel, 1978). Zebraic chalcedony occurs frequently in cherts associated with evaporites (Milliken, 1979; Arbey, 1980); it has also been reported in deep-sea cherts (Keene, 1983) and as a cement in Proterozoic cherty iron formations (Simonson, 1987). The second variety is called microflamboyant quartz (Milliken, 1979) and displays a fabric intermediate between quartzine and megaquartz. It has also been called flamboyant lutecite (Folk and Pittman, 1971) and megaquartz with flamboyant spectral extinction (Chowns and Elkins, 1974). Spectral extinction is a form of undulose extinction that results from the intergrowth of numerous quartz prisms with unclear boundaries and slightly different axial orientations, and perhaps some interpenetrative twinnings.

CLASSIFICATION

A classification of siliceous rocks, both descriptive and genetic, is used here for practical purposes. It does not include silcretes and cherts of hydrothermal–volcanic origin, which are beyond the scope of this presentation.

The classification is as follows:

Siliceous rocks of primary biogenic origin
    Ferruginous ribbon radiolarian cherts
    Carbonaceous ribbon radiolarian cherts
    Diatomites, porcelanites, and diatomaceous cherts
    Opaline rocks of the Atlantic and Gulf Coastal Plain

    Spiculites and spiculitic cherts
    Novaculites
Siliceous rocks of primary inorganic origin
    Magadi-type cherts
    Lagoonal and lacustrine cherts
Siliceous rocks of secondary replacement origin
    Syngenetic cherts in chalks and limestones
    Early diagenetic cherts in limestones and dolostones
    Late diagenetic cherts in limestones and dolostones
    Diagenetic cherts in evaporites

SILICEOUS ROCKS OF PRIMARY BIOGENIC ORIGIN

Ferruginous Ribbon Radiolarian Cherts

Depositional Environment. Within the framework of plate tectonics, the association of mostly Mesozoic ribbon radiolarian cherts or radiolarites with ophiolites, reported in many orogenic belts, retained its geotectonic significance. However, it lost its assumed genetic aspect, according to which silica was believed to be derived, by means of unclear processes, from submarine, silica-undersaturated basaltic volcanics. These ferruginous ribbon radiolarian cherts are now interpreted as mineralogically equivalent to the above-mentioned deep-sea cherts, but deposited in a geotectonic context different from that of present-day ocean basins, that is, within orogenic belts. Jenkyns and Winterer (1982) pointed out that no siliceous sediment resembling ribbon radiolarian cherts has been cored from present-day ocean basins. The cherts from the Jurassic, Cretaceous, and Cenozoic beds beneath the oceans of today are all clearly lenticular and surrounded by pelagic carbonates and claystones. No thick, stratigraphically continuous ribbon chert sequence and no chert-shale couplet have been recorded. Although the mineralogy of these rocks is comparable, the repetitively bedded or ribbonlike aspect of Mesozoic radiolarian cherts observed on land remains unique and restricted to orogenic belts. Jenkyns and Winterer (1982) believed that, regardless of the genetic process or combination of processes assumed for the genesis of ribbon radiolarian cherts, they appear to be deep-water deposits formed either in small basins of various types, mainly in arc-related regions for the Great Valley Sequence and possibly the Franciscan of California, or in embryonic oceans dominated by transform faulting for the sequences of the Western Tethys involved in the Alpine orogenic belt.

Because the genesis of ribbon radiolarian cherts is mainly biogenic and not volcanogenic as previously thought, their formation was controlled by local paleo-oceanographic conditions; hence they can be associated both with continental and oceanic crust as shown in the Apennines (Folk and McBride, 1978; McBride and Folk, 1979). However, in the case of the Jurassic ribbon radiolarian cherts of the island of Elba, local hematite and manganese enrichments in the lower portion of the section, which rests on basalt of the oceanic crust, indicate the metasomatic influence of a former spreading axis (Barrett, 1981, 1982).

The typical occurrence of ribbon radiolarian cherts is a bedding style characterized by rhythmically alternating thin beds of chert (often about 10 cm thick) with sharp basal and top contacts and siliceous mudstones or shales (also about 10 cm thick); hence the name of ribbon bedding. However, this bedding style is quite variable in terms of bed thicknesses, chert-to-shale ratios, and lateral variations of bed thicknesses at the scale of a given section and within a basin (Baltuck, 1983).

Mesozoic ribbon radiolarian cherts are chiefly reddish brown in color, but green and green–red mottled ones are frequent, as well as dark gray, black, and even colorless (Davis, 1918). The reddish color is due to minute hematite flecks, indicating deposition in an oxidizing environment; the greenish colors originate from the presence of illite and chlorite, which appear upon diagenetic reduction of hematite; gray to black colors are generated by carbonaceous matters and phosphate (Cayeux, 1929).

A certain relationship occurs between color and preservation of Radiolaria. They are better preserved in red cherts because clay minerals and hematite flecks inhibited diagenetic processes such as solution–reprecipitation and neomorphism, which were more active in lighter green colors where Radiolaria tests appear poorly preserved (Thurston, 1972). Particular colors characterize certain stratigraphic intervals, certain textures, and distinct environments; therefore, they can be used as diagnostic criteria. McBride and Thomson (1970) and Folk and McBride (1978) proposed to standardize color descriptions by using Munsell color numbers and names.

The rhythmic bedding of radiolarian cherts remains controversial. McBride and Folk (1979) summarized four major interpretations (**Fig. 7.1**): (1) diagenetic segregation of silica from an initially homogeneous siliceous mud; (2) alternating phases of rapid and slow production of Radiolaria in surface waters, combined with a constant rate of mud deposition; (3) phases of rapid deposition of Radiolaria by currents during constant slow mud deposition; and (4) phases of rapid deposition of mud by currents during constant deposition of Radiolaria. Primary sedimentary structures in radiolarian cherts seem to indicate redeposition by turbidity currents or other types of bottom currents from adjacent submarine highs. They consist of common graded bedding, parallel lamination, faint cross lamination, starved current ripples, parallel alignment of Radiolaria and sponge spicules, and presence of shale intraclasts (Nisbet and Price, 1974; Robertson, 1977; McBride and Folk, 1979). However,

there are instances where these sedimentary structures indicating redeposition do not occur, while top and bottom contacts of the chert beds with the shales still remain sharp. This situation led some investigators to consider the effects of biogenic productivity or of diagenetic segregation, because some radiolarian cherts show symmetrical geochemical and mineralogical changes and symmetrical variation in Radiolaria abundance from the middle part of a given chert bed toward its top and bottom (Mizutani and Shibata, 1983; Sano, 1983; Steinberg et al., 1983; Murchey, 1984). In summary, the present consensus is that rhythmic bedding of ribbon radiolarian cherts cannot be ascribed to a single process because interferences between depositional features and diagenetic changes have obscured the picture (Jenkyns and Winterer, 1982; Baltuck, 1983; Hein and Karl, 1983; Jones and Murchey, 1986; Hesse, 1990a). However, major and rare earth element data from ribbon radiolarian cherts (Murray et al., 1992) appear consistent with a dominantly early diagenetic origin of chert–shale couplets. These data are also incompatible with many depositional theories relying on turbidity currents or other transport mechanisms and on biogenic productivity fluctuations related or not to Milankovitch cyclicity.

The conclusions of the above-mentioned authors is that early diagenetic migration of labile SiO_2 from protoshale to protochert is largely responsible for the chert–shale couplets, and that such a process of diagenetic fractionation has not received appropriate attention as a viable mechanism.

Petrography. In thin section, a typical red radiolarian chert (**Plate 27.F**) consists of a very fine grained siliceous cement colored by minute hematite flecks intimately associated with clay minerals. In this red semitranslucent to opaque cement are scattered minute areas of clear silica, circular, elliptical, tetrahedral, or rodlike in outline, which represent the remains of Radiolaria, often with preserved internal lattice structure. Closer examination of radiolarian remains shows that they consist of microquartz granules, 5 to 30 μ in size, that have replaced original tests. Their internal filling, besides hematitic microquartz cement, can be one or several interfering spherulites of chalcedony, a single quartz crystal, or even hematitic red clay. All degrees of destruction may be observed until only faint outlines of Radiolaria consist of irregular spots of slightly coarser, clear microquartz mosaic scattered in the fine hematitic cement.

Radiolaria vary greatly in abundance (50% to 75%). They may be quite sparsely scattered throughout the hematitic cement or so abundant as to be packed in reciprocal contact, reducing the cement to an interstitial filling. This close packing could be of depositional origin, resulting from winnowing and current deposition, or diagenetic and produced by pressure solution and stylolitization (Baltuck, 1983). The two latter processes carried to an extreme could

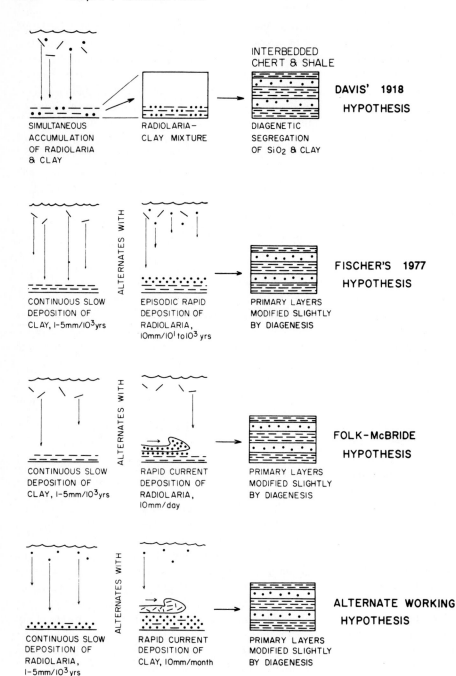

FIGURE 7.1 Diagram of four hypotheses of the origin of ribbon radiolarian chert–shale rhythms. From McBride and Folk (1979). Reprinted by permission of the Society of Economic Paleontologists and Mineralogists.

eventually destroy the radiolarian molds and change the rock into a ferruginous microquartzite of variable opacity. However, the most common situation is not only a perfect preservation of the Radiolaria skeletons, but also a general absence of compactional effects, indicating that complete lithification preceded compaction with the exception of a very few cases where flattened skeletons occur. In general, Radiolaria skeletons appear completely unsorted in terms of size, but laminae occur in which Radiolaria of particular size and abundance are concentrated and still others where abundance varies symmetrically from the center of a given bed, upward and downward.

Radiolaria can be associated with some accessory constituents, such as silicified, thin-shelled pelagic pelecypods; phosphatic remains of unknown origin; siliceous sponge spicules; as well as nonskeletal constituents that appear to be transported argillaceous fecal pellets and shale intraclasts.

The interstitial cement of red radiolarian cherts consists of a mosaic of microquartz with a crystal size ranging from 5 to 20 μ and generally uniform in a given bed. The quartz mosaic that gives a typical pinpoint extinction under crossed nicols is intimately mixed with numerous shapeless flecks of hematite, minute radiolarian debris, and clay minerals. In some instances, hematite may be so abundant that the cement becomes practically an opaque red groundmass in which no details can be observed. In most instances, hematite flecks tend to hide the presence of clay minerals, which are also evenly scattered throughout the microquartz cement. However, distinct elliptical concentrations and lamellae of clay minerals occur that could represent, respectively, either burrow fillings flattened by compaction or real current laminae.

The mineralogical composition of the clay mineral fraction varies little and was reviewed by Steinberg et al. (1977a, b) and Jenkyns and Winterer (1982). The assemblage is mainly illite and chlorite, with illite generally of the muscovite type, sometimes with a tendency toward phengite and a variable crystallinity of chlorite. Smectite and mixed-layers illite/smectite, chlorite/smectite, and illite/chlorite are frequently present. Kaolinite and attapulgite were reported in small amounts. The association illite–chlorite is very characteristic of radiolarian cherts, and both minerals are responsible for the green or mottled green–red color that appears upon diagenetic reduction of hematite. Among other minerals, feldspars, mainly albite, oligoclase, and K-feldspars, occur in many Jurassic radiolarian cherts (Steinberg et al., 1977a). They are too small to be seen under the petrographic microscope and were detected by X-ray diffraction; their positive anomaly of europium indicates that they are not authigenic but of magmatic origin, where such an anomaly occurs as the result of reduction of this element in magmas. However, it is not possible to detect if these feldspars are detrital or express a volcanism contemporaneous with sedimentation (Steinberg et al., 1977b).

Karl (1984) demonstrated in a geochemical study that certain elements could be used for discriminating source components and depositional provinces for Upper Mesozoic and Cenozoic cherts of deep-sea drillings, as well as for cherts of the Franciscan assemblage in California, in spite of their diagenetic evolution. Elements that define unique ratios for identifying chert sediment sources include Al, Fe, Mn, Ni, Ba, Cu, K, Ti, Zr, Pb, V, Y, Mg, and B. Element ratios plotted on Fe–Ba–Ni, Cu–Ba, and Fe–Mg–K discriminant diagrams characterize samples in distinct source fields.

The reddish siliceous mudstones and shales forming the second term of the couplet range from types containing 10% to 40% microquartz to a hematitic groundmass of illite–chlorite. The siliceous mudstones have a weakly developed conchoidal fracture, whereas fissility appears with predominance of clay minerals. Most shales are devoid of structures except for a general mass extinction of clay minerals under crossed nicols. Local evidence of burrowing appears to be shown by concentrations of silt-size quartz and random orientation of mica flakes. Rarity of laminae suggests that the homogeneous aspect of the shales is not necessarily depositional, but could result from intensive bioturbation (McBride and Folk, 1979). Radiolaria are poorly preserved, often flattened by compaction and reduced to barely recognizable clay–microquartz aggregates indicating extensive dissolu-

tion, probably at the sediment–water interface (Barrett, 1981, 1982).

Diagenetic Evolution. Laterally irregular bedding, very common in ribbon radiolarian cherts, is caused by variations in the rate of silica diagenesis. This typical pinching and swelling and knoblike structures of chert beds is a response of early lithified areas to the compaction of the surrounding shales. These areas appear often highly rich in silica and in Radiolaria that are preserved with original spherical forms, whereas outside such areas they might be flattened. These features indicate that induration started immediately after deposition and was associated with flowage between cherts and shales due to differential compaction and fluid-escape phenomena. The large spectrum of deformation of ribbon radiolarian cherts ranges from ductile features, such as diapirlike structures (Folk and McBride, 1976) and chert dikes (Steinitz, 1970), to various types of brittle brecciation, with fillings displaying equigranular and various types of fibrous quartz cements.

The origin of brecciation in chert beds remains controversial (McBride and Folk, 1977, 1979) because numerous processes can be involved from early to late diagenetic stages, such as changes of volume during silica mineralogical evolution; disruption of particular beds by movement of pressured fluids through sediments undergoing lithification under sedimentary overloading (Steinitz, 1970; Murchey, 1984; Paris et al., 1985); and movement of heated high-pressure hydrothermal fluids associated with volcanism (Crerar et al., 1982; Murchey, 1984; Paris et al., 1985). See also Chapter 2 for a discussion of chert pseudobreccias and pseudoconglomerates developed under these peculiar circumstances.

In highly folded belts, tectonic deformation strongly affects ferruginous ribbon radiolarian cherts and may change them beyond recognition (Cayeux, 1929). Generally, Radiolaria become elongate ellipsoids, with their main axes always parallel to bedding. These highly deformed individuals consist of patches of clear microquartz mosaic surrounded by a finer and darker mosaic of hematitic microquartz. The overall aspect is that of a pseudoquartzite with lenticular texture or amygdaloidal pseudoquartzite. Exaggeration of the stretching and flattening process eventually changes the lenses into streaks of microquartz entirely devoid of internal structure. The rocks eventually appear as clear schistose pseudoquartzites in which aggrading neomorphism almost eliminated hematite and thus mimic almost perfectly real tectonized quartzites.

Carbonaceous Ribbon Radiolarian Cherts

Depositional Environment. These cherts share a similar general depositional environment with the ferruginous types except that oxidizing conditions of hematite deposition were replaced by anoxic conditions due probably to restricted circulation on the basin floors.

Carbonaceous ribbon radiolarian cherts are grayish to dull black and have a rather scaly fracture similar to that of quartzites. They owe their typical aspect to the intimate association of microquartz mosaic with carbonaceous flecks, clay minerals (illite–chlorite), and phosphate. They are known under the French name of "phtanites" (Cayeux, 1929) and the German designation of "Lydite" (Heritsch and Heritsch, 1943).

Petrography.   A first glance at carbonaceous radiolarian cherts reveals scattered white spots of chalcedony corresponding to individual Radiolaria set in the darker microquartz groundmass. The original test sometimes appears as a peripheral ring of short chalcedony fibers against which the filling of the interior of very elongate fibers abuts. In other instances, Radiolaria are represented by patches of relatively clear microquartz mosaic, with undulose extinction quite distinct from the finer and darker interstitial cement of microquartz intimately mixed with minute shapeless carbonaceous flecks. The carbonaceous matter (graphite) may represent up to 30% of the groundmass and, together with phosphate (apatite) and minute crystals of pyrite, it expresses the important organic contribution preserved in the anoxic environment of deposition. The carbonaceous matter occurs also in streaks and localized concentrations, filling Radiolaria molds, which interfere with one another and eventually build a very irregular network either parallel to bedding or completely independent of it.

Diagenetic Evolution.   Carbonaceous ribbon radiolarian cherts share the same diagenetic features as the above-described ferruginous types. Tectonic deformation leads to the almost complete elimination of the carbonaceous pigment and the generation of clear schistose pseudoquartzites.

Diatomites, Porcelanites, and Diatomaceous Cherts

Depositional Environment.   The Miocene Monterey Formation of California is used here as a typical example of siliceous sedimentation controlled by diatoms (Garrison et al., 1981; Pisciotto and Garrison, 1981; Isaacs et al., 1983). It is a deep-marine episode of a late Cenozoic cycle of basin formation and filling associated with wrench-faulting along the California margin. Monterey sedimentation occurred in basin-floor, slope, shelf, and ridge-top environments, which were at times anoxic. In most places, the Monterey Formation shows a lower calcareous facies, a middle transitional phosphatic facies, and a thick upper siliceous facies consisting of diatomaceous rocks and related diagenetic porcelanites and cherts. The widespread siliceous facies consists of rapidly deposited diatom ooze resulting from high plankton productivity caused by climatic cooling and intensified upwelling during Late Miocene (Isaacs et al., 1983).

Petrography.   The petrographic descriptions of Bramlette (1946), which still remain unsurpassed, illustrated the diagenetic derivation of the cherty phase from softer diatomites. These descriptions were enhanced recently by X-ray diffraction, SEM and TEM studies, and isotopic analyses (Pisciotto, 1981a).

Diatomites are soft and porous rocks, gray-yellowish, and extremely light. By addition of clay minerals, diatomites grade into diatomaceous mudstones and shales. Most of the diatomites of the Monterey Formation are finely laminated, but thickness and distinctness of the laminae are highly variable. Under the microscope, the laminae result from regularly alternating light-colored layers made of opaline aggregates of diatoms and dark-colored layers in which clay minerals and irregularly shaped calcite granules predominate over diatom debris. In light-colored layers, diatom shells may be entirely preserved, highly packed, but with their axis parallel to bedding. In places, individuals have been deformed by compaction. However, more frequently, the diatom shells have been broken into minute debris to build a felty mass. The porosity of pure diatomites, which consist of opal-A, is microporosity within and between diatom debris. Among other constituents are Radiolaria, silicoflagellates, small foraminifers, sponge spicules, and fish scales, the latter more frequent in anoxic environments. Detrital minerals and glaucony are very rare constituents; hydrocarbons and carbonaceous shapeless flecks occur in varying amounts, and a few diatoms contain very minute altered shards of volcanic glass.

The term porcelanite designates an argillaceous porous siliceous rock, less hard, less dense, and less vitreous than chert. It contains 25% to 50% clay minerals and is one of the diagenetic products of diatomaceous shales (Bramlette, 1946; Jones and Murchey, 1986). The microporosity of porcelanites gives them a dull aspect resembling that of unglazed porcelain, hence the reason for its name. Porcelanites are commonly not laminated, but rather thin bedded. Mineralogically, the Monterey Formation shows two types of porcelanites, the first one with opal-CT and the second with microquartz forming the uniform groundmass intimately mixed with clay minerals in which are scattered a few calcite-filled molds of foraminifers. Microporosity results from molds of small diatoms and silicoflagellates.

Diatomaceous cherts occur as distinct beds and as concretionary or nodular masses, hard and vitreous, but still displaying traces of bedding planes and laminae. Cherts deriving from pure diatomites are mineralogically of two types like the porcelanites, with opal-CT or with microquartz forming the groundmass, which can still contain up to 25% clay minerals. Chalcedony usually fills the larger pore spaces such as the cavities of foraminifers and the interior of the largest diatoms. Megaquartz may form the central part of veinlets bordered by chalcedony.

Diagenetic Evolution. The mineralogical duality described above for porcelanites and cherts of the Monterey Formation was fully recognized only recently (Isaacs et al., 1983). Although the general burial diagenetic sequence is biogenic opal-A to diagenetic opal-CT to diagenetic quartz, it has been observed that not all rocks containing diagenetic quartz are vitreous, brittle, and "chert-looking." Conversely, many glassy, brittle, and hard chert rocks contain only opal-CT as the diagenetic silica mineral. Petrographic studies, X-ray diffraction, and chemical analyses showed that these differences in textures and physical properties derive from original compositional differences and not from the degree of diagenetic silica remobilization, and that the presence of clay minerals in the Monterey Formation, predominantly mixed-layer illite/smectite, is a critical factor in the generation of crystallization features (Isaacs et al., 1983).

The lithological succession due to burial diagenesis can be summarized as follows: pure diatomite to opal-CT chert, to quartz chert; diatomaceous shale to opal-CT to porcelanite to quartz porcelanite. In this succession, most of the changes of physical properties occur during opal-CT formation, such as marked increase in density, hardness, and brittleness and marked decline in porosity, whereas the generation of diagenetic quartz has a much smaller effect on all these properties (Isaacs et al., 1983). The two-phase transformation, which occurred by solution and in-place precipitation of silica, is associated with abrupt reductions of 15% to 39% porosity resulting from compaction. Isaacs et al. (1983) estimated that opal-CT rocks became abundant at 40° to 50°C or at burial depths of 750 to 1,100 m, whereas quartz rocks became well developed around 80°C or at burial depths of 1,500 to 2,400 m.

Calcium-rich dolomite, at times ferroan, is the main diagenetic carbonate in the Monterey Formation and apparently developed at various times during burial (Pisciotto, 1981b; Isaacs et al., 1983). Reworked and redeposited dolomite nodules indicate very early formation (Bramlette, 1946; Surdam and Stanley, 1981). At greater depths of burial, dolomite formed within zones of increasing organic-matter maturation, as suggested by syntaxial overgrowths on detrital dolomite in dolomitic sandstones as well as by wide ranges in carbon and oxygen isotopic compositions (Pisciotto, 1981b; Surdam and Stanley, 1981). Still later phases of dolomite formation consist of dolosparite cements filling entirely or partially fractures in dolomitic mudstone breccias (Redwine, 1981; Roehl, 1981).

Reservoir Properties. The Monterey Formation is considered the main source rock for oil in California and at the same time a major reservoir of its own derived hydrocarbons. The abundance of organic matter ranges from 1% to 25% by weight and reaches an average of 12% in the weakly siliceous organic shale units. Thermal diagenesis of this abundant organic matter, which is predominantly kerogene, produced both a highly asphaltic fraction and oil at temperatures lower than 100°C (Surdam and Stanley, 1981). Porcelanites, diatomites, and less commonly cherts contain substantial microporosity. The average values are 60% for diatomaceous rocks with opal-A, 30% for porcelanites with opal-CT, and 25% for quartz porcelanites (Isaacs, 1981). The porosity–depth pattern is characterized by two abrupt reductions in porosity (of 10% to 35% porosity) corresponding to the two silica phase transformations, opal-A to opal-CT and opal-CT to quartz (Isaacs, 1981). The dolomites are also microporous, mostly as intercrystalline voids and non-fabric-selective vugs.

Apparently, the storage capacity of Monterey rocks is mainly the result of microporosity, but permeability was generated by fracturing in a situation similar to that of chalk reservoirs. Indeed, the best and most prolific reservoirs are located in phosphatic and calcareous mudstones and shales that underwent secondary dolomitization and thus became naturally fractured reservoirs. Other reservoirs correspond to unusual types of dolomitic breccias resulting from open extension fractures partially cemented by dolosparite cement with remaining large open voids. These breccias show an exploded texture in which the puzzlelike fragments are separated by a veinlike or dikelike matrix that was apparently an injected slurry of water, oil, and smaller dolostone fragments. Although the real cause of brecciation remains disputed, Roehl (1981) and Redwine (1981) considered that regional tectonic stresses induced repeated periods of rock dilation, followed by natural hydraulic fracturing.

Opaline Rocks of the Atlantic and Gulf Coastal Plain

Depositional Environment. The interpretation of ribbon radiolarian cherts and diatomites as biogenic deposits of planktonic origin related to upwelling in open oceanic environments or along continental margins of the past can also be extended to continental shelves and even to inner shelf environments, as shown by Calvert (1983) for the Namibian shelf today.

The interpretation of normal shallow-water biogenic origin for the fine-grained Cenozoic opaline claystones and associated clastic and carbonate rocks of the Atlantic and Gulf coastal plains, involved in transgressive–regressive cycles, was demonstrated by Wise and Weaver (1973) and Weaver and Wise (1974). The origin of these rocks had been previously attributed to assumed volcanic sources in the Gulf Coast or in the Caribbean areas. These earlier interpretations postulated volcanic ash accumulations or problematic direct precipitation from silica-rich bottom waters circulating above these ashes in restricted coastal environments.

Petrography. The silica-rich opaline claystones in the Paleogene rocks of the Atlantic and Gulf coastal plains are light- to dark-gray in color, thick bedded, and present a conchoidal fracture. Their mineral suite is dominated by authigenically precipitated opal-CT, montmorillonite, and occasional admixtures of zeolites of the heulandite family (clinoptilolite). SEM studies (Wise and Weaver, 1973) showed that the common matrix of these claystones consists of the typical lepispheres of opal-CT, 3 to 12 μ in diameter, wherever free-growth morphologies are developed; if not, the lepispheres coalesce, forming a dense felty mass. This matrix contains numerous molds or opal-CT-replaced tests of diatoms, with the most perfectly developed lepispheres as lining or partial filling; Radiolaria; and sponge spicules, all of which indicate an open marine environment of shelf depth. These microfossils, by in-place dissolution and reprecipitation, generated the opal-CT lepisphere matrix. Although the biogenic origin of this matrix appears clearly demonstrated, the presence of zeolites may require a minor volcanic contribution, and this interpretation might also apply to some of the montmorillonite associated with them. However, the bulk of the clay minerals appears detrital in origin (Wise and Weaver, 1973).

In the southern portion of the Atlantic Coastal Plain, Carver (1980) reported the various types of cementation of opaline siltstones and carbonates, deposited in shallow marine environments and subsequently weakly buried. Indeed, X-ray diffraction of opaline cherts invariably indicates opal-C with an immature d(101) spacing, indicating that the sediments were never buried deeply or submitted to temperatures in excess of 100°C for any significant length of time. The biogenic opal is well preserved as skeletons of Radiolaria and diatoms, but predominantly as siliceous sponge spicules. Intact spicules with lepispheric or massive opal filling are most common.

An interesting aspect of these rocks is the occurrence of primary opal and opal-cemented carbonates and sandstones that show no evidence of former presence of biogenic opal (Carver, 1980). Furthermore, opaline cherts are to some extent impure, containing fragments of sponge spicules, fine detrital grains, and opal clasts "floating" in the massive opal matrix that is optically structureless. According to Carver (1980), the presence of detrital material dispersed within an opaline matrix indicates that opal is primary and was deposited directly from seawater as a bottom sediment. The presence of opal intraclasts and fecal pellets indicates that opal was located at the sediment–water interface, soft enough to be ripped up into clasts or ingested by organisms. In some opaline cherts, quartz grains deeply penetrate into rounded opal clasts, suggesting that opal remained in a soft condition for a certain length of time after burial. SEM examination of this primary opal shows lepispheres only in voids and often very crowded, to the extent of forming a felty mass of cristobalite blades. In some instances, opal displays a faint birefringence probably produced by parallel to subparallel orientation of cristobalite crystals.

Besides this unusual case of primary depositional opal, most of it occurs as massive cement or rim cement throughout the entire spectrum of sandstones, siltstones, spiculites, and bioclastic carbonates (Carver, 1980). Massive cement is structureless, whereas the rim cement is banded and faintly birefringent. The bands are due to different hues of the brownish transmitted-light color of opal. Rim cements follow the boundaries of clastic grains or bioclasts, the latter after an initial phase of pressure solution. Rim cements represent a gradual filling by opal of interstitial spaces; when these cements are bound by free surfaces, they are often botryoidal and faintly birefringent. The birefringent elements are elongate and oriented perpendicularly to substrates or radiate at a high angle. Complete fillings of interstitial cavities are complex and often show the successions opal–chalcedony–microquartz or opal–chalcedony–opal, with a final filling of wavellite. Alternating bands of length-fast and length-slow chalcedony in pore fillings suggest, as pointed out by Carver (1980), frequent changes in pore water chemistry (temporary enrichment in sulfate and Mg ?) during the last stages of diagenesis. But, as discussed above, the real meaning of the two types of chalcedony remains disputed.

Opal also replaced tests of foraminifers, bryozoans, gastropods, and pelecypods after they underwent an early phase of pressure solution (Carver, 1980). When the opal replacement is incomplete, the inner parts of the tests are replaced by chalcedony or microquartz; in all instances, preservation of original structures is perfect. In summary (Carver, 1980), opal-C is the predominant form of precipitated silica, and the process ranges from primary synsedimentary to early diagenetic and shallow burial, after an early phase of pressure solution of bioclasts. Opal shows no evidence of progressive recrystallization to chalcedony or microquartz. Chalcedony and microquartz generally postdate opal cement as cavity-filling or carbonate replacement, but the succession of the various phases is more complex. The origin of silica obviously combines a biogenic synsedimentary dissolution–reprecipitation for claystones with an early diagenetic origin for coarser clastic and carbonate rocks during which it may have been derived from the above-described associated opaline claystones (Wise and Weaver, 1973; Weaver and Beck, 1977; Carver, 1980).

Spiculites and Spiculitic Cherts

Depositional Environment. Concentrations of siliceous sponges occur in a variety of marine (occasionally freshwater) environments. They originate either from essentially in place accumulations in shallow-water and shoreline conditions associated with clastic and carbonate rocks (Cayeux, 1929; Cavaroc and Ferm, 1968; Lane, 1981) or in basinal concentrations by turbidity currents associated with dark organic-rich shales and carbonates (Newell, 1957). In most instances, silica is biogenic in origin and, by diagenetic dissolution and reprecipitation, accounts for the formation of spiculitic cherts, which generally represent only a small fraction of the various host rocks.

Petrography. Detailed petrographic descriptions of arenaceous spiculites, pure spiculites, and spiculitic cherts from shallow-water environments were provided by Cayeux (1929).

Arenaceous spiculites occur as transitional rocks between glauconitic arenites with opal–chalcedony cement and pure spiculites. They are gray–yellow in color and have a granular aspect and high microporosity, and consequently relatively low specific gravity. Under the microscope, the groundmass consists of an intimate association of opal and chalcedony with fine crystalline residual calcite and clay minerals in which are scattered detrital grains of quartz, pelletoidal to lobate glaucony, and muscovite. Glaucony may also replace sponge spicules, fill internal cavities of small foraminifers, or be scattered as flecks in the groundmass. Small amounts of pyrite are also present. Organic fragments are always in smaller proportion than detrital minerals; they are predominantly siliceous sponge spicules, Radiolaria, diatoms, a few small foraminifers, and bioclasts of pelecypods and pelmatozoans. The preservation of sponge spicules presents numerous varieties in which opal, chalcedony, glaucony, and pyrite, whether alone or associated in all proportions, occur within a given individual. Opal, the main constituent of the siliceous portion of the matrix, appears with three major aspects: a homogeneous, slightly birefringent mass enclosing scattered clay minerals; numerous minute globules containing isolated spots of chalcedony embedded in a groundmass of chalcedony; and, finally, larger globules with concentric structure associated with chalcedony spherulites with well-developed radiating fibers. Chalcedony occurs also by itself, developing around sponge spicules, but at the expense of the groundmass, and forming incipient cherty nodules consisting of irregular sheaves of fibers with hazy boundaries that can eventually evolve into distinct nodules.

Pure spiculites (Cayeux, 1929) are almost entirely made up of a felty concentration of siliceous sponge spicules. Their interstitial cement is insignificant and shows a few grains of detrital quartz and glaucony. Spicules are strongly comminuted and preserved as opal or chalcedony. Their central canal appears often enlarged by diagenetic dissolution, whereas their external boundaries are marked by rows of globular opal. The interstitial cement is mainly formed by undifferentiated massive and slightly birefringent opal, with a few scattered clay minerals, occasionally globular. Scattered spots of chalcedony occur within the opal, which also contains irregularly shaped concentrations of small calcite crystals, indicating a possible replacement of a precursor carbonate cement. When chalcedony is more com-

pletely developed, it is always accompanied by a partial destruction of globular opal, particularly along the margins of the sponge spicules. This evolution leads, as in the case of arenaceous spiculites, to the formation of cherty zones with hazy boundaries.

Well-developed spiculitic cherts (**Plate 28.A**) are vitreous, of a bluish color, and characterized by a large predominance of chalcedony over opal (Cayeux, 1929). The abundant remains of siliceous sponge spicules can only be detected through the particular orientation or size of the chalcedonic fibers. At times, only the axial canal of the spicules has been preserved as a single or double streak of opal or glaucony extending across chalcedonic fibers. Occasionally, the spicules were entirely dissolved after having acted as centers for fibroradiated zones of chalcedony. In other instances, spicules were preserved as streaks of opal globules or as shapeless opaline residues, which are almost beyond interpretation were it not for the surrounding petrographic context. These cherts may eventually consist only of a groundmass of interfering chalcedony spherulites throughout which are scattered the faint remains of sponge spicules reduced to rodlike bodies of different lengths representing their central canals. These bodies consist either of finer-grained, darker microfibrous chalcedony or a combination chalcedony–glaucony or intimately mixed chalcedony–opal. Some of the rodlets may be hollow and colored brown by iron oxides. In other varieties, the fan-shaped chalcedony develops on such a large scale that the obscure traces of siliceous sponges are reduced to the darker points at the center of irregularly interfering chalcedony spherulites.

Novaculites

Depositional Environment. The term novaculite is applied to a very dense, structureless, and light-colored bedded chert, consisting essentially of microquartz. It is thick bedded with rare shale partings, and typically represented by the Caballos Novaculite of Texas and the Arkansas Novaculite, both of Devonian–Mississippian age (Goldstein and Hendricks, 1953; Goldstein, 1959; McBride and Thomson, 1970).

Novaculite beds and associated cherts display an array of sedimentary and diagenetic structures that have been the subject of a long-lasting controversy concerning their environment of deposition. Interpretations range from deep marine to shallow marine, with occasional subaerial exposure, and the question is as yet unresolved (McBride and Thomson, 1970; Folk, 1973; Folk and McBride, 1976; McBride and Folk, 1977; Jones and Knauth, 1979).

Petrography. A typical novaculite appears under the microscope as a well-developed pure microquartz mosaic that is either uniform or has a characteristic clotted or pelletoidal texture under crossed nicols. This groundmass contains traces of clay minerals and silt-size detrital quartz granules. Remains of siliceous sponge spicules and Radiolaria are sparsely scattered within the microquartz mosaic. The grains of microquartz are very uniform in size and range from less than 1 to 5 µ. These grains are so minute that they barely give the polarization colors of quartz; however, their distinct outlines and the absence of fibrous texture indicates that neither opal nor chalcedony is present, except in a few cases where spherulitic spots of chalcedony have been described.

Among the microquartz mosaic of some novaculites are little cavities with rhombic outlines along the margins of which quartz grains form a lining with preferred orientation. These cavities indicate leaching of an original carbonate mineral, probably dolomite or ferroan dolomite; some of the cavities have been subsequently filled by iron oxides.

Petrographic and SEM observations by Jones and Knauth (1979) revealed in some samples of Arkansas Novaculite, unaffected by regional metamorphism, significant preservation of depositional features such as the morphology of Radiolaria, including radial spines and monaxonic siliceous sponge spicules with central canals, as well as early diagenetic textures, the most important being relict opal-CT lepispheres now replaced by microquartz. These relict lepispheres fill Radiolaria tests, axial canals of sponge spicules, former microvoids, and even the inside of a phosphatic conodont. When a former void around the lepispheres was infilled with fibrous chalcedony, the microspherulitic form is particularly well preserved. Lepispheres were also observed under SEM and occur as microspherules 10 to 45 µ in size. These lepispheres are larger than those of opal-CT reported in Cenozoic deep-sea cherts, which are about 10 µ, but similar to those described by Meyers (1977) in Mississippian cherts. As mentioned above, these lepispheres are probably aggregates, rather than single individuals. Jones and Knauth (1979) found that these samples of Arkansas Novaculite show textures identical to those observed during the typical diagenesis of deep-sea siliceous oozes where the transformation sequence is opal-A to opal-CT to quartz, as discussed above. These authors also stated that isotopic temperatures of 21° to 29°C suggest chertification at shallow burial depths. In a particular sample, the novaculite contained dolomite rhombs and well-preserved siliceous sponge spicules. It also showed areas of microbrecciation in which the dolomite rhombs were truncated by the breccia clasts, indicating dolomitization before brecciation in place. Fractures are filled with quartzine, whose disputed significance is well known.

In samples of Arkansas Novaculite affected by regional metamorphism, Jones and Knauth (1979) observed a groundmass consisting of structureless and massive microquartz mosaic with no evidence of opal-CT lepispheres. Preservation of microfossils is poor. Radiolaria and siliceous sponge spicules occur as "ghosts," which in plane-polarized light are simply outlined by streaks of carbonaceous organic

matter and are shown under crossed nicols by darker patches of cryptocrystalline quartz surrounded by coarser microquartz mosaic.

Novaculites appear to have been originally siliceous biogenic oozes, but their environment of deposition remains unclear. This is particularly true when novaculites submitted to thermal metamorphism had their microquartz mosaic changed to aggregates of coarse, polygonal triple-point, euhedral quartz crystals, 100 µ or more in size, under temperatures estimated at 760°C (Keller et al., 1977, 1985).

SILICEOUS ROCKS OF PRIMARY INORGANIC ORIGIN

Magadi-type Cherts

Depositional Environment. Magadi-type cherts result from the conversion to chalcedony of two chemically precipitated hydrous sodium silicates called magadiite ($NaSi_7O_{13}(OH)_3 \cdot 3\ H_2O$) and kenyaite ($NaS_{11}O_{20.5}(OH)_4 \cdot 3\ H_2O$) in alkaline lakes under semiarid climates (Eugster, 1967, 1969; Hay, 1968). Although this conversion is in reality a replacement process that takes place in less than a few hundred years (Eugster, 1969), that is, geologically speaking, instantaneous, these cherts are classified, for convenience purposes, as being inorganically primary precipitates.

Eugster (1967, 1969) and Hay (1968) provided the first descriptions of magadiite precipitated from the alkaline waters of Lake Magadi and nearby Lake Natron (Kenya), in the East African rift valley, during the Late Pleistocene when the lake levels were more than 10 m higher than at present. Today, the brines of Lake Magadi contain high amounts of dissolved silica, up to 2,400 ppm SiO_2, reached mainly by evaporation concentration in the semiarid climate of the area. As a consequence of high pH levels resulting from precipitation of trona, which removes considerable quantities of bicarbonate, the brines appear to be slightly undersaturated with respect to amorphous silica. It is not clear how magadiite and kenyaite precipitate periodically in a varvelike pattern interbedded with silts and clays. The process requires a periodical lowering of pH in a chemically stratified lake. This could happen without dilution near the pycnocline, where oxidation of methane produces CO_2 (Eugster, 1969), or by dilution at the interface between silica-rich brines and freshwater entering the floor or margins of the lake during seasonal rain and runoff (Eugster, 1969; Hay, 1968).

Conversion of magadiite to chalcedony occurs by dehydration and loss of sodium directly or through an intermediate kenyaite phase (Eugster, 1969). It can take place both in near-surface environments and at somewhat greater depths within lacustrine sediments. Magadiite alters spontaneously to chalcedony in saline environments with a pH of 9.5 to 10.0 (O'Neil and Hay, 1973), but the conversion can be aided by rainfall and runoff. The cherts form undulating, thin lenticular or platy beds and nodules with characteristic reticulated and cracked surfaces due to dehydration. The rate of conversion was estimated by Eugster (1969) and Hay (1968) at a few hundred years, because in some sequences of gravels, alternating with continuous chert beds, the gravels contain angular chert chips, indicating that the conversion was fully completed before transportation and before the next overlying bed of magadiite was deposited.

Petrography. The platy and nodular cherts of Lake Magadi and Lake Natron (Tanzania) localities (Hay, 1968) show an opaque white rim over a dense, homogeneous translucent chert core whose color varies from yellow–brown to dark gray; some mottling and color zoning may be present. The groundmass, which may contain crystals and molds of trona, is a mosaic of irregularly shaped individuals of equant microquartz ranging in size from 2 to 10 μ. The crystals have a wavy extinction, which under crossed nicols tends to blur their outlines. However, some nodules show microfibrous chalcedony, length fast and length slow, forming spherulites or irregular zones scattered within the equant microquartz groundmass (Hay, 1968). In reality, compared with other cherts, not a single characteristic appears to be unique to Magadi-type chert, not even the surface reticulation patterns (Schubel and Simonson, 1990). These authors, however, pointed out that the only possible exception is a rectilinear or gridwork orientation of the microquartz crystals, which appears inherited from the precursor magadiite, displaying a similar extinction pattern due to the presence of 10- to 20-μ spherical aggregates of platelike crystals. This pseudomorphism implies a direct volume-for-volume replacement of magadiite by quartz. However, void-filling patches of chalcedony and the relatively high density of these cherts indicate that silica was added during and after the conversion of magadiite to chert. As mentioned above, this conversion occurred in multiple stages, with early episodes in the brines and within the sediments, and late episodes in contact with dilute meteoric waters.

Among the ancient examples of chert formations interpreted as lacustrine Magadi-type cherts and described below, only two cases, from the lacustrine Pliocene Gila Conglomerate of New Mexico and from the Rome Formation of Oregon (Sheppard and Gude, 1986), display the typical gridwork orientation of the chalcedony crystals. It is probable that such an orientation was destroyed by diagenesis in older cases, stressing the fact that this structure alone is perhaps not even a sufficient character by itself and should be used in context.

Among the ancient chert formations interpreted as lacustrine Magadi-type cherts are those of the Cambrian of the Officer Basin in Australia (White and Youngs, 1980) and of the Middle Devonian Old Red Sandstone of northern Scotland (Parnell, 1986). This author described chert nodules replacive within organic-rich, carbonaceous siliciclastic laminae and micronodular dololutites of a former stratified lake. These cherts exhibit soft-sediment deformation, including injection along shrinkage fractures within the nodules. The nodules were nucleated along vertical cracks in the host sediment and show polygonal or parallel-aligned patterns. Precipitation of the assumed magadiite precursor seems to have occurred from alkaline groundwaters rather than in the lake water column. Petrographically, the nodules consist entirely of a microquartz mosaic containing numerous stellate quartz pseudomorphs, up to 1 mm across, of an evaporitic mineral, which could have been trona and which cuts displacively across the chertified remains of organic-rich laminae. The micronodular fabric of the dololutite is also preserved in some chert nodules in spite of total replacement by microquartz.

Vugs parallel to primary lamination developed during silicification and are filled by quartz mosaic or chalcedony. Larger vugs are lined with chalcedony overlays, followed by final infilling by mosaic quartz or zebraic chalcedony. Spherulitic chalcedony also progressively replaces the original fabric. It is both length fast and length slow, and a mixture of both types occurs within single spherulites. Some of the quartz mosaic displays a highly undulose extinction, making the boundaries of the crystals difficult to define; it is actually microflamboyant quartz as defined by Milliken (1979).

Other instances of assumed Magadi-type cherts were described in the Jurassic and Eocene to Pleistocene in the western United States (Surdam et al., 1972; Eugster and Surdam, 1973). Sheppard and Gude (1986) reviewed all occurrences in the coterminous United States. Their petrographic study showed that the dense Magadi-type cherts consist mainly of an aggregate of interlocking, irregular, fine grains of chalcedony forming a gridwork pattern, ranging in size from less than 2 μ to about 40 μ, with an average of 25 μ. All the chalcedony shows anomalous extinction and is length slow. Its mean index of refraction is between 1.53 and 1.54, distinctly lower than that of equant microquartz. Some cherts contain veinlets and irregular patches of coarser fibrous chalcedony and of spherulitic chalcedony. Spherulites reach 250 μ in diameter and are also length slow. Some other chert nodules display abundant, but vague, ellipsoidal bodies of chalcedony visible under crossed nicols. These bodies, 0.03 to 1.4 mm in longest dimension, typically consist of a very fine grained core (grain size less than 2 μ) and a coarser fibrous rim. Some ellipsoidal bodies show a vague concentric structure within their cores. The fibrous chalcedony rims of the bodies are length fast, in contrast to all the other observed chalcedony. Sheppard and Gude (1986) interpreted these bodies as relics of original magadiite, because this mineral occurs in the lacustrine deposits described by Eugster (1967) and Hay (1968) in spherulites of about the same size as the ellipsoidal bodies in the cherts.

In the Magadi-type cherts described by Sheppard and Gude (1986), crystal molds of assumed evaporitic minerals are filled entirely or partially by chalcedony, microquartz, megaquartz, and calcite. Often, molds are lined by a thin rim of length-slow chalcedony, followed inward by microquartz, and finally by megaquartz, which may have euhedral crystal faces projecting into unfilled cavities. This diagenetic sequence is of general occurrence, and calcite always postdates megaquartz. X-ray diffraction studies by Sheppard and Gude (1986) revealed in the rind of the cherts, in addition to chalcedony, calcite, clinoptilolite, silhydrite, and opal-CT, whereas similar studies by Hay (1968) and Eugster (1969) on the Pleistocene nodules of Lakes Magadi and Natron recognized, in addition to quartz, magadiite, kenyaite, clinoptilolite, and erionite. Apparently, magadiite and other hydrous sodium silicate minerals are absent from cherts older than late Pleistocene; they obviously reacted during diagenesis to form silica or aluminosilicate phases.

Magadi-type cherts, being indicative of alkaline, lacustrine depositional environments, represent a key for the search of precipitated evaporitic minerals and for diagenetic minerals formed by the reaction of volcanic glass with the brines, in particular zeolites such as analcime, chabazite, clinoptilolite, erionite, mordenite, and phillipsite, some of which are of potential economic interest (Sheppard and Gude, 1986).

Lacustrine Cherts

Depositional Environment. In ephemeral playa lakes associated with the Coorong Lagoon of South Australia, where magnesium-rich fluids dolomitize their carbonate substratum (see Chapter 6), Peterson and Von der Borch (1965) reported direct inorganic precipitation of silica. The dominant precipitates in these lakes are dolomite, magnesite, and magnesian calcite. The surface of the sediment, which dries completely during the summer, contains plates of desiccated sediment related to mudcracks. The plates, about 1 cm thick and 10 cm across, are internally complex, indicating penecontemporaneous fragmentation and reworking. Their uppermost part, a few millimeters thick, has a porcelaneous texture and represents the cherts formed from a precipitated silica gel.

These playa lakes undergo fluctuations in level and in pH that are seasonally controlled. When the lakes are high, their alkaline waters have very high pH values, ranging between 9.5 and 10.2, resulting from the photosynthetic activity of phytoplankton. Under these conditions, detrital grains of quartz, feldspars, and clay minerals are partially dissolved and release silica, and the waters become supersaturated with respect to amorphous silica. This source of silica is well shown under the microscope, where detrital grains appear deeply pitted with complex and irregular shapes and ragged fringes and are surrounded by a halo of fine-grained silica and carbonates. Evaporation of lake waters with associated decay of phytoplankton material, hypersalinity, and a drop of pH to 6.5 to 7.0 leads to precipitation of an opal-A–cristobalite gel consisting, under SEM, of minute globular units. This gel coats the desiccation plates, but mainly impregnates the unconsolidated carbonate mud. It is assumed to mature subsequently into a typical chert.

Opaline cherts were described in Lake Bogoria in the Kenya rift, which is an alkaline lacustrine environment (Renaut and Owen, 1988), and differ from the Magadi-type cherts commonly associated with these conditions. These cherts cement beach gravels and also form crusts up to 15 cm thick, but are in fact intimately associated with hot springs. They are white to cream in color, typically porcelaneous, with a dense slightly vitreous interior. Under the microscope, they show an opal groundmass with abundant and perfectly preserved diatoms; enclosed laths of K-feldpsar and lithoclasts of trachyte are marginally replaced by opal. SEM examination (Renaut and Owen, 1988) shows that most diatoms have preserved intraparticle porosity, whereas other areas of microporosity might be concentrations of decayed organic matter. Many pores contain discoidal bodies, 1 to 3 μ in diameter, irregular filamentous masses, and sheaflike mucilage. The disks resemble coccoid cyanobacteria; the other organic matter could also be cyanobacterial or bacterial in origin.

Renaut and Owen (1988) stressed that the abundance of diatoms indicates a subaqueous precursor deposited during a Holocene higher lake level and with fresher conditions than at present. Silica certainly originated from sublacustrine hot springs and was probably a product of hydrothermal alteration of local volcanic rocks. Silica was precipitated around the orifices of hot springs as a soft gelatinous precursor of silica or sodium silicate. X-ray diffraction indicates opal-A with traces of opal-CT, which could represent an early diagenetic transformation, eventually leading to microquartz. According to Renaut and Owen (1988), precipitation was due to rapid cooling and perhaps a decrease of pH or microbial actions. Lithification by agglomeration and cementation of opal particles began before subaerial exposure. In summary, this type of chert, formed by sublacustrine hydrothermal action, is bound to be localized and related to rift faulting. It is transitional to cherts of volcanic origin, some of which will be discussed in relation to the genesis of Precambrian banded-iron formations (see Chapter 9).

Petrography. Ancient analogues of playa lake cherts were described in the Triassic rift grabens of the eastern United States (Wheeler and Textoris, 1978). These cherts occur as beds up to 60 cm in thickness. They are dense, dark gray, and consist of medium crystalline chalcedony of only length-fast type and of microquartz. The silica is assumed to have precipitated as an opal-A–cristobalite gel.

SILICEOUS ROCKS OF SECONDARY REPLACEMENT ORIGIN

These siliceous rocks occur predominantly as chert nodules, but also as tabular layers and as entirely chertified formations that developed mostly as replacement of carbonate and evaporite rocks. This replacement was accompanied in some instances by a minor amount of pore-filling textures.

The most often investigated types of replacement chert pertain to carbonate rocks. Three problems are related to this replacement, which are of critical importance: (1) the constraints of the geochemical environment of silicification; (2) the source or sources of silica; and (3) the relative timing of silicification with respect to the diagenetic phases of the carbonate host rock. Many aspects of these three problems remain disputed, while a certain amount of confusion pertains to the question of timing introduced by a general and indiscriminate use of the terms "very early," "early," and "late replacement." To avoid further confusion, three environments of silicification need to be defined.

Syngenetic replacement characterizes silicification that occurred under phreatic submarine conditions, essentially penecontemporaneous with carbonate sedimentation. This replacement ranges from the water–sediment interface to several meters into the sediment, within the reach of storm activity, and perhaps in some cases involving early dewatering and initial compaction. This type of chertification is amply demonstrated by the intraformational reworking and fracturing of fully indurated cherts within unconsolidated carbonate sediments.

Early diagenetic replacement characterizes silicification that occurred outside the marine environment when porous carbonate rocks at various stages of lithification were submitted to the diagenetic mixing zone between the vadose phreatic freshwater lense and the underlying marine phreatic zone (see Chapter 5 and **Fig. 5.5**).

Late diagenetic replacement characterizes silicification that occurred when completely lithified carbonate rocks reached the burial environment and received a silica influx from associated rocks, derived from pressure solution of quartz grains, compaction, or smectite–illite conversion of shales, a situation discussed above in the section on silica cementation of sandstones (see Chapter 1). Late diagenetic replacement occurred also when carbonate rocks underwent subsequent cycles of uplifting and folding, and groundwater circulation brought silica, which was deposited in systems of joints and faults.

Unquestionably, the three above-mentioned types of silicification grade into each other in time. Indeed, many cherts show that their history may span the entire spectrum from early diagenetic to deep burial, which can be reconstructed if sufficient textural and mineralogical relicts were preserved in them.

Syngenetic Cherts in Chalks

Genetic Environment. The Cretaceous chalk of Western Europe is predominantly a white, pelagic, nannofossil calcilutite probably deposited as a low-Mg calcite mud in a continental shelf setting. It is extremely bioturbated and generally displays few traces of original bedding planes. However, it contains numerous and rhythmically spaced streaks of black carbonaceous chert (flint) nodules or tabular layers that have been extensively studied for many years (Cayeux, 1929; Sieveking and Hart, 1986). These chert nodules and tabular beds can be traced laterally for tens and even hundreds of kilometers, showing that they originated from processes of regional to basinal significance. In detail, chert morphology is complex, displaying a complete transition from individual vague nodules to complex, well-defined, digitated and coalescing shapes, indicating that chert replaced many types of burrow networks, and to tabular continuous chert beds in which burrows are barely visible because silicification involved the entire chalk sediment (Bromley and Ekdale, 1986).

The origin of black chert in chalk was poorly understood for many years, except for the fact that it appeared to represent replacement of the chalk in layers more or less parallel to the sea floor. Recently, the geochemical study of Clayton (1986) provided for the first time a large-scale understanding of the process.

From a geochemical viewpoint, the black cherts of the chalk were formed by reprecipitation in pore waters of the sediment of silica dissolved from siliceous microorganisms, mainly siliceous sponges. This precipitation was accompanied by partial replacement of the mud with elimination of calcium carbonate, and by other effects on the mineralogy and chemistry of other mineral phases present in the chalk at that time. Clayton (1986) undertook a detailed trace element geochemical investigation precisely along these lines to discover what kinds of chemical signatures would be left that could allow understanding of the process of chertification in the chalk. He stressed the fact that cherts tend to form at any important change of the chalk fabric and that tabular cherts develop even in the absence of burrowing structures. Therefore, according to Clayton (1986), burrows are not the cause of silicification, but are probably porosity and permeability contrasts between burrowing structures and host chalk, whereas discontinuities in the chalk, such as hardgrounds, determined the location of the process or modified the pattern of silicification caused by some other process. Clayton (1986) studied an exception to the above rule, a peculiar and complex type of chert that is a cylindrical nodule formed around a cemented central chalk core that in its middle contains a pyritized, glauconitized, and often phosphatized vertical burrow of *Bathichnus paramoudrae*. This type of nodule, called a *paramoudra structure,* is unique because silicification occurred as a ring around the chalk core, which was obviously, and exceptionally, the cause of silicification; thus it provides a rare insight into the geochemistry of silica replacement of chalk. According to Clayton (1986) and given here in a very simplified way, anaerobic bacterial sulfate reduction occurred in the central burrow of paramoudras and released large excesses of dissolved sulfide (H_2S or HS^-), which diffused outward toward more oxic conditions. Oxidation of the sulfide at the oxic–anoxic boundary substantially reduced local pH, causing dissolution of the chalk and chemical "seeding" of dissolved silica, as well as formation of chert with an annular geometry.

The geochemical results obtained by Clayton (1986) in his study of the paramoudras can be extended to the following summary of the general conditions of chert formation in the chalk. Aerobic bacterial degradation of organic matter in the upper portion of the chalk sediment gradually depleted pore waters of dissolved oxygen so that, at a certain depth, oxygen concentration became too low to sustain aerobic bacteria. However, oxidation of organic matter continued deeper into the sediment by using sulfate, making the environment sulfate reducing. Under these sulfate-reducing conditions, if iron had been abundant, pyrite would have formed; but since this was not the case, the former sulfate migrated in the form of H_2S upward toward the more oxic conditions. This corresponded to the generation within the chalk sediment of a redox boundary, separating an upper oxic environment from an underlying anoxic environment. Along this boundary, sulfide-oxidizing bacteria reoxidized H_2S to SO_4^{2-}, liberating hydrogen ions in the process. These released hydrogen ions led to an intensive but local calcite dissolution, and the high concentration of dissolved carbonate ions liberated in this way represented a suitable seeding agent for silica from saturated solutions. In essence, calcite dissolution and silica precipitation occurred simultaneously. The source of silica in chalks presents no major problem because it contains abundant siliceous sponge spicules, diatoms, and Radiolaria. Dissolution of biogenic opal-A in the chalk pore waters increased silicon concentration beyond the solubility of opal-CT, which was the first silica phase to precipitate from this highly concentrated and metastable solution upon introduction of the seeding agent represented by a high concentration of dissolved carbonate ions.

In summary, the replacement of chalk by silica straddled the redox boundary within the unconsolidated sediment. Clayton (1986) was able to explain that the chert morphology is determined by the geometry of the H_2S–O_2 mixing zone. In homogeneous and bedded chalk, where burrows were absent or poorly defined, widespread mixing took place along a regular zone, and tabular chert was generated. In places where burrows were well defined and introduced greater permeability, heterogeneous mixing was largely confined to the burrow systems, and rows of complicated digitate–nodular cherts were formed, reflecting the shapes of the various burrow systems. Higher up in the sediment, above the mixing zone, only local concentrations of organic matter were able to initiate local sulfate reduction conditions, and hence silicification on a small scale generated simple nodules of chert around patches of decaying sponges, and the vertical, organic-rich *Bathichnus* burrows triggered the columnar cherts called paramoudra structures. Furthermore, systems of conjugated joints associated with dewatering of the chalk sediment and very early compaction also provided ideal local mixing zones for H_2S and O_2, generating oblique and intersecting sheets of chert.

On the basis of the dispersion of clay minerals within the cherts, Clayton (1986) suggested a porosity of 75% to 80% at the time the silica structure was able to resist compaction, that is, within the top 10 m or so of the sediment. Also, the height of the tallest paramoudra structure represents the minimum depth since it was formed in general oxidizing conditions, that is, above the redox boundary. Therefore, he suggested that bedded and nodular cherts formed 5 to 10 m inside the unconsolidated chalk, but he also considered much shallower depths. Clayton (1986) noted the absence of reworked cherts and related it to the growth history of the chert nodules. As mentioned above, the main process was an intense precipitation of opal-CT lepispheres following dissolution of the skeletal calcite of the chalk. Submarine reworking of the nascent chert at this stage would disaggregate the lepispheres and disperse them even if they were already forming a kind of self-supporting framework together with the carbonate. Clayton (1986) assumed that the chert became firmly cemented later upon precipitation of an interstitial opal-CT–chalcedony phase, and apparently only during subsequent "burial" diagenesis did opal-CT recrystallize to its present microquartz mineralogy.

If this mineralogical and textural evolution is correct and common to many types of biogenic cherts, the time frame during which it is supposed to have occurred in the chalk does not appear realistic. Indeed, Cayeux (1929, 1941) described in the Turonian chalk of the Paris basin and in the cliffs of the Channel tabular beds and rows of black chert laterally interrupted and replaced by chert breccias. These intraformational breccias consist of completely indurated and angular fragments of black chert with conchoidal fracturing. The fragments display all the features of the nonfragmented tabular or nodular cherts, including their peripheral patina. Cayeux felt that these monogenic breccias demonstrated the synsedimentary origin of completely indurated cherts within the unconsolidated chalk mud. He also said that submarine fracturing of these fragile bodies resulted from a "mysterious" dynamic action; this was the only possible interpretation at a time when catastrophic, short-lived, high-energy events such as storms and related tempestite deposits were not yet understood. These observations also indicate that the redox boundary proposed by Clayton (1986) might have been much closer to the sediment–water interface than he suggested, at least within the range of reworking of the chalk mud by storms. Further confirmation of the geologically instantaneous nature of silicification was provided by Deflandre (1934), who described in the black cherts of the chalk of the Paris basin perfectly preserved remains of Hystrichospheres and Dinoflagellates; staining indicated that they were "mummified" as cellulose, an organic substance that in seawater would completely decay within 12 hours after death of these microorganisms.

Additional evidence of the geologically instantaneous induration of chert nodules in chalk is also provided by their being involved in synsedimentary slumping (**Fig. 7.2**) and other depositional structures (Cayeux, 1929, 1941; Kennedy and Juignet, 1974).

The regular repetition of chert bands at 50-cm to 2-m intervals that is so characteristic of the chalk is probably the reflection of the rhythmicity of its own sedimentation, as suggested by many authors (Kennedy and Garrison, 1975). Therefore, it seems logical that bands of chert nodules and tabular cherts would also reflect this cyclicity. According to Clayton (1986), each individual hiatus in sedimentation (bedding planes, erosional surfaces, hardgrounds, and so on) would have led to an extensive period with a stationary redox boundary, a few meters below or less, generating a tabular chert or a row of nodules. After this episode, relatively rapid chalk sedimentation may have restricted the development of a stable redox stratification, and the next generation of chert could only develop during the next hiatus. The well-developed and laterally continuous chert bands, which can extend for tens to hundreds of kilometers laterally, may represent basinwide breaks in sedimentation, while discontinuous bands could express local interruptions.

Petrography. The cherts of the chalk (**Plate 28.B**) are generally of an evenly distributed black color, with occasional mottled appearance. Banding may be present, parallel to bedding or concentric. Waxy luster is typical and the con-

choidal fracture corresponds to a complete absence of grain. In thin section (Cayeux, 1929), the black cherts appear almost clear and colorless or with a faint brownish color due to very fine carbonaceous flecks. Within this translucent groundmass, the shapes of small foraminifers, shell bioclasts, and sponge spicules, normally found also in the enclosing chalk, may be seen faintly but faithfully outlined. However, some of the finest skeletal microstructures have been obscured. Under crossed nicols, the microquartz mosaic appears extremely fine and becomes coarser only in the areas occupied by microfossils. Among them, siliceous sponge spicules are predominant. They are often devoid of an axial canal and consist always of chalcedony fibers oriented at random or radially; debris of foraminifers display similar features. Most bioclasts of pelecypods and echinoderms have lost their internal microstructure and are represented by patches of irregularly crystallized chalcedony.

These chert nodules rarely consist entirely of dark or pure black material; almost every one of them shows breaks in the homogeneity, which appear as cloudy patches. These light-colored patches are chalk residues either completely silicified or only along their periphery with unaltered chalk preserved in the center. Other light-colored areas correspond to imperfectly silicified bioclasts.

Practically all black cherts grade into the enclosing chalk by means of a gray-whitish transitional zone called patina that belongs more to the nodule than to the chalk, as clearly shown upon fracturing. This transitional zone is an original feature of cherts and does not represent any weathering effect, as the poorly chosen name patina may suggest (Cayeux, 1929). Indeed, no relationship exists between the thickness of the transitional zone and the age of the nodule or its distance to the present topographic surface. It is also significant that the features of the transitional zone are not

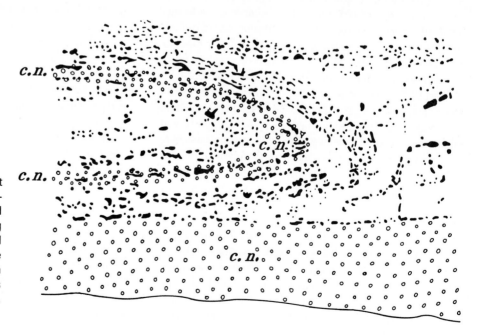

FIGURE 7.2 Stringers of chert and bed of nodular chalk (c.n.) arranged in a kind of recumbent fold produced by submarine slumping and underlain by a horizontal bed of nodular chalk (c.n.). The whole structure is overlain by a chalk with horizontal stringers of chert. Cliffs west of Etretat, Normandy, France. From Cayeux (1929).

found along later compaction or tectonic fractures intersecting nodules. The texture of the transition zone is usually fine grained, porcelaneous, and relatively porous. Microscopic investigation reveals a confused groundmass of microquartz and chalcedony veiled by a concentration of amorphous impurities and numerous chalky residues at various stages of silicification. This heterogeneous texture indicates that chert nodules grew from the center toward the periphery. This represents actually a zone of incomplete silicification, with concentration of material rejected by the process and made partially porous by subsequent minor decalcification.

Syngenetic Cherts in Limestones and Dolostones

Genetic Environment. Carozzi and Gerber (1978) described a type of synsedimentary chert conglomerate in the Lower Burlington Limestone (Middle Mississippian) of eastern Missouri where completely indurated and fragile chert nodules were penecontemporaneously fragmented within an unconsolidated crinoidal calcarenitic sediment by the action of a short-lived and localized high-energy event interpreted as the touchdown of a tornado system (see also Chapter 2). It appears as if unusual events of catastrophic nature and penecontemporaneous with sedimentation can provide the indisputable demonstration of the syngenetic origin of some cherts in carbonate rocks, as shown above for chalks as well.

Other instances of syngenetic cherts were described by Banks (1970) in two tidal flat dolostone members of the Mississippian Leadville Limestone of west-central Colorado. The syngenetic origin is established by the fact that laminae of the enclosing carbonate wrap around chert nodules in patterns indicating differential compaction of carbonate sediment around indurated chert. Furthermore, abraded and angular clasts of syngenetic chert occur in wave-formed intraformational breccias. Banks (1970) concluded that these cherts were not only formed before any appreciable lithification of the carbonate sediment, but also behaved as brittle rock almost immediately upon their formation. According to Banks (1970), these syngenetic cherts display two textural and color varieties. The first is a black chert consisting mainly of cryptocrystalline to microcrystalline quartz with minor amounts of chalcedony. The black color is caused by the inclusion of carbonaceous material. Enclosed dolomite is anhedral, but etched and embayed by the surrounding microquartz mosaic, which may also contain ghosts of the laminae of the host dolostone. The second variety is white to blue–white and gray in color; it is also called "sand chert" because its occurrence is limited to arenaceous dolostones and to sand-grain marker beds in dolostones. These light-colored cherts consist predominantly of chalcedony with subordinate local areas of microquartz. Scattered in the groundmass are rhombohedra of dolomite and grains of detrital quartz with radial chalcedonic overgrowths. In the sand-

grain marker beds of the dolostones, all stages of chert formation can be observed, from several grains of quartz welded together by chalcedony to real nodules of sand chert.

Banks (1970) stressed the fact that eolian sand is abundant in the dolomitic rocks containing syngenetic chert and that it appears as the dominant if not the only source of silica available, particularly since the quartz grains are heavily replaced marginally by dolomite. One could add to Banks's statement that silt-size eolian quartz dust should also be considered because it is volumetrically more important than sand grains and could be of even greater importance as a source. Biogenic silica, such as siliceous sponge spicules, Radiolaria, or any recognizable form of siliceous microfossils, has not been found in the dolostones or in the syngenetic cherts. Clearly, the hypersaline waters of the tidal flats dissolved detrital quartz and reprecipitated silica in the form of nodules at the sediment–water interface, where they could be intraformationally reworked by waves and currents.

Dietrich et al. (1963) described in the Cambro-Ordovician Knox Group of Virginia conditions where tidal-flat dolomitization was at times and in places interrupted by the influx of silica. The original carbonate sediment was a pelletoidal mud that underwent initial dolomitization at the sediment–water interface, which is shown by zoned dolomite rhombs with a variety of zonal structures. The deposition of silica in the form of nodules interrupted the dolomitization process. These cherts obviously replaced the calcium carbonate portion of the initial sediment because they show ill-defined relics of pelletoidal texture and of bedding laminations. A few nodules show shrinkage cracks from subaerial exposure, and locally the resulting fragments were imbricated by intraformational wave action before deposition of the overlying carbonate sediment. The zoned dolomite rhombs within the nodules were spared by silicification; therefore, silicification selectively replaced calcite and preserved whatever initial dolomite had developed before that time. In some instances, silicification prevented further dolomitization, in others, dolomitization resumed after silicification and resulted in the partial or total conversion of the remaining unsilicified carbonate into a mosaic of small, interlocking, clear and unzoned anhedral crystals. These early and mutually interrupted diagenetic episodes of silicification and dolomitization did not always have a simple set sequence; they overlapped in time and space, but they represent two aspects of early diagenesis in very shallow, hypersaline marine waters. Comparable conditions of marine silicification preceding early dolomitization were also described by Gao and Land (1991) for chert nodules from the Arbuckle Group (Early Ordovician) of southwest Oklahoma.

A syngenetic origin of chert nodules in the Middle Ordovician Hardy Creek Limestone of southwest Virginia (Harris, 1958) was also proposed on the basis of the observation of an apparent correlation between the shape and spacing of chert nodules and the arrangement and spacing of desiccation cracks surrounding them, as visible on bedding planes. Chert nodules are centered in desiccation polygons, and the shapes of the polygons appear directly controlled by the shapes of the chert nodules, whose generation clearly preceded desiccation processes.

Another example of syngenetic chert formation was reported by Coniglio (1987) in the Cambro-Ordovician Cow Head Group of western Newfoundland. This group contains abundant Radiolaria and siliceous sponge spicules, indicating a biogenic source of silica. There uncompacted burrows and uncompacted graptolites in silicified shales, including both completely silicified beds and nodules, indicate that chertification was very close to the sediment–water interface.

In summary, limestones and dolostones provide abundant data in favor of syngenetic chertification similar to that occurring in the chalk and often with a similar rhythmical repetition as in the Keokuk–Burlington Formation of the Middlewest. These conditions suggest a similar geochemical setting as described in the chalk by Clayton (1986), which remains to be investigated.

Petrography. Cherts of syngenetic origin in limestones are not petrographically very different from those described in the chalk, except that they can be of various colors besides black and that they generally contain a greater number of incompletely silicified bioclasts or patches of the original host limestone because of its lesser porosity–permeability than the chalk. The contact between nodule and host rock is usually sharp, with a thin intermediate patina. In general, a mosaic of microquartz predominates over areas of length-fast chalcedony, while megaquartz concentrates in replaced bioclasts. Microquartz may occur also as interparticle and intraparticle cement, particularly within debris of pelmatozoans. When interparticle pores are sufficiently large, successive cement overlays of chalcedony occur, at times botryoidal, which are followed during final occlusion by megaquartz.

Early Diagenetic Cherts in Limestones and Dolostones

Genetic Environment. Knauth (1979) proposed a geochemical model according to which many nodular cherts in limestones have formed in the diagenetic mixing zone between the vadose freshwater phreatic lense and the underlying marine phreatic zone, an environment where the Dorag-type dolomitization also occurs (see Chapter 5 and **Fig. 5.5**).

Under these conditions, vadose phreatic freshwater, after having dissolved opal-CT of the skeletons of siliceous sponges, Radiolaria, and diatoms scattered in the carbonate rocks through which they circulated, became mixed with marine phreatic waters and produced solutions highly supersaturated with respect to quartz and undersaturated with respect to calcite and aragonite. This situation

is an ideal diagenetic environment for rapid chertification of porous carbonate rocks under shallow lithification conditions.

In this process of chertification, which still awaits testing in ancient and modern settings, it is not yet demonstrated that opal-CT is the first silica phase replacing the carbonate; but Knauth's hypothesis is strongly supported by available oxygen and hydrogen isotopic studies of nodular cherts in carbonates throughout the geological column (Hesse, 1990b). The proposed model has interesting implications for the geometry of chert nodules. Indeed, they are not evenly distributed in limestones but tend to be concentrated along definite bands parallel to bedding. Such bands could then represent the position of the mixing zone within the porous carbonates and even indicate the thickness of the zone, provided it was not restricted by strong variations of relative porosity and permeability during the hydrologic flow process (Knauth, 1979).

Furthermore, in cases of transgressive–regressive evolution of the carbonate platform, the shape and extent of the chertified zone would vary accordingly. Because the model proposed by Knauth (1979) is an aspect of the Dorag-type dolomitization model, the question can be raised about the relationship between dolomitization and chertification. The proposed model accounts for the generation of rhomb-bearing cherts that contain unreplaced dolomite rhombs, while the surrounding rock may be a limestone or a dolostone. Obviously, it is a question of relative timing and of successive occurrence of solutions supersaturated in quartz or in dolomite.

In most instances, the well-developed and zoned rhombs of dolomite or ferroan dolomite dispersed in a microquartz groundmass indicate a limestone host rock that underwent incipient dolomitization interfering with or just preceding chertification (**Plate 28.C**). These dolomite rhombs, in spite of their general resistance to silicification, may at times show evidence of selective replacement by silica, mostly along cleavage planes. In such a case, the central part of individuals is replaced by microquartz, or the latter forms a dark, rhomb-shaped zone separating the core from the marginal portion, both being in optical continuity. The incipient replacement may also be shown by corrosion of the obtuse angles of the rhombs, whereas the rest of the crystals retain their sharp outline. Finally, some rhombs may be dissolved into rhombic molds.

Geeslin and Chafetz (1982) described in the Ordovician Aleman Formation of southwestern New Mexico, which is a stromatolitic dolostone with rhythmic chert beds, an example of early chertification that appears to fit quite well Knauth's model. Microquartz is the predominant constituent of cherts, followed by chalcedony and minor amounts of megaquartz. Pelletoidal and laminated textures are common within chert beds, as well as abundant packed remains of siliceous sponge spicules. Associated with these layered cherts are botryoidal white quartzose nodules generally called "cauliflower nodules," which represent cherts after evaporitic nodules. They consist mostly of microquartz, megaquartz, and microflamboyant quartz that contain rectangular relics of former anhydrite crystals replaced mainly by megaquartz and occurring either in a parallel manner as a typical pile-of-brick texture of primary anhydrite or as smaller laths in sheaflike aggregates.

Siliceous sponge spicules are considered by Geeslin and Chafetz (1982) the most likely source of silica and the mixing zone model proposed by Knauth (1979) as the environment of diagenesis. The following suite of events was suggested: (1) penecontemporaneous to early diagenetic evaporite formation in an intertidal-sabkha environment with accompanying dolomitization; (2) vadose silicification and dissolution of anhydrite nodules, because there is absence of gypsum hydration textures indicating that the nodules were submitted to a diagenetic water-saturated environment; and (3) phreatic mixing zone of silicification and dolomitization with algal mats being particularly susceptible to silicification. Many nodules have a central cavity filled with a geopetal dolomite matrix indistinguishable from the host rock, a feature indicating that the nodules were silicified before lithification of the carbonate. Additional evidence in favor of this early chertification consists of fractures found in some nodules that have a conchoidal shape. Fractured pieces of chert nodules or chert beds are separated by several millimeters to several centimeters of dolomite matrix; therefore, breakage took place probably by syneresis in the brittle material while the surrounding sediment was still unlithified. Final lithification of the dolostone took place after the cherts were brittle and fully formed.

The origin of silica for early diagenetic cherts in limestones and dolostones seems to be adequately accounted for by the redistribution of the biogenic silica, particularly from siliceous sponge spicules within the carbonate rocks themselves (Meyers, 1977; Geeslin and Chafetz, 1982). This interpretation fits the hypothesis of Knauth (1979) who postulated that chert nodules received their silica from pore waters that circulated through large volumes of porous carbonates at various stages of initial lithification, which now appear devoid of silica. Furthermore, in the mixing zone, waters were supersaturated in silica; hence the siliceous sponge spicules occurring in the corresponding limestones were not dissolved but incorporated into the precipitating silica. This explains the striking concentration of such spicules in nodules, compared with their absence or small number in underlying and overlying carbonates. However, other instances were discussed above where silica originated from the diagenetic dissolution of detrital grains (Banks, 1970), an interpretation also applied to syngenetic cherts.

Petrography. Early diagenetic chertification in limestones is extremely widespread and has been the subject of numerous studies. It would be redundant to describe an endless series of examples because the final result is, in most cases, the preservation to a high degree of preexisting depositional and diagenetic textures of the host linestone by a combination of microquartz, length-fast chalcedony, and megaquartz. Among these quartz varieties, microquartz tends to be a replacement of fine textures of carbonate mud and of skeletal remains, whereas chalcedony and mosaics of megaquartz replace preferentially larger bioclasts and various types of cavity-filling calcite cements (**Plate 28.D, E**). A characteristic feature of replacement is the abundance of undigested carbonate inclusions readily visible and giving, under crossed nicols, a brownish tinge to quartz. Many spectacular features of replacement are displayed by oolitic calcarenites with sparite cement (Choquette, 1955; Swett, 1965) or redeposited ooids (Hesse, 1987). Most silicified ooids show a concentric arrangement of carbonate inclusions within concentric bands of microquartz and megaquartz; other ooids with cores of detrital quartz grains show euhedral overgrowths on them, displaying the same pattern of concentric carbonate inclusions (**Plate 28.F**). Incompletely silicified ooids show that replacement can be centrifugal or centripetal.

Grain-supported limestones with a well-established sequence of types of interstitial cements indicate that chalcedony and megaquartz can also occur as cements by themselves developed between generations of carbonate cements. In such cases (Meyers, 1977), the various types of quartz cements are totally devoid of carbonate residues.

Whenever silicification is a replacement process, it appears to have a highly selective character; that is, certain skeletal or nonskeletal constituents are totally or partially replaced, while others remain unaffected. The process is highly complex and certainly a function of the original mineralogy and texture of the constituents and of the porosity–permeability conditions preceding silicification. Meyers (1977) found in Mississippian limestones a crude order of decreasing susceptibility to replacement by microquartz, which is as follows: carbonate mud, bryozoan bioclasts, thin-shelled brachiopods, thick-shelled brachiopods, and syntaxial calcite overgrowths on crinoids, the last three being only partially replaced or unaffected. However, different orders of susceptibility to replacement by silica have been described (Jacka, 1974; Hatfield, 1975; Hesse, 1987). The question should be dealt with case by case, because it is further complicated by the reversible nature of carbonate–silica replacements (Walker, 1962; Hesse, 1987). Indeed, Walker (1962) documented petrographically several types of such replacement reversals: carbonate minerals located along and adjacent fractures in cherts; transections of authigenic siliceous textures such as microquartz, banded chalcedony, and megaquartz by carbonate minerals; and occurrence within carbonate minerals of inclusions inherited from chert such as botryoidal zones of chalcedony. Multiple replacement reversals were also demonstrated. Walker (1962) showed in the

Oneota Formation (Ordovician) of southern Wisconsin two or more reversals of replacement with a diagenetic history as follows: (1) deposition of oolitic calcarenite; (2) partial replacement of carbonate by silica with formation of chert nodules; (3) partial dolomitization of chert nodules transecting their internal textures; and (4) partial silicification of dolomite with chert pseudomorphs of dolomite rhombohedra. Reversals of chert–carbonate replacement occurred between stages 2 and 3 and between 3 and 4; additional replacement reversals may have occurred, but textural complexities make interpretations questionable beyond the second stage of silicification.

Silicification and Early Carbonate Diagenesis.
Several studies make it possible to relate the relative time of silicification with the various types of cements and diagenetic features displayed by host carbonates. As expected, silicification, showing an extremely versatile geochemical behavior, seems to have involved the entire time span of early diagenetic processes that carbonates underwent before deep burial.

Meyers (1977) and Meyers and James (1978), on the basis of cement stratigraphy in the Mississippian limestones of New Mexico, established that pore-filling, length-fast chalcedony precipitated before and after early carbonate cements, both related to two unconformities and interpreted as freshwater phreatic in origin. Additional evidence for such an interpretation consists of angular fragments of cherts reworked in conglomerates overlying the unconformities, which indicate that the cherts were fully indurated and fragile constituents before the unconformities occurred. Meyers and James (1978) assumed that this chertification occurred under a very thin overburden of a few meters to a few tens of meters.

Another example of the relationship between silicification and unconformity was described by Namy (1974) in the Marble Falls Group (Lower Pennsylvanian) of central Texas. Silica associated with the unconformity occurs as a thin crust of silicified oolitic calcarenite just beneath the eroded top of an oncolitic pavement, which is a widespread regional unconformity. The silicified oolitic crust shows a gradual decrease downward in silica content. In the uppermost part of the crust, ooids are replaced by microquartz, and the cement is microquartz and fibrous chalcedony of pore-filling origin. In the middle of the crust, cement and ooid concentric rings consist of microquartz, whereas most of the ooid cores remain calcitic. Near the base of the crust, ooids are entirely calcitic, but the cement is microquartz.

A critical observation of genetic significance (Namy, 1974) is that in the middle and lower part of the silicified crust the silica cement postdates a thin rim cement of micrite and finely crystalline sparite around the ooids. Several inches below the crust, the interstitial cement of the ooids is cavity-filling sparite, with no traces of earlier cement phases

preserved. The thin rim cement of micrite and microsparite can be interpreted as representing stabilized incipient submarine phreatic cements that were arrested when the oolitic shoal was exposed slightly above sea level in a beachrock situation, where mixing marine phreatic and freshwater vadose waters led to conditions favorable to precipitation of replacive and pore-filling silica cement, as well as to the dissolution features visible in corroded oncolites and bioclasts at the base of the silicified crust. Silica is assumed to have originated from dissolution of siliceous sponge spicules washed over the shoal by storms or high tides (Namy, 1974). In summary, silicification occurred during an episode of mixing and subaerial exposure between marine phreatic carbonate cementation and freshwater phreatic carbonate cementation.

Silicification taking place beneath the water table in the freshwater phreatic environment was documented by McBride (1988) in pebbles and cobbles of Cretaceous carbonate rocks in Eocene to Neogene fluvial conglomerates and lag deposits of west Texas. This silicification, which occurred after deposition, is supported by oxygen isotopic values of silica indicating that it precipitated from meteoric waters and by the absence of conchoidal fracture surfaces and crescent impact marks typical of many fluvially transported chert clasts. Silicification appears on well-rounded carbonate pebbles either as thin rinds of microquartz or thick rinds of radially spherulitic chalcedony with small amounts of microquartz; but some silicification occurs as internally alternating limestone and chertified bands, conformable with the external shape of the pebbles. Thin-section examination shows that the replacive silica is microquartz, spherulitic chalcedony, a small amount of lutecite, and quartz with relict spherulitic texture. X-ray diffraction shows quartz as the predominant silica phase, but traces of opal-CT occur in a few samples, although only ghosts of lepispheres appear under SEM.

Chalcedony is clearly of replacement origin because within a given pebble it cuts across the boundaries of components of the limestone and includes residues of calcite of the same texture as the unreplaced limestone. According to McBride (1988), chertification was accomplished by groundwaters rich in silica, probably derived from hydrolysis of volcanic ash in Neogene volcanic tuffs. It must have occurred fairly rapidly because it is unlikely that adequate hydrologic and chemical conditions could have prevailed for more than a few million years. The process and its rapid rate are comparable in a broad sense to more superficial and vadose silcretization processes, where durations can be on the order of 30,000 years (Thiry and Millot, 1987; Thiry et al., 1988).

It is also possible to constrain even further the timing of silicification in terms of its relationship with aragonite conversion to low-Mg calcite and of high-Mg calcite conversion to low-Mg calcite, and by using very detailed petrographic evidence.

Jacka (1974), in his study of the Getaway carbonate turbidites (Middle Permian) of the Delaware basin, concluded that silicification took place before aragonite conversion to low-Mg calcite and also before stabilization of high-Mg calcite to low-Mg calcite. Indeed, he observed that aragonitic shells were pseudomorphosed in the silicified portions of the fossils, but destroyed in the calcitized areas. He also found microdolomite inclusions in the silicified portions of former high-Mg calcite shell fragments, but not in the unsilicified parts. Jacka (1974) assumed that exsolution of dolomite from high-Mg calcite took place during dissolution of the calcite and its replacement by silica, that is before its stabilization to low-Mg calcite. This is not in agreement with subsequent studies by Lohmann and Meyers (1977), who interpreted the occurrence of microdolomite inclusions in low-Mg calcite as an exsolution formed when a high-Mg calcite precursor stabilized to low-Mg calcite. However, Richter (1972), using carbonate inclusions in authigenic quartz developed in skeletal constituents of limestones from Paleozoic to Jurassic age, found that silicification postdated the conversion of aragonite to low-Mg calcite and predated the high-Mg calcite stabilization. Indeed, authigenic quartz in original aragonite and low-Mg calcite contained inclusions of low-Mg calcite, whereas quartz in original high-Mg calcite had high magnesium content.

Choquette (1955) gave petrographic proofs that silicification in the State College Oolite took place during and after aragonite dissolution. Hesse (1987), in his studies on Middle Cretaceous carbonate-bearing turbidites of the Eastern Alps, found that silicification postdated aragonite conversion to low-Mg calcite and also stabilization of high-Mg calcite. In the latter situation, he observed rows of microdolomite crystals in quartzine that continued, with the same optical orientation, into unsilicified portions of echinoderm bioclasts.

In summary, in early replacement chertification, an almost complete spectrum of time relationship occurs between silicification and carbonate diagenetic environments, as well as the various phases of calcium carbonate cements, a situation stressing the versatility of silica behavior.

Late Diagenetic Cherts in Limestones and Dolostones

Genetic Environment. A very fine line separates early and late diagenetic cherts as defined above because, in many instances of early cherts (Meyers, 1977; Hesse, 1987; Maliva and Siever, 1989a), a certain amount of burial or sufficient depth of burial for intergranular pressure solution and mechanical grain deformation of carbonate sands was said to be involved, separating successive phases of silicification; but the amount of this burial was not always reliably documented. Consequently, it is difficult to tell when carbonates really left the freshwater phreatic environment to penetrate

into a fully developed deep burial environment. Under deep-burial conditions, porosity–permeability conditions are low, circulation of fluids extremely slow, and replacive silica has different origins such as pressure solution of quartz grains, compaction of shales, smectite–illite conversion, and so on. Only detailed petrographic and geochemical studies on a regional basis, which are still missing, could characterize these new conditions and provide criteria to define real late-burial chertification in limestones and dolostones.

The designation of "late" also comprises chertification under a completely different setting, during a subsequent tectonic cycle. In such a case described by Banks (1970), chertification took place at the upper part of the freshwater phreatic environment within a folded, fractured, and karstified carbonate terrane undergoing seasonal evaporative concentration and precipitation of silica derived from underlying sandstones.

Petrography. As mentioned above, little petrographic data are available on deep-burial cherts because they may also have been included among early cherts formed under some burial.

The late cherts found in folded limestones described by Banks (1970) are black to dark brown in color and consist of cryptocrystalline to microcrystalline quartz masses, nodules, stringers, and layers. Nodular masses of minute, double-terminated euhedral crystals of quartz also occur in association with late cherts. They formed after complete induration of the carbonate host, as shown by the following features: layers of chert cut bedding planes at angles of 1° to 5°; these layers also bifurcate along strike and may cut bedding at angles up to 90°; other layers closely follow stylolitic zones and breccias, solution channels, and tectonic fractures; laminae of the host carbonate do not bend around these nodules, suggesting differential compaction, in fact many laminae abut against nodules or continue undisturbed through them. The most convincing evidence of late formation is the occurrence of *feeders,* which are minute joints and fractures lined with quartz or doubly terminated euhedral quartz crystals that connect the major cherty masses, expressing the microfracturation along which silica-bearing solutions circulated.

Diagenetic Cherts in Evaporites

Genetic Environment. Numerous studies on cherts developed in evaporites were undertaken following the recognition by Folk and Pittman (1971) that quartzine and lutecite are assumed to be characteristic of evaporite mineral replacement or at least, as mentioned above, of the presence of diagenetic solutions rich in sulfate and magnesium outside evaporitic conditions. Studies of examples ranging from Precambrian to Cretaceous confirmed that the hypersaline and particularly high pH nature of some evaporitic environments allow high concentrations of silica in solution to develop, with consequent replacement of evaporites, generally as nodules replacing anhydrite (Siedlecka, 1972, 1976; Chowns and Elkins, 1974; Tucker, 1976; Milliken, 1979; Geeslin and Chafetz, 1982).

Petrography. Milliken (1979) summarized the typical petrographic features of chert replacing anhydrite nodules in the Mississippian of southern Kentucky and northern Tennessee. The diagnostic spherulitic length-slow chalcedony or quartzine and small amounts of lutecite are accompanied by megaquartz with a very strong undulose extinction and euhedral megaquartz with a cubic appearance that is not a pseudomorph after cubical minerals, but corresponds to stubby crystals that consist of either a short prism terminated by two sets of rhomb faces that form a quasi-hexagonal pyramid or a prism terminated by three faces of a single rhomb. The near 90° angle between the faces of the single rhomb gives to these crystals the cubic appearance. These crystals occur adjacent to internal cavities of the nodules. Other quartz forms are microflamboyant quartz; randomly fibrous microcrystalline quartz, which is unusual because of its low birefringence and the presence of randomly distributed small

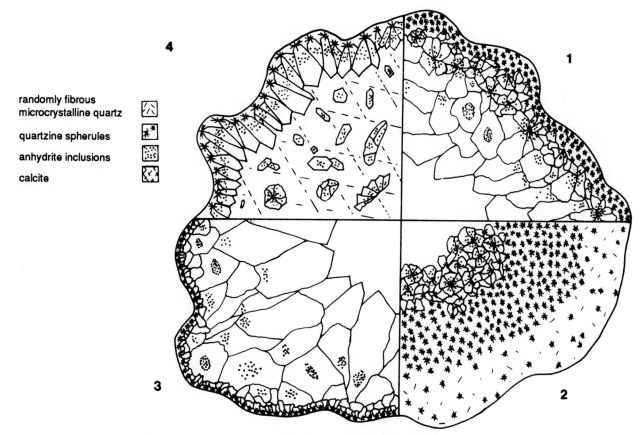

randomly fibrous
microcrystalline quartz

quartzine spherules

anhydrite inclusions

calcite

FIGURE 7.3 Examples of common variants of chert replacing evaporites shown as a composite nodule. **1.** Typical nodule. Thin rim of spherulitic quartzine followed inward by megaquartz of increasing crystal size with frequent anhydrite inclusions. **2.** Nodule with a large proportion of fibrous quartz. Thick rim of spherulitic quartzine followed inward by strongly undulose megaquartz in central part of nodule. Megaquartz may show crude radial or fanlike undulosity and may display a nucleus of spherulitic quartzine or microflamboyant quartz with abundant anhydrite inclusions. **3.** Nodule with very thin rim of spherulitic quartzine followed by a large proportion of megaquartz with anhydrite inclusions. **4.** Nodule margin of spherulitic quartzine followed inward by a rim of palisade megaquartz and large secondary calcite-replaced center, with isolated inclusions of quartz crystals indicating incomplete initial replacement of anhydrite by chert. From Milliken (1979). Reprinted by permission of the Society of Economic Paleontologists and Mineralogists.

laths with undulose, inclined extinction, which Milliken (1979) thought to be related to lutecite; and, finally, zebraic chalcedony as overlays on euhedral quartz in the center of a few nodules, where it appears as a cavity filling devoid of evaporite inclusions. Naturally, anhydrite inclusions in megaquartz, quartzine, and microflamboyant quartz are the most convincing proofs of replacement.

Milliken (1979) recognized a recurring sequence of quartz fabrics in the chert nodules she investigated from the edge toward the center, although no single nodule presents the complete sequence. A frequent simple succession consists of a rim of spherulitic quartzine followed toward the center by megaquartz. The composite section is as follows: (1) isolated quartzine spherulites in randomly fibrous microcrystalline quartz; (2) interlocking spherulites of quartzine, lutecite, or microflamboyant quartz; (3) spherulites of quartzine, lutecite, or microflamboyant quartz grading into megaquartz on spherulite edges, most of the megaquartz being of "cubic" type; (4) mosaic of megaquartz spherulites and undulose megaquartz; (5) simple quartz crystals (normal euhedral or "cubic"), spherulites of quartzine, microflamboyant quartz, or megaquartz isolated in secondary cavity filling of calcite or dolomite, which replaced anhydrite when silicification did not reach completion; (6) overlays of fibrous quartz (zebraic, chalcedonic, lutecitic, or quartzinic) resting on euhedral quartz terminations; and (7) breccias of the various types of evaporite-replacement quartz described above in a secondary cavity-filling calcite or dolomite. These breccias occurred when nodules partially collapsed after silicification, leaving at times the rim unbroken.

The relative importance of the various fabrics varies greatly from one nodule to another (**Fig. 7.3**). In most nodules, the rim begins with fabrics 2 or 3; if the internal cavity is large, fabric 4 may be missing and 5, 6, or 7 may follow 3. None of the nodules has both fabric 6 and a filling of secondary carbonate minerals. Finally, fabric 7 may follow 3 or fabric 7 alone may be present.

Diagenetic Evolution. The silicification history of these nodules was also investigated by Milliken (1979) by means of oxygen-isotope analyses from which temperatures and compositions of the replacing fluids were derived. The earliest microcrystalline quartz was interpreted as formed in fluids similar to seawater at a temperature in the range of 25° to 30°C, given the fact that the investigated area was located during the Pennsylvanian at 10° latitude S. Quartzine probably formed in waters of composition intermediate between seawater and modern meteoric water, at a slightly higher temperature (40°C) than did microcrystalline quartz. Megaquartz was formed probably in phreatic freshwater. Zebraic chalcedony, which follows in the diagenetic sequence, formed at a somewhat lower temperature in the same phreatic environment, perhaps near to the surface.

In summary, silicification of these anhydrite nodules

was a continuous process extending over a long period of time, ranging from marine phreatic, mixing marine–freshwater, and freshwater, and freshwater phreatic diagenetic environments, and from near-surface temperatures to shallow burial temperatures no higher than 40°C.

TYPICAL EXAMPLES

Plate 27.F. Ferruginous radiolarian chert. Microcrystalline chalcedony cement colored red by minute hematite flecks intimately associated with clay minerals. The uncompacted Radiolaria appear as circular to elliptical patches of clear microquartz mosaic, often with preservation of the lattice structure on their periphery; rare individuals are cement filled. Jurassic, Pennine nappes, Zermatt, Valais, Switzerland.

Plate 28.A. Banded spiculitic chert. Groundmass of dark pelletoidal arenaceous and argillaceous calcisiltite partially replaced by cryptocrystalline chalcedony, with relicts of pellets, ostracod shells, and monaxonic sponge spicules. The two lighter-colored intercalated layers correspond to zones of concentration of monaxonic sponge spicules, where silicification reached maximum development, and coarser crystallinity forming an irregular mosaic of fibrous chalcedony and microquartz enclosing the spicules consisting of single crystals of quartz with well-preserved central canals. Kinkaid Formation, Upper Mississippian, Bunscombe, Illinois, U.S.A.

Plate 28.B. Chertified calcilutite (chalk). The groundmass consists of bioturbated, pelletoidal, cryptocrystalline chalcedony with opaque pyrite concentrations. It shows randomly scattered relicts of monaxonic sponge spicules, echinoid spines, and thin pelecypods replaced by clear mosaics of microquartz. Chalk, Upper Cretaceous, Dieppe, France.

Plate 28.C. Rhomb-bearing chert. Originally, a calcisiltite with scattered, sand-size bioclasts. It appears as a groundmass of cryptocrystalline chalcedony with relicts of bioclasts of crinoids, bryozoans, sponge spicules, pelecypods, and ostracods replaced by mosaics of microquartz. Scattered perfect rhombs of zoned dolomite. Bird Spring Group, Upper Pennsylvanian, Arrow Canyon Range, Clark County, Nevada, U.S.A.

Plate 28.D. Chertified biocalcarenite. Grain-supported framework of bioclasts of crinoids, bryozoans, pelecypods, and brachiopods replaced by mosaics of homogeneous microquartz or by mosaics of megaquartz displaying rims of bladed crystals and center portions of coarser equant mosaic simulating a cavity-filling texture. Calcisiltite groundmass with detrital quartz grains is replaced by cryptocrystalline chalcedony showing opaque relicts of smaller bioclasts. Jeffersonville Limestone, Middle Devonian, Falls-of-the-Ohio, Indiana, U.S.A.

Plate 28.E. Chertified biocalcarenite. Grain-supported

framework of bioclasts of crinoids, bryozoans, brachiopods, ostracods, and thin pelecypods replaced either by dark cryptocrystalline chalcedony or irregular mosaics of microquartz and megaquartz. The interstitial sparite cement was also completely replaced by mosaics of megaquartz reproducing some of the original cavity-filling texture (bladed marginal and equant central), which has variably interfered with the silicification of the bioclasts. Burnt Bluff Group, Middle Silurian, Oakfield, Wisconsin, U.S.A.

Plate 28.F. Chertified oolitic calcarenite. Grain-supported framework of normal ooids with concentric and fibroradiated structure developed around pelletoidal, bioclastic, or detrital quartz nuclei. Most of the ooids were replaced by cryptocrystalline chalcedony containing abundant unreplaced original calcite; other ooids were entirely replaced by homogeneous cryptocrystalline chalcedony; still others had their detrital quartz nuclei enlarged secondarily into large subhedral crystals showing relics of original structure. Interstitial sparite cement was totally replaced by mosaics of megaquartz reproducing the original cavity-filling texture, that is, bladed marginal and equant central. State College Oolite, Upper Cambrian, State College, Pennsylvania, U.S.A.

REFERENCES

ARBEY, F., 1980. Les formes de la silice et l'identification des évaporites dans les formations silicifiées. *Bull. Centre Rech. Expl. Prod. Elf-Aquitaine,* 4, 309–365.

BALTUCK, M., 1983. Some sedimentary and diagenetic signatures in the formation of bedded radiolarite. In A. Iijima, J. R. Hein, and R. Siever (eds.), *Siliceous Deposits in the Pacific Region.* Developments in Sedimentology 36, Elsevier Scientific Publishing Co., New York, 299–315.

BANKS, N. G., 1970. Nature and origin of early and late cherts in the Leadville Limestone, Colorado. *Geol. Soc. Amer. Bull.,* 81, 3033–3048.

BARRETT, T. J., 1981. Chemistry and mineralogy of Jurassic bedded chert overlying ophiolites in the North Apennines, Italy. *Chem. Geol.* 34, 289–317.

———, 1982. Stratigraphy and sedimentology of Jurassic bedded chert overlying ophiolites in the North Apennines, Italy. *Sedimentology,* 29, 353–373.

BRAMLETTE, M. N., 1946. The Monterey Formation of California and the origin of its siliceous rocks. *U. S. Geol. Survey Prof. Paper 212,* 57 pp.

BROMLEY, R. G., and EKDALE, A. A., 1986. Flint and fabric in the European chalk. In G. de G. Sieveking and M. B. Hart (eds.), *The Scientific Study of Flint and Chert.* Cambridge University Press, New York, 71–82.

BROWN, G., CATT, J. A., HOLLYER, S. E., and OLLIER, C. D., 1969. Partial silicification of chalk fossils from the Chilterns. *Geol. Magazine,* 106, 583–586.

CALVERT, S. E., 1983. Sedimentary geochemistry of silicon. In S. R. Aston (ed.), *Silicon Geochemistry and Biogeochemistry.* Academic Press, New York, 143–186.

CAROZZI, A. V., and GERBER, M. S., 1978. Synsedimentary chert breccia: a Mississippian tempestite. *J. Sed. Petrology,* 48, 705–708.

CARVER, R. E., 1980. Petrology of Paleocene–Eocene and Miocene opaline sediments, southeastern Atlantic Coastal Plain. *J. Sed. Petrology,* 50, 569–582.

CAVAROC, V. V., Jr., and FERM, J. C., 1968. Siliceous spiculites as shoreline indicators in deltaic sequences. *Geol. Soc. Amer. Bull.,* 79, 263–271.

CAYEUX, L., 1929. *Les roches sédimentaires de France. Roches siliceuses.* Mém. Carte. Géol. Dét. France, Imprimerie Nationale, Paris, 774 pp.

——— 1941. *Causes anciennes et causes actuelles en géologie.* Masson et Cie, Paris, 81 pp. See also L. Cayeux, 1971. *Past and Present Causes in Geology,* translated and edited by A. V. Carozzi. Hafner Publishing Company, New York, 162 pp.

CHOQUETTE, P. W., 1955. A petrographic study of the "State College" siliceous oolites. *J. Geology,* 63, 337–347.

CHOWNS, T. M., and ELKINS, J. E., 1974. The origin of quartz geodes and cauliflower cherts through silicification of anhydrite nodules. *J. Sed. Petrology,* 44, 885–903.

CLAYTON, C. J., 1986. The chemical environment of flint formation in Upper Cretaceous chalks. In G. de G. Sieveking and M. B. Hart (eds.), *The Scientific Study of Flint and Chert.* Cambridge University Press, New York, 45–54.

CONIGLIO, M., 1987. Biogenic chert in the Cow Head Group (Cambro-Ordovician), western Newfoundland. *Sedimentology,* 34, 813–823.

CRERAR, D. A., NAMSON, J., CHYI, M. S., WILLIAMS, L., and FEIGENSON, M. D., 1982. Manganesiferous cherts of the Franciscan Assemblage. I. General geology, ancient and modern analogues, and implications for hydrothermal convection at oceanic spreading centers. *Economic Geol.,* 77, 519–540.

DAVIS, E. F., 1918. The radiolarian cherts of the Franciscan Group. *Univ. Calif. Publ. Geol.,* 11, 235–432.

DEFLANDRE, G., 1934. Sur les microfossiles d'origine planctonique conservés à l'état de matière organique dans les silex de la craie. *Acad. Sci. Paris, Comptes Rendus,* 119, 797–799.

DIETRICH, R. V., HOBBS, C. R. B., JR., and LOWRY, W. D., 1963. Dolomitization interrupted by silicification. *J. Sed. Petrology,* 33, 646–663.

EUGSTER, H. P., 1967. Hydrous sodium silicates from Lake Magadi, Kenya: precursors of bedded chert. *Science,* 157, 1177–1180.

———, 1969. Inorganic bedded cherts from the Magadi area, Kenya. *Contributions Mineralogy Petrology,* 22, 1–31.

———, and SURDAM, R. C., 1973. Depositional environment of the

Green River Formation: a preliminary report. *Geol. Soc. Amer. Bull.,* 84, 1115–1120.

FOLK, R. L., 1973. Evidence of peritidal deposition of Devonian Caballos Novaculite, Marathon basin, Texas. *Amer. Assoc. Petroleum Geologists Bull.,* 57, 702–725.

———, 1975. Third-party reply to Hatfield: Discussion of Jacka, A. D., 1974, Replacement of fossils by length-slow chalcedony and associated dolomitization. *J. Sed. Petrology,* 45, 952.

———, and MCBRIDE, E. F., 1976. The Caballos Novaculite revisited. Part I. Origin of Novaculite members. *J. Sed. Petrology,* 46, 659–669.

———, and ———, 1978. Radiolarites and their relation to subjacent "oceanic" crust in Liguria, Italy. *J. Sed. Petrology,* 48, 1069–1102.

———, and PITTMAN, J. S., 1971. Length-slow chalcedony: a new testament for vanished evaporites. *J. Sed. Petrology,* 41, 1045–1058.

FRONDEL, C., 1978. Characters of quartz fibers. *Amer. Mineralogist,* 63, 17–27.

GAO, G., and LAND, L. S., 1991. Nodular chert from the Arbuckle Group, Slick Hills, SW Oklahoma: a combined field, petrographic, and isotopic study. *Sedimentology,* 38, 857–870.

GARRISON, R. E., 1974. Radiolarian cherts, pelagic limestones, and igneous rocks in eugeosynclinal assemblages. In K. J. Hsü and H. C. Jenkyns (eds.), *Pelagic Sediments: On Land and under the Sea.* Internat. Assoc. Sedimentologists, Special Publ. 1. Blackwell Scientific Publications, London, 367–399.

———, DOUGLAS, R. G., PISCIOTTO, K. E., ISAACS, C. M., and INGLE, J. (eds.), 1981. *The Monterey Formation and Related Siliceous Rocks of California.* Soc. Econ. Paleontologists and Mineralogists, Pacific Section Special Publ., Los Angeles, 327 pp.

GEESLIN, J. H., and CHAFETZ, H. S., 1982. Ordovician Aleman ribbon cherts: an example of silicification prior to carbonate lithification. *J. Sed. Petrology,* 52, 1283–1293.

GOLDSTEIN, A., Jr., 1959. Cherts and novaculites of Ouachita facies. In H. A. Ireland (ed.), *Silica in Sediments.* Soc. Econ. Paleontologists and Mineralogists, Special Publ. 7, 135–149.

———, and HENDRICKS, T. A., 1953. Siliceous sediments of Ouachita facies in Oaklahoma. *Geol. Soc. Amer. Bull.,* 64, 421–442.

HARRIS, L. C., 1958. Syngenetic chert in the Middle Ordovician Hardy Creek Limestone of southwest Virginia. *J. Sed. Petrology,* 28, 205–208.

HATFIELD, C. B., 1975. Replacement of fossils by length-slow chalcedony and associated dolomitization: discussion. *J. Sed. Petrology,* 45, 951–952.

HAY, R. L., 1968. Chert and its sodium silicate precursors in sodium carbonate lakes of East Africa. *Contributions Mineralogy Petrology,* 17, 255–274.

HEIN, J. R., and KARL, S. M., 1983. Comparisons between open-ocean and continental margin chert sequences. In A. Iijima, J. R. Hein, and R. Siever (eds.), *Siliceous Deposits in the Pacific Region.* Developments in Sedimentology 36, Elsevier Scientific Publishing Co., New York, 25–43.

HERITSCH, F., and HERITSCH, H., 1943. Lydite und ähnliche Gesteine aus den karnischen Alpen. *Mitt. Alpenl. Geol. Ver. Oesterreich,* 34, 127–164.

HESSE, R., 1987. Selective and reversible carbonate–silica replacements in Lower Cretaceous carbonate-bearing turbidites of the Eastern Alps. *Sedimentology,* 34, 1055–1077.

———, 1990a. Origin of chert: diagenesis of biogenic siliceous sediments. In I. A. McIlreath and D. W. Morrow (eds.), *Diagenesis.* Geological Association of Canada, Geoscience Canada, Reprint Series 4, 227–251.

———, 1990b. Silica diagenesis: origin of inorganic and replacement cherts. In I. A. McIlreath and D. W. Morrow (eds.), *Diagenesis.* Geological Association of Canada, Geoscience Canada, Reprint Series 4, 253–275.

ISAACS, C. M., 1981. Porosity reduction during diagenesis of the Monterey Formation, Santa Barbara coastal area, California. In R. E. Garrison, R. G. Douglas, K. E. Pisciotto, C. M. Isaacs, and J. C. Ingle (eds.), *The Monterey Formation and Related Siliceous Rocks of California.* Soc. Econ. Paleontologists and Mineralogists, Pacific Section, Special Publ., Los Angeles, 257–271.

———, PISCIOTTO, K. E., and GARRISON, R. E., 1983. Facies and diagenesis of the Monterey Formation, California: a summary. In A. Iijima, J. R. Hein, and R. Siever (eds.), *Siliceous Sediments in the Pacific Region.* Developments in Sedimentology 36, Elsevier Scientific Publishing Co., New York, 247–282.

JACKA, A. D., 1974. Replacement of fossils by length-slow chalcedony and associated dolomitization. *J. Sed. Petrology,* 44, 421–427.

JENKYNS, H. C., and WINTERER, E. L., 1982. Palaeoceanography of Mesozoic ribbon radiolarites. *Earth Planet. Sci. Letters,* 60, 351–375.

JONES, D. L., and KNAUTH, L. P., 1979. Oxygen isotopic and petrographic evidence relevant to the origin of the Arkansas Novaculite. *J. Sed. Petrology,* 49, 581–597.

———, and MURCHEY, B., 1986. Geologic significance of Paleozoic and Mesozoic radiolarian chert. *Ann. Rev. Earth Planet. Sci.,* 14, 455–492.

JONES, J. B., and SEGNIT, E. R., 1971. The nature of opal. 1. Nomenclature and constituent phases. *J. Geol. Soc. Australia,* 18, 57–68.

KARL, S. M., 1984. Sedimentologic, diagenetic, and geochemical analysis of Upper Mesozoic ribbon cherts from the Franciscan assemblage at the Marin Headlands, California. In M. C. Blake, Jr. (ed.), *Franciscan Geology of Northern California.* Soc. Econ. Paleontologists and Mineralogists, Pacific Section, Los Angeles, 71–88.

KASTNER, M., 1980. Length-slow chalcedony: the end of the new testament. *EOS, Trans. Amer. Geophys. Union,* 61, 399.

KEENE, J. B., 1983. Chalcedonic quartz and occurrence of quartzine (length-slow chalcedony) in pelagic sediments. *Sedimentology,* 30, 449–454.

KELLER, W. D., STONE, C. G., and HOERSCH, A. L., 1985. Textures of Paleozoic chert and novaculite in the Ouachita Mountains of Arkansas and Oklahoma, and their geological significance. *Geol. Soc. Amer. Bull.,* 96, 1353–1363.

———, VIELE, G. W., and JOHNSON, C. H., 1977. Texture of Arkansas Novaculite indicates thermally-induced metamorphism. *J. Sed. Petrology,* 47, 834–843.

KENNEDY, W. J., and GARRISON, R. E., 1975. Morphology and genesis of nodular chalks and hardgrounds in the Upper Cretaceous of southern England. *Sedimentology, 22*, 311–386.

———, and JUIGNET, P., 1974. Carbonate banks and slump beds in the Upper Cretaceous (Upper Turonian–Santonian) of Haute-Normandie, France. *Sedimentology, 21*, 1–42.

KNAUTH, L. P., 1979. A model for the origin of chert in limestone. *Geology, 7*, 274–277.

LANE, G., 1981. A nearshore sponge spicule mat from the Pennsylvanian of west-central Indiana. *J. Sed. Petrology, 51*, 197–202.

LOHMANN, K. C., and MEYERS, W. J., 1977. Microdolomite inclusions in cloudy prismatic calcites: a proposed criterion for former high magnesium calcites. *J. Sed. Petrology, 47*, 1078–1088.

MALIVA, R. G., and SIEVER, R., 1988. Pre-Cenozoic nodular cherts: evidence for opal-CT precursors and direct quartz replacement. *Amer. J. Science, 288*, 798–809.

———, and ———, 1989a. Chertification histories of some Late Mesozoic and Middle Palaeozoic platform carbonates. *Sedimentology, 36*, 907–926.

———, and ———, 1989b. Nodular chert formation in carbonate rocks. *J. Geol., 97*, 421–433.

———, KNOLL, A. H., and SIEVER, R., 1989. Secular change in chert distribution: a reflection of evolving biological participation in the silica cycle. *Palaios, 4*, 519–532.

MCBRIDE, E. F., 1988. Silicification of carbonate pebbles in a fluvial conglomerate by groundwater. *J. Sed. Petrology, 58*, 862–867.

———, and FOLK, R. L., 1977. The Caballos Novaculite revisited. Part II: chert and shale members and synthesis. *J. Sed. Petrology, 47*, 1261–1286.

———, and ———, 1979. Features and origin of Italian Jurassic radiolarites deposited on continental crust. *J. Sed. Petrology, 49*, 837–868.

———, and THOMSON, A., 1970. The Caballos Novaculite, Marathon region, Texas. *Geol. Soc. Amer. Special Paper 122,* 129 pp.

MEYERS, W. J., 1977. Chertification in the Mississippian Lake Valley Formation, Sacramento Mountains, New Mexico. *Sedimentology, 24*, 75–105.

———, and JAMES, A. T., 1978. Stable isotopes of cherts and carbonate cements in the Lake Valley Formation (Mississippian), Sacramento Mts., New Mexico. *Sedimentology, 25*, 105–124.

MILLIKEN, K. L., 1979. The silicified evaporite syndrome. Two aspects of silicification history of former evaporite nodules from southern Kentucky and northern Tennessee. *J. Sed. Petrology, 49*, 245–256.

MIZUTANI, S., and SHIBATA, K., 1983. Diagenesis of Jurassic siliceous shale in central Japan. In A. Iijima, J. R. Hein, and R. Siever (eds.), *Siliceous Deposits in the Pacific Region.* Developments in Sedimentology 36, Elsevier Scientific Publishing Co., New York, 283–297.

MURCHEY, B., 1984. Biostratigraphy and lithostratigraphy of chert in the Franciscan complex, Marin Headlands, California. In M. C. Blake, Jr. (ed.), *Franciscan Geology of Northern California.* Soc. Econ. Paleontologists and Mineralogists, Pacific Section, Los Angeles, 51–70.

MURRAY, R. W., JONES, D. L., and BUCHHOLZ TEN BRINK, M. R.,

1992. Diagenetic formation of bedded chert: evidence from chemistry of the chert–shale couplet. *Geology, 20*, 271–274.

NAMY, J. N., 1974. Early diagenetic cherts in the Marble Falls Group (Pennsylvanian) of central Texas. *J. Sed. Petrology, 44*, 1262–1268.

NEWELL, N. D., 1957. Paleoecology of Permian reefs in the Guadalupe Mountains area. In H. S. Ladd (ed.), *Treatise on Marine Ecology and Paleoecology.* Geol. Soc. Amer. Memoir 67, vol. 2, 407–436.

NISBET, E. G., and PRICE, I., 1974. Siliceous turbidites: bedded cherts as redeposited ocean ridge-derived sediments. In K. J. Hsü and H. C. Jenkyns (eds.), *Pelagic Sediments: On Land and under the Sea.* Internat. Assoc. Sedimentologists Special Publ. 1, Blackwell Scientific Publications, London, 351–366.

O'NEIL, J. R., and HAY, R. L., 1973. $^{18}O/^{16}O$ ratios in cherts associated with the saline lake deposits of East Africa. *Earth Planet. Sci. Letters, 19*, 257–266.

PARIS, I., STANISTREET, I. G., and HUGHES, M. J., 1985. Cherts of Barberton greenstone belt interpreted as products of submarine exhalative activity. *J. Geol., 93*, 111–129.

PARNELL, J., 1986. Devonian Magadi-type cherts in the Orcadian Basin, Scotland. *J. Sed. Petrology, 56*, 495–500.

PETERSON, M. N. A., and VON DER BORCH, C. C., 1965. Chert: modern inorganic deposition in a carbonate-precipitating locality. *Science, 149*, 1501–1503.

PISCIOTTO, K. E., 1981a. Diagenetic trends in the siliceous facies of the Monterey Shale in the Santa Maria region, California. *Sedimentology, 28*, 547–571.

———, 1981b. Review of secondary carbonates in the Monterey Formation, California. In R. E. Garrison, R. G. Douglas, K. E. Pisciotto, C. M. Isaacs, and J. C. Ingle (eds.), *The Monterey Formation and Related Siliceous Rocks of California.* Soc. Econ. Paleontologists and Mineralogists, Pacific Section, Special Publ., Los Angeles, 273–283.

———, and GARRISON, R. E., 1981. Lithofacies and depositional environments of the Monterey Formation, California. In R. E. Garrison, R. G. Douglas, K. E. Pisciotto, C. M. Isaacs, and J. C. Ingle (eds.), *The Monterey Formation and Related Siliceous Rocks of California.* Soc. Econ. Paleontologists and Mineralogists, Pacific Section, Special Publ., Los Angeles, 97–122.

RAD, U. VON, and RÖSCH, H., 1974. Petrography and diagenesis of deep-sea cherts from central Atlantic. In K. J. Hsü and H. C. Jenkyns, *Pelagic Sediments: On Land and under the Sea.* Internat. Assoc. Sedimentologists, Special Publ. 1, Blackwell Scientific Publications, London, 327–347.

REDWINE, L. E., 1981. Hypothesis combining dilation, natural hydraulic fracturing, and dolomitization to explain petroleum reservoirs in Monterey Shale, Santa Maria area, California. In R. E. Garrison, R. G. Douglas, K. E. Pisciotto, C. M. Isaacs, and J. C. Ingle (eds.), *The Monterey Formation and Related Siliceous Rocks of California.* Soc. Econ. Paleontologists and Mineralogists, Pacific Section, Special Publ., Los Angeles, 221–248.

RENAUT, R. W., and OWEN, R. B., 1988. Opaline cherts associated with sublacustrine hydrothermal springs at Lake Bogoria, Kenya Rift valley. *Geology, 16*, 699–702.

RICHTER, D. E., 1972. Authigenic quartz preserving skeletal material. *Sedimentology,* 19, 211–218.

ROBERTSON, A. H. F., 1977. The origin and diagenesis of cherts from Cyprus. *Sedimentology,* 24, 11–30.

ROEHL, P. O., 1981. Dilation brecciation—A proposed mechanism of fracturing, petroleum expulsion, and dolomitization in the Monterey Formation, California. In R. E. Garrison, R. G. Douglas, K. E. Pisciotto, C. M. Isaacs, and J. C. Ingle (eds.), *The Monterey Formation and Related Siliceous Rocks of California.* Soc. Econ. Paleontologists and Mineralogists, Pacific Section, Special Publ., Los Angeles, 285–315.

SANO, H., 1983. Bedded cherts associated with greenstones in the Sawadani and Shimantogawa Groups, southwest Japan. In A. Iijima, J. R. Hein, and R. Siever (eds.), *Siliceous Rocks in the Pacific Region.* Developments in Sedimentology 36, Elsevier Scientific Publishing Co., New York, 427–440.

SCHUBEL, K. A., and SIMONSON, B. M., 1990. Petrography and diagenesis of cherts from Lake Magadi, Kenya. *J. Sed. Petrology,* 60, 761–776.

SHEPPARD, R. A., and GUDE, A. J., 1986. Magadi-type chert—A distinctive diagenetic variety from lacustrine deposits. In F. A. Mumpton (ed.), *Studies in Diagenesis. U.S. Geol. Survey Bulletin 1578,* 335–345.

SIEDLECKA, A., 1972. Length-slow chalcedony and relicts of sulphates—evidences of evaporitic environments in the Upper Carboniferous and Permian beds of Bear Island, Svalbard. *J. Sed. Petrology,* 42, 812–816.

———, 1976. Silicified Precambrian evaporite nodules from northern Norway: a preliminary report. *Sed. Geol.,* 16, 161–175.

SIEVEKING, G. de G., and HART, M. B. (eds.), 1986. The scientific study of flint and chert. Cambridge University Press, New York, 190 pp.

SIMONSON, B. M., 1987. Early silica cementation and subsequent diagenesis in arenites from four Proterozoic iron formations of North America. *J. Sed. Petrology,* 57, 494–511.

STEINBERG, M., BONNOT-COURTOIS, C., and TLIG, S., 1983. Geochemical contribution to the understanding of bedded chert. In A. Iijima, J. R. Hein, and R. Siever (eds.), *Siliceous Deposits in the Pacific Region.* Developments in Sedimentology 36, Elsevier Scientific Publishing Co., New York, 193–210.

———, DESPRAIRIES, A., FOGELSANG, J.-F. MARTIN, A., CARON, D., and BLANCHET, R., 1977a. Radiolarites et sédiments hypersiliceux océaniques: une comparaison. *Sedimentology,* 24, 547–563.

———, FOGELSANG, J.-F., COURTOIS, C., MPODOZIS, C., DESPRAIRIES, A., MARTIN, A., CARON, D., and BLANCHET, R., 1977b. Détermination de l'origine des feldspaths et des phyllites présents dans des radiolarites mésogéennes et des sédiments hypersiliceux océaniques par l'analyse des terres rares. *Bull. Soc. Géol. France,* Série 7, XIX, 735–740.

STEINITZ, G., 1970. Chert "dike" in Senonian chert beds, southern Negev, Israel. *J. Sed. Petrology,* 40, 1241–1254.

SURDAM, R. C, and STANLEY, K. O., 1981. Diagenesis and migration of hydrocarbons in the Monterey Formation, Pismo syncline, California. In R. E. Garrison, R. G. Douglas, K. E. Pisciotto, C. M. Isaacs, and J. C. Ingle (eds.), *The Monterey Formation and Related Siliceous Rocks in California.* Soc. Econ. Palentologists and Mineralogists, Pacific Section, Special Publ., Los Angeles, 317–327.

———, EUGSTER, H. P., and MARINER, R. H., 1972. Magadi-type chert in Jurassic and Eocene to Pleistocene rocks, Wyoming. *Geol. Soc. Amer. Bull.,* 83, 2261–2266.

SWETT, K., 1965. Dolomitization, silicification, and calcitization patterns in Cambro-Ordovician oolites from Northwest Scotland. *J. Sed. Petrology,* 35, 928–938.

THIRY, M., and MILLOT, G., 1987. Mineralogical forms of silica and their sequence of formation in silcretes. *J. Sed. Petrology,* 57, 343–352.

———, BERTRAND AYRAULT, M., and GRISONI, J.-C., 1988. Ground-water silicification and leaching in sands: example of the Fontainebleau Sand (Oligocene) in the Paris Basin. *Geol. Soc. Amer. Bull.,* 100, 1283–1290.

THURSTON, D., 1972. Studies of bedded cherts. *Contributions Mineralogy Petrology,* 36, 329–334.

TUCKER, M. E., 1976. Replaced evaporites from the Late Precambrian of Finnmark, Arctic Norway. *Sed. Geol.,* 16, 193–204.

WALKER, T. R., 1962. Reversible nature of chert–carbonate replacement in sedimentary rocks. *Geol. Soc. Amer. Bull.,* 73, 237–242.

WEAVER, C. E., and BECK, K. C., 1977. Miocene of the S.E. United States: a model of chemical precipitation in a peri-marine environment. *Sed. Geol.,* 17, 1–234.

WEAVER, F. M., and WISE, S. W., JR., 1972. Ultramorphology of deep sea cristobalitic chert. *Nature,* 237, 56–57.

———, and ———, 1974. Opaline sediments of the southeastern coastal plain and horizon A: biogenic origin. *Science,* 184, 899–901.

WHEELER, W. H., and TEXTORIS, D. A., 1978. Triassic limestone and chert of playa origin in North Carolina. *J. Sed. Petrology,* 48, 765–776.

WHITE, A. H., and YOUNGS, B. C., 1980. Cambrian alkali playa-lacustrine sequence in the northeastern Officer Basin, South Australia. *J. Sed. Petrology,* 50, 1279–1286.

WILSON, R. C. L., 1966. Silica diagenesis in Upper Jurassic limestones of southern England. *J. Sed. Petrology,* 36, 1036–1049.

WISE, S. W., JR., and WEAVER, F. M., 1973. Origin of cristobalite-rich Tertiary sediments in the Atlantic and Gulf Coastal Plain. *Gulf Coast Assoc. Geol. Soc. Trans.,* 23, 305–323.

———, and ———, 1974. Chertification of oceanic sediments. In K. J. Hsü and H. C. Jenkyns (eds.), *Pelagic Sediments: On Land and under the Sea.* Internat. Assoc. Sedimentologists, Special Publ. 1. Blackwell Scientific Publications, London, 301–326.

CHAPTER 8

PHOSPHORITES

INTRODUCTION

Sedimentary phosphatic rocks, or phosphorites, occur throughout the geological column from Precambrian to Recent. The term phosphorite is theoretically reserved for rocks containing 50% apatite (18% P_2O_5), but no general agreement exists among authors. The main constituents of phosphorites are varieties of apatite, in particular carbonate fluorapatite called francolite with a simplified formula $Ca_{10}(PO_4, CO_3)_6F_{2-3}$, which has more than 1% fluorine by weight. The partial replacement of F by OH can vary, but when the proportion of F by weight is less than 1%, the corresponding variety is a carbonate hydroxyl apatite called dahllite whose simplified formula is $Ca_{10}(PO_4, CO_3OH)_6(OH, F)_2$. The calcium ion may be substituted by sodium, strontium, and magnesium and by a variety of trace elements typical of marine planktonic origin, such as uranium, vanadium, thorium, and molybdenum. Both francolite and dahllite are best identified by X-ray diffraction techniques and chemical analysis because they are, under the microscope, fibrous, weakly birefringent, and almost identical optically.

Petrographically, the term *collophane* is routinely and informally used for descriptive purposes when sedimentary apatites appear amorphous to microcrystalline, and when their particular chemical composition has not been established.

As for the origin of cherts, oceanographic studies have brought fundamental data to the understanding of Subrecent to Recent marine phosphorites. Present-day marine phosphorite formation is very restricted and almost an exception to uniformitarian principles when compared with older occurrences. Many phosphorites found today on continental shelves are reworked residual deposits formed in pre-Holocene times, and in some cases the original formation dates back to the Miocene, which is the last peak phase of the periodical development of phosphorites during geological time (Bentor, 1980a). Baturin (1982) described in great detail the origin, composition, and distribution of the phosphorites encountered today on the sea floor throughout the world. The problem of their formation is highly complex from a geochemical and sedimentological viewpoint and, consequently, the debate continues. Bentor (1980b) outlined the numerous unsolved problems. An important one pertains to dissolved oceanic phosphorus and the solubility of carbonate fluorapatite. According to Bentor (1980b), phosphorus is at present a rare trace element in oceanic waters, which contain an average of 70 ppb P. Near-surface waters are everywhere strongly depleted by biological uptake, with values down to a few parts per billion, and even less to the extent that biological reproduction can be frequently inhibited. In the zone of organic regeneration, that is at 200 to 400 m depths, the concentration of dissolved phosphorus increases sharply to values of 50 to 100 ppb. From there downward to the ocean floor, phosphorus content remains constant or decreases slightly.

Regional differences in the phosphorus content of deep-ocean waters are strong, for instance, 60 to 100 ppb in the Pacific and Indian Ocean, but only 35 to 85 ppb in the Atlantic. Semienclosed oceanic basins have very low phosphorus contents: deep waters of the Red Sea contain only 15 ppb and in the Mediterranean, 6 ppb. The reason for this situation is that both enclosed seas are replenished by surface waters only, inhibiting the influx of phosphorus-richer deeper waters from connecting oceans.

The dominant phosphate species in the ocean is HPO_4^{2-}, which, in the critical range of pH 7 to 8, accounts for 85% to 90% of all dissolved phosphorus and 44% of it is complexed with organic matter (Bentor, 1980b). But how do these levels of phosphorus concentration in ocean water relate to solubility of carbonate fluorapatite? This critical question remains as yet unsolved. It is not known if seawater is or is not saturated with respect to carbonate fluorapatite. There is, however, some evidence that seawater is probably undersaturated, as shown by phosphatic skeletal debris found corroded on the sea floor.

The situation is entirely different for interstitial waters in sediments rich in organic matter. As described below, in Recent anoxic sediments of the Peru shelf, 1,400 ppb P were found (Burnett, 1977) and 2,500 ppb P in similar sediments off Namibia by Baturin (1982). These values, which are up to 100 times higher than those of the open ocean, certainly correspond to oversaturation with respect to apatite. This indicates that interstitial waters within reducing sediments are the environment of formation of phosphorites today and most probably in the past as well (Bentor, 1980b).

For Subrecent to Recent marine phosphorites, it is possible to state a certain number of associated genetic factors (Slansky, 1980, 1986; Baturin, 1982). They are oceanic upwelling, which is latitudinally controlled; slow rate of sedimentation; direct precipitation of apatite within interstitial waters; replacing of host sediments by apatite; association of both precipitation and replacement processes; penecontemporaneous or early phases of intraformational reworking and winnowing processes of variable importance and duration; and, finally, a general context of transgressive–regressive cycles or at least sea-level ocillations. Following uniformitarian principles, with the reservation that present-day marine phosphatization is volumetrically negligible and upwelling temporarily not so active as during other past periods, it is possible to recognize most of the above-mentioned genetic factors among ancient phosphorites (Sheldon, 1981). In fact, ancient phosphorites do not appear to be a geochemically anomalous phenomenon when compared to modern oceanic conditions such as the Peru margin (Filippelli and Delaney, 1992).

The most important factor of phosphatization is oceanic upwelling of cold, oxygen-poor but nutrient-rich waters. These waters contain today 50 to 100 ppb P and rise along tropical to subtropical arid coasts within 50° of the equator, characterized by prevailing winds blowing from land to ocean. For ancient phosphorites, these latitudinal and climatic conditions were essentially the same and can be rec-

ognized starting from Early Proterozoic times when continents drifted in such situations as a function of plate tectonics (Cook and McElhinny, 1979). At the shelf edge, these upwelling waters undergo mixing with oxygenated waters in the photic zone, causing a high phytoplanktonic productivity in surface waters. This high productivity requires so much oxygen that the water layers immediately below the photic zone become rapidly oxygen deficient and constitute an anoxic environment that preserves a great amount of the dead phytoplankton falling on the sea floor under the upwelling site. Any nutrient still associated with organic matter is brought back to the surface instead of being dispersed into the open ocean; therefore, the system becomes a self-perpetuating trap of organic matter (Demaison and Moore, 1980). The conditions of aridity mentioned above, corresponding to a very slow rate of sedimentation, lead to a deposition of organic-rich dark, fine-grained siliciclastic and argillaceous sediments, together with siliceous muds formed by the tests of Radiolaria and diatoms. These sediments deposited below the oxygen-deficient zone, the precursors of deep-water black shales rich in hydrocarbons, also contain pyrite due to the reduction of sulfates by anaerobic bacteria and, naturally, to the high concentrations of phosphorus in their interstitial waters reaching conditions of oversaturation with respect to apatite (Demaison and Moore, 1980). As pointed out by Slansky (1980, 1986), there is no present or past evidence in favor of direct precipitation of apatite from seawater as previously assumed by many authors.

In spite of the above-mentioned generalized statements, it is obvious that numerous geochemical and depositional aspects of the formation of marine phosphorites remain unclear. Among the geochemical ones is the effect of Mg^{2+} ions, which are common inhibitors of direct phosphate precipitation in the interstitial waters of anoxic sediments, probably because magnesium competes with calcium for sites in the apatite lattice. A decrease of the effect of magnesium upon apatite precipitation could be obtained either by absorption of magnesium in associated clay minerals, by incipient submarine dolomitization, or during diagenetic replacement by apatite. Controversies on depositional conditions pertain to the causes of the well-established periodicity of major phosphate accumulations, which reach peaks in the Cambrian, Ordovician, Permian, Cretaceous–Early Cenozoic, and Miocene, apparently in response to hypothetical and complex global oceanic processes (Sheldon, 1980). Another subject of discussion concerns the suggested depositional models of marine phosphorites capable of explaining their baffling and simultaneous lateral gradations or close associations not only with cherts, organic-rich shales, and carbonates, but also with evaporites and even red beds (**Fig. 8.1**). This situation is shown, for instance, by the Permian Phosphoria Formation of the northern Rocky Mountains (McKelvey et al., 1959; Sheldon, 1963, 1987). A crude correlation appears also to exist between marine phosphorites and major iron deposits. In that respect, Hite (1978) rejected

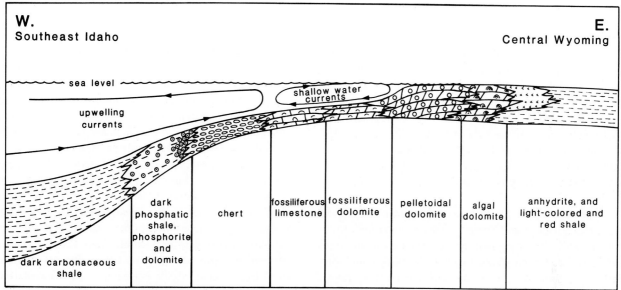

FIGURE 8.1 Ideal depositional model of Phosphoria Formation (Permian), northern Rocky Mountains. From McKelvey et al. (1959).

the concept of oceanic upwelling and proposed that apatite, silica, dolomite, and iron-rich minerals could be precipitated by the mixing of warm brines from evaporite basins on continents with cold oxygenated seawater. This mixing is assumed to have taken place in and at mouths of marine accessways to Mediterranean-type evaporite basins where high-density brines reflux to the ocean. In this hypothesis, continents supply phosphorus and other trace elements found in phosphorites and iron ores as the result of reactions between corrosive brines and detrital minerals washed into the evaporite basins. One major objection to this hypothesis is the fact that many marine phosphorites are not related to on-land evaporitic basins.

The economical importance of phosphorites as basic fertilizers required for high productivity of the world's agricultural lands has prompted the publication of important monographs in which regional stratigraphy, field observations, megascopic descriptions, depositional models, and economic considerations largely outweighed the sedimentary petrographic aspect, which therefore, with a few rare exceptions, remains very sketchy (Cook and Shergold, 1986a; Notholt et al., 1989; Burnett and Riggs, 1990; Notholt and Jarvis, 1990).

CLASSIFICATION

There is no general agreement on the petrographic designation of the particles forming a large proportion of marine phosphorites. Definitions suggested here are based on their size and shape.

The smallest particles, called phosphatic spherules, are defined as tiny granules (50 to 100 μ), spherical or slightly

ellipsoidal in shape, and with no internal structure. They are considered to be organic in origin, possibly bacteria or extremely small fecal pellets, although their great regularity in shape does not fit this latter possible origin well.

Phosphatic pellets are elongate to ellipsoidal bodies (length to width ratio = 2:1), generally ranging in size from 0.05 to 0.5 mm. Their origin is extremely variable. They could be either fecal or lithic pellets or products of direct precipitation or could correspond to complete phosphatization of microfossils. This variability of origin requires that they be interpreted case by case as a function of the general petrographic context in which they occur.

Superficial, normal, and composite phosphatic ooids, as well as pisoids, occur in marine phosphorites. Ooids generally range in size from 0.1 mm to 2 mm, and to greater than 2 mm for pisoids. Their nuclei consist of pellets, small intraclasts, and in some cases grains of detrital minerals and bioclasts such as bone fragments.

Ellipsoidal bodies of collophane ranging in size from a few millimeters to 1 cm with a characteristic internal fluidal texture are generally considered coprolites of unknown organisms. The term coprolite, as pointed out by Slansky (1980, 1986), is to be used carefully, because they occur in Proterozoic to Holocene phosphorites, which means that the potential variety of unknown organisms that produced them is great.

Intraclasts represent intraformationally reworked fragments of earlier semiconsolidated to consolidated phosphorites. Their shape is extremely variable depending upon their original composition and amount of transport. They range in size from 0.1 mm to greater than 3 mm.

Phosphatic nodules are spheroidal to ellipsoidal bodies of an average size of several centimeters that are assumed to have been formed essentially in situ as concretionary masses. They may consist of pure collophane or contain a variety of smaller phosphatic constituents and bioclasts.

Other constituents of marine phosphorites are various types of bioclasts such as gastropods, corals, bryozoans, sponge spicules, foraminifers, which can be partially phosphatized, phosphatic brachiopods, and fragments of fish bones and scales. Finally, Cayeux (1941) recognized the ubiquitous presence of bacteria in phosphorites of all ages and their probable great genetic importance. Coccoid cyanobacteria were recently demonstrated to be important contributors by Soudry and Champetier (1983), Krajewski (1984), and Soudry (1987).

The petrographic classification of marine phosphorites used here (Black and Carozzi, 1988) is modified from Slansky (1980, 1986) and derived from the classification of limestones (see Chapter 5 and **Fig. 5.4**). It utilizes two basic parameters: grain size of the predominant (> 10%) phosphatic figured constituents, and nature of matrix and/or cement. Phosphalutite (0.01 to 0.03 mm), phosphasiltite (0.03 to 0.064 mm), phospharenite (0.064 to 2 mm), and phospharudite (> 2mm) are further qualified by prefixes such as bio-, pel-, oo-, intra-, and extra- to describe bioclasts, pellets, ooids, intraclasts, and extraclasts.

Phosphalutites and phosphasiltites may contain up to 10% sand-size components of organic or inorganic origin. Phospharenites containing 10% to 30% sand-size constituents in a matrix of phosphalutite or phosphasiltite are called matrix supported; those with more than 30% sand-size constituents and possessing an interstitial matrix as above and/or a cement of precipitated collophane (> 0.01 mm), defined as *microsphatite* by Slansky (**Plate 29.A**), are called grain supported. Nonphosphatic constituents that form part of the matrix (quartz, glaucony, clay minerals) and/or part of the cement (carbonates, silica) are designated according to their mineralogy, grain or crystal size, and relative abundance, such as grain-supported oophospharenite with 10% quartz–glaucony matrix, or grain-supported pelphospharenite with 20% sparite cement. Phospharudites contain more than 30% granule- (2 to 4 mm) or pebble- (4 to 64 mm) size constituents with interstitial material of all previous phosphorite types and/or phosphatic and nonphosphatic cements.

Cook and Shergold (1986b) introduced for phosphorites a modified version of Dunham's classification of limestones (see Chapter 5 and **Fig. 5.3**). It includes a first group of nonpelletoidal phosphorites divided into mudstone phosphorites and wackestone phosphorites, a second group of pelletoidal phosphorites divided into packstone phosphorites and grainstone phosphorites, a third group of boundstone phosphorites, and, finally, phoscrete as the equivalent of crystalline carbonate. This classification suffers the same poor terminology, ambiguities, and imprecisions previously pointed out for the carbonate classification from which it was derived.

SUBRECENT TO RECENT MARINE PHOSPHORITES

Authigenic Precipitated Phosphorites

Burnett (1974, 1977), and Burnett et al. (1980) investigated the phosphorite deposits from offshore Peru and Chile, which occur mainly as small pellets and nodules associated with laminated, organic-rich diatomaceous oozes, which express the extremely high organic productivity of the surface waters off the west coast of South America. These deposits are of critical importance for the understanding of the genesis of marine phosphorites because they are one of only two areas, the other being the South West Africa shelf near Walvis Bay, where phosphorites have been proved by radiometric dating to be forming from 150,000 years ago through episodic phases until today (Burnett and Veeh, 1977; Burnett et al., 1980). Petrographically (Burnett, 1977), the main mineral component of the irregularly shaped nodules is a cryptocrystalline, optically anisotropic variety of apatite, often called collophane. Microprobe and X-ray diffraction observations confirmed that this material is a fluorine-rich variety of apatite, probably francolite. Other mineral phases identified optically include quartz, feldspars (plagioclase and orthoclase), glaucony, dolomite, opaques, and calcite occurring as foraminifer tests. Skeletal remains of opaline silica (diatoms) and phosphatic fish bones were also observed. Phosphatic pellets scattered within nodules are structureless, with often a nucleus of quartz, feldspar, or glaucony grains. SEM examinations (Burnett, 1977) of freshly fractured surfaces of nodules and small pellets suggest that apatite formed authigenically as a direct chemical precipitate rather than by replacement. Indeed, apatite crystals are variable in size, but generally on the order of 1μ in length. Most crystals are prismatic, many are double terminated, and their basal sections are hexagonal. They occur as individual rods or associated in rosettes on a substrate that is most commonly the opaline silica of diatom frustules, although at times it is a feldspar crystal or fish-bone apatite. It is not known whether this association implies that silica is a nucleation surface for apatite growth or whether it is simply a function of the relative abundance of these two major constituents of the sediment (Burnett, 1977). It is possible that small undetected fragments of calcite may have served as "seeds" for epitaxial growth of apatite. Similar crystals of precipitated apatite were also reported from the same areas by Baturin (1982).

After inorganic precipitation of apatite in the interstitial waters of the sediment, some process of concentration of apatite into indurated pellets, ooid-like concretions, and nodules is necessary. According to Burnett (1977), it was probably brought about by winnowing and reworking processes, possibly in response to changes of the sedimentary environment caused by tidal or storm waves and currents, but mainly by eustatic sea-level fluctuations, or tectonism. Indeed, removal by currents of the fine, unphosphatized particles of the sediment from the sea floor leaves a natural concentrate of pellets and nodules at various stages of induration and phosphatization, which can be submitted to further precipitation processes.

Loughman (1984) described the Aramachay Formation (Lower Jurassic) onland Peru, which is a sequence of 300 m of organic-rich mudstones grading upward into 100 m of phosphatic siltstones, cherts, and arenites. He interpreted this facies succession of phosphatic authigenesis as representing the first ever described fossilized upper margin of an oceanic oxygen-minimum zone comparable with that observed on the present-day Peruvian shelf.

Diagenetic Replacive Phosphorites

In similar deposits offshore Peru, Manheim et al. (1975) found clear petrographic evidence of replacement of carbonate tests of Holocene benthonic foraminifers in the interstitial waters of organic-rich sediments. SEM studies show a steplike and gradual phosphatization of entire foraminiferal tests, with the ultimate result being phosphatic pellets having retained the general shape of the tests, but with total destruction of the fine skeletal structure. At very high magnification, francolite forms minute spherulitic crystals. Manheim et al. (1975) assumed that benthonic foraminifers proliferated intermittently due to variations of the upwelling and during decrease or disappearance of anoxic conditions, to be phosphatized when upwelling was active, accumulation of planktonic matter high, and anoxic conditions well established.

Cenozoic limestones almost entirely phosphatized by replacive francolite were described along the South African continental margin by Parker and Siesser (1972). These phosphatized, relatively deep water limestones are grain-supported to mud-supported calcarenites consisting typically of entire or fragmented tests of planktonic foraminifers, which predominate over benthonic ones. Other constituents are echinoid plates and spines, bryozoans, and pelecypod bioclasts. All this skeletal material is invariably calcium carbonate and not replaced by any phosphatic minerals as described in other localities (Dietz et al., 1942). This situation results probably from the larger surface area and microporosity of the carbonate mud compared to bioclasts, which makes it more susceptible to replacement by phosphate-rich interstitial solutions. The matrix of these calcarenites is an intimate association, indistinguishable under the microscope, of collophane, calcilutite, and sometimes goethite identified by X-ray diffraction. Silt-size angular grains of quartz and feldspars are scattered throughout the matrix.

Comparisons with unphosphatized Upper Middle Miocene limestones of similar texture and biogenic composition from the Agulhas Bank leave no doubt that these pelagic limestones had their carbonate mud replaced by francolite under submarine conditions by interstitial phosphate-rich solutions (Parker and Siesser, 1972).

In addition to phosphatized limestones, the South African continental margin displays glauconitic and glauco-conglomeratic phosphorites mostly developed along the eastern margin of the Agulhas Bank during Late Cenozoic and Early Pleistocene times (Parker, 1975). The glaucophosphorites consist of glaucony grains (20% to 60%), microfossil shells and bioclasts (1% to 30%), quartz grains (5% to 20%), and minor amounts of macrofossil shell debris (1% to 7%). These constituents are set in a matrix of collophane and calcilutite, forming a mud-supported to grain-supported texture. Glaucony grains vary in size from fine to coarse sand, and some are lobate. Pyrite occurs as inclusions in glaucony grains and as flecks in the matrix. Microfossils are thin-walled foraminifers associated with pelecypod bioclasts. All have remained calcitic in the same way as those previously described by Parker and Siesser (1972) in the phosphatized limestones of the same general area. Quartz is usually present as angular grains, ranging in size from silt to very fine sand. Collophane pellets also occur, ranging from completely homogeneous to faintly zoned due to concentration of inclusions toward the center. Other accessory constituents are zircon, feldspar, garnet, and tourmaline. The matrix is an intimate association of collophane, calcilutite, and mineral blebs of organic matter or iron oxides such as goethite. These inclusions give the matrix a turbid aspect in plane-polarized light. Dahllite rings may surround glaucony and quartz grains, while pores within the rock are often filled with secondary sparite cement.

The glaucoconglomeratic phosphorites (Parker, 1975) consist of variably phosphatized limestone pebbles forming up to 50% of the rock set in a matrix very similar to the above-described glaucophosphorites. The general texture is predominantly pebble supported, but matrix-supported varieties exist. The pebbles are equant to extremely irregular; they consist of foraminiferal debris and angular silt-size quartz grains set in a collophane–calcilutite matrix with finely divided goethite. They are identical to the above-described phosphatized pelagic limestones (Parker and Siesser, 1972). Some of the glaucoconglomeratic phosphorites are bedded into units of different composition separated by erosional discontinuities, with truncations, cavities, and borings filled at time by unphosphatized calcilutite. The intraformational nature of these extremely poorly sorted conglomeratic phosphorites is evident, but the nature and shape of the pebbles indicate neither high-energy nor long-distance transport by traction processes.

According to Parker (1975), the glaucophosphorites were formed after a transgression when the zone of biogenic deposition migrated from the shelf edge across the shelf itself in response to a rising sea level. Planktonic foraminifers and carbonate mud were deposited over coarser glauconitic

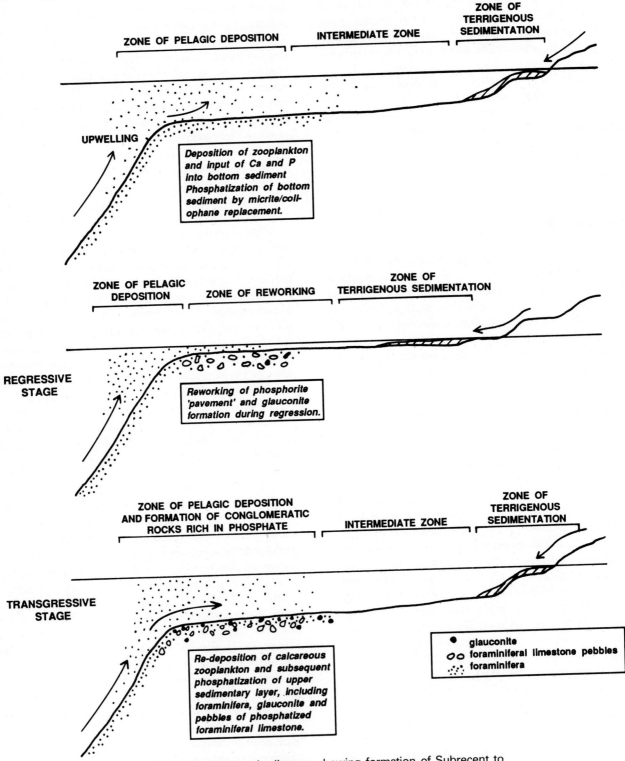

FIGURE 8.2 Schematic diagram showing formation of Subrecent to Recent conglomeratic and nonconglomeratic phosphorites during transgressive–regressive cycles along western margin of South Africa. From Birch (1979). Reprinted by permission of the Society of Economic Paleontologists and Mineralogists.

and quartzose mid-shelf sediments, mixed together to a high degree by bioturbation and subsequently phosphatized by replacement within their interstitial waters. The glaucoconglomeratic phosphorites of the steeper parts of the Agulhas Bank formed when the glaucophosphorites became unstable and slid downslope in the form of mud flows or turbidity currents. During their movement, these flows broke up the semiconsolidated materials, which were incorporated into the flowing material, generating the poorly sorted conglomeratic phosphorites commonly found today near shelf edges. However, conglomeratic phosphorites occur also in the shallow parts of the shelf, and Parker (1975) proposed that, during phases of low stand of sea level, abnormally strong currents due to tides and storm waves would be responsible for their formation, inducing extensive transport accompanied by winnowing effects. Birch (1979), in a new study of the same phosphorites, stressed the difficulty of visualizing such a process, given the fact that these glaucoconglomeratic phosphorites appear rather as extremely poorly sorted lag deposits, indicating very little transport. He proposed an alternative hypothesis for the formation of both conglomeratic and nonconglomeratic phosphorites during transgressive–regressive cycles (**Fig. 8.2**). This hypothesis is based on the fact that, on the flat-lying shelf of southwest Africa, both phosphatized pelagic limestones and glaucoconglomeratic phosphorites were recovered from the same stations. This implies that the two types of sediments formed at the same general place at different times under different environmental conditions. During transgressive and regressive cycles, sedimentation zones migrated back and forth across the shelf, leading to overlap of zones. It seems very likely that the phosphatized foraminiferal bedrock, formed during a transgression, was broken up almost in place by a following regression, and that these new conditions of water depth and change of upwelling were favorable for the formation of glaucony. During the subsequent transgression, renewed carbonate deposition and further phosphatization took place when the zone of deposition moved back across the shelf. Therefore, the pebbles of previously phosphatized limestones and glaucony grains were already in place as detrital constituents of the phosphatic rocks that were formed during the second phase of mineralization. This interpretation of the phosphatic conglomerates as winnowed lag deposits eliminates the necessity of the transport of any constituents for long distances over flat-shelf topography. It explains also the frequently observed repeated episodes of submarine phosphatization.

Association of Authigenic Precipitated and Diagenetic Replacive Phosphorites

Birch (1980) described conditions along the western margin of South Africa where, during Late Miocene to Early Pleistocene, authigenic and diagenetic processes of phospha-

tization were apparently active simultaneously under the action of two types of upwelling (**Fig. 8.3**). The first process corresponds to the previously described association of phosphatized pelagic limestones and glaucophosphorites, in part conglomeratic, located in open shelves and offshore topographic highs close to the shelf edge. The deposition of calcareous plankton was related to diffuse, divergent upwelling of water rich in nutrients. Transgressive–regressive cycles were a critical factor in the formation of glauconitic and conglomeratic facies. The second process formed authigenic massive or layered, pelletoidal, mud-supported phosphorites in shallow, restricted, lagoonal or estuarine environments along the margins of the hot, arid hinterland. Intense, wind-induced upwelling of nutrient-rich waters led to the development of siliceous phytoplankton, which upon decay generated apatite precipitation in the interstitial waters of the shallow-water muds. Phosphate pellets were probably formed at the water–sediment interface around suitable nuclei of either detrital origin or produced by intraformational reworking by periodic storms. According to Birch (1980), authigenic phosphatization was greatly facilitated by solar heating in the shallow lagoonal environment, by reduced terrigenous influx, and possibly by microorganism activity.

A mixing of authigenic and diagenetic forms of phosphorites may occur during transgressive–regressive cycles of large amplitude, which can lead to superposition of the two depositional environments. The model of phosphatization by a dual mechanism appears applicable to other phosphogenic provinces such as the Moroccan continental shelf and onshore phosphorites (Birch, 1980).

ANCIENT MARINE PHOSPHORITES

Among the works dealing with the petrography of ancient marine phosphorites, a few typical examples were selected to avoid needless repetition and to illustrate the major subdivisions of the proposed classification.

Phospharudites

Phospharudites display an extraordinary petrographic variety because they may consist of cobble-size constituents formed by any combination of the above-mentioned types of phospharenites cemented or not by calcitic and/or phosphatic cements. However, most frequently, phospharudites are unconsolidated gravel lag deposits, indicating very little transport and reworking; actually, their cobble-size constituents are of controversial origin and are called *nodules*. This term means that they were formed as concretions essentially in place by discontinuous episodes of phosphate precipitation and replacement, separated by phases of submarine exposure and minor reworking, or at times by uninterrupted phosphatization with no reworking at all. Therefore, the

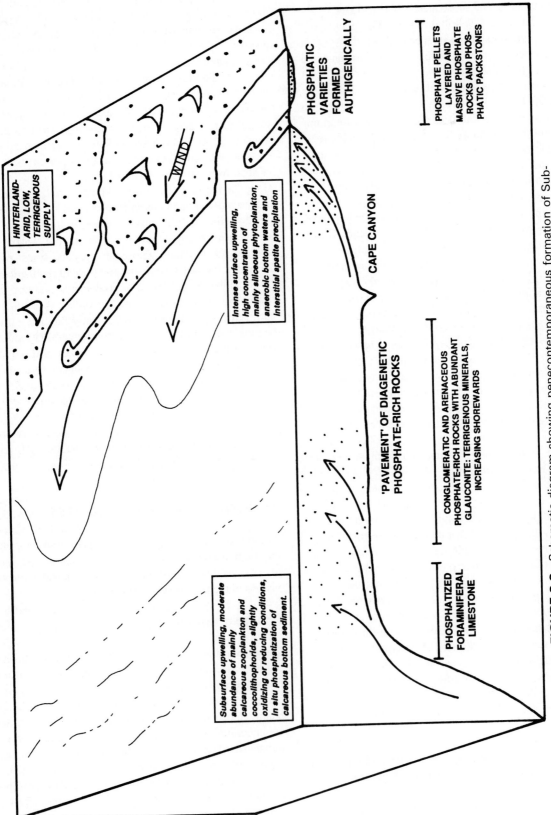

FIGURE 8.3 Schematic diagram showing penecontemporaneous formation of Subrecent to Recent phosphorites by authigenic and diagenetic processes along the southwestern margin of South Africa. From Birch (1980). Reprinted by permission of the Society of Economic Paleontologists and Mineralogists.

The following labels appear within the figure:

HINTERLAND-ARID, LOW, TERRIGENOUS SUPPLY

WIND

Intense surface upwelling, high concentration of mainly siliceous phytoplankton, anaerobic bottom waters and interstitial apatite precipitation

Subsurface upwelling, moderate abundance of mainly calcareous zooplankton and coccolithophorids, slightly oxidizing or reducing conditions, in situ phosphatization of calcareous bottom sediment.

PHOSPHATIC VARIETIES FORMED AUTHIGENICALLY

PHOSPHATE PELLETS LAYERED AND MASSIVE PHOSPHATE ROCKS AND PHOSPHATIC PACKSTONES

CAPE CANYON

'PAVEMENT' OF DIAGENETIC PHOSPHATE-RICH ROCKS

CONGLOMERATIC AND ARENACEOUS PHOSPHATE-RICH ROCKS WITH ABUNDANT GLAUCONITE: TERRIGENOUS MINERALS, INCREASING SHOREWARDS

PHOSPHATIZED FORAMINIFERAL LIMESTONE

spectrum of environmental and diagenetic conditions is very broad, as shown in Subrecent to Recent deposits (Bremner, 1980; Marshall and Cook, 1980), and only a few typical examples of phosphatic nodules are described below.

Very characteristic phosphatic nodules, probably of Pleistocene age, occur on banks, escarpments, and walls of submarine canyons off Southern California (Dietz et al., 1942). These nodules, which vary in shape from thin slabs to nodular masses, display protuberances, cavities, and perforations, leading to complex morphologies. In general, they are hard and dense, with a smooth, glazed surface consisting of a thin discolored layer of collophane or of manganese oxide obscuring the internal structure.

In thin section, collophane is the principal constituent, whereas anisotropic francolite replaces bioclasts and other calcareous debris originally entrapped in the nodules. Francolite also occurs as fibrous concentric rings around grains, forming superficial and normal ooids. The texture of the nodules varies strongly; some appear as pieces of homogeneous collophane and francolite with no visible enclosed skeletal or mineral constituents; however, most of them show a layered structure, and many have phospharenitic or phospharuditic textures. The layers are irregular and nonconcentric and contain phosphatic pellets and ooids similar to those described in phospharenites, as well as foraminifer tests, bone fragments, and bioclasts of corals and bryozoans, either unaltered or phosphatized. Grains of detrital minerals are extremely abundant in some nodules and consist of quartz, feldspars, micas, and pyroxenes. Glaucony is abundant either in rounded grains with a coating of manganese oxide, or as filling of foraminiferal tests, or as fine disseminated granules.

The external morphology and the internal composition of these nodules indicate that they resulted from direct discontinuous precipitation of phosphate within the interstitial waters of the sediments deposited on the walls of submarine canyons undergoing oceanic upwelling. At times, the nodules were exposed at the sediment–water interface, where they were slightly reworked in place, perforated, and encrusted.

Other phosphatic nodules occurring within quartz–glaucony arenites with calcite cement of the Albian of the Paris Basin (Cayeux, 1939, 1941a, 1950) are concretions generated essentially in place, which differ from the host rock only by the fact that phosphate has extensively replaced the calcite cement and most of the bioclasts as shown by abundant calcitic inclusions in the replacive collophane. However, these nodules show light petrographic and faunistic differences when compared with the host rock, indicating that a limited amount of reworking took place within the environment of phosphatization. The small reworking is also shown by fissures, borings, and encrustations. Several cycles of reworking are also revealed by large nodules enclosing smaller ones. The fissures are usually filled by a different type of phosphate, and its detailed petrographic and micropalentological features indicate that the transport of the nodules was either coastward or seaward; these contradictory directions of short transport may be observed within the same deposit.

Kennedy and Garrison (1975) described in the Cenomanian of southeast England phosphatic and glauconitic nodules that are mainly whole or fragmentary molds of fossils. The origin of these nodules is a complex history involving the following stages: infilling of shells with carbonate mud; shallow burial; prefossilization and cementation of fossil infillings, probably by high-Mg calcite; dissolution of aragonitic shell material; sea-floor erosion and submarine exposure of nodules; phosphatization and sometimes glauconitization, replacing carbonate mud and cement; and boring and encrustation by various organisms. Many nodules display the evidence of several repetitions of this succession of events, as shown by the variable degree of abrasion and phosphatization of the nodules and by successive generations of borings and encrusting organisms.

Phosphate nodules within Pennsylvanian black and gray shales of the epicontinental seas of the midcontinent area of the United States display an example of concretionary phosphatization entirely devoid of any kind of intraformational reworking (Kidder, 1985). These nodules range from spheroidal to platy in shape and have an average size of about 3 cm. The enclosing shales bend around the nodules, indicating that the latter were fully cemented before compaction and lithification of the shales. The nodules rarely show any internal structure except for a few instances of crude concentric and horizontal lamination. They consist of pellets, microfossils (mainly Radiolaria), and occasional megafossils held together by various types of interstitial cements. The pellets are subspheroidal, 100 to 150 μ in diameter, and consist of finely crystalline apatite. Their origin is multiple: they are either fecal pellets containing debris of radiolarian skeletons and scattered flecks of organic matter or phosphatized entire radiolarian tests, or they show phosphatic envelopes developed around centers of calcite cement, in which case they are interpreted as having formed by dissolution of the central part of a phosphatic pellet, followed by calcite cementation; some of them appear in fact fractured and rehealed. The phosphate pellets can be surrounded by rims of apatite cement, indicating that they acted as nucleation sites for cementation. The microfossils are mainly Radiolaria, with preservation ranging from perfect to poor. In silicified nodules, early apatite rim cement surrounds Radiolaria, indicating that they also acted as nucleation sites for precipitation of the apatite cement. The megafossils enclosed in the nodules are nautiloids, fish-bone fragments, inarticulate brachiopods, sponge spicules, and vascular plant material.

Apatite and calcite are the dominant cements; silica (microquartz and chalcedony) occurs also as cement and in some nodules replaces bioclasts; pyrite is replacive and present in many nodules. The apatite cement usually forms a thin (up to 25 μ) fibroradiated isopachous rim cement around all types of grains, including entire Radiolaria, pellets, concentrations of organic matter, and large bioclasts. This early cementation is of greatly variable importance and can consist of several distinct superposed layers. Calcite cement occurs as large irregular crystals, often including flecks of pyrite, but more commonly as intergranular poikilotopic cement. Temporal relationships between the various cements indicate the following diagenetic sequence (Kidder, 1985): apatite rim cement along margins of intragranular and intergranular voids, followed in some cases by a slight dissolution; replacive pyrite, whenever present, followed apatite and preceded calcite; silica and calcite represent the last stages of cementation, although they rarely coexist within a single nodule and their timing is difficult to establish; one single case was found of calcite cement crosscutting silica. Both calcite and silica occur within voids lined by apatite rim cement, and microquartz can be seen grading into chalcedony.

In summary, the pathway of cementation, which, after apatite rim cement, is followed in places by replacive pyrite, bifurcates to calcite or to microquartz–chalcedony in the final stages depending on local microenvironmental conditions (Kidder, 1985).

Phospharenites

Phospharenites are among the most common types of phosphorites and consist of pelletoidal, intraclastic, oolitic–pisolitic, and bone types displaying a complete range of textures from pressure solution to mud- and grain-supported, with phosphatic and/or carbonate interstitial matrix or cement.

Pelphospharenites. In pelphospharenites (Cayeux, 1939, 1941a, 1950; Lowell, 1952; Soudry and Nathan, 1980), pellets appear as ovoidal, subrounded to well-rounded bodies, sometimes broken or molded against each other, with well-developed reciprocal interpenetration, indicating a semiconsolidated state at the time of initial pressure solution (**Plate 29.B**). Pellets are usually of a brownish color, but may become almost black by concentration of organic matter, probably graphitic. This organic matter, together with minute pyrite flecks, may be either irregularly distributed or associated in clusters or form ill-defined nuclei or incomplete peripheral rings. However, most pellets are devoid of internal structures and consist of cryptocrystalline collophane; but this appearance is deceptive and results from the converging effects of authigenic and diagenetic processes combined with rounding by abrasion. The various possible origins can only be detected by searching for incomplete stages of replacement or precipitation within a given pelletoidal phospharenite.

The most frequent type of pellet is formed by precipitated phosphate around a skeletal or mineral grain. It appears as a small body of massive collophane centered around an entire test of pelagic or benthonic foraminifer. Such tests may still remain calcitic, whereas their internal cavities can be sparitic or filled by collophane or glaucony. All stages of phosphatization of the foraminiferal test and its internal filling can be observed until total disappearance leads to a homogeneous collophane pellet. However, nuclei of detrital grains of quartz and feldspar remain as testimonies of the process. Precipitation and replacement by collophane were obviously followed by a minor amount of reworking and abrasion since the thickness of the phosphatic material around respective nuclei varies appreciably. In these pellets, vague concentric zoning or halos may be visible, which could indicate periodic precipitation of apatite around the nuclei separated by films of organic matter.

Pellets are also generated by phosphomicritization of bone fragments (Soudry and Nathan, 1980). Typically, bone debris appear as anisotropic prismatic and angular shards, with sharp edges produced by the crushing action of scavengers, and consist of francolite with well-preserved typical bony structure. The process of phosphomicritization, controlled by endolithic algae, leads also to a final isotropic phosphate material and is similar to algal micritization in the carbonate marine phreatic environment. Two stages of the process are recognizable: (1) a brown isotropic envelope is formed by successive perforations along the edges of the bone fragment; and (2) micritization gradually spreads toward the center, destroying the original bony structure. This phosphomicritization is accompanied by a dulling of the edges and a rounding of the bone debris due to mechanical action, and the final product is again a homogeneous pellet (Soudry and Nathan, 1980).

Other pellets are clearly small intraclasts displaying the original texture of the semiconsolidated mud that was reworked, such as microfossils, smaller pellets, minute grains of detrital minerals, laminations, and so on. The best criterion for this type of pellet is to find the internal constituents in part truncated by the margins of the pellet. Phosphomicritization can also destroy most of these original features and again generate collophane pellets almost totally devoid of internal structures, except for the detrital mineral granules. Other larger intraclasts can break up during transport into their individual constituents, for instance liberating microfossils and pellets that may in turn undergo the same evolution.

Finally, some relatively large phosphatic pellets are true aggregates consisting of phosphatic pellets, bioclasts, and small phosphatic intraclasts cemented together by acicular francolite (Soudry and Nathan, 1980). Phosphomicritization can also destroy such an assemblage and change it into a homogeneous body of collophane.

Phosphatic chalks belong also to the group of pelphospharenites, having aquired their pelletoidal texture by addition of various types of phosphatic pellets (Cayeux, 1939, 1941a, 1950; Jarvis, 1980a, b, 1992). Phosphatic chalks have a gray to brownish color and a somewhat coarse texture, which leads to their easy disaggregation. Most of the pellets display organic nuclei that are predominantly foraminifers, still calcareous but with their internal cavities filled by collophane stained by organic matter. The foraminifers are usually surrounded by an envelope of pure and hyaline phosphate. When phosphatization replaces foraminifers, it usually begins with their internal partitions and subsequently involves the entire test until total destruction. Contrary to the amorphous filling of the cavities, the limpid envelopes react under crossed nicols and appear to be francolite with a fine texture consisting of concentric layers with a few radial divisions. These observations indicate two phases of phosphatization. The material that fills the cavities originated from replacement of the original chalky matrix as shown by enclosed tiny calcitic residues. The envelope is a concretionary precipitated product. Other pellets are devoid of organic remains. They are either the product of the complete evolution of the preceding type, during which all organic traces were destroyed by phosphatization, or are small intraclasts rounded by mechanical abrasion and phosphatized in the same manner as the filling of foraminiferal cavities. These pellets may also be coated by an envelope of limpid phosphate. The matrix containing these various types of phosphatic pellets has the extremely fine grained character of the calcilutite forming the normal chalk and strongly contrasts with the pelletoidal constituents and other partially phosphatized bioclasts it may include. Phosphatization of the matrix is very limited and revealed only by a yellowish color and a certain opacity.

In summary, as pointed out by Cayeux (1939, 1941a, 1950), phosphatic chalks are detrital rocks by their phosphatic constituents and pelagic by their matrix. The phosphatic elements seem to have been generated in agitated conditions over submarine hardgrounds and shoals resulting from synsedimentary uplifts and gentle folds with large radius of curvature, without any relations to shorelines. The phosphatic pellets were subsequently distributed by currents into adjacent submarine depressions where normal chalk sedimentation continued uninterrupted (Cayeux, 1941b).

Intraphospharenites. Intraphospharenites display a wide range of debris of all types derived from reworked phosphorites (**Plate 29.C**). These debris are characterized by their edges truncating all their internal lithic or skeletal constituents and by shapes ranging from extremely irregular to moderately rounded, indicating a limited transport distance. Intraphospharenites are poorly sorted deposits and may contain fragments of nonphosphatized materials. Interstitial cements are either cavity-filling calcite or collophane precipi-

tated and/or replacive. Incipient replacement of calcite by apatite is shown by a yellowish color of the calcite.

Bone-Biophospharenites. Bone-biophospharenites form relatively thin but widespread units that correspond to major or minor unconformities and submarine diastemic surfaces. They express the accumulation of fish bones produced by mass mortality of fish, which occasionally occurs in areas of oceanic upwelling as a result of poisoning by phytoplankton blooms. In the Phosphoria Formation, bone phospharenites, or bone beds (**Plate 29.D**), are used as stratigraphic markers (McKelvey et al., 1959).

Bone-biophospharenites, often bituminous, appear mostly as microbreccias containing bone fragments, shark teeth, fish scales, and coprolites. The contribution of fish to bony phospharenites is striking. However, it does not suggest that fish population was necessarily more abundant in that environment, but mass killing followed by phosphatic sedimentation under anoxic conditions led to unusual conditions of preservation. Fossil bones consist almost exclusively of collophane with a small percent of organic matter. They can be divided into two groups characterized by the absence or presence of typical bony microstructure. When the microstructure has not been preserved, bone fragments are still easily identifiable by their peculiar morphology, which is totally different from that of associated grains of detrital minerals. Bone fragments appear always as sharp-angled shards or as crudely prismatic bodies with subrectangular sections; these shapes result from fragmentation by scavengers. Some debris can be identified only under crossed nicols, which reveal the bony structure not apparent in plane-polarized light. Fragments without bony structure usually polarize with a dark bluish interference color, showing small lines more brilliant than the surrounding groundmass and appearing to represent accretion lines parallel to each other or slightly divergent. Other fragments have longitudinal pseudocleavages or a very apparent zonation appearing, under crossed nicols, as a juxtaposition of alternating colored and black streaks.

When the microscopic structure is preserved in bone fragments, the Haversian canals with concentric osseous laminae may often be observed; this is also true for the pores or lacunae, the most characteristic histologic features. Even the minute canaliculi surrounding the lacunae are sometimes visible under high magnification.

In general, the collophane of fossil bones is colored yellowish-brown by organic matter, but certain samples are colorless. The collophane of certain bones may show a distinct pleochroism, as well as a weak double refraction. Around Haversian canals, vascular cavities, and other structures, the collophane often displays, under crossed nicols, an apparent spherulitic texture due to concentric layers and not radial fibers. Several minerals are closely associated with collophane in fossil bones; the most common is calcite with cavity-filling texture in cancelled bones or in medullar cavities of long bones; other cementing minerals are quartz, chalcedony, opal, dolomite, barite, fluorite, pyrite, and collophane.

Oo-Pisophospharenites.

Oolitic and pisolitic phosphorites (**Plate 29.E**) are rather rare varieties (Cayeux, 1939, 1941a, 1950; Swett and Crowder, 1982). Superficial ooids consist of pellets and other types of detrital and skeletal nuclei surrounded by a single concentric layer of hyaline apatite of variable thickness, but not exceeding a few microns. In normal ooids, which display the same kinds of nuclei as superficial ones, the concentric structure is apparent because of the distinct differences in the index of refraction of adjoining bands of apatite. Most of these ooids have the colorless or lighter zone at the periphery, with progressively darker bands inward, but at times the sequence is reversed. A relatively small proportion of ooids display remarkably uniform concentric layers made of fibrous laminae of francolite or alternating francolite and amorphous collophane. In very rare instances (Horton et al., 1980), alternating calcite and collophane concentric layers might indicate replacement of former carbonate ooids, but the general consensus is that phosphatic ooids are of primary precipitated origin developed around nuclei (Swett and Crowder, 1982). However, their general shape is irregularly subcircular to elliptical in thin section, showing that they appear to be concretionary bodies developed almost in place with limited mechanical action; some of them are highly flattened as if they were soft at the time of formation and recall similar shapes observed in hematitic ooids. When phosphatic ooids are developed around cores of detrital grains, they show the general inverse proportionality of the thickness of phosphate coatings to detrital core, indicating some energy control of the coating process (Swett and Crowder, 1982). However, under no circumstances do phosphatic ooids either reach the perfect spherical shape that characterizes the mature mechanical evolution of carbonate ooids or build appreciable accumulations with the structures of submarine dunes.

Composite ooids have been reported (Slansky, 1980, 1986), which have nuclei consisting of two or three smaller superficial ooids cemented together by collophane and coated by several common concentric rings.

Early Diagenetic Cementation of Phospharenites.

Most of the above-described phospharenites are predominantly grain supported and cemented either by cavity-filling calcite or by various fabrics of precipitated collophane. In that respect, it appears appropriate to review early diagenetic phosphate cement fabrics as classified by Krajewski (1984) in the Albian condensed glauconitic limestones of the Tatra Mountains, which are of general application. The cement consists of crypto- to microcrystalline carbonate fluorapatite, pale yellow to dark brown in color due to dispersed iron oxides, pyrite, and organic matter. Krajewski (1984) recognized six micromorphological types of phosphatic cement on the basis of general aspect, relation to walls of pores and cavities, and presence and morphology of microbial fabrics associated with the cements, which he interpreted as coccoid cyanobacteria.

Phosphate rim cement envelopes are developed around pellets and skeletal grains of all types as thin uniform coatings, generally 5 to 20 μ thick, but reaching in places 100 μ. Whenever grains are pressure welded, the envelopes may join each other without interruptions, a situation observed in many bone phospharenites. The rim envelopes may contain dispersed microorganic traces visible as dark oval to irregular concentrations of organic matter, a few microns in size. Toward the central parts of the coated grains, where, particularly, bioclasts with original structural microporosity exist, rim envelopes grade into infillings of intraparticle porosity. Advanced stages of rim cement envelopes display darker microstreaks of impurities parallel to grain surfaces, expressing successive phases of cement accretion, a situation previously described in some pelletoidal phospharenites that may represent a transition toward phosphatic ooids (Lowell, 1952; Swett and Crowder, 1982).

Phosphate infillings of intraparticle porosity also involve microboring systems of these grains and show the same microorganic traces observed in rim cements. At times, when internal cavities are particularly large, phosphate cementation proceeds from the walls toward the centers, which may remain empty or are filled by other cements such as calcite. The filling of microboring systems is nothing else but the initial stage of the previously described phosphomicritization (Soudry and Nathan, 1980). The rim cement coating the walls of the large cavities within bioclasts or nodules consists of a thin uniform layer, 5 to 20 μ thick, occasionally reaching 100 μ, with relatively rare microorganic traces. Whenever the latter increase in number locally, the rim cement grades into a palisade fabric (Krajewski, 1984). Some of these internal rim cements show a radial arrangement of apatite crystallites aligned with their *c* axes perpendicular to the substrate surface. Multiple phosphate rim cements occur also and consist of a variable number of superposed layers of various colors due to variations in their iron oxides or pyrite content. These multiple rim cements can fill the large cavities completely or leave central spaces to other cement minerals; they also show microorganic traces consisting of coccoidal concentrations of organic matter arranged as an incipient string of beads transitional toward the formation of a palisade fabric. Krajewsli (1984) found phosphatic cement with palisade fabric closely associated with rim cement, most often within ammonite chambers devoid of internal sediment. Phosphatic palisade cement develops directly over internal surfaces of walls and septas and is covered by phosphatic rim cement. The palisade fabric consists of numerous juxtaposed, elongate, microorganic traces aligned with their longer axes perpendicular to the substrate surface. These microorganic traces are dark, oval, elongate, or irregular concentrations of organic matter, 5 to 40 μ in size, arranged as strings of beads and coated by isotropic carbonate fluorapatite. The complexity of this fabric ranges from a single string of beadlike bodies to a crust consisting of many juxtaposed strings and eventually to highly entangled associations. They represent the best expression of the important role of fossilized coccoid cyanobacteria in the genesis of early phosphate cements (Cayeux, 1941b; Krajewski, 1984).

Finally, phosphate cement may occur as cluster masses representing variable concentrations of microorganic traces, ranging from a single cluster to many, always coated by isotropic carbonate fluorapatite. These clusters occur as interstitial cement of phospharenites and within larger internal cavities of constituents.

Phosphalutites and Phosphasiltites

Phosphalutites and phosphasiltites appear as beds and nodules, usually of dark color and dull appearance (Cayeux, 1939, 1941a, 1950), consisting of isotropic and undifferentiated matrix of collophane, clay minerals, and calcite with local slightly birefringent patches of francolite. Pyrite occurs as irregular grains and flecks scattered irregularly throughout the matrix, together with minute grains of detrital quartz, feldspar, and glaucony. Clay minerals may represent in places an appreciable proportion of the matrix but are difficult to detect within the collophane matrix. Calcite represents relics of a former carbonate mud extensively phosphatized. Most phosphalutites and phosphasiltites contain numerous spicules of siliceous sponges of monaxonic and tetraxonic types, generally broken and irregularly distributed in a manner suggesting intense bioturbation action (**Plate 29.F**). Their central canal, very much enlarged by dissolution, is filled by homogeneous phosphate identical to that of the matrix. Other portions of the spicules are silicified by quartz and chalcedony or silicified and phosphatized at the same time. Phosphatization of the spicules gives rise to amorphous globular bodies or to extensions of the matrix, which penetrate into the spicules at right angles to their external surface and may, in some instances, reach the enlarged internal canal. Some individuals may be entirely replaced by undifferentiated collophane similar to that of the matrix or by francolite displaying a fibrous aspect. Tests of pelecypods and brachiopods, often associated with the sponge spicules, display the same types of phosphatization; small debris of pelmatozoans show their cellular network well preserved, but of darker color than the matrix due to the concentration of pyrite flecks.

Phosphalutitic material occurs also as interstitial matrix of quartz arenites and glauconitic quartz arenites, consisting of brownish isotropic collophane mixed with clay minerals and frequent calcite relics, indicating phosphatiza-

tion of a precursor calcite cement. Cambrian varieties of these arenites may contain concentrations of tests of inarticulated brachiopods such as *Lingula,* consisting originally of francolite.

DIAGENETIC EVOLUTION OF ANCIENT MARINE PHOSPHORITES

The petrographic study of the various types of ancient marine phosphorites has shown the complexity of their formation, which often involves repeated episodes of phosphatization by precipitation and/or replacement, submarine exposure, and perforations, a complexity even shown by apparently homogeneous-looking nodules. Pelphospharenites, because of the sand size of their constituents, are the ideal field for petrographic studies attempting to unravel in detail the successive stages of the early diagenetic evolution of marine phosphorites. Cook (1970) undertook such a study on the Meade Peak Member of the Phosphoria Formation. He found that phosphatization, calcitization, and silicification were particularly well developed, but that less important late diagenetic processes such as ferruginization, feldspathization, fluoritization, and oxidation also occurred. He established the diagenetic sequence under the microscope by using all available types of contact between minerals, in particular caries structures (in which one mineral shows scalloped-shaped penetrations in the other, resembling filled dental cavities), pseudomorphic textures, transecting relationships, parallel orientation patterns, and automorphic textures. Cook (1970) developed a seven-stage early diagenetic sequence involving repeated phases of phosphatization, calcitization, and silicification (**Fig 8.4**), which is as follows:

Stage 1. Phosphatization. This stage takes place in interstitial waters leading to the generation by precipitation of phosphatic pellets and phosphatization of calcareous bioclasts (gastropods) and of siliceous bioclasts (sponge spicules). Argillaceous and silty rocks contain irregular patches and wisps of collophane, with gradational boundaries indicating a formation by diagenetic replacement of silica and silicates, which appears volumetrically more important than the phosphatization of carbonates (Cook, 1970).

Stage 2. Calcitization. This stage is rarely developed and involves the replacement of apatite by calcite. It appears, for instance, as fine-grained calcite transecting growth lines of a phosphatic brachiopod bioclast, and as irregular patches of calcite within primary collophane pellets accompanied by destruction of their preexisting internal structures. Calcite is not cavity filling because it lacks all the typical fabrics of such an origin.

Stage 3. Phosphatization. This second episode of phos-

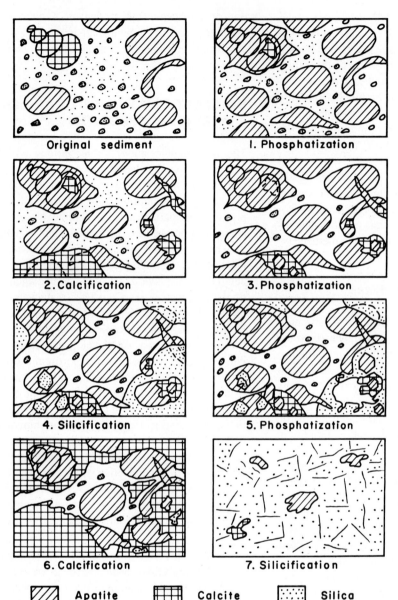

FIGURE 8.4 Schematic representation of the diagenetic sequence of pelletoidal phosphorites of the Meade Peak Member, Phosphoria Formation (Permian), showing repeated episodes of phosphatization, calcitization, and silicification. From Cook (1970). Reprinted by permission of the author and the Geological Society of America.

phate generation is predominantly phosphatization of calcite. It is distinguished from the earlier episode by the fact that it is colorless compared to the brownish original phosphate, but it is mainly characterized by its well-developed crystallinity, represented by hexagonal crystals of apatite replacing stage 2 calcite.

Stage 4. Silicification. This stage involves replacement of stage 2 calcite and stage 3 apatite by quartz crystals with well-developed hexagonal forms; during this stage, hexagonal overgrowths are also generated around detrital quartz grains. In other instances, silicification forms pseudomorphs after preexisting structures. This diagenetic silica is probably derived from earlier reactions within the sediment when detrital and biogenic silica as well as silicates were replaced by phosphates (Cook, 1970).

Stage 5. Phosphatization. This is a relatively minor stage of phosphatization, which occurs as irregular branching fingers of apatite penetrating stage 4 quartz, but also as better developed crystalline forms replacing that same quartz. This phosphatization is clearly distinct from those of stages 1 and 3 because its apatite combines caries structures and euhedral crystallinity in its replacement of stage 4 quartz.

Stage 6. Calcitization. This diagenetic phase is well developed and may involve dolomitization in some cases. It is distinguished from earlier stage 2 calcitization by cross-cutting relationships to stage 4 quartz and stage 5 apatite. It also tends to be coarser grained than stage 2 calcite and commonly replaces both quartz and apatite, with destruction of phosphate pellets and preexisting structures. In conditions where silica and phosphate coexist, dolomite rhombs preferentially replace silica.

Stage 7. Silicification. This final stage is a coarse silicification highly destructive of all preexisting carbonates and phosphates. This feature distinguishes it from stage 4 silicification, which is finer grained and commonly preserves earlier structures. Transecting veins and parallel orientation patterns demonstrate replacement of stage 6 calcite. Most of this silica forms a quartz mosaic, but locally cavity-filling textures indicate some solution of carbonates before stage 7 silicification.

Other stages of late diagenesis were observed by Cook (1970), such as feldspathization of collophane represented by a well-developed prismatic habit of feldspars replacing a collophane groundmass. Ferruginization appears as a late diagenetic event in which iron oxides, probably hematite and goethite, appearing as fine dust or well-developed hexagonal crystals, replace more commonly quartz and dolomite rather than calcite or apatite. Oxidation is also a late, near-surface diagenetic process changing ferrous to ferric oxides, but mainly removing carbon organic matter. This process leads to a discoloration of phosphatic pellets along their margins forming a vague concentric zoning or to a radial penetration along tiny fractures. Finally, rare euhedral crystals of fluorite occur in some samples, which might have derived from replacement of fluorapatite by calcite.

Cook (1970) considered that the repeated phases of diagenetic phosphatization, calcitization, and silicification he observed occurred soon after deposition within the sediments as reactions to fluctuations in pH. He concluded that given the right physicochemical conditions during early diagenesis, phosphate is a relatively mobile constituent. This situation is amply demonstrated by detailed petrography of phosphorites of all ages.

TYPICAL EXAMPLES

Plate 29.A. Quartz–glaucony arenite with microsphatite cement. Grain-supported framework of subangular quartz grains and pelletoidal to lobate glaucony grains. Interstitial cement is homogeneous microsphatite with disseminated pyrite flecks and clay minerals. Upper Cambrian, Central Mineral Region, Texas, U.S.A.

Plate 29.B. Pelphospharenite consisting entirely of ellipsoidal flattened pellets of amorphous to weakly birefringent collophane, interlocked by advanced pressure solution. The very reduced interstitial material is a film of opaque pyrite flecks and clay minerals emphasizing the grain contacts. Phosphoria Formation, Permian, Kelley, Wyoming, U.S.A.

Plate 29.C. Intraphospharenite consisting of a grain-supported framework of a large variety of subangular to subrounded intraclasts of pelphospharenites, oophospharenites, phosphasiltites, and phosphalutites in slight pressure-solution contacts. Interstitial cement is sparite stained yellow by partial apatite replacement. Phosphoria Formation, Permian, Dry Ridge, Idaho, U.S.A.

Plate 29.D. Bone phospharenite (bone bed) consisting of a grain-supported framework of angular splinters of fish bones and scales made of collophane and francolite, often replaced by secondary calcite and quartz. Simple or composite envelopes of microsphatite surround many debris and are shared by adjacent ones, indicating a diagenetic concretionary origin in place. However, a few pellets of collophane and some phosphatic ooids are scattered among the bioclasts. The reduced interstitial cement is sparite. Tectonic veinlets of late calcite cement intersect the rock (upper half of picture). Fish Scale Bed, Meade Peak Member, Phosphoria Formation, Dry Ridge, Idaho, U.S.A.

Plate 29.E. Oophospharenite consisting of a grain-supported framework of well-developed circular to elliptical phosphatic ooids, sometimes broken, with phosphalutite or pyritic nuclei. Interstitial material is an association of coarse anhedral dolomite crystals, microsphatite cement, and pyrite flecks. Galena Group, Middle Ordovician, Guttenberg, Iowa, U.S.A.

Plate 29.F. Spiculitic phosphalutite consisting of a bioturbated groundmass of collophane with minute pyrite flecks in which are scattered abundant monaxonic sponge spicules replaced by single, clear calcite crystals. Minute bioclasts of crinoids, bryozoans, ostracods, and small foraminifers are distributed throughout the groundmass. Top of Muncie Creek Shale, Iola Formation, Upper Pennsylvanian, Union Station, Kansas City, Missouri, U.S.A.

REFERENCES

BATURIN, G. N., 1982. *Phosphorites on the Sea Floor. Origin, Composition, and Distribution.* Developments in Sedimentology 33, Elsevier Scientific Publishing Co., New York, 343 pp.

BENTOR, Y. K. (ed.), 1980a. *Marine Phosphorites—Geochemistry, Occurrence, Genesis.* Soc. Econ. Paleontologists and Mineralogists, Special Publ. 29, 249 pp.

———, 1980b. Phosphorites. The unsolved problems. In Y. K. Bentor (ed.), *Marine Phosphorites—Geochemistry, Occurrence, Genesis.* Soc. Econ. Paleontologists and Mineralogists, Special Publ. 29, 3–18.

BIRCH, G. F., 1979. Phosphatic rocks of the western margin of South Africa. *J. Sed. Petrology,* 40, 93–110.

———, 1980. A model of penecontemporaneous phosphatization by diagenetic and authigenic mechanisms from the western margin of southern Africa. In Y. K. Bentor (ed.), *Marine Phosphorites—Geochemistry, Occurrence, Genesis.* Soc. Econ. Paleontologists and Mineralogists, Special Publ. 29, 79–100.

BLACK, N. R., and CAROZZI, A. V., 1988. A carbonate phosphatization model: Phosphorites of basal Maquoketa Group (Late Ordovician) of eastern Missouri and eastern Iowa, U.S.A. *Archives Sci. Genève*, 41, 303–335.

BREMNER, J. M., 1980. Concretionary phosphorite from SW Africa. *J. Geol. Soc. London,* 137, 773–786.

BURNETT, W. C., 1974. *Phosphorite Deposits from the Sea Floor off Peru and Chile: Radiochemical and Geochemical Investigations Concerning Their Origin.* Hawaii Inst. Geophysics Report HIG-74-3, 164 pp.

———, 1977. Geochemistry and origin of phosphorite deposits from off Peru and Chile. *Geol. Soc. Amer. Bull.,* 88, 813–823.

———, and RIGGS, S. R., (eds.), 1990. *Phosphate Deposits of the World.* Volume 3. *Neogene and Modern Phosphorites.* Cambridge University Press, London, 464 pp.

———, and VEEH, H. H., 1977. Uranium-series disequilibrium studies in phosphorite nodules from the west coast of South Africa. *Geochim. Cosmochim. Acta,* 41, 755–764.

———, VEEH, H. H., and SOUTAR, A., 1980. U-series, oceanographic and sedimentary evidence in support of Recent formation of phosphate nodules off Peru. In Y. K. Bentor (ed.), *Marine Phosphorites—Geochemistry, Occurence, Genesis.* Soc. Econ. Paleontologists and Mineralogists, Special Publ. 29, 61–71.

CAYEUX, L., 1939, 1941a, 1950. *Les Phosphates de chaux sédimentaires de France (France métropolitaine et d'outremer). Etude des gîtes minéraux de la France.* Service Carte géol. France, Imprimerie Nationale, Paris, vol. I, 350 pp, vol. II, 310 pp, vol. III, 360 pp.

———, 1941b. *Causes anciennes et causes actuelles en géologie.* Masson & Cioe, Paris, 81 pp. See also L. Cayeux, 1971, *Past and Present Causes in Geology,* translated and edited by A. V. Carozzi, Hafner Publishing Company, New York, 162 pp.

COOK, P. J., 1970. Repeated diagenetic calcitization, phosphatization, and silicification in the Phosphoria Formation. *Geol. Soc. Amer. Bull.,* 81, 2107–2116.

———, and McELHINNY, M. W., 1979. A re-evaluation of the spatial and temporal distribution of sedimentary phosphate deposits in the light of plate tectonics. *Economic Geol.,* 74, 315–330.

———, and SHERGOLD, J. H. (eds.), 1986a. *Phosphate Deposits of the World.* Volume 1. *Proterozoic and Cambrian Phosphorites.* Cambridge University Press, London, 386 pp.

———, and ———, 1986b. Proterozoic and Cambrian phosphorites—an introduction. In P. J. Cook and J. H. Shergold (eds.), *Phosphate Deposits of the World.* Volume 1. *Proterozoic and Cambrian Phosphorites.* Cambridge University Press, London, 1–8.

DEMAISON, G. J., and MOORE, G. T., 1980. Anoxic environments and oil source bed genesis. *Amer. Assoc. Petroleum Geologists Bull.,* 64, 1179–1209.

DIETZ, R. S., EMERY, K. O., and SHEPARD, F. P., 1942. Phosphorite deposits on the sea floor off south California. *Geol. Soc. Amer. Bull.,* 53, 815–848.

FILIPPELLI, G. M., and DELANEY, M. L., 1992. Similar phosphorus fluxes in ancient phosphorite deposits and a modern phosphogenic environment. *Geology,* 20, 709–712.

HITE, R. J., 1978. Possible genetic relationships between evaporites, phosphates, and iron-rich sediments. *Mountain Geologist,* 14, 97–107.

HORTON, A., IVIMEY-COOK, H. C., HARRISON, R. K., and YOUNG, B. R., 1980. Phosphatic ööids in the Upper Lias (Lower Jurassic) of central England. *J. Geol. Soc. London,* 137, 731–740.

JARVIS, I., 1980a. The initiation of phosphatic chalk sedimentation—the Senonian (Cretaceous) of the Anglo-Parisian Basin. In Y. K. Bentor (ed.), *Marine Phosphorites—Geochemistry, Occurrence, Genesis.* Soc. Econ. Paleontologists and Mineralogists, Special Publ. 29, 167–192.

———, 1980b. Geochemistry of phosphatic chalks and hardgrounds from the Santonian to Early Campanian (Cretaceous) of northern France. *J. Geol. Soc. London,* 137, 705–721.

———, 1992. Sedimentology, geochemistry, and origin of phosphatic chalks: the Upper Cretaceous deposits of NW Europe. *Sedimentology,* 39, 55–97.

KENNEDY, W. J., and GARRISON, R. E., 1975. Morphology and genesis of nodular phosphates in the Cenomanian Glauconitic Marl of southeast England. *Lethaia,* 8, 339–360.

KIDDER, D. L., 1985. Petrology and origin of phosphate nodules from the Midcontinent Pennsylvanian epicontinental sea. *J. Sed. Petrology,* 55, 809–816.

KRAJEWSKI, K. P., 1984. Early diagenetic phosphate cements in the Albian condensed limestone of the Tatra Mountains, Western Carpathians. *Sedimentology,* 31, 443–470.

LOUGHMAN, D. L., 1984. Phosphate authigenesis in the Aramachay Formation (Lower Jurassic) of Peru. *J. Sed. Petrology,* 54, 1147–1156.

LOWELL, W. R., 1952. Phosphatic rocks in the Deer Creek–Wells Canyon area, Idaho. *U.S. Geol. Survey Bull. 982-A,* 52 pp.

MANHEIM, F., ROWE, G. T., and JIPAD, D., 1975. Marine phosphorite formation off Peru. *J. Sed. Petrology,* 45, 243–251.

MARSHALL, J. F., and COOK, P. J., 1980. Petrology of iron- and phosphorus-rich nodules from the E. Australian continental shelf. *J. Geol. Soc. London,* 137, 765–771.

McKELVEY, V. E., WILLIAMS, J. S., SHELDON, R. P., CRESSMAN, E. R., CHENEY, T. M., and SWANSON, R. W., 1959. The Phosphoria, Park City and Shedborne Formations in the western phosphate field. *U.S. Geol. Survey Prof. Paper 313-A,* 47 pp.

NOTHOLT, A. J. G., and JARVIS, I. (eds.), 1990. *Phosphorite Research and Development.* Geol. Soc. London, Special Publ. No. 32, 326 pp.

———, SHELDON, R. P., and DAVIDSON, D. (eds.), 1989. *Phosphate Deposits of the World.* Volume 2. *Phosphate Rock Resources.* Cambridge University Press, London, 566 pp.

PARKER, R. J., 1975. The petrology and origin of some glauconitic and glauconglomeratic phosphorites from the South African continental margin. *J. Sed. Petrology,* 45, 230–242.

———, and SIESSER, W. G., 1972. Petrology and origin of some phosphorites from the South African continental margin. *J. Sed. Petrology,* 42, 434–440.

SHELDON, R. P., 1963. Physical stratigraphy and mineral resources of Permian rocks in western Wyoming. *U.S. Geol. Survey Prof. Paper 313-B,* 273 pp.

———, 1980. Episodicity of phosphate deposition and deep ocean circulation—A hypothesis. In Y. K. Bentor (ed.), *Marine Phosphorites—Geochemistry, Occurrence, Genesis.* Soc. Econ. Paleontologists and Mineralogists, Special Publ. 29, 239–247.

———, 1981. Ancient marine phosphorites. *Ann. Rev. Earth Planet. Sci.,* 9, 251–284.

———, 1987. Association of phosphorites, organic-rich shales, chert, and carbonate rocks. *Carbonates and Evaporites,* 2, 7–14.

SLANSKY, M., 1980. *Géologie des phosphates sédimentaires.* Bureau Rech. Géol. Minières, Mémoire 114, 92 pp.

———, 1986. *Geology of Sedimentary Phosphates.* Studies in Geology, North Oxford Academic Publishers, Ltd., distributed by Elsevier Science Publishing Co., New York (revised and updated translation of M. Slansky, 1980, by P. Cooper), 210 pp.

SOUDRY, D., 1987. Ultra-fine structure and genesis of the Campanian Negev high-grade phosphorites (southern Israel). *Sedimentology,* 34, 641–660.

———, and CHAMPETIER, Y., 1983. Microbial processes in the Negev phosphorites (southern Israel). *Sedimentology,* 30, 411–423.

———, and NATHAN, Y., 1980. Phosphate peloids from the Negev phosphorites. *J. Geol. Soc. London,* 137, 749–755.

SWETT, K., and CROWDER, R. K., 1982. Primary phosphatic oolites from the Lower Cambrian of Spitsbergen. *J. Sed. Petrology,* 52, 587–593.

CHAPTER 9

IRONSTONES

INTRODUCTION

Iron occurs in all sedimentary rocks essentially as an accessory component reaching only a few percent. But when this content exceeds 15% or reaches even more than 30%, the rock has an anomalously high iron content and becomes an ore of great commercial interest. Any lower limit of iron content is an arbitrary decision, and hence the designation of these iron-rich rocks as ironstones or iron-formations has generated an endless debate aggravated by the lack of communication between authors studying Precambrian iron-formations and those concerned with Phanerozoic ironstones (Trendall, 1983a; Young, 1989a). For reasons of convenience of presentation, the term ironstone is used as chapter title as in the case of sandstone or limestone, and, following the suggestions of Young (1989a), the terms Precambrian ironstone-formations (with a hyphen) and Phanerozoic ironstones are used.

The most common iron minerals in sedimentary rocks, regardless of their depositional, diagenetic, or metamorphic origin (James, 1954) are oxides (hematite, magnetite, goethite, and limonite), carbonate (siderite), silicates (berthierine, chamosite, glaucony, stilpnomelane, greenalite, grunerite, and minnesotaite), and sulfides (pyrite and marcasite).

The question of the origin of ironstones is difficult because they cannot be studied according to uniformitarian principles. Indeed, the iron deposits forming today in subaqueous conditions, although covering a large spectrum of depositional environments, are most often volumetrically insignificant and cannot be related to the Precambrian ironstone-formations and Phanerozoic ironstones. For instance, bog ores are forming today in mid- to high-latitude lakes and swamps under tundra conditions in North America, Europe, and Asia. These deposits range from concretionary forms to accumulations of earthy soft material consisting of goethite and manganese oxides. Groundwater becomes acidic beneath accumulating humic matter and contains iron in solution as bicarbonate; when rising at the surface of swamps and marshes, these waters, in oxygenated conditions, precipitate ferric hydroxides. These iron deposits are actually phreatic-vadose crusts and potential indicators of paleowater tables (Borchert, 1960).

Any short review of Phanerozoic and present-day iron deposits in marine environment has to stress a mineralogical nomenclature problem (Young, 1989a) concerning the distinction between chamosite and berthierine. International standards clearly recommend that the term chamosite be used for a 2:1 trioctahedral chlorite (1.4 nm repeat) with Fe^{2+} as the dominant divalent octahedral cation (with an end member formula of $(Fe_5^{2+}Al)(Si_3Al)O_{10}(OH)_8$), whereas the term berthierine should be reserved for an Fe-rich 1:1 type layer silicate of the serpentine group (0.7 nm repeat) having appreciable tetrahedral Al. The chemical composition of berthierine was reviewed by Brindley (1982). Although the identification of these two important minerals is often difficult, efforts should be made to distinguish them whenever possible.

Berthierine represents a relatively shallow water aspect of iron formation in present-day marine conditions. It has been found in muds at depths of up to 150 m in the pro-delta areas of the Niger and Orinoco, the Malacca Straits, and on the shelf off Sarawak. In these instances, it forms either dark-green fecal pellets of mud with minute detrital quartz grains and with no internal concentric rings except an outer coating of goethite, infillings of skeletal voids, or replacement of calcareous bioclasts (Porrenga, 1965, 1967). Berthierine is diagenetically precipitated from ferrous iron-rich pore waters at the expense of a kaolinite precursor and under bacterial activity. Subsequent partial dehydration and burial may convert berthierine into chamosite found in older deposits (Brindley, 1982). Berthierine occurs also in Recent, organic-rich, sandy muds in Loch Etive, a hydrographically restricted sea loch of western Scotland (Rohrlich et al., 1969). The pellets are dark brown to green, faintly birefringent in thin section. Their shape is spherical to ovoidal, and their size varies from 50 to 200 μ. The organic-rich mud contains scattered silt-size grains of various detrital minerals. Many pellets show a faint concentric structure and a thin anisotropic envelope, which displays a clear extinction cross under crossed nicols. This indicates that small crystal plates are arranged tangentially and form a pale yellow envelope up to 20 μ thick. A similar thin envelope surrounds fine sand-size grains of detrital quartz and, in some instances, replaces partially polycrystalline quartz grains. The environmental conditions, although different from those reported by Porrenga (1965, 1967), also indicate an early diagenetic origin within the interstitial waters of organic-rich sediments in restricted marine conditions. The pellets of Loch Etive do not display any goethite rim, indicating no winnowing or reworking from their original position and related oxidation during such processes.

Nodules and crusts of ferromanganese oxides represent the deep-water aspect of iron formation in present-day marine conditions. They are widely distributed on the ocean floors of today, where they seem to form independently from submarine volcanic–hydrothermal activity by direct or indirect precipitation within interstitial waters. Bacterial action is perhaps involved, and the growth rates are extremely slow (Horn, 1972; Glasby, 1977; Jenkyns, 1978; Sarem and Fewkes, 1979; Baturin, 1988).

Of critical importance for both Recent marine and ancient deposits are inquiries about the source, transport, and deposition of iron. Two sources of iron, continental weathering or contemporaneous submarine hydrothermal volcanic processes, can be considered. The question has been highly debated and reviewed (Lepp, 1975). It appears at present that continental weathering under humid tropical climate generating iron-rich lateritic soils is the most widely accepted interpretation. However, in spite of numerous geochemical studies, the conditions under which iron is transported by groundwater of fluvial systems to its final deposition in assumed marine environments remains unclear. In freshwater and seawater, iron in true solution reaches only a few parts per million; the bulk of it is transported as insoluble ferric hydroxides stabilized by colloidal organic matter and adsorbed at the surface of clay minerals. Upon reaching the marine environment, the colloidal suspensions flocculate, and iron is released in the interstitial pore waters of the sediment and reprecipitated during early diagenesis in a variety of iron minerals according to the particular environment of deposition. But even in a rather simplified approach as this, it becomes clear that detrital input, associated with the iron, should be at a minimum, whereas other depositional and early diagenetic factors should be active to concentrate the iron in sufficient amount to generate an ore. At this stage the problem becomes a sedimentological one, leading to depositional models that, in the complete absence of modern analogs, are of highly speculative nature. As pointed out by Garrels (1987), an entire book could be dedicated to this subject.

Attempts at understanding the origin of ancient ironstones have led to a succession of geochemical speculations even more farfetched than those presented for the origin of phosphorites, for which at least a few Subrecent to Recent examples provide some realistic constraints to the proposed models. A direct consequence of these theoretical preoccupations, combined with economical studies of the ironstones, has led to a relatively small number of objective detailed petrographic studies, most of them are biased in favor of demonstrating a given genetic speculation.

Nevertheless, iron-rich sedimentary rocks are generally divided into two major groups: Precambrian ironstone-formations, mainly banded but also pelletoidal, among which the Superior-type is the most widespread, and Phanerozoic minette-type oolitic and bioclastic ironstones including the Paleozoic Clinton-type and the Mesozoic original minette type. These two major groups represent highly different sets of mineralogical, geochemical, and sedimentological conditions, although some similarities exist. The two types succeeded each other in time and do not form at present, which raises another question: how do they relate to the fundamental problem of evolution of the geochemistry of the earth's surface and of composition of the atmosphere?

CLASSIFICATION

The nomenclature and classification of ironstones is as yet not systematic, although several authors (Dimroth, 1968, 1976, 1979a, b; Dimroth and Chauvel, 1973; Chauvel and Dimroth, 1974; Beukes, 1980a, b; Lougheed, 1983) recognized that the basic principles of carbonate rock classification presented by Folk (1962) could be adapted with some modifications to ironstones. Even the most refined and comprehensive attempt by Beukes (1980b) falls short of its goal, because it is very difficult, it not impossible, to devise a general classification for rocks so different in texture and mineralogy. Precambrian ironstone-formations are predominantly banded deposits of chert, siderite, hematite, magnetite, pyrite, and iron silicates, and in minor amounts arenitic (granular) pelletoidal and oolitic deposits. Phanerozoic ironstones are mostly berthierine, chamosite, hematite, and goethite oolitic deposits with interstitial fossiliferous siderite and calcite cement or matrix.

To use a nomenclature inspired from that of carbonates led to an extended controversy, because according to Kimberley (1979, 1980a, b) and Dimroth (1968, 1979a, b) Precambrian and Phanerozoic ironstones were in fact carbonate rocks replaced by silica and iron compounds during early diagenesis through an eluviation–replacement process (of enormous scale, at least for the Precambrian). The origin of iron and silica was considered as yet unknown, unless perhaps derived from leaching of interbedded argillaceous and pyroclastic materials. But Dimroth and Chauvel (1973) temporarily disavowed this enigmatic large-scale epigenesis, stating that the textures of limestones and arenitic ironstone deposits were indeed similar but not identical. Some Precambrian ironstone-formations and Phanerozoic oolitic ironstones contain constituents with shrinkage cracks, indicating an initial gellike condition. Their ooids in particular are predominantly ellipsoidal and often highly distorted bodies, indicating that they were soft at the time of deposition, a feature never encountered in any limestones. With a few possible exceptions among Phanerozoic ironstones, the wholesale replacement of original calcareous ooids by iron minerals is at present rejected by most authors (Bradshaw et al., 1980; Adeleye, 1980). The only basic similarity between pelletoidal–oolitic ironstones and calcarenites are textures typical of shallow marine environment with wave and current transportation and erosion.

At long last, the terms described below for Precambrian banded ironstone-formations and Phanerozoic oolitic ironstones seem to have reached a consensus. They are meant to be more descriptive than genetic, although, in the case of Precambrian banded ironstone-formations, which have undergone appreciable metamorphism, the distinction between depositional and late diagenetic features is complex and requires a case-by-case approach because of the development of fabric-destructive new iron silicates (Klein, 1973, 1983; Klein and Bricker, 1977).

In the Precambrian banded ironstone-formations, particularly of Western Australia, three major types of banding were described (Trendall and Blockley, 1970; Ewers and Morris, 1981; Gole, 1981). Macrobanding refers to the major alternations between the two generally contrasting lithologies of the banded ironstone-formations and mixed shales or other rock types. Such macrobands have variable thicknesses, ranging for instance between 0.50 m to more than 5 m. Mesobanding refers to banding at the scale usually meant by the term *banded ironstone-formation*, that is, with a thickness of 10 to 50 mm. Mesobands can consist either of almost pure chert or of cryptocrystalline aggregates of several iron minerals, or be internally banded by alternations of regularly repeated layers of even thickness ranging from 0.2 to 2 mm (average 1 mm) called *microbands* (Trendall, 1965). Microbands (**Plate 30.A**) are apparently couplets of iron-rich and chert-rich laminae that at first glance have a striking similarity to annual varves (**Fig. 9.1**). However, Ewers and Morris (1981) found that the iron-rich laminae of the microbands are themselves composite layers consisting of several (3 to 20) alternating microlayers of hematite (or other iron minerals) and chert, whereas the iron-poor laminae are crudely layered mixtures of chert and ankerite or siderite (**Figs. 9.2, 9.3**). Hence, Trendall (1983a, b) in order to emphasize the

FIGURE 9.1 Core specimen of banded iron-formation from the Precambrian Dales Gorge Member, Brockman Iron Formation, Western Australia, showing relatively thin, dark, hematite-rich mesobands alternating with relatively thick, lighter, varved mesobands consisting of regularly repetitive pairs or microbands of hematite overlain by chert, believed to represent annual varves (see Figs. 9.2 and 9.3 for detailed view). Photograph courtesy of Richard C. Morris, CSIRO, Division of Exploration Geoscience, Wembley, W. A.

NON
VARVED

VARVED
MESOBAND
WITH
FOUR
COMPLETE
MICROBANDS
OR
VARVES
OF HEMATITE
OVERLAIN BY
CHERT + ANKERITE

NON
VARVED
MESOBAND
WITH
LATE
DIAGENETIC
MAGNETITE
AT
BOTTOM

1.0 mm

FIGURE 9.2 Examples of complex repetitive pattern of a varved mesoband consisting of four complete microbands or varves. Ideally, each varve consists of a finely laminated zone of alternating hematite and chert followed by a chert zone with enrichment in diagenetic ankerite (or siderite). Precambrian Dales Gorge Member, Brockman Iron Formation, Western Australia. Plane-polarized light. Photomicrograph courtesy of Richard C. Morris, CSIRO, Division of Exploration Geoscience, Wembley, W.A.

NON VARVED

VARVED
MESOBAND
WITH
ELEVEN
COMPLETE
MICROBANDS
OR
VARVES
OF
HEMATITE OVERLAIN BY
CHERT + ANKERITE

1.0 mm

HEMATITE
MESOBAND
50

FIGURE 9.3 Examples of complex repetitive pattern of a varved mesoband consisting of 11 complete microbands or varves. Ideally, each varve displays the same structure as illustrated in Fig. 9.2. Precambrian Dales Gorge Member, Brockman Iron Formation, Western Australia. Plane-polarized light. Photomicrograph courtesy of Richard C. Morris, CSIRO, Division of Exploration Geoscience, Wembley, W.A.

very complex nature of these seasonally controlled annual layers (varves), changed the term of microbands to *aftbands* and called their internal laminae *micron bands*. The origin of these finer structures within the aftbands is not yet understood.

On the other hand, a longer-term cyclicity for the mesobands is also required, which should be in some manner related to that of the aftbands. A cyclicity of aftbands of about 23.3 years was observed (Trendall, 1983b), and its expansion could account for that of the mesobands, although the question still remains unclear.

Dimroth (1968) created the petrographic term *femicrite* as an equivalent for carbonate micrite to describe cryptocrystalline iron minerals forming the matrix of Precambrian ironstone-formations. Mesobands in which femicrite is the predominant constituent are called felutite mesobands. They show microscale cross-laminations, scouring, washouts, and graded bedding, indicating a mechanical origin and deposition as an original fine ooze, with properties similar to those of an argillaceous mud whose originally very small particles were destroyed by subsequent aggrading neomorphism or metamorphism. Femicrite can also occur as an interstitial matrix in mud- and grain-supported fabrics of sand-size allochthonous constituents such as intraclasts, pellets, and ooids.

Dimroth (1968) called femicrite mesobands showing aftbands ferhythmites. A typical feature of ferhythmites is the complete lack of cross-cutting relationships between iron mineral aftbands, as well as at times their crinkled character; both characteristics bear striking similarities, as mentioned above, with varved deposits, in particular carbonate–anhydrite ones. A special kind of ferhythmite consists of femicrite concentrated along algal laminae of hemispheroidal stromatolites, which are otherwise chertified (**Plate 30.B**). All these fabrics are interpreted as evidence of the in situ precipitation of iron minerals through highly controlled and abrupt changes of chemical or biochemical processes (Beukes, 1973).

Chert is the next important constituent of Precambrian banded ironstone-formations; its possible multiple origin presents a major problem. It can be interpreted as matrix chert deposited as a silica gel, as cement chert (comparable to sparite), or possibly as replacement chert (Beukes, 1980b). Matrix chert may form relatively continuous layers and appears generally clean, with occasionally streaks of dust of iron minerals appearing as thin laminae. Cement chert occurs as chalcedony with fibers oriented perpendicular to the margins of particulate constituents. Chalcedony sometimes recrystallized into coarse-bladed quartz with *c* axes oriented in the same way as for chalcedonic fibers, but generally recrystallization generated very finely crystalline quartz mosaic (Dimroth, 1968). In reality, matrix chert and cement chert are very difficult to recognize microscopically except when unequivocal cavity-filling fabrics can be observed. At any rate, the generation of chert implies, like that of iron, abrupt periodical changes of chemical or biochemical processes, which in this case involve a larger proportion of silica.

The particulate constitutes are pellets, intraclasts, ooids, pisoids, oncoids, and particular kinds of concavoconvex shardlike bodies (Dimroth, 1968; Gross, 1972; Beukes, 1980b).

Ironstone pellets range in diameter from 0.1 to 0.3 mm. They are spheroidal, flat ovoidal, or irregularly shaped with long axes often parallel to bedding (**Plate 30.C**). The pellets consist of a variety of types of femicrite, including microcrystalline quartz together with one or several iron minerals. They have generally no internal structure and appear to originate from intraformational reworking of semiconsolidated muds. They seem to have been still soft at the time of deposition. In the Precambrian, pellets, previously called "granules," do not seem to be of fecal origin in the absence of infauna or scavengers (LaBerge, 1964; Dimroth and Chauvel, 1973; Beukes, 1980b). For the Phanerozoic pellets, fecal and lithic origins were certainly associated.

Intraclasts are fragments of ironstones or cherts (in the Precambrian) with diameters ranging from 0.1 mm to 10 cm and above. They have internal structures demonstrating a derivation through intraformational reworking of penecontemporaneous sediments (**Plate 30.B, C**). Small intraclasts are spheroidal to ovoidal, larger ones discoidal. Large intraclasts commonly show internal laminations or pelletoidal textures truncated by the sharp external boundaries of the intraclasts. Some intraclasts are slightly deformed, but, in general, they appear to have been well consolidated at the time of deposition. Many show cracks, which could represent syneresis or subaerial desiccation; this is particularly true for the Precambrian chert intraclasts (Gross, 1972; Beukes, 1980b).

Ironstone ooids vary in size between 0.5 and 5 mm, averaging between 1 and 2 mm; pisoids with diameters above 5 mm are extremely rare. Most ooids have cores consisting of structureless pellets of femicrite of variable composition, detrital quartz grains, chert and microcrystalline quartz (in the Precambrian), and a variety of bioclasts in the Phanerozoic. The concentric layers consist of any combination of different iron minerals, depending on the environment of formation. In the Precambrian, these layers can alternate with chert concentric laminae. Some ooids show shrinkage cracks, generally separating core from concentric rings; smaller cracks in the core are irregular, whereas those in the layers are radial and concentric. Double and composite ooids are common. Many ooids display appreciable deformation, with or without preferred orientation, which seems to indicate that the concentric rings consisted of a kind of iron or silica gel accreted at the surface of pelletoidal nuclei. Apparently, ooids reached their deformed and flattened shapes before the gel solidified and were later reworked and concentrated in particular beds by the mechanical action of waves and currents.

Concavoconvex bodies called "shards," with internally deformed concentric structures and contraction cracks, were described by Dimroth (1968) in the Precambrian ironstone-formations of Canada. Their size and internal features indicate that they are probably welded fragments of peeled-off concentric rings of ooids. The term shard used in this sense is confusing and should be avoided because real volcanic glass shards preserved as ferrostilpnomelane as well as altered pyroclastic rocks were described in the Proterozoic Brockman ironstone-formation in the Hamersley Range of Western Australia (LaBerge, 1966a, b).

Ironstone pisoids (Beukes, 1980b) share many of the properties of ooids. Their concentric rings are similar in composition, but their cores are more variable in nature, ranging from pellets of femicrite to small intraclasts. Many pisoids have flattened or kidney-shaped forms, indicating a plastic state at deposition. Boundaries between some pisoids and adjacent matrix or cement can be gradational rather than sharp, indicating again that pisoids are frequently concretionary bodies developed essentially in place.

Typical ironstone oncoids consist of stromatolitic colonies with irregularly concentric laminae coating a variety of pelletoidal and intraclastic nuclei. They have often been reported in relation with hemispheroidal stromatolitic colonies (Walter et al., 1976).

The above-mentioned particulate iron-rich constituents in chert matrix or cement of mesobands could also be characterized by terms such as peloidstones and ooidstones, but with no implied relation with Dunham's (1962) terms for carbonate rocks of wackestones, packstones, or grainstones. Beukes (1980b) reviewed all the detailed aspects of the nomenclature of ironstones, with particular emphasis on Precambrian ironstone-formations.

The nomenclature of Phanerozoic ironstones is somewhat simpler. They consist of hematite, goethite, berthierine, and chamosite ooids, pellets, intraclasts and, variably, iron-replaced bioclasts set in a carbonate, argillaceous, sideritic, limonitic, or hematitic matrix or cement. According to Young (1989a), Dunham's classification (1962) was used as a basis for the nomenclature of Phanerozoic ironstones by adopting terms such as mudironstones or oolitic packironstones. In our opinion, such nomenclature suffers from the same poor terminology, ambiguities, and imprecisions previously pointed out for the original classification of carbonates (see Chapter 5) and its application to phosphorites (see Chapter 8). The modified classification of Folk (1962), as described above, could be used with some reservations, because in some instances it would be difficult to decide whether the interstitial materials is a precipitated cement or a detrital matrix.

A classification of Phanerozoic ironstones could also be approached by adapting the classification of carbonate rocks proposed in this book (see Chapter 5 and **Fig. 5.4**), as was done for dolostones. Such a classification would use, for instance, felutite, fesiltite, mud-supported fecalcarenite, grain-supported fecalcarenite, and so on. If the composition of the felutite or fesiltite consists of only one iron mineral, the prefix "fe" is replaced by the name of the mineral; if several iron minerals are present, the term of felutite is kept and its constituents are described in order of decreasing importance, such as, for instance, siderite siltite with scattered chamosite ooids and pelecypod bioclasts; or chamosite–siderite–kaolinite fesiltite with scattered hematite ooids. In the case of arenitic deposits, one example would be grain-supported oolitic fearenite consisting of hematite ooids with cavity-filling calcite–chamosite cement.

PRECAMBRIAN IRONSTONE-FORMATIONS

Introduction

The problematic character of the study of the Precambrian ironstone-formations was described in the following words: "The conclusion cannot be avoided that the iron-formations of the pre-Cambrian, not only of the Lake Superior region, but of the world, were the unique result of some special combination of conditions that has not since been repeated" (Leith et al., 1935, p. 23).

Indeed, ironstone-formations, whether banded or granular, occur throughout the Archean and the Early Proterozoic in the cratonic shields of major continents, and most of them have undergone a certain degree of metamorphism (Gole and Klein, 1981). Both deposits show striking similarities but also differences expressing the contrasts in patterns of tectonism and continental growth. Archean banded ironstone-formations are not as thick and as laterally extensive as the major Proterozoic ones, but Archean sequences were more highly involved in tectonic deformation, and it is not known whether they might have been originally more extensive before their deformation.

Opinions on the average temperature of the earth at the time of formation of the Precambrian ironstone-formations vary from hotter to cooler than today, but most authors assume climatic conditions similar to those of today in an attempt to applying a partial uniformitarian approach (Garrels, 1987). However, the atmosphere was considered anoxic or with very low oxygen content—indeed a higher CO_2 content would facilitate leaching and transport of the iron—but somehow genetically related to an increase of the oxygen toward its present-day level (Garrels et al., 1973; Towe, 1983; Garrels, 1987).

Precambrian ironstone-formations display sedimentary textures of two contrasting types. First, the most abundant consists of the above-described various types of banding, indicating quiet deposition beyond the photic zone (120 m) and certainly below wave base at depths greater than 200 m away from disturbances of any kind. The second type, arenitic or "granular facies," is pelletoidal–oolitic–oncoidal–intraclastic associated with cross-bedding, cut and fills, desiccation, and synaeresis features, indicating shallow agitated waters with current action and clearly established organic participation (stromatolites). Paleogeographic reconstructions are yet too sketchy to establish how these two distinct physical environments are related vertically and horizontally within a single depositional system. Poor outcrop conditions are mainly responsible for this situation, but many Archean and Proterozoic banded ironstone-formations do not seem to have shallow-water lateral equivalents that might represent shorelines, although some Archean ones are interbedded with siliciclastic turbidites interpreted as prograding submarine fans (Barrett and Fralick, 1985). Nevertheless, the great economical importance of Precambrian ironstone-formations for the world's economy generated an enormous amount of data with an equivalent number of hotly disputed genetic issues that are far from being settled (UNESCO, 1973; Mel'nik, 1982; Trendall and Morris, 1983; Kimberley, 1983, James, 1992) and are briefly discussed below.

Genetic Environments

Various fundamental and closely related aspects of this subject remain problematic: the source of iron and silica, the general physiographic context of the basins of deposition, the general and striking absence of terrigenous clastic influx, the mode of precipitation of iron and silica, and finally, the various depositional models.

Sources of Iron and Silica. No agreement exists on the source of iron and silica (Hesse, 1990). James (1954, 1966) derived the iron from chemical weathering of continents, and Garrels (1987) assumed stream waters similar to present-day ones derived from average basalt and ultramafic rock groundwater, except for concentration of ferrous iron comparable to those of magnesium and calcium due to the low oxygen pressure of the atmosphere at the time. In the absence of organisms controlling silica concentrations, he assumed that the concentration of silica was limited to the inorganic solubility of amorphous silica.

Other authors considered volcanism as the source of iron. Goodwin (1973) stressed the close association of individual Archean ironstone-formations of the Canadian Shield with upper pyroclastic phases of predominantly tholeiitic to calcoalkaline, and mafic to silicic volcanic sequences. These pyroclastic phases contain a great assortment of tuffs, breccias, and lava flows of andesitic–rhyolitic composition, expressing a highly explosive volcanism of island-arc type. In the Hamersley Basin of Western Australia, Trendall and Blockley (1970) found a close association between intense explosive silicic volcanism in and around the basin and its banded ironstone-formations. They mentioned bands 1 to 5 cm thick of stilpnomelane containing relics of shards, which indicates that they represent subaerial fallout tuffs. These facts, combined with calculations indicating that continental weathering was an inadequate source, led Trendall and Blockley (1970) and Trendall (1973) to consider peri-

odic episodes of explosive volcanism as the source for iron, perhaps combined with quiet marginal fumerolic activity.

For the same Hamersley Basin, Morris and Horwitz (1983) visualized deposition on a submarine volcanigenic platform or bank, protruding into or marginal to an ocean. Upwelling of marine bottom currents resulted in precipitation, under anoxic conditions, of iron, silica, and other components, representing the pulsating output of a large oceanic rift or hot spot. The currents generally persisted during sedimentation of the entire Hamersley Group, but could be temporarily interrupted or diverted by eustatic changes of sea level, growth of barrier reefs, and oscillating emergence and submergence of intervening volcanic islands. Ash emission from the latter combined with chemical precipitation could be responsible for the intercalated "shales," which would be essentially of pyroclastic origin. According to Morris and Horwitz (1983), a deep-ocean environment is precluded by the geological setting.

On the other hand, Holland (1973) and Drever (1974) assumed upwelling ferrous iron from the deep ocean deposited in oxidized form over open or restricted shelves together with supersaturation with respect to amorphous silica.

Silica concentration in Precambrian seawater could have been much higher than today, 20 to 200 ppm according to Holland and Malinin (1979), since no organisms were available to use it for making skeletons. The origin of silica is generally considered continental as assumed for iron, but a submarine or subaerial volcanic contribution has to be considered. Hughes (1976) described Late Precambrian chert beds with preserved vitroclastic texture of silicic composition, indicating submarine accumulation and transport of contemporaneous subaerial volcanics. Simonson (1987) documented petrographically early silica cementation in several Early Proterozoic arenitic ironstone-formations. In them, pellets and ooids are set in an interstitial cement of either chalcedony forming spherulites and fibrous rims, or chert in irregular to regular mosaics, or megaquartz crystals randomly oriented or arranged in cavity-filling fabrics, with crystal size increasing toward pore centers. As in another example of Early Proterozoic siliciclastic formation, Simonson (1985b) assumed precipitation of silica at the sediment–water interface by thermal waters rising through both kinds of arenitic sediments because of a drastic decrease in silica solubility with decreasing temperatures. This mechanism would be even more effective if the geothermal gradient were assumed to have been steeper along Early Proterozoic passive margins than it is today (Simonson, 1987). This process could also be extended to the chert bands of the banded ironstone-formations, although their cyclicity would be difficult to explain.

General Physiographic Context. Since the works of Trendall (1968, 1972, 1973, 1983b) on the Hamersley Basin of the northern part of western Australia, the physiographic context of the basins of deposition of banded ironstone-formations has been interpreted as isolated bodies of freshwater or marine water or barred basins with variously restricted connection to the sea, rather than open oceans with broad shelves and extensive epeirogenic adjacent seas. However, this approach cannot be overgeneralized and certainly varied from area to area. A more recent tendency was in favor of an open marine platform environment (Gross, 1980; Beukes, 1983; Simonson 1985a) or even, as mentioned above, an oceanic volcanogenic platform comparable to the present-day Bahama Platform (Morris and Horwitz, 1983).

Modes of Precipitation of Iron and Silica. The chemical precipitation of iron in marine basins, whether restricted or not (Cloud, 1973), was advanced by many authors since the initial studies of banded ironstone-formations. James (1954) distinguished early diagenetic facies: oxides, silicates, carbonates, and sulfides. Goodwin (1973) considered them to be organized in a somewhat depth-related sequence, that is, oxides–silicates–carbonates–sulfides with increasing depth and distance from shoreline. These assumptions relied on rather rudimentary paleotectonic and paleogeographic data and, at present, are considered as purely descriptive, expressing at best effects of regional metamorphism (Klein, 1983; Morris and Horwitz, 1983; Trendall, 1983b; Morris and Trendall, 1988).

As of today, the causes of precipitation of iron remain uncertain. Trendall (1983b) assumed for the Hamersley Basin of western Australia that it was caused by the oxidation of ferrous iron in basin water when oxygen developed during photosynthesis by algae floating either at or below the water surface. The chemical evidence presented by Morris (1973), in particular with respect to phosphorus, demonstrated significant biological activity during deposition of the banded ironstone-formations. The precipitation of silica, however, could be simply the result of evaporative conditions; in such a case, both processes would be independently related to annual insolation with ample possibilities for minor differences in the contribution of the two constituents (Trendall, 1983b).

The question of the organic origin in which certain bacteria would be responsible for depositing iron compounds, whereas other microorganisms would be silica precipitators, also remains open. But if it were answered, the question of marine versus freshwater would still remain open. LaBerge (1973) described in cherts the widespread occurrence of spheroidal structures ranging from 5 to 40 µ in diameter and consisting of organic matter pigmented by extremely fine hematite, siderite, pyrite, and fine iron silicates such as greenalite, stilpnomelane, and minnesotaite. He also described similar spheroidal structures in siderite, which is the least disputed among primary precipitated iron-bearing minerals. Klemm (1979) described silica microspherules, covered or not with sievelike plates, in the chert matrix of

banded ironstones from the Transvaal Supergroup of South Africa. Today, all these structures, which could be inorganic spherulites of gelatinous material, are disputed and classified among pseudomicrofossils (Walter and Hofmann, 1983). Nevertheless, granular ironstone-formations contain stromatolitic oncoids and larger stromatolite colonies (Walter and Hofmann, 1983), which still could be marine or freshwater.

Depositional Models

In terms of depositional models, the hypothesis of freshwater basins remains a reasonable alternative. Hough (1958) concluded that present-day Lake Michigan approaches the conditions required for the formation of Precambrian banded ironstone-formations. He visualized the surrounding region at a very mature geomorphic development and subtropical to warm-temperate climate with light to moderate rainfall. Iron and silica were the main products of weathering, alternatively produced by leaching of soils during the cooler and the warmer periods of the year. Hough (1958) did not consider the alternate seasonal delivery of iron and silica to the basin of deposition as the primary cause of the banding of ironstone-formations, but said that it coincided with the periods of deposition of these materials as controlled by processes within the basin of deposition. The latter was a large freshwater lake, low in nutrients, and hence with low organic productivity. It was of sufficient depth to allow development of density stratification of the water during the summer.

In summer, iron would be oxidized in the epilimnion of the lake, but on settling into the hypolimnion isolated from the atmosphere under slightly reducing and acid conditions, iron would go back into solution in a ferrous state and only silica would precipitate. During the winter, when water was uniformly cold and undergoing complete circulation under oxidizing and alkaline conditions, iron would be oxidized to the ferric state throughout the lake and would precipitate on the bottom. According to Hough (1958), the highest amount of iron probably precipitated during a relatively short period of time at the beginning of the cold season when the summer's accumulation of iron in solution was first exposed to oxygen-saturated waters at the start of the complete circulation. Hough (1958) felt that silica could have been deposited more or less constantly throughout the year and would have complicated the annual varved pattern of deposition of winter bands of iron minerals and summer bands of silica. He also thought that any minerals common to banded ironstone-formations, such as hematite, siderite, and magnetite, could be deposited in such an environment because their stability fields include the ranges of pH and Eh occurring in this type of lake. In summary, Hough's (1958) interpretation of Precambrian banded ironstone-formations calls for simple or complex aftbanding in the form of annual varves deposited in a deep lake. The tectonic context and the amount of subsidence required for their deposition and preservation were not considered at that time.

The interpretation of aftbanding as simple or complex annual varves suggests an evaporitic environment as the best analog. At first, this environment was assumed to have been a playa–lake complex (Eugster and Chou, 1973) following the discovery that a magadiite gel could be an attractive precursor for the chert bands; but in such an interpretation, many other important aspects remain unexplained.

Detailed studies of the Archean and Proterozoic banded ironstone-formations in Australia, particularly the Hamersley Group of Western Australia (Trendall and Blockley, 1970; Trendall, 1972, 1973, 1983b; Ewers and Morris, 1981; Gole, 1981) revealed that aftbands seem to be the result of annual season control of the primary precipitation, while aftbanded chert mesobands and nonaftbanded chert could reflect a 25-year environment cycle. There seems to be no doubt that aftbands are chemical evaporative varves deposited and correlatable over hundreds of square kilometers, if not across the entire basin. As pointed out by Trendall (1973, 1983b), this annual deposition of a thin basinwide skin of precipitates with a clocklike regularity, even if these precipitates can be even more complex inside, as mentioned above, occurred for at least 1 million years and affords a picture of exceptional stability. Direct seasonal effects on the basin environment triggered the precipitation mechanism and led apparently to an annual increase of silica and iron concentrations above their permissible solubilities by evaporation. The Hamersley basin appears to have had an ovoidal shape and an area of about 100,000 km^2. According to Trendall (1973, 1983b), it was probably almost completely enclosed except for at least a partial connection with the open ocean, which would contribute to maintain the delicate balance between evaporation and water influx and also the stable water level, features that are difficult to visualize without some oceanic control. The perfect preservation of annual aftbanding, which was certainly in the form of delicate gelatinous precipitates, indicates negligible bottom currents and no disturbance of the basin floor by waves or storms. This points to a minimum depth of 50 to 200 meters. However, a few exceptions to completely quiet conditions on the basin floor occur (Simonson and Goode, 1989). They consist of thin layers and lenses of ferruginous intraclastic, pelletoidal, and oolitic chert arenites that could indicate short-lived, high-energy events such as turbidites originating from the margins of the basin and/or bottom return flow induced by storms. The surrounding land was flat, under desert climate, and drainage was negligible; iron and silica, as mentioned above, were provided by volcanism. Annual evaporation rates, estimated on the order of 300 to 500 cm/year (Garrels, 1987) with marked seasonal variations that define the aftbands, indicate a low latitude off the equator, perhaps at one of the tropics, but no reliable paleomagnetic estimates of latitudes are as yet available (Trendall, 1983b). Subsidence of the basin was probably 16,000 to 8,000 years/m.

Considering the hypothesis of an evaporative barred basin similar to that assumed for Phanerozoic evaporites (see Chapter 10), Trendall (1973) reached the conclusion that the transition near the end of Precambrian times, from evaporitic iron-chert varves to carbonate–saline varves, could mark a rapid transitional change of oceanic composition of fundamental importance in the surface geochemical evolution of the earth.

Garrels (1987) developed a geochemical model for explaining the aftbanding of the banded ironstone-formations of Australia and concluded that they are varved precipitates resulting from the evaporation of stream waters leaching mafic rocks and draining into large nonmarine basins under seasonal temperatures of 40°C or more. He calculated the results of evaporating such waters and thought that he could theoretically duplicate aftbanding, which he felt was strikingly similar to the Permian Castile carbonate–anhydrite varves (see Chapter 10). However, the proposed geochemical model contains many discrepancies. In particular, it does not account for the internal smaller-scale layering of the aftbands themselves, which led to an extended discussion (Morris and Trendall, 1988; Garrels, 1988).

Petrography

A complete description of the bewildering petrographic varieties of Precambrian banded and arenitic ironstone-formations is beyond the scope of this book. It requires a case-by-case approach, because these deposits are a direct function of the effects of metamorphism on various depositional associations of iron-rich minerals and chert. Consequently, only statements of general significance are given in this section.

The simplest way to approach these rocks, as shown by James (1954), is by means of associations of minerals: the chert–siderite association; the chert–magnetite–hematite association in which hematite can be secondary after magnetite, and vice versa; the chert–silicate association, including greenalite, stilpnomelane, minnesotaite, and various types of iron-rich chlorites; and the sulfide association, including pyrite and marcasite massive or scattered in slates. Mesobanding and aftbanding were described above and involve, in simple or complex manners, chert with carbonate, oxides, and silicates. Abrupt and gradational vertical and lateral changes of iron minerals within aftbands are common and may express depositional, diagenetic, or metamorphic effects.

In arenitic ironstone-formations, the variations of texture and composition of femicrite, as well as pellets, ooids, pisoids, intraclasts, and oncoids set in a matrix or cement of chert and iron minerals, do not require further elaboration. The effects of metamorphism are extremely complex. Among iron silicates, only greenalite appears to be considered of primary depositional origin, whereas minnesotaite, and stilpnomelane are metamorphic transformations of greenalite, and ribeckite and crocidolite are probably derived from hematite and magnetite. Eventually, under high grades of metamorphism, oxides, silicates, and carbonates are replaced by a range of iron-rich pyroxenes and amphiboles such as grunerite, accompanied by a new generation of hematite, whereas the original chert turns into mosaics of megaquartz. The final product, besides relicts of banding or ghosts of particulate constituents, bears little resemblance to the original sediment.

PHANEROZOIC OOLITIC IRONSTONES

Introduction

Phanerozoic oolitic ironstones or minette-type oolitic ironstones (Young and Taylor, 1989) developed on a considerably smaller scale than their Precambrian predecessors and suffered little or no metamorphism.

Phanerozoic oolitic ironstones consist predominantly of ooids and pisoids of silica-rich, aluminous goethite or hematite, of Al-rich berthierine–chamosite, or of a combination of these mixed with kaolinite. In addition to the enrichment of the major elements Fe, Al, and Si, rather high contents of the trace elements P, V, Cr, Zr, Th, and often also Co, Ni, As, and Zn are characteristic of minette-type oolitic ironstones (Siehl and Thein, 1989). Associated with the ooids are bioclasts and intraclasts partially or totally replaced by hematite and chamosite. The interstitial matrix or cement consists usually of berthierine, chamosite, siderite, and calcite. Evidence is widespread of bioturbation, intraformational reworking, and typical high-energy, shallow-water depositional structures, indicating that, regardless of their origin, the ooids were dispersed and concentrated at a short distance from their original sites of formation. These oolitic ironstones essentially consist of the reorganization of two very common products of weathering processes, iron oxides and clay minerals, but they build a very distinctive sedimentary facies for which there is no single modern analog. Consequently, they have also generated controversies about their origin that have lasted even longer than those pertaining to Precambrian banded ironstone-formations.

The controversies involve three major aspects: (1) the climatic and plate tectonics context of the major peaks of development of oolitic ironstones during the Phanerozoic; (2) whether the ferruginous ooids are recycled from continental latosols, marine mechanical, marine early diagenetic, burial diagenetic, or replaced calcareous ooids; and (3) whether the depositional models are large-scale shallowing-upward sequences ending with deltaic and barrier bar conditions with formations of ooids, and hence belonging to transgressive–regressive cycles; autocyclic systems of intertidal or subtidal migrating sand waves in a regressive context; or concentrations of ooids in offshore conditions either by storms that swept coastal ooid banks or by turbidity currents, as in the case of the Middle Silurian Clinton-type ironstones (Hunter, 1970; Bayer, 1989).

The general geotectonic and temporal distribution of Phanerozoic oolitic ironstones has been strongly debated in recent years (Van Houten and Karasek, 1981; Van Houten and Bhattacharyya, 1982; Van Houten and Purucker, 1984; Van Houten, 1985, 1986; Maynard, 1986; Van Houten and Arthur, 1989; Young, 1989b). A general agreement was reached on a few major points.

The general geotectonic framework appears to correspond to periods of continental breakup and dispersal, or subdued orogeny, with accumulations of oolitic ironstones in three kinds of cratonic basins: (1) foredeep along the interior side of mobile belts at times of reduced deformation and curtailed clastic influx, (2) intracratonic basins characterized by extended stability, and (3) along cratonic margins at times of divergence or initial convergence of lithospheric plates.

The temporal distribution reveals at least three cyclic patterns of decreasing amplitude recorded also in common clastic deposits, as well as a significant association with black shales. Two major plate-controlled periods of oolitic ironstone and black shale development, which lasted between 150 and 170 Ma, correspond to the Ordovician–Devonian and the Jurassic–Paleogene; both time spans were marked by extensive epicontinental seas and oceanic anoxia. These conditions presumably led to intensified chemical activity, such as the CO_2 content of the atmosphere producing warmer climate (greenhouse), higher rainfall, and increased rate of chemical weathering. Within these generally favorable episodes, the Ordovician peak corresponds to a first-order maximum of the global sea level curve, whereas that of the Jurassic corresponds to a minimum. Furthermore, paleogeographic conditions were contrasted (**Fig. 9.4**) and are not fully understood. On the one hand, the Ordovician oolitic ironstones accumulated around the northwestern margin of assembled Gondwana and Armorica during highstand of sea level, widespread tectonic activity, volcanism, absence of land plants, and in southern latitudes with no paleosoils and with an ice cap at the end of the period. On the other hand, Jurassic oolitic ironstones developed in middle northern latitudes on the unstable European part of assembled Laurasia at a time of low global sea level, mild moist climate with abundant vegetation, mild tectonic instability, and no continental glaciation (Van Houten, 1985).

During the two long, favorable episodes, the development of oolitic ironstones and black shales seems to have been influenced by a hypothetical quasiperiodic (about 32 Ma) pattern of global climate changes accompanied by recurrent, relatively high sea-level stands and expansions of oxygen-depleted waters (Van Houten and Arthur, 1989). These authors also recognized episodes of fluctuating coastal onlaps, one to several million years long, reflecting smaller variations of sea level, sediment influx, and/or subsidence.

A striking feature at the lowest cyclicity level was the

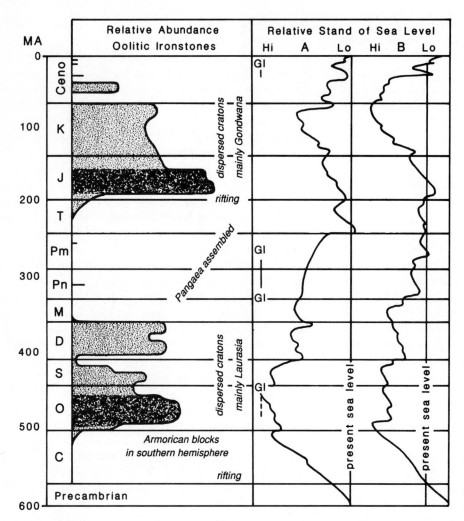

FIGURE 9.4 Phanerozoic record of ferric oxide–chamosite oolitic ironstones. Major Ordovician and Jurassic episodes are shown by darker shading. GI: widespread glaciation. Estimated sea-level curves: **A** after Hallam (1984); **B** after Vail et al. (1977). From Van Houten (1985). Reprinted by permisison of the author and the Geological Society of America.

generally moderate clastic influx, typical of oolitic ironstone deposition, showing a superposition of shoaling-upward regressive sequences, ending with hiatus and unconformity corresponding to the oolitic ironstone deposits that were in turn overlain by rapid transgressive muds, often under anoxic conditions. The duration of the regressive phase was estimated to range between 250,000 and 500,000 years, on the order of the 400,000 year Milankovitch cycles (Van Houten, 1986). Also characteristic was the general physiography of inland seas or continental margins with broad embayments, where lagoons and barrier bars were associated and easily modified by transgressions and regressions. Finally, local factors were active, inducing the local and lenticular characters of the oolitic ironstones (Van Houten and Arthur, 1989).

Origin of Iron-Rich Ooids

Two major origins are presently favored: either generation of these ooids as microconcretions within quiet lagoonal conditions with subsequent short-distance mechanical reworking and concentration, or reworking and fluvial transport of pedogenic ferruginous ooids into the marine environment,

where they undergo mineralogical changes during early and late diagenesis. The total replacement of original calcareous ooids, defended for many years, appears today as an exceptional occurrence.

Kearsley (1989) undertook a detailed study of the various types of ferruginous ooids using a variety of techniques (transmitted light microscopy, secondary electron, backscattered electron, and X-ray emission imagery). He presented an elaborate combined mineralogical and fine textural classification consisting of three major classes further subdivided into a total of 15 minor categories.

Class A consists of Fe oxide–hydroxide ± kaolinite-dominated ooids. These ooids are black, dark brown, red or orange in color and extremely widespread. They consist of varying proportions of goethite, hematite, and occasionally magnetite with common berthierine–chlorite, kaolinite, and bauxite. The iron oxides–hydroxides form complete oolitic coatings or occur as discrete laminae between more phyllosilicate-rich layers. They represent either the primary mineral phase grown in situ or mechanically accreted or a secondary alteration product from an iron-rich phyllosilicate.

Class B consists of berthierine–chlorite ± kaolinite–

dominated ooids. These ooids are pale green, gray, or white in color. They normally consist of berthierine or chlorite, kaolinite, silica, minor iron oxide, calcite, and francolite. They have two possible origins: either by transformation of iron oxyhydroxides–kaolinite ooids or from primary growth from gel, pore, or surface fluids.

Among the pure berthierine ooids, which are extremely widespread and often highly deformed, some show well-developed, alternating concentric laminae consisting of either small, randomly oriented platelets or coarser crystals with a poorly defined tangential fabric or forming lenticular masses with more random or even subradial orientation. Subradial fabric indicates growth from an enveloping fluid, whereas the precipitated coarser crystallites may have been reoriented periodically to a subtangential fabric by mechanical grain rolling.

Class C consists of siderite, calcite, apatite, pyrite-dominated ooids. This class contains a great variety of mineral and textural combinations produced by replacement or displacement of earlier phyllosilicate ooid constituents. Calcite replacement of the ooid cortex is common in sideritic ironstones, at times with preservation of precursor sideritic textures. The ooid may become a single crystal of clear calcite with uniform extinction. The lack of the distinctive fine structures of primary calcite or aragonite ooids prevents any confusion.

The major conclusion of the fundamental work of Kearsley (1989) is that the most common origin of iron-rich ooids is by growth of pure berthierine from an enveloping fluid producing a random to subradial fabric of crystallites that are periodically reoriented by rolling into concentric laminae with subtangential platelets. Although these berthierine ooids withstood rolling and abrasion, they will be flattened easily by compaction and may even break in a brittle manner. The berthierine framework is sufficiently microporous and micropermeable to react to fluids during early diagenesis in the shallow marine environment, in particular when accumulated in shoals in oxidizing conditions. In this manner, coarser laminae may be altered into iron oxyhydroxides and siderite according to changing chemical conditions.

According to Kearsley (1989), primary berthierine ooids could be changed during early diagenesis into iron hydroxides and siderite, but the reverse has also been strongly supported, that is, goethite or hematite ooids changed to berthierine–chamosite by diagenetic processes. Indeed, the formation of iron-rich ooids and pisoids, consisting mainly of aluminum-rich goethite, in tropical weathering profiles and from a variety of substrates, including glauconitic argillaceous sandstones, has been amply demonstrated (Nahon et al., 1980; Ambrosi and Nahon, 1986; Teyssen, 1984, 1989; Siehl and Thein, 1978, 1989). After erosion, reworking, fluvial transportation, and final redeposition, these ooids, submitted to reducing diagenetic processes below groundwater

level, may convert to berthierine–chamosite if the primary bulk composition of the silica-rich, aluminous goethite ooids corresponds to that of the resulting high-Al berthierine. This replacement may affect entire ooids or be differential and result in alternating concentric layers of unchanged goethite and neoformed berthierine (Siehl and Thein, 1989). These authors reported cases of diagenetic profiles with an upper section of brown goethite ores and a lower one below groundwater level with berthierine and siderite, as well as deep burial situations where goethite ooids are entirely changed to berthierine, with preservation of all the sedimentological features indicating the original agitated and oxidizing depositional environment. However, geochemical objections were raised against this interpretation (Maynard, 1986). Contradictions are also apparent in the interpretation of SEM studies of the internal fabric of ironstone ooids. According to Bhattacharrya and Kakimoto (1982), the internal fabric of ironstone ooids is tangential–concentric, indicating a mechanical accretion of detrital kaolinite and hydrated iron oxide minerals, and not radial–concentric as in pedogenic ooids and pisoids. However, Siehl and Thein (1989) found that a concentric orientation can also develop during in situ growth and that it appears independent of the mode of formation of the ooids. In summary, the above-discussed duality of origin still remains unsolved (Myers, 1989), as well as the demonstration that the major oolitic ironstones were generated from reworked pedogenic components.

The association of goethite–hematite ooids that are said to have developed in agitated oxidizing conditions and of berthierine–chamosite ooids generated in quieter reducing conditions can also be explained from a sedimentological viewpoint. They could have formed independently under adjacent shallow marine environments with distinct chemical and physical conditions, where iron, alumina, and associated trace elements recombined differently. James and Van Houten (1979) reached such conclusions in a study of Miocene ironstones of Colombia, in which they described goethitic and chamositic ooids with different textural properties. On the one hand, goethite ooids are spherical, well sorted and multilayered. Most of them consist of structureless goethite, rarely with a small amount of chamosite; their nuclei are made of structureless goethite or chamosite, quartz grains, ooid fragments, shell bioclasts, fish teeth, and fecal pellets. Many of the goethite ooids show two stages of formation represented by a multilayered or structureless inner part with a distinct goethite coating and an outer part consisting of many thin, concentric lamellae. On the other hand, chamositic ooids have a large range of size, and many display no distinct nucleus. Most of them have very irregular and twisted shapes, reflecting intense plastic deformation.

According to James and Van Houten (1979), the plastically deformed chamositic ooids probably formed in situ in the quiet conditions of continuously submerged embayments during abandonment stages. In contrast, the better sorted, spheroidal goethitic ooids developed under locally higher energy conditions, where they underwent reworking, breakage, and renewed growth; they were finally redeposited in abandoned embayments when minor transgressions covered exposed and weathered mudflats.

This review would be incomplete without mentioning the hypothesis held for many years that Phanerozoic ironstone ooids represent carbonate ooids completely ferruginized on a large scale by groundwater leachates from overlying soil horizons. This massive introduction of iron was proposed by many authors (Cayeux, 1909, 1922; Kimberley, 1979, 1980a and b; Dimroth, 1976, 1979a, b). Numerous petrographic, geochemical, sedimentological, and stratigraphic objections have been presented against this hypothesis, which is presently rejected (Bradshaw et al., 1980; Van Houten and Bhattacharyya, 1982; Maynard, 1986).

Depositional Models

Phanerozoic oolitic ironstones, as briefly mentioned above, belong to shoaling-upward sequences expressing transgressive–regressive cycles associated with prograding shorelines. The oolitic ironstones are located at the top of the coarsening upward sequence, which is most often siliciclastic, sometimes mixed siliciclastic–carbonate, or even entirely carbonate. This top location corresponds to the standstill position with minimum clastic influx, considered the highest energy and associated with erosional features, reworking surfaces, condensed zones, and diastems (**Fig. 9.5**). However, the oolitic ironstones can extend into the basal portion of the overlying transgressive sequence when reworking continued with decreasing intensity. At any rate the lagoon–barrier bar relationship is critical (Sheldon, 1970; Bayer, 1989) because it allows ooids to be formed in lagoons and subsequently accumulated in barrier bars or even transported seaward and deposited as tempestites or turbidites. In a more detailed fashion, this general model is viewed by Van Houten and Bhattacharyya (1982) as a microtidal embayed coast along which the sequence begins with muddy, open-shelf bioclastic sediments overlain by the front of a small prograding delta. The waning of detrital influx by delta abandonment develops, for instance, low-energy mud flats where ooids are formed and concentrated by gentle local currents into thick barrier bars, which may interfere with sheeted shoreline sandbodies. The association of ooid development with reduced or interrupted influx of siliciclastic material explains cases where ooids very rarely have detrital grains as core, or none at all (Brookfield, 1971; Talbot, 1974). It appears reasonable that plastic ooids formed in the gently agitated shallow-water lagoons could be transported by currents in cross-bedded bars without too much mechanical fracturing. Highly distorted ooids, characteristic of very low energy environments, are limited to early stages of accretion (Knox,

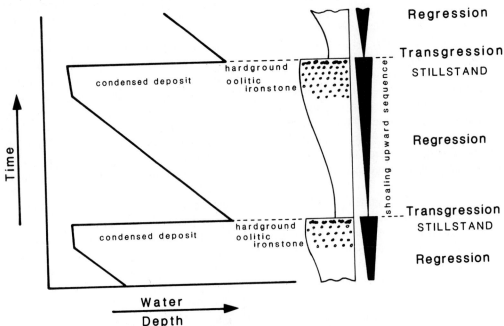

FIGURE 9.5 Facies model of oolitic ironstone-bearing, shoaling-upward detrital sequence emphasizing relation to prolonged regression and relatively rapid transgression. From Van Houten and Bhattacharyya (1982). Reprinted by permission of the authors and *Annual Review of Earth and Planetary Sciences.* Copyright Annual Review, Inc.

1970); therefore, some rolling by currents and reworking by scavengers did occur, and short-distance transportation from original place of formation to final place of deposition appears unquestionable. Interruptions in the process of accumulation, perhaps related to shifting currents, permitted burrowing of the deposits and formation of thin hardgrounds by precipitation of ferric oxide cement. Terminal layers of the oolitic ironstones, before being overlaid by transgressive marine muds, can be at times ferruginous, phosphatic, and intraclastic.

As pointed out by Sheldon (1970), the lagoonal environment is an excellent model not only from a mechanical but also a geochemical viewpoint. Indeed, it can vary within a short time interval from saline to brackish waters, as well as from oxidizing to reducing conditions, and permit the formation of the numerous above-described types of goethite and Fe-chloritic ooids with their complex concentric layers of different iron and clay minerals. This environment could also become at times a site for carbonate deposition, calcitic or sideritic, and, finally, the adjacent barrier bars are ideal places for size sorting and accumulation of completed ooids that grew in the lagoons. This model applies quite well to many Phanerozoic oolitic ironstones, in particular to the Silurian Clinton type of the United States and some Jurassic minette types of Europe (Bayer, 1989).

A different model, still involving an embayed coastline, but with a higher energy context and the influx through rivers of pedogenic ferruginous ooids from continental latosols, was proposed by Teyssen (1984, 1989) and Siehl and Thein (1978, 1989) for some types of minette ironstones of Lorraine and Luxembourg. This autocyclic model consists of large-scale sandwaves developed by tidal currents associated with subtidal shoals with channels. The cyclic repetition results from the migration of the sandwaves over deeper-water muddy deposits, located seaward in front of the sandwaves, and within the general context of a slowly prograding shoreline. The reworked continental origin of the ooids would account for quartz grains, fragments of ooids, and debris of iron crusts, rather than bioclasts, in the cores of most investigated ooids. However, this depositional model for minette-type ironstones seems to be an exception rather than the rule.

Petrography

Among the numerous varieties of Phanerozoic oolitic ironstones, only the major ones are briefly described below, based largely on accounts of authors of the past whose petrographic descriptions remain today of outstanding precision and quality, regardless of the subsequent changes of opinion concerning the environments of ooid formation.

Oolitic Hematite–Chamosite Ironstones. In most Ordovician types of the United States (Mayville and Neda ironstones), well-stratified deposits (Hawley and Beavan, 1934; Kelley, 1951) show ooids as small oblate spheroids flattened parallel to bedding (a "flaxseed" texture). The ooids consist of numerous concentric rings of goethite and hematite surrounding a nucleus of bioclastic or detrital mineral nature (**Plate 30.D**). Four types of nuclei are common: bioclasts of corals and bryozoans highly replaced by iron oxides, quartz predominating over other minerals, numerous intraclasts of structureless or oolitic ironstones, and broken ooids. The matrix between the ooids is microcrystalline hematite with minor amounts of goethite and clay minerals. Recrystallization of the hematite matrix may spread into the ooids and partially destroy their concentric structure. A small amount of the tiny crystals of calcite and dolomite may be scattered in the matrix, together with detrital minerals, organic matter, and a few phosphatic micronodules.

The Lower Ordovician Wabana ironstone of Newfoundland (Hayes, 1915) consists of ooids of hematite or of alternating hematite and chamosite layers. In many individuals, hematite obscures chamosite and apparently replaces it; some ooids are made of concentric layers of chamosite alone. Nuclei are quartz grains, small bioclasts, and minute patches of structureless hematite, chamosite, or siderite. Interstitial spaces are filled by microcrystalline hematite with a small amount of chamosite, but generally by a mosaic of quartz, especially by siderite with well-developed rhombohedral texture. Siderite is highly replacive of all other constituents and tends to destroy partially or totally original depositional fabrics. It is obviously the last-formed component in the marine environment.

The Middle Silurian oolitic hematitic ironstones called Clinton type (Alling, 1947; Hunter, 1970; Bayer, 1989) were formed in higher energy than the Ordovician types; they are characterized by an oolitic texture developed around bioclasts, and more rarely around detrital quartz grains. The ooids consist either of hematite concentric rings or of alternating hematite and chamosite, in which case the hematite appears darker and duller. Associated constituents are bioclasts of crinoids and bryozoans and intraclasts. The general fabric is grain supported with cavity-filling calcite, often stained by hematite flecks and quartz, the latter cement leading often to syntaxial overgrowths on detrital quartz grains. The evolution of the bioclasts forming the cores of most ooids shows that they were mechanically abraded and replaced by hematite before being surrounded by alternating concentric rings of hematite and of chamosite, which formed when the composition of the marine environment changed, becoming enriched in silica and aluminum. After final reworking and redistribution under higher-energy conditions, calcite cement precipitated, associated in place with chert (now quartz) representing silica, which had not been bound by iron compounds.

However, the alternation of chamosite and hematite concentric layers in ooids, generally interpreted as resulting from fluctuating chemical conditions or synsedimentary oxidation of chamosite into hematite, can be challenged by the local occurrence of ooids, which appear to have been originally chamositic and were altered to hematite under late burial diagenesis (Cotter, 1992).

Oolitic Goethite–Hematite–Chamosite Ironstones. These ironstones form most of the minette-type ironstones of the Jurassic of Central Europe (**Plate 30.E**) and are by far the most complex Phanerozoic oolitic ironstones (Cayeux, 1909, 1922; Siehl and Thein, 1978, 1989; Teyssen 1984, 1989). Their ooids consist of goethite, hematite, berthierine, and chamosite associated with bioclasts, partially or completely replaced by the above-mentioned iron minerals, and intraclasts. The ooids are of variable shape, ranging from spheroidal to very ellipsoidal, and often broken; their nuclei are quartz grains or minute fragments of structureless, intraclastic ferruginous mud. Goethite ooids are dense and opaque, showing little internal structure; other ooids consist of hematite layers, alternating layers of goethite and berthierine or chamosite, or entirely of chamosite, in which case they are highly distorted, indicating an original gellike plastic state. The matrix or cement is generally berthierine, chamosite, or siderite in high-quality ores and calcite in lean ores. However, siderite can be extremely widespread as interstitial material and, as in previous examples, highly replacive of all other constituents. Intercalated among the high-energy oolitic beds are bands of felutite and fesiltite deposited during episodes of low energy and consisting of a mixture of berthierine, chamosite, and siderite.

Oolitic Chamosite–Siderite Ironstones. Examples of these rocks belong to the Northampton Sand Ironstone of the Jurassic of Great Britain (Hallimond, 1925; Taylor, 1949). Ellipsoidal chamosite ooids have an average size of 0.3 to 0.4 mm. Broken and highly distorted ooids are often very abundant, and the relative proportion of ooids and sideritic groundmass varies widely. The concentric structure of the ooids due to the tangential arrangement of chamosite platelets is beautifully displayed by alternating lighter and darker green layers; under crossed nicols, a perfect extinction cross is clearly displayed (**Plate 30.F**). Nuclei consist either of fragments of structureless green chamosite mud or of crystalline chamosite, with prominent fibrous cleavage and well-marked pleochroism. Large chamosite pellets, similar in size and shape to normal ooids, are mixed with them and appear as rolled fragments of chamositic mud.

The composition of the chamositic ooids is extremely variable. It reflects either the incorporation of abundant kaolinite in certain concentric layers, giving them a brownish to grayish color, or the partial replacement of some layers by goethite with all transitions to a complete penecontemporaneous oxidation of chamosite to goethite, or the presence of concentric layers of yellowish siderite. All the mineralogical variations express the synsedimentary variability of the chemical conditions during formation of the ooids. However, certain ooids show an irregular central replacement by siderite, which is late diagenetic, and may spread to the entire body and destroy all traces of concentric structure. This replacive siderite is in optical continuity with the colorless one of the matrix, which displays many varieties of marginal replacement of the ooids. The siderite of the matrix is fine to coarsely granular, rarely spherulitic, and individual rhombs may be zoned. Siderite is also highly replacive of associated bioclasts such as pelecypod shells and echinoderm fragments. In addition to the predominant siderite, the matrix contains irregular patches of chamosite, detrital quartz, glaucony grains, mica flakes, heavy minerals, and phosphatic pebbles.

Oolitic Chamosite Ironstones. Chamosite ooids of rather unusual type were described from the Callovian type-locality of Chamoson in the Swiss Alpes (Déverin, 1945, 1948; Delaloye and Odin, 1988). These ooids show numerous and clearly developed fine concentric rings of chamosite surrounding relatively small and poorly outlined nuclei consisting of pelmatozoan bioclasts completely replaced by chamosite. The ooids are dispersed within a calcareous matrix formed by strongly comminuted crinoid and echinoid debris, pelecypods, brachiopods, ostracods, and sponge spicules. Two different interpretations were proposed to explain these peculiar chamosite ooids. Déverin (1945, 1948) assumed that the replaced bioclast was progressively and structurally modified or, so to speak, digested by chamosite from the exterior toward the interior. The total volume of the bioclast was supposed to remain constant in time; only the thickness of the concentric coatings would increase by centripetal rearrangement and recrystallization of the chamosite, which would eventually by replacement totally destroy the original crinoidal core, leaving only relicts of its original microstructure. According to Déverin (1945, 1948), replacement of crinoid fragments by iron-bearing solutions took place in a relatively quiet marine reducing environment, interrupted at times by periods of agitation and reworking, which would help shaping the ooids and disperse them throughout the sediment. As the chamosite coatings grew thicker, chemical exchanges between seawater and bioclastic nuclei decreased, explaining the parallel decrease of the stratified character of the envelopes toward the nuclei. Delaloye and Odin (1988) proposed a different mechanism. First, the pores of the microstructure of the bioclasts were filled with chamosite; then the chamosite replaced the carbonate skeleton, which protected the earlier internal crystallization of chamosite. The totally replaced bioclast acted then as a nucleus around which authigenic chamosite precipitated, concentrically accreting the body into an ooid. This type of accretion resulted in an increase in the size of the ooid compared to that of the original bioclast, a common feature of the development of ooidal texture regardless of the accepted concretionary or mechanical mode of accretion. In relation to the ooids from Chamoson, Odin et al. (1988) discussed the various opinions concerning the role of green marine clays in the genesis of Mesozoic oolitic ironstones.

TYPICAL EXAMPLES

Plate 30.A. Banded ironstone consisting of alternating light-colored mesobands of needles and sheaves of minnesotaite with interstitial fine chert and scattered minute siderite rombs, and opaque mesobands showing alternating continuous aftbands of magnetite and discontinuous ones of minnesotaite and fine chert. Negaunee Iron Formation, Marquette Range Supergroup, Proterozoic, Marquette County, Michigan, U.S.A.

Plate 30.B. Chertified stromatolite showing a framework of superposed thin, dark cyanobacterial laminae replaced by weathered minnesotaite and thicker laminae of microquartz mosaic with scattered minnesotaite crystals. Trapped within the stromatolite is a large subrounded intraclast consisting of magnetite and microquartz (lower left of picture). Negaunee Iron Formation, Marquette Range Supergroup, Proterozoic, Marquette County, Michigan, U.S.A.

Plate 30.C. Pelletoidal ironstone consisting of a grain-supported framework of subrounded elongated pellets and intraclasts formed by an intimate association of magnetite and fine chert. Interstitial cement is a mixture of granular quartz, minnesotaite, and siderite. Negaunee Iron Formation, Marquette Range Supergroup, Proterozoic, Marquette County, Michigan, U.S.A.

Plate 30.D. Oolitic hematite with hematitic matrix. The grain-supported framework consists of normal and composite hematite ooids with characteristic ellipsoidal shape flattened parallel to bedding and appreciable pressure-solution contacts. Nuclei are small and mostly crinoid bioclasts replaced partially by hematite or pellets of dark hematitic siltstone. Interstitial dark groundmass is hematitic siltstone; clear patches are pull-out artifacts from thin-section preparation. Neda Formation, Maquoketa Group, Upper Ordovician, Dodge County, Wisconsin, U.S.A.

Plate 30.E. Oolitic hematite with calcite cement. Grain-supported framework consisting of superficial, normal, composite, and broken hematite ooids with characteristic ellipsoidal shape, but oriented at random. Nuclei of ooids are either dark pellets of massive hematitic siltstone or minute and partially hematitized crinoid bioclasts. Partially hematitized crinoid bioclasts are scattered among the ooids. Moderate pressure solution occurs between all constituents. Interstitial yellowish-green cement is cavity-filling sparite frequently replaced by chamosite. Clear patches are pull-out artifacts from thin-section preparation. Minette Ore, Aalenian, Homecourt, France.

Plate 30.F. Oolitic chamosite with siderite matrix. Mud-supported normal chamosite ooids with numerous concentric rings and characteristic extinction cross. Nuclei are very small and consist of structureless green chamosite mud or of a fragment of crystalline chamosite. Interstitial matrix is sideritic siltstone deeply altered to iron oxides, with scattered minute grains of detrital quartz; in places (lower-right corner of picture) the matrix is replaced by aggregates of clear siderite crystals. Northampton Sand Ironstone, Jurassic, Great Britain.

REFERENCES

ADELEYE, D. R., 1980. Origin of oolitic iron formations. Discussion. *J. Sed. Petrology,* 50, 1001–1003.

ALLING, H. L., 1947. Diagenesis of the Clinton hematite ores of New York. *Geol. Soc. Amer. Bull.,* 58, 991–1018.

AMBROSI, J. P., and NAHON, D., 1986. Petrological and geochemical differentiation of lateritic iron crust profiles. *Chem. Geol.,* 57, 371–393.

BARRETT, T. J., and FRALICK, P. W., 1985. Sediment redeposition in Archean iron formations: examples from the Beardmore–Geraldton greenstone belt, Ontario. *J. Sed. Petrology,* 55, 205–212.

BATURIN, G. N., 1988. *The Geochemistry of Manganese and Manganese Nodules in the Ocean.* Sedimentology and Petroleum Geology Series, D. Reidel Publishing Company, Dordrecht, 342 pp.

BAYER, U., 1989. Stratigraphy and environmental patterns of ironstone deposits. In T. P. Young and W. E. G. Taylor (eds.), *Phanerozoic Ironstones.* Geol. Soc. London, Special Publ. 46, 105–117.

BEUKES, N. J., 1973. Precambrian iron-formations of southern Africa. *Economic Geol.,* 68, 960–1004.

———, 1980a. Lithofacies and stratigraphy of the Kuruman and Griquatown iron-formations, northern Cape Province, South Africa. *Geol. Soc. Africa Trans.,* 83, 69–86.

———, 1980b. Suggestions towards a classification and nomenclature for iron-formations. *Geol. Soc. Africa Trans.,* 83, 285–290.

———, 1983. Paleoenvironmental setting of iron-formations in the depositional basin of the Transvaal Supergroup, South Africa. In A. F. Trendall and R. C. Morris (eds.), *Iron-formation, Facts and Problems.* Developments in Precambrian Geology 6, Elsevier Scientific Publishing Co. New York, 131–209.

BHATTACHARYYA, D. P., and KAKIMOTO, P. K., 1982. Origin of ferriferous ooids: an SEM study of ironstone ooids and bauxite pisoids. *J. Sed. Petrology,* 52, 849–857.

BORCHERT, H., 1960. Genesis of marine sedimentary iron ores. *Institution Mining Metallurgy Trans., London,* 69, 261–279.

BRADSHAW, M. J., JAMES, S. J., and TURNER, P., 1980. Origin of oolitic ironstones. Discussion. *J. Sed. Petrology,* 50, 295–304.

BRINDLEY, G. W., 1982. Chemical composition of berthierines. *Clays and Clay Minerals,* 30, 153–155.

BROOKFIELD, M., 1971. An alternative to the "clastic trap" interpretation of oolitic ironstone facies. *Geol. Magazine,* 108, 137–143.

CAYEUX, L., 1909, 1922. *Les minerais de fer oolithique de France. I. Minerais de fer primaires,* 344 pp. II. *Minerais de fer secondaires,* 1052 pp. Étude des gîtes minéraux de la France, Service Carte Géol. France, Imprimerie Nationale, Paris.

CHAUVEL, J.-J., and DIMROTH, E., 1974. Facies types and depositional environment of the Sokoman Iron Formation, Central Labrador Trough, Quebec. *J. Sed. Petrology,* 44, 299–327.

CLOUD, P., 1973. Paleoecological significance of banded iron-formations. *Economic Geol.,* 68, 1135–1143.

COTTER, E., 1992. Diagenetic alteration of chamositic clay minerals to ferric oxide in oolitic ironstone. *J. Sed. Petrology,* 62, 54–60.

DELALOYE, M. F., and ODIN, G. S., 1988. Chamosite, the green marine clay from Chamoson: a study of Swiss oolitic ironstones. In G. S. Odin (ed.), *Green Marine Clays.* Developments in Sedimentology 45, Elsevier Publishing Co., New York, 7–28.

DÉVERIN, L., 1945. *Les minerais de fer oolithique des Alpes et du Jura.* Mat. Carte Géol. Suisse, Série Géotechnique, Livr. 13, vol. 2, 115 pp.

———, 1948. Oolithes ferrugineuses des Alpes et du Jura. *Bull. Suisse Minéralog. Pétrog.,* 28, 95–102.

DIMROTH, E., 1968. Sedimentary textures, diagenesis, and sedimentary environments of certain Precambrian ironstones. *Neues Jahrb. Geologie Paläontologie Abhdl.* 130, 247–274.

———, 1976. Aspects of sedimentary petrology of cherty iron-formation. In K. H. Wolf (ed.), *Handbook of Strata-bound and Stratiform Ore Deposits.* II. *Regional studies and specific deposits,* vol. 7, *Au, U, Fe, Mn, Hg, Sb, W, and P deposits.* Elsevier Scientific Publishing Co., New York, 203–254.

———, 1979a. Models of physical sedimentation of iron formations. In R. G. Walker (ed.), *Facies Models.* Geological Association of Canada, Geoscience Canada, Reprint Series 1, 175–182.

———, 1979b. Models of diagenetic facies of iron formations. In R. G. Walker (ed.), *Facies Models.* Geological Association of Canada, Geoscience Canada, Reprint Series 1, 183–189.

———, and CHAUVEL, J.-J., 1973. Petrography of the Sokoman Iron Formation in part of the Central Labrador Trough, Quebec, Canada. *Geol. Soc. Amer. Bull.,* 84, 111–134.

DREVER, J. I., 1974. Geochemical model for the origin of Precambrian banded iron formations. *Geol. Soc. Amer. Bull.,* 85, 1099–1106.

DUNHAM, R. J., 1962. Classification of carbonate rocks according to depositional texture. In W. E. HAM (ed.), Classification of carbonate rocks. Amer. Assoc. Petroleum Geologists, Memoir 1, 108–121.

EUGSTER, H. P., and CHOU, I.-M., 1973. The depositional environments of Precambrian banded iron formations. *Economic Geol.,* 68, 1144–1168.

EWERS, W. E., and MORRIS, R. C., 1981. Studies of the Dales Gorge Member of the Brockman Iron Formation, Western Australia. *Economic Geol.,* 76, 1929–1953.

FOLK, R. L., 1962. Spectral subdivision of limestone types. In W. E. Ham (ed.), Classification of carbonate rocks. *Amer. Assoc. Petroleum Geologists,* Memoir 1, 62–84.

GARRELS, R. M., 1987. A model for the deposition of the microbanded Precambrian iron formations. *Amer. J. Sci.,* 287, 81–106.

———, 1988. Reply to a discussion: A model for the deposition of the microbanded Precambrian iron formations. *Amer. J. Sci.,* 288, 669–673.

———, PERRY, E. A., and MACKENZIE, F. T., 1973. Genesis of Precambrian iron-formations and the development of atmospheric oxygen. *Economic Geol.,* 68, 1173–1179.

GLASBY, G. P. (ed.), 1977. *Marine Manganese Deposits.* Elsevier Oceanography Series 15, Elsevier Scientific Publishing Co., New York, 523 pp.

GOLE, M. J., 1981. Archean banded iron-formations, Yilgarn Block, Western Australia. *Economic Geol.,* 76, 1954–1974.

———, and KLEIN, C., JR., 1981. Banded iron-formations through much of Precambrian time. *J. Geol.,* 89, 169–183.

GOODWIN, A. M., 1973. Archean iron-formations and tectonic basins of the Canadian shield. *Economic Geol.,* 68, 915–933.

GROSS, G. A., 1972. Primary features in cherty iron-formations. *Sedimentary Geol.,* 7, 241–261.

———, 1980. A classification of iron formations based on depositional environments. *Canadian Mineralogist,* 18, 215–222.

HALLAM, A., 1984. Pre-Quaternary sea-level changes. *Ann. Rev. Earth Planet. Sci.,* 12, 205–243.

HALLIMOND, A. F., 1925. *Iron Ores: Bedded Ores of England and Wales. Petrography and Chemistry.* Great Britain Geol. Survey Spec. Reports Mineral Resources, 29, 139 pp.

HAWLEY, J. E., and BEAVAN, A. P., 1934. Mineralogy and genesis of the Mayville iron ore of Wisconsin. *Amer. Mineralogist,* 19, 493–514.

HAYES, A. O., 1915. *Wabana Iron Ores of Newfoundland.* Geol. Survey Canada, Memoir, 78, 163 pp.

HESSE, R., 1990. Silica diagenesis: origin of inorganic and replacement cherts. In I. A. McIlreath and D. W. Morrow (ed.), *Diagenesis.* Geological Association of Canada, Geoscience Canada, Reprint Series 4, 253–275.

HOLLAND, H. D., 1973. The oceans: a possible source of iron in iron formations. *Economic Geol.,* 68, 1169–1172.

———, and MALININ, S. D., 1979. The solubility and occurrence of non-ore minerals. In H. L. Barnes (ed.), *Geochemistry of Hydrothermal Ore Deposits,* 2nd ed., Wiley-Interscience, New York, 461–508.

HORN, D. R. (ed.), 1972. *Ferromanganese Deposits on the Ocean Floor.* Lamont–Doherty Geological Observatory, Columbia University. Office for the International Decade of Ocean Exploration, N.S.F., Washington D.C., 293 pp.

HOUGH, J. L., 1958. Fresh-water environment of deposition of Precambrian banded iron formations. *J. Sed. Petrology,* 28, 414–430.

HUGHES, C. J., 1976. Volcanogenic cherts in the Late Precambrian Conception Group, Avalon Peninsula, Newfoundland. *Canadian J. Earth Sci.,* 13, 512–519.

HUNTER, R. E., 1970. Facies of iron sedimentation in Clinton Group. In G. W. Fisher, F. J. Pettijohn, J. C. Reed, Jr., and K. N. WEAVER (eds.), *Studies of Appalachian Geology, Central and Southern.* Wiley-Interscience, New York, 101–121.

JAMES, H. E., Jr., and VAN HOUTEN, F. B., 1979. Miocene goethitic and chamositic ooids, northeastern Columbia. *Sedimentology, 26,* 125–133.

JAMES, H. L., 1954. Sedimentary facies of iron-formations. *Economic Geol.,* 49, 235–291.

———, 1966. Chemistry of iron-rich sedimentary rocks. *U. S. Geol. Survey Prof. Paper 440-W,* 61 pp.

———, 1992. Precambrian iron-formations: nature, origin, and mineralogic evolution from sedimentation to metamorphism. In K. H. Wolf and G. V. Chilingarian (eds.), *Diagenesis, III.* Developments in Sedimentology 47, Elsevier Scientific Publishing Co., New York, 543–590.

JENKYNS, H. C., 1978. Pelagic environments. In H. G. Reading (ed.), *Sedimentary Environments and Facies.* Blackwell Scientific Publications, Oxford, 314–371.

KEARSLEY, A. T., 1989. Iron-rich ooids, their mineralogy and microfabric: clues to their origin and evolution. In T. P. Young and W. E. G. Taylor (eds.), *Phanerozoic Ironstones.* Geol. Soc. London, Special Publ. 46, 141–164.

KELLEY, V. C., 1951. Oolitic iron deposits of New Mexico. *Amer. Assoc. Petroleum Geologists Bull.,* 35, 2199–2228.

KIMBERLEY, M. M., 1979. Origin of oolitic iron formations. *J. Sed. Petrology,* 49, 111–132.

———, 1980a. Origin of oolitic iron formations—Reply. *J. Sed. Petrology,* 50, 299–302.

———, 1980b. Origin of oolitic iron formations—Reply. *J. Sed. Petrology,* 50, 1003–1004.

———, 1983. Constraints on genetic modeling of Proterozoic iron formations. In L. G. Medaris, Jr., C. W. Byers, D. M. Mickelson, and W. C. Shanks (eds.), *Proterozoic Geology: Selected Papers from an International Proterozoic Symposium.* Geol. Soc. Amer. Memoir 161, 227–235.

KLEIN, C., JR., 1973. Changes in mineral assemblages with metamorphism of some banded Precambrian iron-formations. *Economic Geol.,* 68, 1075–1088.

———, 1983. Diagenesis and metamorphism of banded iron-formations. In A. F. Trendall and R. C. Morris (eds.), *Iron-formation: Facts and Problems.* Developments in Precambrian Geology 6, Elsevier Scientific Publishing Co., New York, 417–469.

———, and BRICKER, O. P., 1977. Some aspects of the sedimentary and diagenetic environment of Proterozoic banded iron-formation. *Economic Geol.,* 72, 1457–1470.

KLEMM, D. D., 1979. A biogenetic model of the formation of the banded iron formation in the Transvaal Supergroup/South Africa. *Mineral. Deposita,* 14, 381–385.

KNOX, R. W., 1970. Chamosite ooliths from the Winter Gill Ironstone (Jurassic) of Yorkshire, England. *J. Sed. Petrology,* 40, 1216–1225.

LABERGE, G. L., 1964. Development of magnetite in iron-formations of the Lake Superior region. *Economic Geol.,* 59, 147–161.

———, 1966a. Altered pyroclastic rocks in iron-formation in the Hamersley Range, Western Australia. *Economic Geol.,* 61, 147–161.

———, 1966b. Pyroclastic rocks in South African iron-formations, *Economic Geol.,* 61, 572–581.

———, 1973. Possible biological origin of Precambrian iron formations. *Economic Geol.,* 68, 1098–1109.

LEITH, C. K., LUND, R. J., and LEITH, A., 1935. Pre-Cambrian rocks of the Lake Superior region. *U.S. Geol. Survey Prof. Paper 184,* 34 pp.

LEPP, H., (ed.), 1975. *Geochemistry of Iron.* Benchmark Papers in Geology No. 18. Dowden, Hutchinson and Ross, Inc., Stroudsburg, Pa., Halsted Press, 464 pp.

LOUGHEED, M. S., 1983. Origin of Precambrian iron-formations in the Lake Superior region. *Geol. Soc. Amer. Bull.,* 94, 325–340.

MAYNARD, J. B., 1986. Geochemistry of oolitic iron ores, an electron microprobe study. *Economic Geol.,* 81, 1473–1483.

MEL'NIK, Y. P., 1982. *Precambrian Banded Iron-formations. Physicochemical Conditions of Formation.* Developments in Precambrian Geology 5, Elsevier Scientific Publishing Co., New York, 310 pp.

MORRIS, R. C., 1973. A pilot study of the phosphorus distribution in parts of the Brockman Iron Formation. *West Australia Geol. Survey Annual Rept. 1972,* 75–81.

———, and HORWITZ, R. C., 1983. The origin of the iron-formation-rich Hamersley Group of Western Australia—deposition on a platform. *Precambrian Res.,* 21, 273–297.

———, and TRENDALL, A. F., 1988. Discussion. A model for the deposition of the microbanded Precambrian iron formations. *Amer. J. Sci.,* 288, 664–669.

MYERS, K. J., 1989. The origin of Lower Jurassic Cleveland Ironstone Formation of north-east England: evidence from portable gamma-ray spectrophotometry. In T. P. Young and W. E. G. Taylor (eds.), *Phanerozoic Ironstones.* Geol. Soc. London, Special Publ. 46, 221–228.

NAHON, D., CAROZZI, A. V., and PARRON, C., 1980. Lateritic weathering as a mechanism for the generation of ferruginous ooids. *J. Sed. Petrology,* 50, 1287–1298.

ODIN, G. S., KNOX, R. W., GYGI, R. A, and GUERRAK, S., 1988. Green marine clays from the oolitic ironstone facies: habit, mineralogy, environment. In G. S. ODIN (ed.), *Green Marine Clays.* Developments in Sedimentology 45, Elsevier Scientific Publishing Co., New York, 29–52.

PORRENGA, D. H., 1965. Chamosite in Recent sediments of the Niger and Orinoco deltas. *Geologie en Mijnbouw,* 44, 400–403.

———, 1967. Glauconite and chamosite as depth indicators in the marine environment. *Marine Geol.,* 5, 495–501.

ROHRLICH, V., PRICE, N. B., and CALVERT, S. E., 1969. Chamosite in the Recent sediments of Loch Etive, Scotland. *J. Sed. Petrology,* 39, 624–631.

SAREM, R. K., and FEWKES, R. H., 1979. *Manganese Nodules and Methods of Investigation.* IFI/Plenum Data Company, New York, 723 pp.

SHELDON, R. P., 1970. Sedimentation of iron-rich rocks of Llandovery age (Lower Silurian) in the southern Appalachian basin. In W. B. N. Berry and A. J. Boucot (eds.), *Correlation of the*

North American Silurian Rocks. Geol. Soc. Amer. Special Paper 102, 107–112.

SIEHL, A., and THEIN, J., 1978. Geochemische Trends in der Minette (Jura, Luxemburg/Lothringen). *Geologische Rundschau,* 67, 1052–1077.

———, and———, 1989. Minette-type ironstones. In T. P. Young and W. E. G. Taylor (eds.), *Phanerozoic Ironstones.* Geol. Soc. London, Special Publ. 46, 175–193.

SIMONSON, B. M., 1985a. Sedimentological constraints on the origin of Precambrian iron-formations. *Geol. Soc. Amer. Bull.,* 96, 244–252.

———, 1985b. Sedimentology of cherts in the Early Proterozoic Wishart Formation, Quebec-Newfoundland, Canada. *Sedimentology,* 32, 23–40.

———, 1987. Early silica cementation and subsequent diagenesis in arenites from four Early Proterozoic iron formations of North America. *J. Sed. Petrology,* 57, 494–511.

———, and GOODE, A. D. T., 1989. First discovery of ferruginous chert arenites in the early Precambrian Hamersley Group of Western Australia. *Geology,* 17, 269–272.

TALBOT, M. R., 1974. Ironstones in the upper Oxfordian of southern England. *Sedimentology,* 21, 433–450.

TAYLOR, J. H., 1949. *Petrology of the Northampton Sand Ironstone Formation.* Great Britain Geol. Survey Memoir, 111 pp.

TEYSSEN, T. A., 1984. Sedimentology of the Minette oolitic ironstones of Luxembourg and Lorraine: a Jurassic subtidal sandwave complex. *Sedimentology,* 31, 195–211.

———, 1989. A depositional model for the Liassic Minette ironstones (Luxemburg and France), in comparison with other Phanerozoic oolitic ironstones. In T. P. Young and W. E. G. Taylor (eds.), *Phanerozoic Ironstones.* Geol. Soc. London, Special Publ. 46, 79–92.

TOWE, K. M., 1983. Precambrian atmospheric oxygen and banded iron formations: a delayed ocean model. *Precambrian Res.,* 20, 161–170.

TRENDALL, A. F., 1965. Progress report on the Brockman Iron Formation in the Wittenoom–Yampire area. *West. Australia Geol. Survey Annual Report 1964,* 55–65.

———, 1968. Three great basins of Precambrian banded iron formation deposition: a systematic comparison. *Geol. Soc. Amer. Bull.,* 79, 1527–1544.

———, 1972. Revolution in earth history. *Geol. Soc. Australia J.,* 19, 287–311.

———, 1973. Iron-formations of the Hamersley Group of Western Australia: type example of varved Precambrian evaporites. In UNESCO, *Genesis of Precambrian Iron and Manganese Deposits,* Proceedings of the Kiev Symposium, 20–25 August 1970. UNESCO, Earth Sciences, Paris, 257–270.

———, 1983a. Introduction. In A. F. Trendall, and R. C. Morris (eds.), *Iron-formation: Facts and Problems.* Developments in Precambrian Geology 6, Elsevier Scientific Publishing Co., New York, 1–12.

———, 1983b. The Hamersley Basin. In A. F. Trendall and R. C. Morris (eds.), *Iron-formation: Facts and Problems.* Developments in Precambrian Geology 6, Elsevier Scientific Publishing Co., New York, 69–129.

———, and BLOCKLEY, J. G., 1970. *The Iron Formations of the Precambrian Hamersley Group, Western Australia, with Special Reference to the Associated Crocidolite.* Western Australia Geol. Survey Bull., 119, 366 pp.

———, and MORRIS, R. C. (eds.), 1983. *Iron-formation: Facts and Problems.* Developments in Precambrian Geology 6, Elsevier Scientific Publishing Co., New York, 558 pp.

UNESCO, 1973. *Genesis of Precambrian Iron and Manganese Deposits.* Proceedings of the Kiev Symposium, 20–25 August 1970. UNESCO, Earth Sciences, Paris, 382 pp.

VAIL, P. R., MITCHUM, R. M., and THOMPSON, S., 1977. Seismic stratigraphy and global changes of sea level, Part 4. In C. E. Payton (ed.), *Seismic Stratigraphy: Applications to Hydrocarbon Exploration.* Amer. Assoc. Petroleum Geologists, Memoir 26, 83–97.

VAN HOUTEN, F. B., 1985. Oolitic ironstones and contrasting Ordovician and Jurassic paleogeography. *Geology,* 13, 722–724.

———, 1986. Search for Milankovitch patterns among oolitic ironstones. *Paleoceanography,* 1, 459–466.

———, and ARTHUR, M. A., 1989. Temporal patterns among Phanerozoic oolitic iron ores and oceanic anoxia. In T. P. Young and W. E. G. Taylor (eds.), *Phanerozoic Ironstones.* Geol. Soc. London, Special Publ. 46, 33–50.

———, and BHATTACHARYYA, D. P., 1982. Phanerozoic oolitic ironstones—geologic record and facies model. *Ann. Rev. Earth Planet. Sci.,* 10, 441–457.

———, and KARASEK, R. M., 1981. Sedimentologic framework of Late Devonian oolitic iron formations, Shatti Valley, West-Central Libya. *J. Sed. Petrology,* 51, 415–427.

———, and PURUCKER, M. E., 1984. Glauconitic peloids and chamosite ooids—favorable factors, constraints, and problems. *Earth-Sci. Rev.,* 20, 211–243.

WALTER, M. R., and HOFMANN, H. J., 1983. The palaeontology and palaeoecology of Precambrian iron-formations. In A. F. Trendall and R. C. Morris (eds.), *Iron-formation: Facts and Problems.* Developments in Precambrian Geology 6, Elsevier Scientific Publishing Co., New York, 373–400.

———, GOODE, A. D. T., and HALL, W. D. M., 1976. Microfossils from a newly discovered Precambrian stromatolitic iron formation in Western Australia. *Nature,* 261, 221–223.

YOUNG, T. P., 1989a. Phanerozoic ironstones: an introduction and review. In T. P. Young and W. E. G. Taylor (eds.), *Phanerozoic Ironstones.* Geol. Soc. London, Special Publ. 46, ix–xxv.

———, 1989b. Eustatically controlled ooidal ironstone deposition: facies relationships of the Ordovician open-shelf ironstones of Western Europe. In T. P. Young and W. E. G. Taylor (eds.), *Phanerozoic Ironstones.* Geol. Soc. London, Special Publ. 46, 51–64.

———, and TAYLOR, W. E. G. (eds.), 1989. *Phanerozoic Ironstones.* Geol. Soc. London, Special Publ. 46, 251 pp.

C H A P T E R   1 0

EVAPORITES

INTRODUCTION

Evaporites are chemical rocks consisting of minerals precipitated from aqueous solutions when they gradually change to concentrated brines through the action of evaporative processes taking place generally at the air–water interface. The major evaporite minerals are gypsum ($CaSO_4 \cdot 2H_2O$), anhydrite ($CaSO_4$), and halite ($NaCl$). They can be associated with numerous other complex potassium and magnesium chlorides, such as sylvite and carnallite, as well as sulfates such as polyhalite, kainite, and kieserite to name only a few (Stewart, 1963; Garrett, 1970; Holser,1979). These "potash salts" will not be discussed in detail here.

The usual source for extensive evaporite deposits is seawater, but many evaporite basins could have received important amounts of inflowing continental brines (Hardie, 1984). Furthermore, freshwater and saline groundwater plays a major role in evaporite deposition in saline lakes under continental conditions. Mineralogy alone cannot provide a distinction between marine and nonmarine evaporites, among which gypsum and halite often predominate, and other associated minerals are also nondiagnostic; however, geochemical characteristics provide some clues (Warren, 1989). Furthermore, recent studies (Cody, 1991) showed that chemical sedimentation under supersaturated conditions in marine and nonmarine environments and crystal morphologies of evaporite minerals, in particular gypsum, support the occurrence of widespread and powerful crystallization inhibition by extremely small, substoichiometric concentration (< 1 mg/l) of soluble organic substances. The most powerful inhibitor molecules appear to be polycarboxylic acids, polyphenols, polyphosphates, organic phosphate esters, and hydrolyzed proteins.

Evaporites range from hydrothermal and silica pseudomorphs of typical forms of gypsum crystals in the Early Archean (about 3.5 billion years) in Western Australia (Dunlop, 1978; Buick and Dunlop, 1990), to the actual preservation of anhydrite in the Proterozoic (Chandler, 1988), and of all the spectrum of minerals in the Phanerozoic record. Evaporites by themselves have a great economical importance as industrial material. In petroleum geology, they are equally important as seals of many carbonate reservoirs; structural traps through salt diapirism; and source rocks, together with their associated carbonates, when stratification of brines occurred and reducing conditions in bottom waters of evaporitic basins preserved the organic matter produced by surface phytoplankton from destruction, thus allowing its maturation under adequate burial conditions (Kirkland and Evans, 1981; Schreiber, 1988; Warren, 1989; Melvin, 1991).

Aside from these important economic aspects, ancient evaporites are paleoclimatic indicators if they were restricted, as today, to arid areas of low latitudes, that is, between 10° and 40° north and south of the equator, characterized by subtropical high pressures, relatively low humidity, and evaporation largely exceeding rainfall and runoff. Naturally, exceptions occur, such as gypsum crystals and thin crusts scattered in glaciolacustrine Pleistocene deltaic sequences, where they are formed diagenetically as dispersed, finely crystalline pyrite reacts with calcium-rich oxidizing groundwater (Bain, 1990).

Only in the last two decades has the study of evaporites changed from chemical, mineralogical, and textural analysis of chemical precipitates and their subsequent burial metamorphism, almost equating them to metamorphic rocks (Stewart 1963; Borchert and Muir, 1964; Braitsch, 1971), to the understanding that evaporites are sedimentary rocks, albeit deposited under peculiar environments, but still susceptible of being interpreted in terms of depositional–diagenetic models (Kirkland and Evans, 1973; Dean and Schreiber, 1978; Handford et al., 1982; Kendall, 1984; Schreiber, 1986; Warren, 1989). However, the investigation of evaporites is complex and difficult, and hence a subject of basic controversies. First, the conditions of precipitation of gypsum and anhydrite in evaporating seawater remain unsolved in spite of several studies (Posnjak, 1938, 1940; Hardie, 1967). Under conditions of very high temperature and very high brine concentrations, anhydrite precipitates, accompanied by metastable gypsum, and perhaps facilitated by the presence of certain types of macromolecular organic components (Cody and Hull, 1980). Under less drastic conditions of temperature and brine concentrations, gypsum precipitates as the only original evaporitic sulfate in most depositional environments. Consequently, many former concepts (Ogniben, 1955, 1957) that various textures of gypsum, such as selenitic and alabastrine, are entirely of diagenetic origin from hydration of anhydrite are no longer acceptable. These are today considered products of aggrading neomorphism of gypsum itself, or even of primary depositional origin.

Modes of precipitation of evaporites are quite variable.

They precipitate either subaerially or in brine bodies, both at the air–brine interface and in their middle, after which they settle by gravity at the bottom where they can be remobilized by currents and redeposited elsewhere, even by turbidity currents. Evaporites may also precipitate at the brine–sediment interface or finally, within the sediments. Furthermore, many evaporitic minerals are not strictly depositional chemical precipitates, but early diagenetic minerals developed within nonevaporitic sediments by replacement or displacement, whereas others are early diagenetic replacements of true primary precipitates. This original diversity of formation of evaporite minerals is further complicated by their high susceptibility to large-scale postdepositional changes, resulting from subaerial exposure, groundwater circulation, burial, and tectonic processes (Schreiber et al., 1982). These changes result from their differential solubilities, leading to formation of metastable hydrates, solution–reprecipitation, aggrading or degrading neomorphism (recrystallization), transformation into another mineral phase—the most common transformation being the gypsum–anhydrite–gypsum cycle (**Fig. 10.1**), resulting from the sequence deposition–burial–uplift–erosion (Murray, 1964)—reciprocal complete or incomplete replacements, and, finally, flowage processes

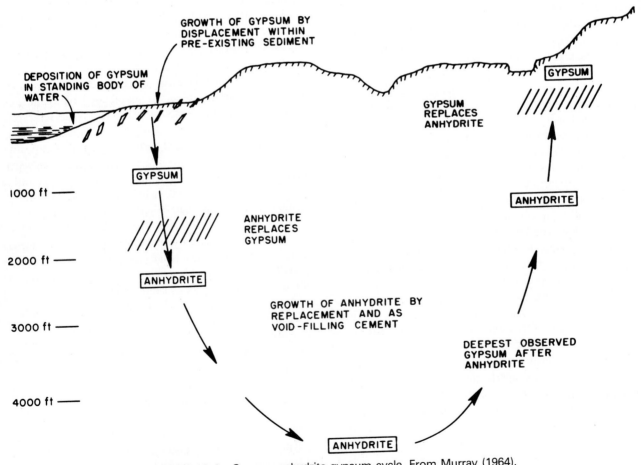

FIGURE 10.1 Gypsum-anhydrite-gypsum cycle. From Murray (1964). Reprinted by permission of the Society of Economic Paleontologists and Mineralogists.

under burial, halokinetic, and tectonic stresses. All these postdepositional changes are very difficult to recognize petrographically, even the simplest one, which, in the gypsum–anhydrite–gypsum cycle, is the regypsification of anhydrite into diagenetic gypsum by the action of groundwater and/or surface weathering (Holliday, 1970).

Two major consequences arise from the above mentioned processes. First, outcrops of unaltered evaporites are relatively rare. Second, studies are generally confined to subsurface materials from cores and mines, although in both instances original depositional features may have been destroyed to an appreciable degree.

An attempt to reconstruct the depositional environments of ancient evaporites does not mean that a strict application of uniformitarian principles is fully warranted (Warren, 1989). The scale of evaporite deposition varied throughout the Phanerozoic, with peaks of development during the Late Cambrian, the Permian, the Jurassic, and the Late Miocene. At such times, evaporites displayed thicknesses and horizontal extents that were two to three times as great as those of modern evaporites. This situation indicates warmer worldwide climates, extensive shallow epeirogenic seas, and tectonically induced basin restrictions. At present, such a combination of factors is not active on a large scale anywhere on the earth's surface. In other words, no modern counterpart exists today on the size and scale of ancient evaporites. Although the detailed and richly rewarding studies of Recent evaporitic settings remain extremely important, they cannot be applied directly to all ancient examples, because marine platform and deep-water basinal evaporites, for instance, are absent today (Kendall, 1984; Warren, 1989).

CLASSIFICATION

With the exception of the practical classification of anhydrite by Maiklem et al. (1969), no satisfactory textural classification of evaporites has been developed. The major reason for this situation is the textural complexity of these rocks, resulting form the above-mentioned combination of numerous depositional and diagenetic fabrics. Whenever megascopic or microscopic crystalline structures are observable, the description of their geometry is relatively easy, and some specific terms were created, such as selenitic gypsum, alabastrine gypsum, felty anhydrite, bacillary anhydrite, and chevron halite. If the depositional fabrics are displayed, they are usually identical to those of other sedimentary rocks, mainly sandstones and limestones, and the textural terms were borrowed or derived from them, such as laminated gypsum, graded-bedded gypsum, clastic gypsum (gypsrudites and gypsarenites), gypsum ooids (gypsolites), nodular anhydrite, laminated anhydrite, massive halite, laminated halite, and halite ooids (halolites). In metasomatic and deformed evaporites, the sedimentary rock terminology is comple-

mented by terms derived from metamorphic rocks as, for instance, porphyroblastic gypsum.

PETROGRAPHY OF MODERN AND ANCIENT EVAPORITIC SETTINGS

Throughout the Phanerozoic, evaporites formed in a wide range of depositional environments, including continental sabkha–playa (fluviolacustrine), coastal supratidal sabkhas, coastal salinas, shallow marine platforms, and deep marine basins. Numerous studies were undertaken of these various environments and led to a number of depositional models to which, whenever possible, data from modern analogs were added. These models were recently reviewed in detail (Kendall, 1984; Schreiber, 1986; Warren, 1989), and it became apparent that certain depositional fabrics, crystal structures, and mineralogical associations could provide clues for particular environments of deposition and in broad terms for their relative depth (**Fig. 10.2**). Therefore, the following brief review of the features of the major depositional environments is combined with a description of their characteristic petrographic features.

Continental Sabkha (Playa) Evaporites

These evaporites are usually precipitated toward the center of enclosed drainage basins, either from ephemeral or perennial saline lakes, or developed within corresponding mud flats, which represent the distal portions of corresponding alluvial fans (Kendall, 1984). These fans consist of the coarser clastic supply so that only the finest mud is spread over the basin, mainly by sheet-floods during storms. Besides these high-energy events, water circulation is limited to the shallow subsurface flow of groundwater toward the center of the basin. During this flow, groundwater undergoes evaporation losses and acquires a concentration gradient.

The mineralogy of these evaporites is complex and a direct function of the chemical composition of the groundwater, which in turn reflects the lithological composition of source areas and the type of weathering they undergo. During the intense evaporation process, saturation with respect to calcium and magnesium carbonates is rapidly reached, with the resulting precipitation of calcite, high-Mg calcite, and dolomite as soft micritic muds on the playa flats, which remain moist because of groundwater discharge. These carbonate muds are transported together with siliciclastic muds toward the center of the playa basin by storm sheet-floods and deposited with an original laminated or cross-laminated structure. These muds, which are not annual varves but reflect only storm periodicity, form a groundmass in which many types of minerals are formed, in a replacive or displacive manner, among which are gypsum, anhydrite, halite, gaylussite, trona, thenardite, and mirabilite, which were in-

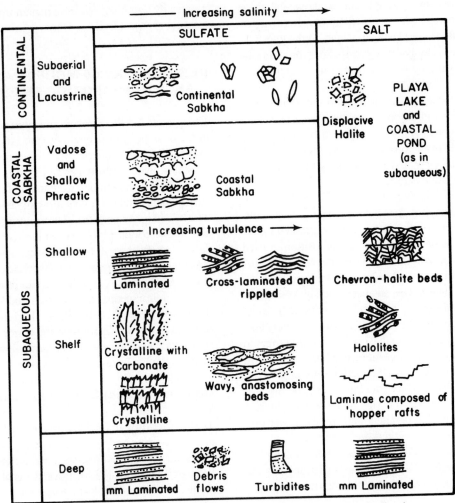

FIGURE 10.2 Summary of the various evaporite textures indicative of particular physical environments. From Warren (1989). Reprinted by permission of Prentice Hall, Englewood Cliffs, N.J.

vestigated in detail by Hardie and Eugster (1970) and Eugster and Hardie (1978). However, many complex processes combine to disrupt continuously and even destroy these structures. They are, for instance, additional groundwater discharge, growth and dissolution of ephemeral evaporite crystals, episodes of surface desiccation with extensive mud cracking, wind deflation concentrating surficial crystals such as gypsum into lag deposits or dunes, dehydration of gypsum into anhydrite, and dissolving effects of rainwater on efflorescent crusts (Hardie et al., 1978). Toward the center of the playa basins, where groundwater becomes increasingly saline, gypsum and halite in particular among other minerals become eventually stable and are preserved in appreciable amounts if the conditions are appropriate. Halite was described in modern continental playa sediments (Handford, 1982; Casas and Lowenstein, 1989) and also in Permian red beds interpreted as continental playas (Smith, 1971). Euhedral to subhedral displacive halite cubes occur widely dis-

persed in the laminated matrix or may develop at such a scale as to interlock in a mosaic, leaving only interstitial polyhedral pockets of matrix. Halite can also occur as coatings around detrital grains and as skeletal hopper crystals.

As the playa lake evaporates further and decreases in size, the ponded brines eventually precipitate crusts of massive halite or trona. Casas and Lowenstein (1989), in a comparison of the petrographic features of modern and Quaternary continental playa halites with some Permian occurrences, showed the greater similarity of their depositional and early diagenetic evolution under shallow burial conditions.

In such a salt pan, single storm sheets consist of a couplet of a thin mud lamina, overlain by a thicker crystalline halite or trona layer. These couplets are also highly modified or destroyed by subsequent salt growth from the underlying groundwater when the lake dries up again.

In the case of perennial saline lakes, laterally continuous layers of similar evaporitic minerals are deposited and separated by siliciclastic mud laminae. Since evaporation is a continuous process in such lakes, the layering does not represent annual varves, but expresses the periodicity of major storms whose influx of mud interrupts chemical precipitation. Although storms could be seasonal, they are most probably irregularly spaced in time.

Supratidal (Coastal Sabkha) Evaporites

On shallow stable shelves under arid climates and low eolian sand influx, the seaward progradation of subtidal and intertidal facies generates broad coastal flats (or sabkhas) that lie just above high tide and extend between coastal lagoons and continental sedimentation.

Depositional and diagenetic processes under complex and delicate hydrological conditions develop this unusual environment, which has been extensively studied in the Persian Gulf and reviewed by many authors (Shearman, 1978; Kendall, 1984; Schreiber, 1986; Warren, 1989). The typical shallowing-upward sequence is described here in a landward direction as a juxtaposition of environmental belts, which is its horizontal expression. It begins with subtidal to low intertidal pelletoidal aragonite facies, overlain by upper intertidal cyanobacterial mats, which eventually turn into a bacterial peat layer as the shoreline progrades and the evaporitic environment begins.

The supratidal portion of the sabkha is divided into three zones: lower, middle, and upper. In the upper portion of the upper intertidal zone and in the lower supratidal zone, syndiagenesis develops to an appreciable extent. Interstitial precipitation of aragonite and of small displacive discoidal–lenticular crystals of gypsum begins within the cyanobacterial mat whereas surface sediments are largely cemented by aragonite, magnesite, and protodolomite. In the most seaward portion of the supratidal zone, the cyanobacterial mat is gradually disrupted by growth of gypsum crystals, which becomes so extensive as to form a layer by itself, called the "gypsum–mush" layer. As the lower supratidal zone progrades, the gypsum crystals become so disruptive that the cyanobacterial mat is almost destroyed and changes into a loose matrix between the crystals. Many gypsum crystals either engulf cyanobacterial filaments as well as carbonate mud and pellets; others develop as poikilotopic crystals, however, without reaching the surface.

In the middle zone of the supratidal environment, evaporation causes salinity of the pore waters to rise sharply. The carbonates are dolomitized; high salinities and high temperatures cause precipitation of ephemeral halite, but the main process is the gradual change of the earlier gypsum into nodular anhydrite, which forms a "chickenwire" structure (**Plate 31.B**) when growing nodules reciprocally interfere and squeeze the interstitial carbonate mud into an irregular reticulated network.

The process of anhydrite formation continues through addition to the initial nodule of crystal pseudomorphs until the gypsum crystals are eventually completely destroyed. Below the gypsum mush layer, where temperatures are lower, gypsum continues to precipitate, but eventually it will be also converted to anhydrite.

The surface aspect of the middle part of the supratidal zone is variable; small anhydrite nodules increase in size and concentration landward, forming nodular layers with an interlocking surface of polygonal dishes, possibly indicating desiccation, and at times covered by eolian sands or planed by storm-driven floods from the lagoon.

The upper part of the supratidal zone is flooded only rarely, every 4 to 5 years, and all its features are syngenetic. They result from the continuous addition of near-surface nodular anhydrite layers, as well as from the alteration of earlier lenticular gypsum to anhydrite nodules with chickenwire structures. These alteration products result from groundwater action, which consists of a mixture of marine and continental components concentrated by rapid evaporation throughout the sabkha section. Isolated gypsum crystals and those in the mush are still involved in the process.

It is clear that in present-day sabkhas the generalized production of anhydrite nodules results from both the change of a gypsum precursor and direct additional anhydrite precipitation (Shearman, 1978). It is interesting to observe that anhydrite nodules are mostly soft and puttylike because of their high natural moisture content. They consist of lathlike cleavage fragments, together with few unbroken rectangular crystals, both of which appear randomly oriented in the center of the nodules, but tend to become, along the edges of the nodules, subparallel to their margins. The nodules grow displacively, and as new crystals develop within the framework of earlier ones, a considerable displacement of the individual constituents occurs. Because of the thin, platy habit of the crystals and the easy, perfect cleavage of anhydrite, rupturing of the crystals is widespread and the entire nodule becomes finally an aggregate of cleavage flakes. This growth by nucleation of new crystals probably results from the reciprocal chipping of cleavage flakes by a process comparable to "collision-breeding" observed when crystals collide in agitated saturated solution of the same salt (Shearman,1978). Anhydrite nodules tend to be aligned in layers, and after their reciprocal interference forming the chickenwire texture, additional growth requires more space and the layers become contorted into ptygmatic, enterolitic folds, and microdiapirs whose formation is facilitated by the moist nature of the nodules and the easy slippage of cleavage fragments. Anhydrite nodules of ancient sabkhas no longer show aggregates of cleavage flakes, but rather a neomorphic felty texture.

In some areas, mixed groundwaters still bring in sulfates and chlorides, which can penetrate deep into the sabkha section and develop large displacive lenticular crystals of gypsum as far down as the buried lagoonal muds and the cyanobacterial mats. Halite is also precipitated there as displacive hopper crystals with extreme skeletal development called "pagoda" halite (Gornitz and Schreiber, 1981).

The surface of the upper supratidal zone is often raised by the intense generation of anhydrite to the extent that the uppermost layers are lifted above the capillary zone into the vadose environment, where they are dried and destroyed by deflation. Appreciable eolian sediments can accumulate at the surface in which secondary gypsum and anhydrite crystals (**Plate 31.A**) also form. Strong deflation can generate rare marine flooding, which deposits thin, ephemeral halite crusts and even displacive halite crystals. But in most modern sabkhas, halite is not an accumulative phase; it is either blown away or dissolved.

Shearman (1978) stressed the need to compare the massive anhydrite generation in sabkhas with the processes occurring in normal seawater undergoing evaporative processes. Normal seawater contains much more sulfate than required to satisfy calcium if precipitated as calcium sulfate, mainly gypsum. Therefore, gypsum precipitates until all or most of the calcium ions are used up, but large amounts of sulfate still remain unused in the residual brine. In sabkhas, the seawater brines are concentrated interstitially within the aragonite mud. Precipitation of gypsum increases the Mg to Ca ratio in brines and promotes dolomitization. In turn, calcium ions liberated by dolomitization combine with the residual sulfate to precipitate more calcium sulfate. The general effect of the sabkha system is to produce twice as much calcium sulfate as would be formed if the same volume of water evaporated in an open basin (Shearman, 1978).

Nevertheless, many aspects of this anhydrite-producing system remain poorly understood. Warren and Kendall (1985) and Warren (1989) discussed at length the pitfalls of applying the present-day marine sabkha model to ancient evaporite sequences without a complete understanding of the implications of the complex hydrology and depositional setting described in the Persian Gulf.

This warning was illustrated by Machel and Burton (1991), who described in the subsurface Late Devonian of Alberta, Canada, gypsum and anhydrite nodules resembling those of sabkhas, but formed during burial of at least several tens to several hundreds of meters by the action of phreatic solutions during or after extensive burial dolomitization of the host rock. The burial and postcompaction origin of these nodules is demonstrated by the following features: they occur within or juxtaposed to dark, bituminous argillaceous seams and pockets, anastomosing bituminous veinlets, and carbonaceous halos, all of which are expressions of pressure-solution seams or stylolitic systems; they replace the host rock and crosscut the boundaries between host rock and stylolitic residues; they contain materials formed during burial, such as authigenic pyrite and dolomite; they show no traces of mechanical compaction; and, finally, the diagenetic texture of anhydrite nodules is felty, bacillary, or blocky, consisting of an irregular mosaic of partially corroded crystals in a finer crystalline matrix. Although anhydrite crystals along nodular margins have a tangential alignment as in sabkha nodules, the crystals are mixed with the surrounding insoluble residues and are partially corroded.

Shallow Marine Evaporites

The deposition of ancient shallow marine evaporites, of which there is no modern equivalent, implied the following conditions: brines saturated with respect to gypsum, halite, or potash–magnesite; potential action of currents and waves causing synsedimentary erosion, transport, and redeposition; significant influence of cyanobacterial activity, and depths of water ranging from a few centimeters to 20 m or more, with a probable average of 5 m. Evaporite precipitation occurred at the air–water interface, at the sediment–water interface, and within the sediments themselves.

Knowledge of ancient shallow marine evaporites was gained mainly from intensive studies undertaken during the 1970s of the Late Miocene (Messinian) of the Mediterranean, both from cores of the Deep Sea Drilling Program (Garrison et al., 1978) and from outcrops in peninsular Italy, Sicily, and Spain (Schreiber et al., 1976). In both situations, burial was not deep enough for gypsum to revert to anhydrite with destruction of depositional fabrics. More recent studies showed that modern shallow-water evaporites, besides those deposited in a few continental basins, are mainly formed in coastal basins or "salinas," which are in fact saline lakes in which the water level is below that of the adjacent ocean and which are replenished by seawater seepage (Arakel, 1980; Kushnir, 1981; Warren, 1982; Warren and Kendall, 1985; Warren, 1989). These studies provided many data confirming the interpretations obtained from assumed ancient shallow-water evaporites and also confirmed that the latter were deposited, with a great depositional diversity, on a scale and in tectonic settings unknown today, such as platforms, rimmed platforms, and ramps.

Under shallow marine conditions, evaporites are deposited more or less in the predictable mineralogical suite of evaporating seawater, that is, gypsum, halite, and potash–magnesia salts. Gypsum displays several textures: laminated, coarsely crystalline or selenitic, coarse clastic (gypsrudites and gypsarenites), and oolitic (gypsolites). Halite is displacive, in crusts, clastic, and oolitic (halolites).

Laminated Gypsum. Laminated gypsum, which can be preceded by swarms of single crystals developed in carbonate muds (**Plate 32.A**), consists of current- or tide-deposited alternating laminae of carbonate micrite or clay minerals and clastic gypsum particles showing normal or reverse graded bedding (**Plate 32.B**). The silt- and sand-size gypsum particles are in fact gypsum crystals, either single or twinned or cleavage fragments, that are generally arranged with their greatest dimension parallel to bedding. These constituents developed under a variety of conditions. They could be either reworked fragments of bottom crusts, crystals precipitated at the air–water interface that sank to the bottom and were reworked, or crystals that grew displacively within the bottom sediment and were also reworked. These gypsum crystals and cleavage fragments eventually develop suturing by reciprocal interpenetration and diagenetic overgrowth, which combine to form an interlocking mosaic. In some laminae, the gypsum crystals apparently precipitated directly from the air–water interface and sank to the bottom without undergoing any appreciable transport. Under such conditions, the crystals displace or poikilotopically enclose organic material and cyanobacterial mats, and their laminations are a function of the structures of the mats. Detailed petrographic examination shows if gypsum did encrust cyanobacterial mats or if the mats acted as trapping agents for the carbonate particles and the gypsum crystals.

The gypsum laminae can display a great variety of sedimentary structures, indicating the action of periodic high-energy events such as cross-bedding, ripple-drift bedding, basal scoured surfaces, and intraformational microbreccias, sometimes graded bedded, which are interpreted as storm deposits when the evaporitic flats were flooded by sediment-laden waters.

Reverse-graded laminae of gypsum (**Fig. 10.3**) were explained in various ways. Ogniben (1955, 1957) suggested that reverse grading was a diagenetic feature if the lamina is interpreted as representing an annual evaporative cycle. The first precipitation was gypsum, but as brines became more concentrated, anhydrite precipitated. The next influx of fresher water would hydrate anhydrite and change it into a more coarsely crystalline gypsum, thus generating the reverse grading. This coarser gypsum is often arranged as a marginal palisade perpendicular to the top of the lamina (microselenite). Schreiber et al. (1976) assumed that no anhydrite was precipitated and that the cyclic dilution of the overlying brine may cause early aggrading neomorphism, that is, growth of coarser selenite crystals at the top of each gypsum lamina, sometimes accompanied by swelling of the top of the bed (**Fig. 10.4**). Garrison et al.(1978) assumed an annual concentration of the brines depositing gypsum in which the earlier bottom crystals are small and numerous and the later fewer and larger, and that this reverse graded bedding is controlled by the original size of the precipitated gypsum crystals. A mechanical interpretation was presented by Hardie and Eugster (1971), who interpreted the reverse grading as highly concentrated coarse gypsum sand sheets deposited during storm surges along shallow offshore banks over underlying gypsum particles. These particles were trapped and bound by cyanobacterial mats expressed by the carbonate laminae. This reverse grading may be subsequently emphasized by early diagenetic aggrading neomorphism and induration during quiet periods between storms.

Laminated gypsum, developed over extensive platforms and ramps, can be persistent for long distances, but not at the scale of those deposited in deep-water basinal conditions discussed below. They are also distinguished by associated cyanobacterial mats and mounds, by clastic textures indicating active bottom currents and storm action, and, finally, by possible exposure features leading to their disruption and subaerial diagenesis. In some instances, laminated

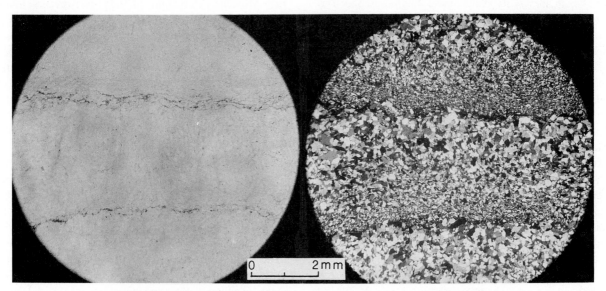

FIGURE 10.3 Reverse graded bedding in primary gypsum laminae separated by thin argillaceous streak. Middle Miocene (Messinian), Sicily. Left: plane-polarized light; right: crossed nicols. Photomicrograph courtesy of Leo Ogniben, Montecatini S.A. (Settore Miniere), Milano, Italy.

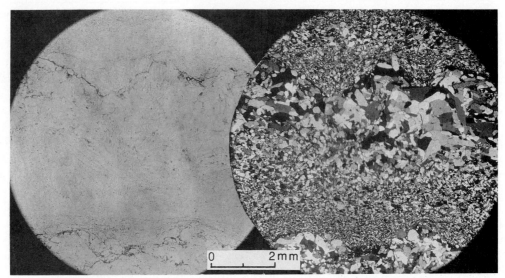

FIGURE 10.4 Reverse graded bedding in primary gypsum lamina produced by aggrading neomorphism to coarser selenite crystals accompanied by swelling of top of lamina. Left: plane-polarized light; right: crossed nicols. Middle Miocene (Messinian), Sicily. Photomicrograph courtesy of Leo Ogniben, Montecatini S.A. (Settore Miniere), Milano, Italy.

gypsum shows a small amount of deformation due to slumping, indicating that the laminae remained in soft conditions for a certain length of time after deposition.

Burial conversion into anhydrite of the above-mentioned depositional or early diagenetic fabrics of gypsum leads to an appreciable loss of original properties due to the development of two diagenetic textures called felty and bacillary, which are readily visible under the microscope. The felty texture (**Plate 31.D**) consists of an association of small and irregular lamellae. Among them are rodlike, spindle-shaped, and rhombic individuals, intergrown in a felty network that includes subrectangular and irregular granules. Many of the lamellar individuals appear curved and show undulatory extinction. The elongation of the lamellae varies much versus crystallographic directions, usually remaining oblique in the irregular laminae and parallel to [010] in the rectangular individuals. The bacillary (**Plate 31.E**) or rodlike structure consists mainly of narrow rods or prisms, elongated parallel to [010], which is the normal habit of anhydrite. However the rods are sometimes elongated obliquely to the crystallographic elements, mainly along the brachydome [011]. The crystals may be aggregated into radiating sheaves. When the boundaries of the individuals are parallel to the cleavages, they are straight; however, they become indented in the crystals elongated obliquely to the crystallographic elements and may even reach a steplike appearance.

Megascopically, the conversion to anhydrite may lead to some convergence with textures developed under coastal sabkha conditions. For instance, slumped and deformed gypsum laminae changed to anhydrite may look like primary displacive anhydrite; but, in general, the geological context, thickness and scale, and mainly the association of textures of shallow marine evaporites prevent such a possible confusion.

Coarsely Crystalline (Selenitic) Gypsum. This type of gypsum occurs as numerous varieties of single and twinned crystals, groupings or stellate clusters, crusts, and superposed beds (Schreiber, 1978). It is best known from the Late Miocene (Messinian) of Italy and was also found recently forming in the Salinas of South Australia (Warren, 1982). Selenitic gypsum in beds (**Plate 32.C, D**) consists mainly of orderly rows of vertically standing, elongate, and generally swallowtail twinned crystals that range in size from a few centimeters to a few meters (Schreiber et al., 1982; Schreiber, 1986; Kendall, 1984). The individual crystals are either vertically aligned in a palisade fabric or arranged into radiating-upward conical clusters (**Fig. 10.5**). Interstitial materials between crystals consist of carbonate micrite, silt- or sand-size gypsum grains, organic matter, or overgrowths on the crystals themselves, which contribute to generate an interlocking crystal mosaic. Swallowtail selenite crystals show, in all cases, faint laminations consisting of anhydrite inclusions, carbonate mud pellets, cyanobacterial filaments, and thin mats. These laminations either pass through the crystals along dissolution surfaces that truncate them or follow the successive positions of the growing crystals. The identical situation observed in similar large crystals growing today in the Salinas of South Australia (Warren, 1982) demonstrated that these large selenite crystals are of primary origin and grew in very shallow water.

Other gypsum crystals associated with the large swallowtail twins display strange forms that are half-twins, with

FIGURE 10.5 Bed of vertically standing swallowtail-twinned selenite crystals of primary origin arranged in radiating-upward clusters. Middle Miocene (Messinian), Sicily. Photograph courtesy of Leo Ogniben, Montecatini S.A. (Settore Miniere), Milano, Italy.

aborted arms often produced by absorption of impurities. Together with these unusual twinned crystals are strangely twisted forms that are not twins, but are crystal splits generating palmate (siva) to fan-shaped clusters of subparallel crystals (**Fig. 10.6**). Splitting occurs at a variety of angles and results from the inclusion of organic matter along curved crystal faces (Orti-Cabo and Shearman, 1977; Schreiber et al., 1982; Schreiber, 1978, 1986).

The various types of crystals of subaqueous gypsum are far from being pure and reveal in thin section a variety of included foreign bodies (Schreiber, 1978). They are cyanobacterial filaments, small spheres, and pellets often concentrated along growth faces, which are interpreted as brine shrimp eggs and fecal pellets; clastic particles include volcanic glass shards, silt-size grains of quartz, feldspars, and other silicates, reworked fragments of gypsum silt and sand, pollen, plant material, reworked foraminifers, diatoms, Radiolaria, and coccoliths. Finally, these primary selenitic crystals also contain small anhydrite inclusions, which according to Ogniben (1957) are not randomly oriented but lie in the [010] plane in zones parallel to [120]. These anhydrite crystals are small and usually euhedral and often display a sawtooth aspect. Their origin is disputed, but they seem to indicate a coprecipitation of anhydrite with primary subaqueous selenitic gypsum under special but unclear conditions.

Coarse selenitic gypsum crystals may be difficult to recognize upon burial conversion to anhydrite. Inclusions of carbonate pellets and of cyanobacterial filaments that defined crystal faces might still appear as relicts within the felty, bacillar, or massive anhydrite mosaic, but inclusions parallel to bedding could simulate laminated gypsum or sabkha anhydrite layers. In fact, Rouchy (1976) stressed the fact that entire beds of large selenite crystals, replaced during burial by anhydrite mosaic, can closely resemble sabkha dis-

FIGURE 10.6 Fan-shaped cluster of subparallel crystals of selenite of primary origin, which are pseudotwins resulting from split twinning. Middle Miocene (Messinian), Cyprus. Photograph courtesy of B. Charlotte Schreiber, Queens College, C.U.N.Y., Flushing, New York.

placement nodules of anhydrite. However, in the case of the Cretaceous Ferry Lake Evaporite of the East Texas Basin, vertical and variably inclined anhydrite nodules, with their outlines preserved by carbonates, organic matter, or clay minerals, appear clearly as replacement products of original, large, primary selenitic gypsum crystals (Loucks and Longman, 1982). Warren and Kendall (1985) confirmed them to be ancient burial equivalents of the subaqueous gypsum crystals forming today in the salinas of Southern Australia (**Fig. 10.7**).

Coarse Clastic Gypsum (Gypsarenites and Gypsrudites). Gypsum sands, pebbly sands, and pebbles consisting of abraded gypsum crystals or cleavage fragments,

associated in places with carbonate mud intraclasts, build distinct intraformational layers called, respectively, gypsarenites and gypsrudites. These beds exhibit all the types of current-laid bedforms (graded bedding, cross-bedding, climbing ripples, oscillation ripples, interference ripples, scour-and-fill structures, load casts, ball-and-pillow structures, and others), indicating current and wave action (Schreiber et al., 1976). Gypsarenites appear almost entirely formed by detrital grains of gypsum; the clastic texture may be seen megascopically by differences in color between grains, but, in many cases, the texture recalls a nondetrital crystalline aggregate. Under the microscope, the detrital texture is evident in spite of the fact that it simulates the appear-

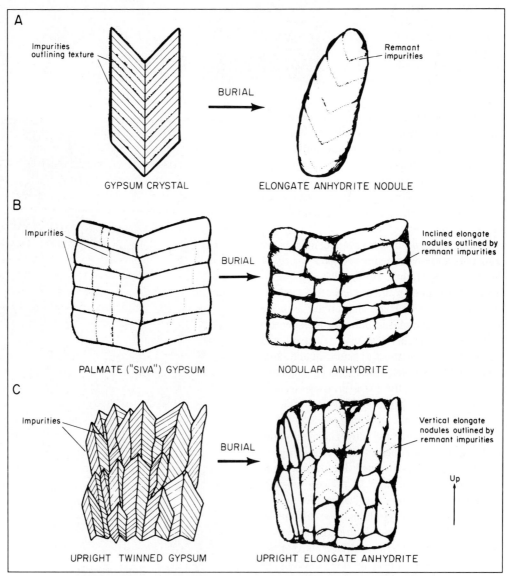

FIGURE 10.7 Schematic diagram showing how some of the various types of primary gypsum are converted into elongate diagenetic anhydrite nodules upon burial. From Warren and Kendall (1985). Reprinted by permission of the American Association of Petroleum Geologists.

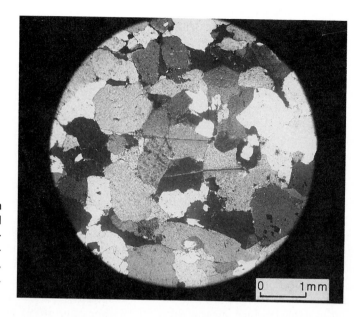

FIGURE 10.8 Gypsarenite with clastic fabric recalling a granitoidal texture. Crossed nicols. Middle Miocene (Messinian), Sicily. Photomicrograph courtesy of Leo Ogniben, Montecatini S.A. (Settore Miniere), Milano, Italy.

ance of a granitoidal rock. The grains are angular, monocrystalline, and interlocking by pressure solution, with no apparent cement (**Fig. 10.8**). The detrital texture is generally emphasized by the occurrence of clastic grains other than gypsum and reworked foraminifers or by a very reduced amount of clay minerals (**Fig. 10.9**).

Coarse clastic gypsum was obviously deposited in a gypsum-saturated environment forming beaches, shoestring sands, sand sheets, offshore bars, and channel fills between more quiet areas where laminae and selenitic gypsum were being formed. In other settings, called "cannibalistic," of the Messinian of the Northern Apennines (Vai and Ricci Lucchi, 1977), autochthonous selenite crystals, growing as far as the shorelines, were removed and carried basinward by slope-controlled flows (fluvial currents and debris flows). Tongues

of clastic gypsum initially advanced from the mouths of torrential streams and bypassed areas where selenitic gypsum continued to grow. Removal of gypsum was intensified by increased subaerial exposure of marginal evaporites during lowering of sea level; eventually, shallow alluvial cones of reworked selenite encroached extensively on the basin area.

The same cannibalistic process is likely to have led to the deposition of turbiditic gypsarenites in deeper basinal conditions described in many locations of the Messinian of the Mediterranean (Parea and Ricci Lucchi, 1972; Ricci Lucchi, 1973; Schreiber et al., 1976; Schlager and Bolz, 1977). The entire Bouma sequence is sometimes fully developed (Schreiber et al., 1976), but generally the turbidites consist only of graded units or have poorly developed parallel laminae in their uppermost portions (Schlager and Bolz,

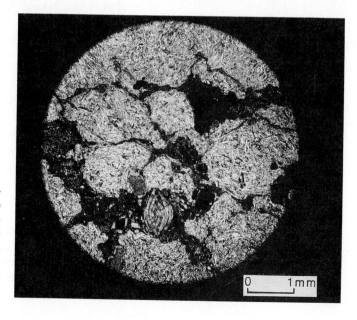

FIGURE 10.9 Gypsarenite replaced by diagenetic felty anhydrite upon burial. Notice reworked foraminifer. Crossed nicols. Middle Miocene (Messinian), Sicily. Photomicrograph courtesy of Leo Ogniben, Montecatini S.A. (Settore Miniere), Milano, Italy.

1977). These turbidites may be entirely formed by gypsum or may show associated carbonate and siliclastic materials. The gypsum-rich turbidites of the Miocene of the Periadriatic Basin described by Parea and Ricci Lucchi (1972) form a thin unit within a thick siliciclastic sequence interpreted as deep-sea fan.

In the Middle Miocene of southern Poland (Peryt and Kasprzyk, 1992), laminated gypsum deposits are accompanied by well-developed redeposition facies of gypsrudites consisting of mechanically abraded clasts of selenitic, laminated, and alabastrine gypsum set in a matrix of gypsarenites. The gypsrudites are intercalated among graded beds consisting of gypsarenites overlain by gypslutites deposited by turbidity currents as incipient Bouma sequences. The gypsrudites show pervasive microfolding related to compressional and extentional strain, combined with extensive microfaulting, indicating semicoherent behavior during mass movements. The stratiform geometry of the gypsrudites and the fact that the intensity of their slumping appears independent of the paleoslope suggest earthquake shocks as the major initial cause of slumping and brecciation.

Collapse and dissolution gypsum breccias were also reported (Garrison et al., 1978; Schreiber, 1978). In the highly interlocked constituents of such breccias, it is possible to recognize twinned selenite crystals broken in various ways, with insoluble residues packed between them. Renewed precipitation develops overgrowths into the interstitial dissolution voids, which can incorporate earlier-formed residues. If the interstitial residues are sufficiently abundant, a new phase of nucleation can begin and completely new twinned selenite crystals may grow. In other situations, if dissolution is carried to an extreme, crystals become isolated, displaced, and broken again, and an unconsolidated residual gypsum breccia is generated.

Oolitic Gypsum (Gypsolites). Gypsum ooids were described among gypsarenites representing shoreline deposits in the Messinian of the southern Apennines (Ciaranfi et al., 1973; Schreiber, 1978). They appear to form dunelike structures with internal cross-bedding. These ooids show a radial overgrowth of fine gypsum crystals forming a rim around the original gypsum nuclei, which are variably rounded by a combination of mechanical abrasion, cleavage-controlled rupturing, and dissolution of edges and corners. The overgrowth rim displays growth layers marked by rows of carbonate inclusions usually also arranged in a radial fashion. High magnification reveals that the crystallites of the rim show distinct growth increments as in primary gypsum crystals along the traces of the [120] planes, whereas the carbonate inclusions are aligned along microdissolution surfaces that represent periods of less saline waters, perhaps associated with cyanobacterial growths expressing carbonate precipitation or trapping. Hardie and Eugster (1971) also re-

ported gypsolites in their study of the Messinian evaporites of Sicily; however, they could have been quiet-water carbonate ooids partially replaced by gypsum. In summary, it seems that gypsum ooids or rim overgrowths formed under high-energy hypersaline conditions within the range of gypsum precipitation. These conditions appear similar to those of halite ooids (halolites) and halite overgrowths being formed today under shallow and highly agitated shoal water under the action of shoreward winds in the Dead Sea (Weiler et al., 1974).

Halite. Halite occurs in shallow waters in three major forms: as crusts that by superposition form bedded halite, as detrital halite, and as displacive skeletal crystals. Halite generally nucleates at the air–water interface, within the water column by brine mixing (Raup, 1970), at the bottom, or within the sediments.

Halite crusts occur in present-day solar salt ponds (Handford, 1990), continental sabkhas or playas (Casas and Lowenstein, 1989) and brine ponds of coastal sabkhas (Shearman, 1970). They were reproduced in the laboratory (Arthurton, 1973) and reported in numerous instances of shallow water environments in the geological column (Wardlaw and Schwerdtner, 1966; Shearman, 1970; Gornitz and Schreiber, 1981; Kendall, 1984; Schreiber, 1986; Hovorka, 1987; Warren, 1989). Halite crusts form by means of several processes. The first is the foundering of rafts of rectangular to cuboidal clear halite crystals, with cube corners pointing down, formed at the air–brine interface in which each crystal undergoes overgrowth after reaching the bottom (**Fig. 10.10A**). The second process consists of bottom-nucleated crystals growing in place and appearing similar in vertical section to the first type. However, horizontal lineation of fluid inclusions within these crystals indicates successive growth faces (**Fig. 10.10B**). The third process consists of the foundering of floating rafts of pyramidal four- or six-sided hopper crystals formed at the air–brine interface (**Fig. 10.11**). When these rafts reach the bottom, they tend to grow on upward-facing faces, developing a chevron-shaped lineation from edge-oriented crystal growth; but in their lower parts they still display relics of the original hopper structure (**Fig. 10.10C**). The fourth process consists of bottom-nucleated crystals growing in place (**Fig. 10.10D**) and having random to corner-oriented aligned individuals. The chevron-shaped growth lineations are similar to those produced by the third process. Cycles of flooding and dissolution may develop truncated surfaces with thin intercalations of gypsum, detrital carbonate or siliciclastic particles, and irregular cavities within the crusts or layers of the chevron-type halite. Subsequent reconcentration of the brines leads to the nucleation of a new layer of halite, which grows upward in the same competitive chevronlike pattern while the cavities are filled by secondary clear halite (Shearman, 1970, 1978). Finally, halite crusts can be produced by mechanical accumu-

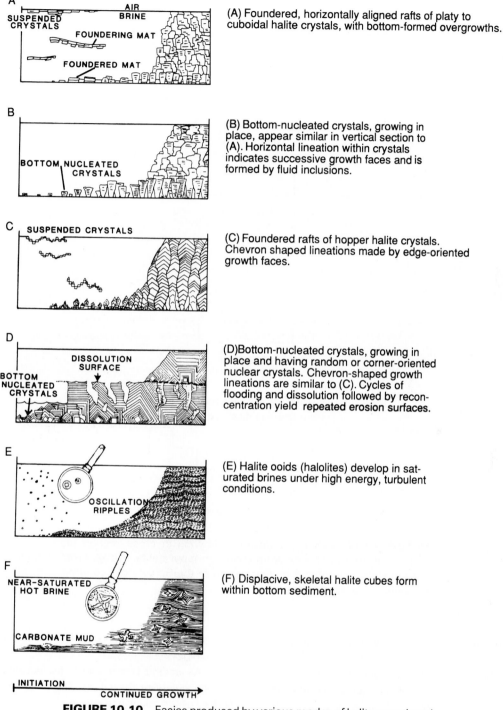

(A) Foundered, horizontally aligned rafts of platy to cuboidal halite crystals, with bottom-formed overgrowths.

(B) Bottom-nucleated crystals, growing in place, appear similar in vertical section to (A). Horizontal lineation within crystals indicates successive growth faces and is formed by fluid inclusions.

(C) Foundered rafts of hopper halite crystals. Chevron shaped lineations made by edge-oriented growth faces.

(D) Bottom-nucleated crystals, growing in place and having random or corner-oriented nuclear crystals. Chevron-shaped growth lineations are similar to (C). Cycles of flooding and dissolution followed by recon-centration yield repeated erosion surfaces.

(E) Halite ooids (halolites) develop in sat-urated brines under high energy, turbulent conditions.

(F) Displacive, skeletal halite cubes form within bottom sediment.

FIGURE 10.10 Facies produced by various modes of halite growth and emplacement. From Gornitz and Schreiber (1981). Reprinted by permission of the Society of Economic Paleontologists and Mineralogists.

lation of halite ooids (halolites), developed in saturated brines under turbulent conditions, displaying oscillation rip-ples and other high-energy bedforms (**Fig. 10.10E**). Halol-ites described as forming today in the Dead Sea (Weiler et al., 1974) are usually spherical and polished, ranging in size from 0.8 to 4 mm. Most halolites display concentric zones of impurities and anhydrite inclusions. A radial orientation of halite crystals was occasionally observed; in other instances, individual halolites consist of single halite crystals as shown by cubic cleavage planes transecting entire grains. The nu-cleus consists generally of corroded halite cubes, but can also be dark clasts of dolostone, limestone, or shale around

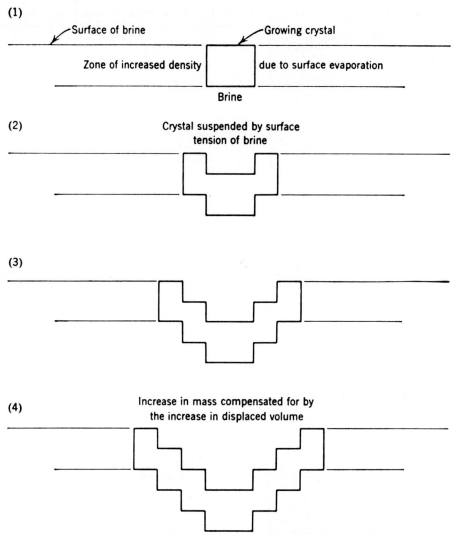

(1)

Surface of brine — Growing crystal

Zone of increased density | due to surface evaporation

Brine

(2)

Crystal suspended by surface
tension of brine

(3)

(4)

Increase in mass compensated for by
the increase in displaced volume

FIGURE 10.11 Schematic diagram showing in cross section the process of growth of a pyramid-shaped hopper halite crystal. From Dellwig (1955). Reprinted by permission of the Society of Economic Paleontologists and Mineralogists.

which halite precipitated. Halolites eventually accumulate on the high parts of beaches, forming low ridges.

Detrital halite is probably more widespread than generally assumed (Kendall, 1984) because it is subjected to recrystallization accompanied by loss of depositional fabrics. It probably derived from the mechanical reworking of halite crusts deposited under quieter conditions and consists of fragments of surface-grown hopper crystals or small cubes, as well as reworked materials from bottom-growing crusts. Crystal growth may continue after deposition of the clastic layers, obscuring their real origin in clear halite.

Displacive halite (**Fig. 10.10F**) was formed under laboratory conditions from solutions by Arthurton (1973) and

Southgate (1982) in experiments that revealed its diversity of morphology.

As previously mentioned, displacive halite occurs also in continental and coastal sabkhas, forming just beneath the surface of soft and water-saturated sediments. Similar conditions occur in shallow marine environments, where sediments are either incorporated in crystals or pushed aside and where phases of subcontemporaneous dissolution are frequent. Simple halite crystals indicate crystal growth in saturated conditions, but generally the crystals have hoppered or skeletal faces and can reach sizes of more than 10 cm (Handford, 1982). Hopper crystals are believed to form as the result of crystal poisoning, but more likely they indicate supersaturation whose increase leads to increasing skeletal development, enclosing large amounts of the surrounding material, such as the "pagoda"-type forms, which in two dimensions appear as the outline of a Thai temple (Gornitz and Schreiber, 1981; Southgate, 1982). Consequently, abundant skeletal displacive halite in shallow marine evaporites is believed to be an indicator of strandline or basin-edge conditions (Warren, 1989). However, Parnell (1983) felt that it could grow subaqueously, although the sequence he investigated displayed evidence of periodic emergence.

Ancient shallow-water halite consists frequently of laminated to bedded halite that in fact represents superposed subaqueous crusts, separated or not by dissolution zones, and showing typically vertical aligned chevron structures (**Fig. 10.12**) and associated pockets of clear halite filling former dissolution cavities (Dellwig, 1955; Wardlaw and Schwerdtner, 1966; Lowenstein, 1982). Kendall (1984) stressed the importance of the presence or absence of fluid-filled inclusions in halite crystals of subaqueous crusts and their use as an indirect indicator of water depth at the time of deposition. Inclusion-rich layers in zoned halite crystals (cloudy halite) develop when brines are supersaturated and crystal growth is rapid. Reduced brine concentration, as the result of halite precipitation, leads to slower, more perfect, and inclusion-free halite layers (clear halite). These cloudy and clear alterations of chevron halite are in fact laminae 0.1 to 0.3 mm thick whose thickness is related to diurnal fluctuations of evaporation rates, which is the only process capable of inducing such rapid changes in brine concentration and brine stability when the brine bodies are of relatively small volume and shallow depth. This situation is in complete contrast with the clear or darker but smaller halite crystals of deep-water evaporites discussed below (Kendall, 1984).

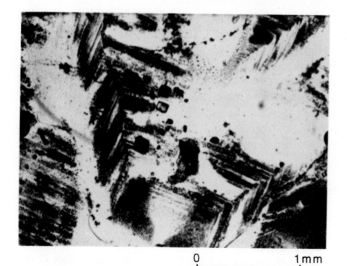

0   1mm

FIGURE 10.12 Pyramid-shaped hopper crystal of halite outlined by brine-filled negative crystals. Plane-polarized light. Photomicrograph courtesy of L. F. Dellwig, Department of Geology, University of Kansas, Lawrence, Kansas.

In a study of a solar salt pond on Bonaire, Netherlands Antilles, Handford (1990) stressed the fact that the various crystal forms and facies (halite pisoids) of primary halite are energy and/or depth dependent, and that some primary features, if preserved in ancient halite deposits, could be used to infer physical energy conditions, subenvironments such as low- and high-energy shorelines, and extremely shallow water depths in ancient evaporite basins.

Finally, cases of halite pseudomorphs after anhydrite-replaced selenitic gypsum or after selenitic gypsum itself have been reported. This replacement appears to characterize environments of deposition of full cycles of evaporites accompanied by appreciable halite precipitation. The origin of this replacement also seems to be incompatible with large amounts of simultaneous replacement of gypsum by anhydrite and is interpreted as a synsedimentary shallow-water thermal-disequilibrium feature (Schreiber and Walker, 1992). The early diagenetic origin of this pseudomorphic replacement by the downward percolation of sabkha brines appears less likely.

Laminated Anhydrite. In the Messinian of Sicily, Ogniben (1957) described laminites of primary anhydrite (**Plate 31.C**) intercalated among halite lenses. They consist of laminae 2 to 2.5 mm thick with typical "pile-of-brick" texture alternating with laminae of carbonate or argillaceous material. The "pile-of-brick" texture appears typically as tightly packed subrectangular crystals of anhydrite. These anhydrite varves also display at times a reverse graded bedding, but less clearly developed than that previously described in laminated primary gypsum (**Fig. 10.13**). This rare instance of primary precipitation of laminated anhydrite related to halite formation must have occurred under conditions of high salinity brines undergoing high-temperature evaporation (Posnjak, 1938, 1940; Hardie, 1967).

Oolitic Anhydrite. This rare occurrence has been reported in argillaceous rocks associated with sandstones themselves cemented by anhydrite (Van Voorthuysen, 1951). Subrounded quartz grains with a thin brown limonite coating form the nuclei of roughly concentric rings of anhydrite. At times the nucleus of quartz has been replaced by a fine to coarse mosaic of calcite. The anhydrite ooids are clearly out of depositional context; they underwent a redeposition process, but, because of the softness of the anhydrite, the distance of transport must have been rather short.

Potash–Magnesia Salts. The origin of these salts is controversial and interpretations range from shallow to deep water (Handford et al., 1982; Kendall, 1984). Schreiber (1986) assumed that they form in shallow water mainly because it is extremely difficult to visualize any evaporative process that would lead to the precipitation of potash–magnesia salts in a water body of large volume for a given sur-

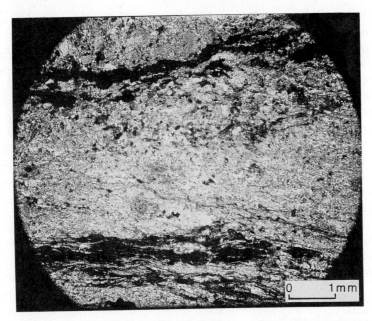

FIGURE 10.13 Reverse graded bedding in lamina of primary anhydrite. Nicols crossed. Middle Miocene (Messinian), Sicily. Photomicrograph courtesy of Leo Ogniben, Montecatini S.A. (Settore Miniere), Milano, Italy.

face area. The salts themselves are extremely diversified, and even if they seem to cover the floor of entire basins, their depocenters may have moved in time, and the whole basin might not have been entirely full of water at the time of deposition. Geochemical studies by Lowenstein (1982) and Hardie (1984) suggested that a large amount of potassic salt was not entirely marine and reworked within depositional basins by meteoric waters, followed by final deposition in shallow lagoons or saline lakes. Many potash–magnesia deposits show upper truncation surfaces that grade laterally into other slightly younger bodies of salt, indicating that some basins apparently displaying continuous fillings of halite and potassic salts were in fact separate and adjacent water bodies of different ages (Schreiber, 1986).

Potash–magnesia salts display structures that appear as varvelike in the Oligocene of the Rhine Valley (Baar and Kühn, 1962). They consist of alternating halite–sylvite laminae in which the latter would represent summer high evaporation. However, in other cases, sylvite and polyhalite are diagenetic replacements of earlier halite, anhydrite, and gypsum, with fabrics showing a complete range from total destruction to excellent preservation of the depositional structures of the precursor minerals (Schaller and Henderson, 1932; Lowenstein, 1982; Harville and Fritz, 1986).

Deep-water Evaporites

In the absence of modern analogues for deep-water evaporites, their interpretations are based essentially on theoretical considerations and the study of ancient examples (Kendall, 1984; Schreiber, 1986; Warren, 1989). To accumulate an appreciable thickness of evaporites in deep water, the brines must be at or near saturation with respect to gypsum, halite, and potassic salts. Crystal growth probably occurs at the air–water interface, and crystals settle through the water column as a pelagic rain. The regular interlaminations of calcite, dolomite, gypsum, and halite reflect variations in the influx of brines and in the rates of temperature and evaporation because large-scale water bodies do not fluctuate under the effect of such short-term variations, in fact they act as buffers (Dean et al., 1975). Some gypsum may grow within the upper layers of the bottom, and some halite may precipitate during brine mixing in a stratified water body (Raup, 1970). Furthermore, evaporite debris flows and turbidites, as described above in the section on shallow-water evaporites and carbonates, accumulate along the margins of the deep basin and contribute to deposition.

The depth of the water of such basins is a subject of debate, and estimates range from about 40 m to 400 or 600 m (Kendall, 1984). At any rate, conditions are such that deposition is generally protected from any action of slumps, waves, currents, and burrowing organisms, so it consists of laminar deposits in which individual laminae range in thickness from 1 mm to more than 1 cm. These rhythmites have been interpreted as representing seasonal or annual increments whose most important characteristic is lateral continuity over great distances to the extent that individual laminae can be correlated across entire basins of deposition.

Sulfate Laminites. The most common type consists of laminated anhydrite (originally gypsum inverted to anhydrite upon burial) alone or in couplets or triplets with calcite, dolomite, or planktonic organic matter that is well preserved in such anaerobic conditions. Anhydrite laminites were described all over the world, for instance in the Permian Castile Formation of Texas and New Mexico (Anderson and Kirkland, 1966; Anderson et al., 1972; Dean and Anderson, 1982), in the Permian Zechstein Group of Germany (Richter-Bernburg, 1960; Anderson and Kirkland, 1966), in the Jurassic Todilto Formation of New Mexico (Anderson and Kirkland, 1966), and in the Middle Devonian Muskeg and Winnipegosis Formations of Western Canada (Davies and Ludlam, 1973; Wardlaw and Reinson, 1971). Gypsum laminae were described in the Late Miocene (Messinian) of Italy and the Mediterranean (Schreiber et al., 1976; Garrison et al., 1978).

Anhydrite laminae are generally thin, ranging from 1 to 10 mm thick. They are typically bound by perfectly smooth flat surfaces, but some may be uneven, crenulated, or display ductile deformations. Individual laminae of uniform thickness or associations of laminae with given thicknesses can be traced over distances of several kilometers. The laminites in the Castile–Lower Salado Formations are 440 m thick and consist of more than 250,000 anhydrite–carbonate couplets (**Figs. 10.14, 10.15**) among which some laminae were traced laterally for more than 115 km (Anderson et al., 1972; Dean and Anderson, 1982).

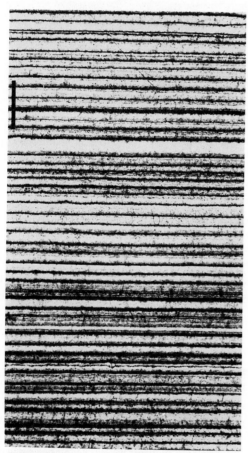

FIGURE 10.14 Typical section of laminites of the Castile Formation (Permian), Delaware Basin, Texas, consisting of dark, organic-rich calcite and light, anhydrite laminae. Plane-polarized light. Bar = 1 cm. From Dean and Anderson (1982). Reprinted by permission of the Society of Economic Paleontologists and Mineralogists.

Further confirmation of these lateral correlations was provided by micromarker beds described by Madsen (1984). They consist of dark gray layers, 0.5 to 20 mm thick, formed by anhydrite, calcite, or dolomite, authigenic albite, and pyrite, with minor amounts of marcasite, fluorite, and mica. These beds may represent sudden clastic influxes, volcanic ash falls, planktonic blooms across the basin, or a slowing down in the rate of sedimentation.

The texture of deep-water anhydrite laminae is burial diagenetic. It is generally felty or bacillary and similar as described above for buried shallow-water laminated gypsum. However, some laminae display a rather uniform texture consisting of an interlocking aggregate of larger subhedral to euhedral crystals set in a denser interlocking mosaic of small crystals. Both crystal sizes of anhydrite have the same subrectangular habit, which should not be confused with the "pile-of-brick" texture typical of the above-described rare occurrences of primary anhydrite. In the Castile Formation (Anderson et al., 1972; Dean and Anderson, 1982), intervals of nodular anhydrite were described with predominantly diagenetic felty texture. They occur among the laminites, but result from an early diagenetic reorganization of the laminites themselves, of which they often contain horizontal relict structures. This nodular anhydrite is often associated with thicker anhydrite laminae or occurs immediately beneath halite layers, indicating that their early diagenetic development was peobably related to locally increased salinity. This situation demonstrates that nodular anhydrite has to be interpreted within its depositional environment and is far from being necessarily diagnostic of supratidal conditions. (Dean et al., 1975).

Laminated Halite. Deep-water halite is invariably laminated and contains anhydrite–carbonate laminae similar to those described above. In turn, halite shows alternating clear transparent and dark gray to black laminae. These variations result from the presence of numerous fluid inclusions in the darker layers, outlining many pyramidal-shaped hopper crystals that grew at the air–water interface, very fine anhydrite crystals, quartz silt, and argillaceous materials. It is not yet clear if the inclusion-free crystals forming the clear halite layers are an expression of precipitation during decreasing, or absence, of clastic influx, or the products of basin-floor growth of inclusion-free clear halite at the partial expense of the hopper-rich layers, or finally a product of burial differential diagenetic fine-grained recrystallization. At any rate, halite layers and laminae have also been traced for many kilometers in a basinwide pattern (Anderson et al., 1972; Dean and Anderson, 1982). In general, laminated halite consists of very minute crystals compared to that forming the thick layers deposited under shallow-water conditions (Schreiber, 1986). The main reason for the fine fabric is due to the fact that most halite crystals that nucleate at the air–water interface and eventually sink to the bottom are

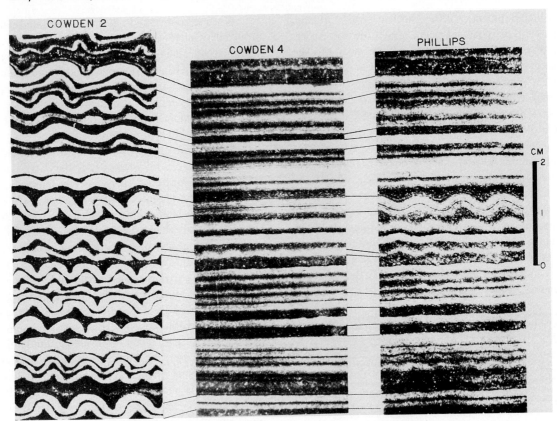

FIGURE 10.15 Correlative core slabs of dark calcite and light anhydrite laminae from the Cowden No. 2, Cowden No. 4, and Phillips No. 1 cores of the Castile Formation (Permian), Delaware Basin, Texas. From Dean and Anderson (1982). Reprinted by permission of the Society of Economic Paleontologists and Mineralogists.

very tiny, except for the occasional rafter hoppers. Whenever bottom-nucleated crystals can form in deep water, they have only a very limited source of ionic supply, mainly derived from depleted density currents; consequently, they also tend to be of small size. Finally, if halite precipitates at the interface of stratified waters of different composition, it also takes the shape of small crystallites, because the zone of precipitation at the interface between these water masses does not provide sufficient components or adequate space for growth of larger crystals (Raup, 1970). Regardless of the possible combinations of origin, halite crystals sink through the water column and generate basinwide clear and dark laminae.

Laminated Potash–Magnesia Salts. The Zechstein (Permian) of the North Sea (Colter and Reed, 1980) shows a great variety of laminated deposits consisting of various associations of anhydrite, halite, polyhalite, and kieserite. Many of the individual evaporitic units, in particular those of polyhalite, can be traced by gamma-ray correlations between wells from deep basinal conditions up into a preexisting marginal carbonate shelf. This procedure showed that

the evaporitic units are arranged in a large-scale foreset pattern, with individual beds thinning both toward the center of the basin and toward the shelf. These conditions indicate not only the primary origin of these potash–magnesia salts, but also afford the means to state that they were deposited in deep-water basinal conditions of at least 200 m depth and over a broad regional extent (Taylor, 1980).

DIAGENETIC EVOLUTION

One major problem presented by the study of evaporites is the distinction between the wide range of primary depositional features, and the extensive spectrum of early to late diagenetic imprints. Several recent reviews were dedicated to this important question (Schreiber et al., 1982; Hardie et al., 1985; Spencer and Lowenstein, 1990). Although no general consensus seems to have been reached on a common terminology, it is clear that textures, fabrics, structures, fluid inclusions, and mineralogical assemblages of evaporites should be classified in three types. These types are either depositional, that is, formed at the time of deposition or deposited in their existing form; postdepositional but preburial, that is, generated by penecontemporaneous or early diagenesis soon after deposition by processes still controlled by the depositional environment (including possible extensive changes in mineralogy but with no changes of primary textures); or burial, that is, formed by late diagenetic or metasomatic processes controlled by subsurface conditions. This approach in three types is more precise than the use of the terms "primary" and "secondary," which do not emphasize sufficiently the fundamental aspect of the timing of the events, although it is clear that some features still can remain of ambiguous origin.

Because of the unusual complexity of evaporites, the preceding portion of this chapter, treating the petrography of modern and ancient evaporitic settings, dealt with the major depositional and diagenetic aspects of these rocks, and hence only a few additional comments are presented here.

Among syndepositional features, Spencer and Lowenstein (1990) stressed the importance of recognizing cements, because recent surficial primary deposits are generally highly porous, and their slightly buried equivalents become rapidly tight. For instance, in the case of halite, petrographic evidence indicates that clear halite cement is obviously intergranular when it is lining pore spaces of a primary crystalline or detrital framework or filling vugs and cavities within a darker chevron-structured halite layer.

Criteria for burial diagenetic features correspond, as previously described, to the destruction of primary depositional features by the development of diagenetic ones such as felty and bacillary anhydrite or selenitic and alabastrine gypsum. However, the most characteristic for anhydrite and halite, according to Spencer and Lowenstein (1990), is the development of sutured mosaic textures in which grain boundaries are sutured in the same way as in carbonate neomorphic pseudosparite. These authors also stressed the occurrence, in buried halite, of polygonal equigranular mosaic textures comparable to the products of experimental annealing in metals where grains tend to optimize their size, shape, and orientation in order to minimize energy, in the manner of bubbles in foam. Halite appears particularly susceptible to the formation of such mosaic texture, which is recognizable, under the microscope, by the fact that the crystals display curved boundaries that meet at triple junctions with angles approaching 120°, with interstitial impurities concentrated along the crystal boundaries. Burial mosaics are also displayed when evaporitic minerals are replacive of others, particularly calcite. The replacement of limestones by anhydrite mosaics (**Plate 31.F**), reaching total destruction of original depositional fabrics, is a large-scale process that has not received proper attention. The same situation occurs with cubical halite versus other evaporitic minerals in particular anhydrite (**Plate 32.F**) and to potash-magnesia salts versus halite.

An unusual aspect of halite diagenesis is the occurrence of very late, ordered, and stoichiometric limpid dolomite along contact between halite and interbedded mudstones and along halite crystal boundaries (Gao et al., 1990). In the Permian example studied by these authors, the limpid dolomite is characterized by a low Sr content, indicating that meteoric waters preserved in the interbedded mudstones were an important source of fluid for the process. The authors suggested that the dolomite precipitated, under favorable sulfate-reduction conditions, during burial compaction of the mudstones from a halite-saturated brine that derived from the mixing of waters released by the mudstones and evaporative brine from the halite.

At the other extreme of replacement processes in the final phase of uplift of the cycle gypsum–anhydrite–gypsum (**Fig. 10.1**), it is important to recall the aspect of diagenetic regenerated gypsum. It occurs in two major forms: porphyroblastic diagenetic gypsum and alabastrine diagenetic gypsum (Ogniben, 1957; Holliday, 1970). The most typical feature of gypsum porphyroblasts is the presence of abundant relics of anhydrite within a single, uniformly extinguishing crystal of gypsum. Porphyroblasts are either euhedral or anhedral, thin and needlelike or thick, isolated or associated in aggregates and rosettes. Although generally isolated, they can replace entire beds of anhydrite. These crystals truncate sharply, but do not disturb the anhydrite fabric, indicating that gypsification took place without volume increase. It appears that these porphyroblasts may have formed before alabastrine gypsum because they are often associated with it and have often embayed and ragged boundaries against it.

Alabastrine gypsum shows a great variety of textures. A first type consists of poorly defined grains with their boundaries virtually nonexistent and with irregular extinction. Individual grains, whenever recognizable, are less than 50 μ in diameter. Under crossed nicols, the rock appears composed of fairly definite areas (superindividuals of Ogniben, 1957), consisting, however, of a number of these poorly defined grains, each area showing also poorly defined boundaries. A second type has a granoblastic texture in which well-defined anhedral and equidimensional grains are intricately interlocked.

When these two types of alabastrine gypsum begin to recrystallize, they become gradually free of anhydrite relics. The first product of this evolution is often an anhedral, equidimensional, granoblastic texture with uniform extinction and grains with straight edges. Further progressive recrystallization leads to a coarser and more variable grain size, with an increasing number of grains becoming subhedral to euhedral.

Porphyroblastic crystals may even develop, cross-cutting the surrounding grains. They are devoid of anhydrite relicts and easily distinguishable from the above-described porphyroblastic secondary gypsum.

In spite of the fact that the unit cell of gypsum is considerably larger than that of anhydrite and that a volume increase is generally believed to occur during gypsification, there is little evidence of such an increase in the numerous studied examples (Holliday, 1970). Most of the secondary gypsification appears to be a volume per volume process with excess sulfate removed in solution. Excess amounts of trace elements, notably strontium and boron, are released from some anhydrite rocks during hydration and occur in alabastrine secondary gypsum as celestite and boron-bearing minerals such as probertite, ulexite, and priceite.

Besides the obvious megascopic deformations of evaporites, exemplified by the syngenetic folds of nodular anhydrite in sabkhas or the burial distortion of deep-water anhydrite laminae, many microscopic features (Shlichta, 1968) indicate the effects of burial deformation, such as deformation twins, slip lines and slip bands, lattice distortion expressed by undulatory extinction and bent cleavages, subdomain grains, and finally, flattening of the grains, leading to foliation and eventually to tectonic deformations related to diapirism and compressional tectonics (Procha, 1968; Schreiber et al., 1982).

As mentioned above, although the issue of timing of events involving the diagenetic evolution of evaporitic minerals is critical, there are still ambiguous features that span the entire spectrum from syndepositional to burial conditions, such as euhedral crystal growth, development of nodular textures, pseudomorphs, coarse crystalline texture, and even cavity-filling cements. Studies of these features within their proper context may afford clues to the timing of their origin. In that respect, the study of fluid inclusions, which are very abundant in evaporitic minerals, particularly halite, can be extremely useful to determine the temperature of formation of the various salts, the composition of the solutions from which they were formed, their relationships to fluctuations of sea level, their complex hydrological systems through which seawater was fed and refluxed, and finally, aspects of their diagenetic evolution (Roedder, 1984a and b; Spencer and Lowenstein, 1990; Bein et al., 1991).

RESERVOIR PROPERTIES

The importance of evaporites in petroleum geology as seals, structural traps, and source rocks was mentioned in the introduction of this chapter. Through burial dissolution, evaporites, in particular halite and potash–magnesia salts, can also become potential reservoirs. An interesting example of intergranular anhydrite dissolution as a cause of secondary porosity was recently described petrographically and with the SEM in dolarenites and quartz arenites (Schenk and Richardson, 1985). In this particular example, as observed under the SEM, the edges of anhydrite crystals adjacent to porosity are characterized by sharp, right-angle projections and reentrants that combine to produce complex needlelike, waferlike, and blocklike forms. This dissolution fringe is highly

controlled by the three prominent anhydrite cleavages, which account for the predominance of right-angle shapes. The subsurface dissolution of anhydrite can be recognized under the petrographic microscope by the following features: isolated remnants of optically continuous anhydrite along dissolution fronts, irregularity of birefringence zonation along crystal edges, and anhydrite cleavages appearing darker adjacent to porosity. According to Schenk and Richardson (1985), incipient dissolution began along the borders between anhydrite and adjacent dolomite or quartz. From these borders, dissolution penetrated anhydrite cleavages, leading to rapid, preferential solution perpendicular to the more prominent cleavages, which, together with intercrystalline boundaries, acted as major conduits. Continued dissolution along cleavages generated the blocky reentrants along the dissolution fringe and, in some instances, separated crystals into several isolated but optically continuous remnants. In the final stage, as the borders of dissolving anhydrite retreated to pore throats, dissolution slowed and became again restricted to intercrystalline boundaries, and the process kept on repeating in adjacent anhydrite-filled intergranular areas. Anhydrite dissolution appears to have been a late burial event, and large amounts of anhydrite were apparently removed by deep recharge of aquifers.

Another aspect of reservoir development occurs when massive lenses of authigenic carbonates (calcite, aragonite, dolomite, and magnesite) develop by large-scale replacement of anhydrite and gypsum mainly by bacterial sulfate reduction (Pierre and Rouchy, 1988). In the Miocene example of Egypt studied by these authors, replacement occurred in two distinct phases: (1) an episode penecontemporaneous with sedimentation or early diagenetic, during which bacteria used planktonic organic matter within anaerobic pore waters of evaporitic marine origin; and (2) a later episode under shallow phreatic conditions associated with halokinetic deformation, when evaporitic sediments were flushed by nonaerated solutions associated with migrating hydrocarbons. This replacement under closed to semiclosed system conditions was accompanied by substantial reduction of volume, generation of substantial secondary porosity at all scales of observation, and collapse structures. Pores range from megascopic, irregulary shaped vugs, a few millimeters to a few centimeters in size, to microporosity visible under the petrographic microscope and the SEM, consisting of irregular voids due to dissolved mineral inclusions, concentric voids and geometric central cavities in zoned crystals, and, finally, triangular pores controlled by crystal lattice. Pierre and Rouchy (1988) interpreted the porosity as developed partly by mineral transformation of sulfate to carbonate accompanied by volume reduction, partly by dissolution of residual or authigenic sulfate by weathering and also by oxidation of native sulfur associated with the carbonates.

If the above-mentioned large-scale replacement of anhydrite and gypsum by calcite is assumed to develop secondary porosity at various scales, it can also generate, under moderate burial diagenesis, large concentrations of nonporous microsparite and sparite, as observed in the Pennsylvanian phylloid algal mounds of Kansas (Dawson and Carozzi, 1986). The origin of this calcite is recognizable, under the microscope, by the pseudomorphic preservation of the typical rectangular pattern of the cleavages of the precursor anhydrite (**Plate 32.E**).

Significant concentrations of very coarsely crystalline sparite showing a nonferroan character, complex cathodoluminescence banding, minor evaporite inclusions and pseudomorphs, and evidence of multiple stages of pore fillings separated by corrosion events have been reported by Scholle et al. (1992) in Permian carbonates of the Guadalupe Mountains and northern and eastern margins of the Permian Basin (west Texas and New Mexico). The blocky sparite fills a variety of porosity types formed by evaporite solution and collapse brecciation. It also postdates all diagenetic features from early marine through dolomitization to late solution seams and in situ fracturation and brecciation. Hence, Scholle et al. (1992) interpreted this blocky sparite as the product of the circulation, during the Cenozoic, of near-surface fresh meteoric solutions that led to evaporite dissolution and calcitization reactions. They described a typical reaction as follows: initial corrosion of felted to bladed anhydrite, one or more stages of hydration–dehydration during conversion to gypsum, dissolution of gypsum, local collapse brecciation, and multistage precipitation of sparite in pores vacated by evaporites.

Subsequent deep-burial dissolution of concentrations of calcite, whether mesogenetic pseudomorphic or late-stage telogenetic, can lead to appreciable secondary porosity, whose real origin and importance have not been appreciated in many instances.

TYPICAL EXAMPLES

Plate 31.A. Calcite pseudomorph after anhydrite. Calcisiltite groundmass showing molds of subrectangular crystals of anhydrite that were first partially filled with geopetal vadose silt and subsequently by cavity-filling freshwater phreatic sparite cement. Cedar Valley Limestone, Middle Devonian, New Jersey Zinc core DDH-6, Macon County, Missouri, U.S.A.

Plate 31.B. Chickenwire felty anhydrite. Interfering micronodules of felty anhydrite are set in a matrix of dark dolosiltite. Each micronodule consists of many anhydrite laths oriented subparallel to its periphery, while having a random arrangement in the center. Joachim Dolomite, Middle Ordovician, Brewer, Perry County, Missouri, U.S.A.

Plate 31.C. Pile-of-brick anhydrite. Laminated primary anhydrite in which its typical subrectangular crystals were deposited subparallel to bedding, simulating the aspect of a wall consisting of superposed bricks. Mulichinco Formation, Lower Cretaceous, Mendoza Basin, Argentina.

Plate 31.D. Felty anhydrite. This secondary texture developed here from conversion of primary gypsum shows a typical wavy and sigmoidal arrangement of sheaves of spindle-shaped crystals. Mulichinco Formation, Lower Cretaceous, Mendoza Basin, Argentina.

Plate 31.E. Bacillary anhydrite. The secondary texture developed here from conversion of primary gypsum shows the characteristic radial arrangement of narrow rodlike crystals or prisms. Mulichinco Formation, Lower Cretaceous, Mendoza Basin, Argentina.

Plate 31.F. Anhydrite replacing ooids. Aggregates of large subrectangular crystals of secondary anhydrite, developed under deep-burial conditions, replaced almost completely a grain-supported oolitic calcarenite consisting of ooids with relatively large pelletoidal nuclei of dark calcisiltite. Mississippian, Renville County, North Dakota, U.S.A.

Plate 32.A. Calcite pseudomorph after gypsum. Pelletoidal dolosiltite groundmass showing a typical lenticular gypsum crystal replaced by a single calcite crystal. Kokomo Dolomite, Upper Silurian, Maumee, Ohio, U.S.A.

Plate 32.B. Laminated primary gypsum. Laminated groundmass of small lenticular crystals of gypsum aligned subparallel to bedding and displaying irregular discontinuous thin intercalations of darker clay minerals. Upward tendency toward a less laminated and coarser mosaic of crystals represents reverse graded bedding. Messinian, Ravanusa, Sicily, Italy.

Plate 32.C. Incipient selenitic gypsum. Irregular mosaic of equicrystalline secondary gypsum derived from anhydrite conversion shows scattered, large porphyroblasts, some of which contain residual anhydrite (right side of picture). Messinian, Serradifalco, Sicily, Italy.

Plate 32.D. Calcite pseudomorph after selenite. Dolomitized pelphosphasiltite displays calcite pseudomorphs after clusters of vertically standing single and twinned crystals of selenite. Rhombs of ferroan dolomite (upper part of picture) are shown by potassium ferricyanide blue staining. Basal Maquoketa Group, Upper Ordovician, Cominco American Incorporated Core CS 1, Jackson County, Iowa, U.S.A.

Plate 32.E. Calcitized anhydrite. This widespread pseudomorphic replacement by freshwater phreatic sparite generally occurs with perfect preservation of the characteristic rectangular cleavage traces of anhydrite. Raytown Limestone, Iola Formation, Upper Pennsylvanian, Iola, Kansas, U.S.A.

Plate 32.F. Replacive cubical halite. Well-developed crystals of halite with characteristic cleavage traces and concentrations of impurities and iron oxides along their margins replaced an argillaceous groundmass with magnesite, bladed anhydrite, minute quartz grains, and pyrite flecks. Reddish patches (left center and upper center of picture) represent another generation of hematitic halite with poorly defined crystal boundaries. Staintondale borehole, Lower Halite Zone of Middle Evaporites, Permian, Yorkshire, Great Britain.

REFERENCES

ANDERSON, R. Y., and KIRKLAND, D. W., 1966. Intrabasin varve correlation. *Geol. Soc. Amer. Bull.,* 77, 241–256.

———, DEAN, W. E., KIRKLAND, D. W., and SNIDER, H. I., 1972. Permian Castile varved evaporite sequence, West Texas and New Mexico. *Geol. Soc. Amer. Bull.,* 83, 59–86.

ARAKEL, A. V., 1980. Genesis and diagenesis of Holocene evaporitic sediments in Hutt and Leeman lagoons, Western Australia. *J. Sed. Petrology,* 50, 1305–1326.

ARTHURTON, R. S., 1973. Experimentally produced halite compared with Triassic layered halite-rock from Cheshire, England. *Sedimentology,* 20, 145–160.

BAAR, A. and KÜHN, R., 1962. Der Werdegang der Kalisalzlagerstätten am Oberrhein. *Neues Jahrbuch Mineralogie Abhandl.,* 97, 289–336.

BAIN, R. J., 1990. Diagenetic, nonevaporitic origin of gypsum. *Geology,* 18, 447–450.

BEIN, A., HOVORKA, S. D., FISHER, R. S., and ROEDDER, E., 1991. Fluid inclusions in bedded Permian halite, Palo Duro Basin, Texas: evidence for modification of seawater in evaporite brine-pools, and subsequent early diagenesis. *J. Sed. Petrology,* 61, 1–14.

BORCHERT, H., and MUIR, R. O., 1964. *Salt Deposits: The Origin, Metamorphism, and Deformation of Evaporites.* University Series in Geology, Van Nostrand Reinhold, New York, 338 pp.

BRAITSCH, O., 1971. *Salt Deposits. Their Origin and Composition. Minerals, Rocks, and Inorganic Materials.* Monograph series of theoretical and experimental studies 4, Springer-Verlag, New York, 297 pp.

BUICK, R., and DUNLOP, J. S. R., 1990. Evaporitic sediments of Early Archaean age from the Warrawoona Group, North Pole, Western Australia. *Sedimentology,* 37, 247–277.

CASAS, E., and LOWENSTEIN, T. K., 1989. Diagenesis of saline pan halite: comparison of petrographic features of modern, Quaternary, and Permian halites. *J. Sed. Petrology,* 59, 724–739.

CHANDLER, F. W., 1988. Diagenesis of sabkha-related, sulfate nodules in the Early Proterozoic Gordon Lake Formation, Ontario, Canada. *Carbonates and Evaporites,* 3, 75–94.

CIARANFI, N., DAZZARO, L., PIERI, P., RAPISARDI, L., and SARDELLA, A., 1973. Stratigraphic characters and tectonic outlines of some Messinian deposits outcropping along the eastern side of the Southern Apennines. In C. W. Drooger (ed.), *Messinian Events in the Mediterranean.* North-Holland Publishing Company, Amsterdam, 178–179.

CODY, R. D., 1991. Organo-cyrstalline interactions in evaporite systems: the effects of crystallization inhibition. *J. Sed. Petrology,* 61, 704–718.

———, and HULL, A. B., 1980. Experimental growth of primary anhydrite at low temperatures and water salinities. *Geology,* 8, 505–509.

COLTER, V. S., and REED, G. E., 1980. Zechstein 2 Fordon evaporites of the Atwick No. 1 borehole, surrounding areas of N. E. England and the adjacent southern North Sea. In H. Füchtbauer and T. M. Peryt (eds.), *The Zechstein Basin with Emphasis on Carbonate Sequences.* Contr. Sedimentology 9, E. Schweizerbart'sche Verlagsbuchhandlung, Stuttgart, 115–129.

DAVIES, G. R., and LUDLAM, S. D., 1973. Origin of laminated and graded sediments, Middle Devonian of western Canada. *Geol. Soc. Amer. Bull.,* 84, 3527–3546.

DAWSON, W. C., and CAROZZI, A. V., 1986. Anatomy of a phylloid algal buildup, Raytown Limestone, Iola Formation, Pennsylvanian, southeast Kansas, U. S. A. *Sed. Geol.,* 47, 221–261.

DEAN, W. E., and ANDERSON, R. Y., 1982. Continuous subaqueous deposition of the Permian Castile evaporites, Delaware Basin, Texas and New Mexico. In C.R. Handford, R. G. Loucks, and G.R. Davies (eds.), *Depositional and Diagenetic Spectra of Evaporites. A Core Workshop.* Soc. Econ. Paleontologists and Mineralogists, Core Workshop No. 3, Calgary, Alberta, Canada, 324–353.

———, and SCHREIBER, B. C., (eds.), 1978. *Marine Evaporites.* Soc. Econ. Paleontologists and Mineralogists, Short Course No. 4, Oklahoma City, Oklahoma, 188 pp.

———, DAVIES, G. R., and ANDERSON, R. Y., 1975. Sedimentological significance of nodular and laminated anhydrite. *Geology,* 3, 367–372.

DELLWIG, L. F., 1955. Origin of Salina salt in Michigan. *J. Sed. Petrology,* 25, 83–110.

DUNLOP, J. S. R., 1978. *Shallow-water Sedimentation at North Pole, Pilbara, Western Australia.* Univ. W. Australia Geol. Dept. and Extension Service, Publ. 2, 30–38.

EUGSTER, H. P., and HARDIE, L. A., 1978. Saline lakes. In A. Lerman (ed.), *Chemistry, Geology, and Physics of Lakes.* Springer-Verlag, New York, 237–293.

GAO, G., HOVORKA, S. D., and POSEY, H. H., 1990. Limpid dolomite in Permian San Andres halite rocks, Palo Duro Basin, Texas Panhandle: characteristics, possible origin, and implications for brine evolution. *J. Sed. Petrology,* 60, 118–124.

GARRETT, D. E., 1970. The chemistry and origin of potash salts. In J. L. Raup and L. F. Dellwig (eds.), *Third Symposium of Salt.* Northern Ohio Geological Society, 1, 211–222.

GARRISON, R. E., SCHREIBER, B. C., BERNOULLI, D., FABRICIUS, F. H., KIDD, R. D., and MÉLIÈRES, F., 1978. Sedimentary petrology and structures of Messinian evaporitic sediments in the Mediterranean Sea. Leg 42 A, Deep Sea Drilling Project. In K. J. Hsu et al. (eds.), *Initial Reports of the Deep Sea Drilling Project,* U. S. Government Printing Office, Washington, D. C., 42, 571–611.

GORNITZ, V. M., and SCHREIBER, B. C., 1981. Displacive halite hoppers of the Dead Sea: some implications for ancient evaporite deposits. *J. Sed. Petrology,* 51, 787–794.

HANDFORD, C. R., 1982. Sedimentology and evaporite genesis in a Holocene continental-sabkha playa basin—Bristol Dry Lake, California. *Sedimentology,* 29, 239–253.

———, 1990. Halite depositional facies in a solar salt pond: a key to interpreting physical energy and water depth in ancient deposits? *Geology,* 18, 691–694.

———, LOUCKS, R. G., and DAVIES, G. R., (eds.), 1982. *Depositional and Diagenetic Spectra of Evaporites. A Core Workshop.* Soc. Econ. Paleontologists and Mineralogists, Core Workshop No. 3, Calgary, Alberta, Canada, 395 pp.

HARDIE, L. A., 1967. The gypsum–anhydrite equilibrium at one atmosphere. *Amer. Mineralogist,* 52, 171–200.

———, 1984. Evaporites: marine or non-marine? *Amer. J. Sci.,* 284, 193–240.

———, and EUGSTER, H. P., 1970. The evolution of closed-basin brines. In B. A. Morgan (ed.), *Fiftieth Anniversary Symposia: Mineralogy and Petrology of the Upper Mantle; Sulfides; Mineralogy and Geochemistry of Non-marine Evaporites.* Mineralogical Soc. Amer., Special Paper 3, 273–290.

———, and ———, 1971. The depositional environment of marine evaporites: a case for shallow water clastic accumulation. *Sedimentology,* 16, 187–220.

———, LOWENSTEIN, T. K., and SPENCER , R. J., 1985. The problem of distinguishing between primary and secondary features in evaporites. In B. C. Schreiber and H. L. Harker (eds.), *Sixth International Symposium on Salt.* The Salt Institute, Alexandria, VA., 1, 11–39.

———, SMOOT, J. P., and EUGSTER, H. P., 1978. Saline lakes and their deposits: a sedimentological approach. In A. Matter and M. E. Tucker (eds.), *Modern and Ancient Lake Sediments.* Internat. Assoc. Sedimentologists, Special Publ. 2. Blackwell Scientific Publications, London, 7–41.

HARVILLE, D. G., and FRITZ, S. J., 1986. Modes of diagenesis responsible for observed succession of potash evaporites in the Salado Formation, Delaware Basin, New Mexico. *J. Sed. Petrology,* 56, 648–656.

HOLLIDAY, D. W., 1970. The petrology of secondary gypsum rocks: a review. *J. Sed. Petrology,* 40, 734–744.

HOLSER, W. T., 1979. Mineralogy of evaporites. In R. G. Burns (ed.), *Marine Minerals.* Mineralogical Soc. America, Short Course Notes 6, 211–294.

HOVORKA, S. D., 1987. Depositional environments of marine-dominated bedded halite, Permian San Andres Formation, Texas. *Sedimentology,* 34, 1029–1054.

KENDALL, A. C., 1984. Evaporites. In R. G. Walker (ed.), *Facies Models,* 2nd ed. Geological Association of Canada, Geoscience Canada, Reprint Series 1, 259–296.

KIRKLAND, D. W., and EVANS, R., (eds.), 1973. *Marine Evaporites. Origin, Diagenesis, and Geochemistry.* Benchmark Papers in Geology. Dowden, Hutchinson and Ross, Inc., Stroudsburg, Pa., 426 pp.

———, and ———, 1981. Source-rock potential of evaporitic environment. *Amer. Assoc. Petroleum Geologists Bull.,* 65, 181–190.

KUSHNIR, J., 1981. Formation and early diagenesis of varved evaporitic sediments in a coastal hypersaline pool. *J. Sed. Petrology,* 51, 1193–1203.

LOUCKS, R. G., and LONGMAN, M. W., 1982. Lower Cretaceous Ferry Lake Anhydrite, Fairway field, East Texas: product of shallow-subtidal deposition. In C. R. Handford, R. G. Loucks, and G. R. Davies (eds.), *Depositional and Diagenetic Spectra of Evaporites. A Core Workshop.* Soc. Econ. Paleontologists and Mineralogists, Core Workshop No. 3, Calgary, Alberta, Canada, 130–173.

LOWENSTEIN, T. K., 1982. Primary features in a potash evaporite district, the Permian Salado Formation of West Texas and New Mexico. In C. R. Handford, R. G. Loucks, and G. R. Davies (eds.), *Depositional and Diagenetic Spectra of Evaporites. A Core Workshop.* Soc. Econ. Paleontologists and Mineralogists, Core Workshop No. 3, Calgary, Alberta, Canada, 276–304.

MACHEL, H. G., and BURTON, E. A., 1991. Burial diagenetic

sabkha-like gypsum and anhydrite nodules. *J. Sed. Petrology,* 61, 394–405.

MADSEN, B. M., 1984. Micromarker beds in the Upper Permian Castile Formation, Delaware Basin, West Texas and southeastern New Mexico. *J. Sed. Petrology,* 54, 1169–1174.

MAIKLEM, W. R., BEBOUT, D. G., and GLAISTER , R. P., 1969. Classification of anhydrite—a practical approach. *Bull. Canadian Petroleum Geol.,* 17, 194–233.

MELVIN, J. M., (ed.), 1991. *Evaporites, Petroleum, and Mineral Resources.* Developments in Sedimentology 50. Elsevier Scientific Publishing Co., New York, 556 pp.

MURRAY, R. C., 1964. Origin and diagenesis of gypsum and anhydrite. *J. Sed. Petrology,* 34, 512–523.

OGNIBEN, L., 1955. Inverse graded bedding in primary gypsum of chemical origin. *J. Sed. Petrology,* 25, 273–281.

———, 1957. Secondary gypsum of the Sulphur Series, Sicily, and the so-called integration. *J. Sed. Petrology,* 27, 64–79.

ORTI-CABO, F., and SHEARMAN, D. J., 1977. Estructuras y fábricas en las evaporitas del Mioceno superior (Messiniense) de San Miguel de Salinas (Alicante, España). *Inst. Investigaciones Geológicas Diputación Provincial Universidad de Barcelona,* 32, 5–54.

PAREA, G. C., and RICCI LUCCHI, F., 1972. Resedimented evaporites in the Periadriatic trough. *Israel J. Earth Sci.,* 21, 125–141.

PARNELL, J., 1983. Skeletal halites from the Jurassic of Massachusetts, and their significance. *Sedimentology,* 30, 711–715.

PERYT, T. M., and KASPRZYK, A., 1992. Earthquake-induced resedimentation in the Badenian (middle Miocene) gypsum of southern Poland. *Sedimentology,* 39, 235–249.

PIERRE, C., and ROUCHY, J. M., 1988. Carbonate replacements after sulfate evaporites in the Middle Miocene of Egypt. *J. Sed. Petrology,* 58, 446–456.

POSNJAK, E.,1938. The system $CaSO_4$–H_2O. *Amer. J. Sci.,* 235A, 247–272.

———, 1940. Deposition of calcium sulfate from sea water. *Amer. J. Sci.,* 238, 559–568.

PROCHA, J. J., 1968. Salt deformation and decollement in the Firtree Point anticline of central New York. *Tectonophysics,* 6, 273–299.

RAUP, O. B., 1970. Brine mixing: an additional mechanism for formation of basin evaporites. *Amer. Assoc. Petroleum Geologists Bull.,* 54, 2246–2259.

RICCI LUCCHI, F., 1973. Resedimented evaporites: indicators of slope instability and deep-basin conditions in Periadriatic Messinian (Apennines Foredeep, Italy). In C. W. Drooger (ed.), *Messinian Events in the Mediterranean.* North-Holland Publishing Company, Amsterdam, 142–149.

RICHTER-BERNBURG, G., 1960. Zeitmessung geologischer Vorgänge nach Warven-Korrelationen. *Geol. Rundschau,* 49, 132–138.

ROEDDER, E.,1984a. *Fluid Inclusions.* Mineralogical Soc. America, Reviews in Mineralogy, 12, 644 pp.

———, 1984b. The fluids in salt. *Amer. Mineralogist,* 69, 413–439.

ROUCHY, J. M., 1976. Sur la genèse de deux principaux types de gypse (finement lité et en chevrons) du Miocène terminal de Sicile et d'Espagne méridionale. *Revue Géogr. Phys. Géol. Dynamique,* série 2, 18, 347–364.

SCHALLER, W. T., and HENDERSON, E. P., 1932. Mineralogy of drill cores from the potash fields of New Mexico and Texas. *U. S. Geol. Survey Bull.,* 833, 124 pp.

SCHENK, C. J., and RICHARDSON, R. W., 1985. Recognition of interstitial anhydrite dissolution: a cause of secondary porosity, San Andres Limestone, New Mexico, and Upper Minnelusa Formation, Wyoming. *Amer. Assoc. Petroleum Geologists Bull.,* 69, 1064–1076.

SCHLAGER, W., and BOLZ, H., 1977. Clastic accumulation of sulphate evaporites in deep-water. *J. Sed. Petrology,* 47, 600–609.

SCHOLLE, P. A., ULMER, D. S., and MELIM, L. A.,1992. Late-stage calcites in the Permian Capitan Formation and its equivalents, Delaware Basin margin, west Texas and New Mexico: evidence for replacement of precursor evaporites. *Sedimentology,* 39, 207–234.

SCHREIBER, B. C., 1978. Environments of subaqueous gypsum deposition. In W. E. Dean and B. C. Schreiber (eds.), *Marine Evaporites.* Soc. Econ. Paleontologists and Mineralogists, Short Course No. 4, Oklahoma City, Okla., 43–73.

————, 1986. Arid shorelines and evaporites. In H. G. Reading (ed.), *Sedimentary Environments and Facies,* 2nd ed. Blackwell Scientific Publications, London, 189–228.

————, (ed.), 1988. *Evaporites and Hydrocarbons.* Columbia University Press, New York, 475 pp.

————, and WALKER, D., 1992. Halite pseudomorphs after gypsum: a suggested mechanism. *J. Sed. Petrology,* 62, 61–70.

————, ROTH, M. S., and HELMAN, M. L., 1982. Recognition of primary facies characteristics and the differentiation of these forms from diagenetic overprints. In C. R. Handford, R. G. Loucks, and G. R. Davies (eds.), *Depositional and Diagenetic Spectra of Evaporites. A Core Workshop.* Soc. Econ. Paleontologists and Mineralogists, Core Workshop No. 3, Calgary, Alberta, Canada, 1–32.

————, FRIEDMAN, G. M., DECIMA, A., and SCHREIBER, E., 1976. Depositional environments of Upper Miocene (Messinian) evaporite deposits of the Sicilian basin. *Sedimentology,* 23, 729–760.

SHEARMAN, D. J., 1970. Recent halite rock, Baja California, Mexico. *Trans. Inst. Mining Metallurgy,* London, 79B, 155–162.

————, 1978. Evaporites of coastal sabkhas. In W. E. Dean and B. C. Schreiber (eds.), *Marine Evaporites.* Soc. Econ. Paleontologists and Mineralogists, Short Course No. 4, Oklahoma City, Okla., 6–42.

SHLICHTA, P. J., 1968. Growth, deformation, and defect structure of salt crystals. In R. B. Mattox (ed.), *Saline Deposits.* Geol. Soc. America, Special Paper 88, 597–617.

SMITH, D. B., 1971. Possible displacive halite in the Permian Upper Evaporite Group of northeast Yorkshire. *Sedimentology,* 17, 221–232.

SOUTHGATE, P. N., 1982. Cambrian skeletal halite crystals and experimental analogues. *Sedimentology,* 29, 391–407.

SPENCER, R. J., and LOWENSTEIN, T. K., 1990. Evaporites. In I. A. McIlreath and D. W. Morrow (eds.), *Diagenesis.* Geological Association of Canada, Geoscience Canada, Reprint Series 4, 141–163.

STEWART, F. H., 1963. Data of geochemistry, 6th ed., Chapter Y, Marine Evaporites. *U. S. Geol. Survey Prof. Paper 440-Y,* 52 pp.

TAYLOR, J. C. M., 1980. Origin of the Werraanhydrite in the U. K. Southern North Sea—a reappraisal. In H. Füchtbauer and T. M. Peryt (eds.), *The Zechstein Basin with Emphasis on Carbonate Sequences.* Contr. Sedimentology 9, E. Schweizerbart'sche Verlagsbuchhandlung, Stuttgart, 91–113.

VAI, G. B., and RICCI LUCCHI, F., 1977. Algal crusts, autochthonous and clastic gypsum in a cannibalistic evaporite basin: a case history from the Messinian of Northern Apennines. *Sedimentology,* 24, 211–244.

VAN VOORTHUYSEN, T. H., 1951. Anhydrite formation in the saline facies of the Munder Mergel (Upper Malm). *Geologie en Mijnbouw N.S.,* 13, 279–282.

WARDLAW, N. C., and REINSON, G. E., 1971. Carbonate and evaporite deposition and diagenesis, Middle Devonian Winnipegosis and Prairie Evaporite Formations of Saskatchewan. *Amer. Assoc. Petroleum Geologists Bull.,* 55, 1759–1786.

————, and SCHWERDTNER, W. M., 1966. Halite–anhydrite seasonal layers in Middle Devonian Prairie Evaporite Formation, Saskatchewan. *Geol. Soc. Amer. Bull.,* 77, 331–342.

WARREN, J. K., 1982. The hydrological setting, occurrence, and significance of gypsum in late Quaternary salt lakes in South Australia. *Sedimentology,* 29, 609–637.

————, 1989. *Evaporite Sedimentology: Importance in Hydrocarbon Accumulation.* Prentice Hall, Englewood Cliffs, New Jersey, Advanced Reference Series, 285 pp.

————, and KENDALL, G. C. ST. C., 1985. Comparison of sequences formed in sabkha (subaerial) and salina (subaqueous) settings: modern and ancient. *Amer. Assoc. Petroleum Geologists Bull.,* 69, 1013–1023.

WEILER, Y., SASS, E., and ZAK, I., 1974. Halite oolites and ripples in the Dead Sea. *Sedimentology,* 21, 623–632.

APPENDIX

THIN SECTION IDENTIFICATION CHART

This chart has been used by the author for the past 30 years in laboratory exercises for advanced undergraduates and graduate courses in sedimentary petrography, as well as in training courses for major oil companies worldwide. Generations of students found this chart extremely useful as a systematic guide or checklist for their search of all observable critical data in a thin section before attempting description and depositional–diagenetic interpretation.

The version of the chart given here has benefited from the input of many students and is suitable for the petrographic study of most sedimentary rocks. It is not all-encompassing, but of sufficient length to be a practical tool; it can be easily modified for particular purposes, but any expansion might defeat its purpose as a didactic tool.

The chart is in essence descriptive, but its final sections deal with interpretative matters such as concise description of the rock to compel a synthesis of the observations, designation of the rock according to any applicable classification, generation of a depositional–diagenetic sequence for the practical application to hydrocarbon exploration, and, finally, interpretation of the depositional environment. Dashed lines are for additional observations and comments.

Sample No. _____
Formation, age _____
Locality _____

Circle appropriate components; fill in parentheses with relative abundance code:
p = predominant, c = common, r = rare.

CLASTS AND GRAINS

LITHIC GRAINS: Intrusives: silicic ( ), intermediate ( ), mafic ( ); volcanics: silicic ( ), intermediate ( ), mafic ( ); glass: silicic ( ), intermediate ( ), mafic ( ), devitrified ( ), pumice ( ), obsidian ( ); volcaniclastics ( ), serpentine ( ), schists ( ), gneisses ( ), quartzites ( ), metaquartzites ( ), arenites ( ), wackes ( ), siltstones ( ), shales ( ), micrite ( ), dolomicrite ( ), sparite ( ), dolosparite ( ), marbles ( ), cherts ( ), coal ( ), bituminous materials ( ), _____

ROUNDING AND OTHER PROPERTIES: well-rounded, rounded, subangular, angular, flat pebbles, desiccation chips, intraclasts, pressure solution, stylolitization, marginal replacement, frayed edges, _____

MONOMINERALIC GRAINS: straight quartz ( ), wavy quartz ( ), fibrous quartz ( ), chert ( ), K-feldspars ( ), plagioclases ( ), biotite ( ), muscovite ( ), chlorite ( ), pyroxenes ( ), amphiboles ( ), olivine ( ), magnetite ( ), hematite ( ), opaques ( ), glaucony: lobate ( ), pelletoidal ( ), vermicular ( ), detrital ( ), zircon ( ), garnet ( ), tourmaline ( ), anhydrite ( ), gypsum ( ), phosphates ( ), micrite pellets ( ), crystals ( ), shards ( ), phenocrysts ( ), _____

TYPES OF GRAIN-TO-GRAIN CONTACTS: point contact, meniscus, straight, concavoconvex, sutured, microstylolitic, marginal replacement, frayed edges, _____

ROUNDING AND OTHER PROPERTIES: well-rounded, rounded, subangular, angular, alteration of feldspars to: vacuolization, quartz, sericite, muscovite, chlorite, calcite, epidote, clay minerals, anhydrite, zeolites, _____

MATRIX

DETRITAL (silt size): clay minerals, sericite, chlorite, muscovite, biotite, epidote, quartz, K-feldspars, plagioclases, pyroxenes, amphiboles, micrite, bioclastic, dolomite, hematite, limonite, pyrite, chamosite, siderite, iron silicates, glaucony, phosphates, bituminous flecks, carbonaceous flecks, anhydrite, gypsum, halite, _____

DIAGENETIC (silt size): quartz, calcite, dolomite, sericite, K-feldspars, plagioclases, epidote, clay minerals (kaolinite, dickite, smectite, illite, chlorite, sericite), zeolites, opaques, _____

ABUNDANCE OF DETRITAL AND/OR DIAGENETIC MATRIX: predominant, common, rare, _____

TEXTURE OF DETRITAL AND/OR DIAGENETIC MATRIX: massive, laminated, cross-laminated, graded bedded, mesobanded, microbanded, aftbanded, bioturbated, fenestral, geopetal, internal sediment slumping, flow, soft-sediment deformation, vitroclastic, welded, _____

CEMENT

PRECIPITATED: aragonite, calcite, dolomite, siderite, opal, microquartz, chalcedonic quartz, zebraic chalcedony, microflamboyant quartz, megaquartz, feldspars, anhydrite, gypsum, halite, polyhalite, phosphates, zeolites, hematite, siderite, chamosite, iron silicates, clay minerals (kaolinite, dickite, smectite, illite, chlorite, sericite), barite, celestine, potash–magnesia salts, _

ABUNDANCE: predominant, common, rare, _

CRYSTALLINITY: amorphous, cryptocrystalline, microcrystalline, microsparite, coarsely crystalline, sparite, poikilotopic, fibrous, palisade, wavy, felty, pseudomorphic, bacillary, selenitic, hopper, _

TEXTURE: interparticle, intraparticle, isopachous, rim, syntaxial overgrowths, cavity filling (bladed, mosaic), infilling of tests, infilling of molds, infilling of fractures, umbrella effects, fenestral, pendulous, meniscus, concretionary, geopetal, banded, _ _ _ _ _ _ _ _ _ _ _ _ _

GENERAL TEXTURE

GRAINS AND MATRIX OR CEMENT RELATIONSHIPS: grain supported, homogeneity or inhomogeneity of packing, displacive texture ("floating grains"), compacted, pressure welded, mud supported, matrix supported, bioaccumulated, bioconstructed, bioturbated, soft-sediment deformation, see also matrix and cement, _ _ _ _ _ _ _ _ _ _ _ _ _ _ _ _ _

TEXTURAL MATURITY (sand-size grains only): high, medium, low, _ _ _ _ _ _ _ _ _ _ _ _

MINERALOGICAL MATURITY (sand-size grains only): high, medium, low, _ _ _ _ _ _ _ _

ORGANISMS

CONSTRUCTING: corals ( ), stromatoporoids ( ), coralline algae ( ), stromatolites ( ), cyanobacterial mats ( ), bryozoans ( ), annelids ( ), calcisponges ( ), _ _ _ _ _ _ _ _ _

ACCUMULATING: pelecypods ( ), gastropods ( ), brachiopods ( ), ostracods ( ), crinoids ( ), plants ( ), large foraminifers ( ), small foraminifers ( ), _ _ _ _ _ _ _ _ _ _

BENTHONIC: large foraminifers ( ), small foraminifers ( ), rotalids ( ), miliolids ( ), textularids ( ), brachiopods ( ), brachiopod spines ( ), pelecypods ( ), gastropods ( ), annelids ( ), bryozoans ( ), corals ( ), echinoids ( ), crinoids ( ), dasyclads ( ), trilobites ( ), ostracods ( ), algal encrustations ( ), oncoids ( ), sponge spicules ( ), bone debris ( ), fish-scale debris ( ), *Chara* ( ), _ _ _ _ _ _ _ _ _ _ _ _ _ _ _ _ _ _ _

PLANKTONIC: radiolaria ( ), diatoms ( ), tintinnoids ( ), conodonts ( ), cephalopods ( ), scaphopods ( ), tentaculites ( ), calcispheres ( ), globigerinids ( ), microfilaments ( ), aptychus ( ), _

BIOCLASTS (sand-size to silt-size skeletal particles mechanically transported and deposited with possible additional fragmentation by scavengers): corals ( ), large foraminifers ( ), small foraminifers ( ), stromatoporoids ( ), coralline algae ( ), stromatolites ( ), cyanobacterial mats ( ), dasyclads ( ), bryozoans ( ), annelids ( ), pelecypods ( ), gastropods ( ), brachiopods ( ), brachiopod spines ( ), plant debris ( ), echinoids ( ), crinoids ( ), trilobites ( ), ostracods ( ), sponge spicules ( ), bone debris ( ), scale debris ( ), cephalopods ( ), scaphopods ( ), tentaculites ( ), _ _ _ _ _ _ _ _ _ _ _ _ _

PROPERTIES OF BIOCLASTS: well rounded, rounded, subangular, angular, with marginal replacements, perforations, micrite envelopes, neomorphosed, pressure solution, stylolitic contacts, _

OTHER CONSTITUENTS

Fecal pellets ( ), lithic pellets ( ), peloids ( ), ooids ( ), pisoids ( ), concretions ( ), nodules ( ), _

DIAGENESIS

NEOMORPHISM: aggrading, degrading, pseudobreccias, clotted texture, devitrification, pseudomicrosparite, pseudosparite, micritization, micrite envelopes, spar-micritization, _

REPLACEMENT: use arrows to connect replacive mineral with replaced mineral. The process is aggrading, degrading, pseudomorphic. aragonite ( ), calcite ( ), dolomite ( ), siderite ( ), microquartz ( ), chalcedonic quartz ( ), zebraic quartz ( ), microflamboyant quartz ( ), megaquartz ( ), K-feldspars ( ), plagioclases ( ), gypsum ( ), anhydrite ( ), halite ( ), potash–magnesia salts ( ), hematite ( ), pyrite ( ), glaucony ( ), chamosite ( ), magnetite ( ), iron silicates ( ), phosphates ( ), clay minerals (kaolinite, dickite, smectite, illite, chlorite, sericite), biotite ( ), muscovite ( ),
zeolites ( ), _

VADOSE PROCESSES: pedological features, bioturbation, flow structures, concretions, ooids, pisoids, crusts, vadose silt, internal sediment, geopetal features, micritization, neomorphism, root casts, oxidation, alteration, _

COMPACTION AND TECTONISM (applies to entire rock): distortion, splitting, interpenetration, stylolitization, fissility, schistosity, cleavage, shearing, Boehm lamellae, wavy extinction, mass extinction, cataclastic, mylonitic, metamorphic, _ _ _ _ _ _ _ _ _ _ _ _ _ _ _

POROSITY

SILICICLASTIC ROCKS

Primary: interparticle, residual interparticle (compaction, pressure solution, overgrowth), _

Secondary: (quartz, silicates, carbonates, sulfates, chlorides) intraparticle dissolution, intercrystalline, molds, inhomogeneity of packing, oversized pores, elongate pores, corroded grains, honeycombed grains, fractured grains, shrinkage of matrix, _ _ _ _ _ _ _ _ _

CARBONATE ROCKS

Fabric selective: interparticle, residual interparticle (compaction, pressure solution, overgrowth), intraparticle, intercrystalline, moldic, moldic enlarged, fenestral, shelter, growth framework, _

Nonfabric selective: fracture, channel, vuggy, shrinkage, _ _ _ _ _ _ _ _ _ _ _ _ _ _ _ _ _ _ _

Fabric or nonfabric selective: breccia, boring, burrow, shrinkage, _ _ _ _ _ _ _ _ _ _ _ _ _

Primary origin: predepositional, depositional, _

Secondary origin: eogenetic, mesogenetic, telogenetic, _

GENERAL FEATURES

Size: micropores ($<\frac{1}{16}$ mm), mesopores ($\frac{1}{16}$ to 4 mm), megapores (4 to 256 mm), _ _ _ _ _

Abundance: predominant, common, rare, _

Environment: predepositional, postdepositional, marine phreatic, mixed marine–freshwater vadose, beachrock, freshwater vadose, freshwater phreatic, mixing, deep burial, _

DESCRIPTION OF ROCK

NAME OF ROCK

DEPOSITIONAL–DIAGENETIC SEQUENCE

ENVIRONMENTAL INTERPRETATION

ENERGY: high, medium, low, _

DEPTH OF WATER: very shallow, shallow, moderate, deep, very deep, _ _ _ _ _ _ _ _ _ _ _

ENVIRONMENT: open marine, marine restricted, brackish, evaporitic, fluvial, lacustrine, continental, karstic, caves, _

TIDES: supratidal, high intertidal, low intertidal, shallow subtidal, deep subtidal, _ _ _ _ _

PROCESSES: traction, debris flow, sandflow, mudflow, slumping, turbidites, gravity, pelagic, bioconstructed, bioaccumulated, in place accumulation, collapse, pyroclastic, hyaloclastic, pedologic, biochemical precipitation, _

ADDITIONAL COMMENTS ON DEPOSITIONAL ENVIRONMENT

SUBJECT INDEX

E

Eastern Australia:
    Paleozoic wackes, 38
Eastern U.S.:
    Ancient lacustrine cherts:
        Triassic rift grabens, 158
Echinoid plate microstructure, 103
Echinoid spine microstructure, 103
Edwards aquifer, Texas:
    dedolomitization, 140
Edwards Group:
    dedolomitization, 140
Egypt, Miocene:
    large-scale replacement of anhydrite by
        calcite, 228
    large-scale replacement of gypsum by
        calcite, 228
Enewetak Atoll:
    seawater dolomitization, 132
England, Southern:
    Cenomanian phosphatic nodules, 181
Eolian processes:
    quartz arenites, 7
Evaporites (*see also* Anhydrite; Gypsum;
        Halite; Potash-magnesia salts;
        Trona):
    burial deformations, 228
    burial diagenetic features, 227–28
    chert replacement types, 166–68
    coastal sabkhas (supratidal) type:
        eolian processes, 214
        generation of chickenwire anhydrite,
            213
        generation of nodular anhydrite,
            213–14
        gypsum-mush, 213
        hopper "pagoda" type halite, 213–14
        mineralogy of evaporites, 213–14
        Persian Gulf, 213–14
        physiographic conditions, 213–14
    continental sabkhas (playas) type:
        mineralogical suite, 211–13
        physiographic condition, 211–13
        storm-stratification, 212–13
    crystallization inhibitors, 209
    deep water type:
        depositional environments, 224
        interpretation of rhythmites, 224
        laminated halite, 225–26
        laminated potash-magnesia salts, 226
        mineralogical suite, 224
        sulfate laminites, 224–25
        turbidity currents, 224
        water depth interpretation, 224
    depositional environments, 209–11
    diagenetic evolution, 209–10, 226–28
    diagenetic terminology, 226–27
    economical importance, 209

    glaciolacustrine Pleistocene, 209
    major minerals, 209
    mineralogical terminology, 211
    mineral phase transformations, 210–11
    neomorphism, 210
    non-uniformitarian approach, 211, 224
    paleoclimatic indicators, 209, 211
    reservoir properties, 228–29
    seals of hydrocarbon reservoirs, 209
    shallow marine type:
        coarse clastic gypsum, 218–20
        coarse crystalline gypsum, 216–18
        Deep Sea Drilling Program, 214
        depositional environments, 214
        gypsarenites, 218–20
        gypsolites, 220
        gypsrudites, 218–20
        halite, 220–23
        laminated anhydrite, 223
        laminated gypsum, 214–16
        mineralogical suite, 214
        oolitic anhydrite, 223
        oolitic gypsum, 220
        physiographic conditions, 214
        potash-magnesia salts, 223–24
        "salinas" basins, 214
        selenitic gypsum, 216–18
        tectonic setting, 214
    significance of fluid inclusions, 228
    source-rocks of hydrocarbons, 209
    stratigraphic range, 209, 211
    structural traps of hydrocarbons, 209
    syndepositional cements, 227
    syndepositional features, 227
    syngenetic deformations, 228
    textural classification, 211
    timing of diagenesis, 226–28

F

Feldspars:
    authigenic:
        in bituminous shales, 70
        in limestones, 101
    basinward variation in shales, 74–75
    chemical composition, 19
    chemical zoning, 19
    ferruginous ribbon radiolarian cherts,
        151
    grains in bituminous shales, 70
    grains in calcareous shales, 70
    grains in carbonaceous shales, 71
    grains in common shales, 69
    grains in siliceous shales, 70
    grains in tillites, 48
    overgrowth in feldspathic arenites,
        17–18
    provenance indicators in feldspathic
        arenites, 19–20

    chemical composition, 19
    chemical zoning, 19
    structural state, 20
    twinning, 20
    replaced by calcite in feldspathic aren-
        ites, 18
    replaced by dolomite in feldspathic aren-
        ites, 18
    replaced by matrix in wackes, 34
    replaced by siderite in feldspathic aren-
        ites, 18
    structural state, 20
    twinning, 20
Feldspathization:
    pelphospharenites, 185–87
Ferricretes:
    relationship to glauconitic facies, 16
Ferry Lake Evaporites:
    anhydrite conversion of selenitic gyp-
        sum, 218
Flexible metaquartzites:
    muscovite, 7
    origin, 7
    texture, 7
Flint, 159–62
Florida:
    tidal flats:
        seawater dolomitization, 133
Fluoritization:
    pelphospharenites, 185–87
Foraminifer test microstructure:
    benthonic, 104
    planktonic, 104
Francolite, definition, 173

G

Galena Group:
    carbonate tempestites, 51
Gastropod shell microstructure, 103
Gaylussite:
    continental sabkhas (playas), 211
Genetic environments:
    cherts in evaporites, 166–67
    early diagenetic cherts in dolostones,
        163–64
    early diagenetic cherts in limestones,
        163–64
    late diagenetic cherts in dolostones, 166
    late diagenetic cherts in limestones, 166
    Phanerozoic oolitic ironstones, 199–
        200, 202–5
    Precambrian ironstone-formations,
        196–98
    syngenetic cherts in chalks, 159–61
    syngenetic cherts in dolostones, 162–63
    syngenetic cherts in limestones, 162–63
Gila Conglomerate:
    Magadi-type cherts, 157

AUTHOR INDEX

Page numbers in regular type are author citations in the text; page numbers in **bold** type are full author references at the end of chapters.